Digital Image Processing Using MATLAB®

Rafael C. Gonzalez
University of Tennessee

Richard E. Woods
MedData Interactive

Steven L. Eddins
The MathWorks, Inc.

ISBN 81-7758-898-2

First Impression, 2006
Second Impression, 2007

Published by Dorling Kindersley (India) Pvt. Ltd., licensees of Pearson Education in South Asia.

Head Office: 482, F.I.E., Pataparganj, Delhi 110 092, India.
Registered Office: 14 Local Shopping Centre, Panchsheel Park, New Delhi 110 017, India.

Printed in India by Baba Barkha Nath Printers.

To Connie, Ralph, and Robert
To Janice, David, and Jonathan
and
To Geri, Christopher, and Nicholas

Contents

Preface

Solutions to problems in the field of digital image processing generally require extensive experimental work involving software simulation and testing with large sets of sample images. Although algorithm development typically is based on theoretical underpinnings, the actual implementation of these algorithms almost always requires parameter estimation and, frequently, algorithm revision and comparison of candidate solutions. Thus, selection of a flexible, comprehensive, and well-documented software development environment is a key factor that has important implications in the cost, development time, and portability of image processing solutions.

In spite of its importance, surprisingly little has been written on this aspect of the field in the form of textbook material dealing with both theoretical principles and software implementation of digital image processing concepts. This book was written for just this purpose. Its main objective is to provide a foundation for implementing image processing algorithms using modern software tools. A complementary objective was to prepare a book that is self-contained and easily readable by individuals with a basic background in digital image processing, mathematical analysis, and computer programming, all at a level typical of that found in a junior/senior curriculum in a technical discipline. Rudimentary knowledge of MATLAB also is desirable.

To achieve these objectives, we felt that two key ingredients were needed. The first was to select image processing material that is representative of material covered in a formal course of instruction in this field. The second was to select software tools that are well supported and documented, and which have a wide range of applications in the "real" world.

To meet the first objective, most of the theoretical concepts in the following chapters were selected from *Digital Image Processing* by Gonzalez and Woods, which has been the choice introductory textbook used by educators all over the world for over two decades. The software tools selected are from the MATLAB Image Processing Toolbox (IPT), which similarly occupies a position of eminence in both education and industrial applications. A basic strategy followed in the preparation of the book was to provide a seamless integration of well-established theoretical concepts and their implementation using state-of-the-art software tools.

The book is organized along the same lines as *Digital Image Processing*. In this way, the reader has easy access to a more detailed treatment of all the image processing concepts discussed here, as well as an up-to-date set of references for further reading. Following this approach made it possible to present theoretical material in a succinct manner and thus we were able to maintain a focus on the software implementation aspects of image processing problem solutions. Because it works in the MATLAB computing environment, the Image Processing Toolbox offers some significant advantages, not only in the breadth of its computational tools, but also because it is supported under most operating systems in use today. A unique feature of this book is its emphasis on showing how to develop new code to enhance existing MATLAB and IPT functionality. This is an important feature in an area such as image processing, which, as noted earlier, is characterized by the need for extensive algorithm development and experimental work.

After an introduction to the fundamentals of MATLAB functions and programming, the book proceeds to address the mainstream areas of image processing. The major areas covered include intensity transformations, linear and nonlinear spatial filtering, filtering in the frequency domain, image restoration and registration, color image processing, wavelets, image data compression, morphological image processing, image segmentation, region and boundary representation and description, and object recognition. This material is complemented by numerous illustrations of how to solve image processing problems using MATLAB and IPT functions. In cases where a function did not exist, a new function was written and documented as part of the instructional focus of the book. Over 60 new functions are included in the following chapters. These functions increase the scope of IPT by approximately 35 percent and also serve the important purpose of further illustrating how to implement new image processing software solutions.

The material is presented in textbook format, not as a software manual. Although the book is self-contained, we have established a companion Web site (see Section 1.5) designed to provide support in a number of areas. For students following a formal course of study or individuals embarked on a program of self study, the site contains tutorials and reviews on background material, as well as projects and image databases, including all images in the book. For instructors, the site contains classroom presentation materials that include PowerPoint slides of all the images and graphics used in the book. Individuals already familiar with image processing and IPT fundamentals will find the site a useful place for up-to-date references, new implementation techniques, and a host of other support material not easily found elsewhere. All purchasers of the book are eligible to download executable files of all the new functions developed in the text.

As is true of most writing efforts of this nature, progress continues after work on the manuscript stops. For this reason, we devoted significant effort to the selection of material that we believe is fundamental, and whose value is likely to remain applicable in a rapidly evolving body of knowledge. We trust that readers of the book will benefit from this effort and thus find the material timely and useful in their work.

Acknowledgments

We are indebted to a number of individuals in academic circles as well as in industry and government who have contributed to the preparation of the book. Their contributions have been important in so many different ways that we find it difficult to acknowledge them in any other way but alphabetically. We wish to extend our appreciation to Mongi A. Abidi, Peter J. Acklam, Serge Beucher, Ernesto Bribiesca, Michael W. Davidson, Courtney Esposito, Naomi Fernandes, Thomas R. Gest, Roger Heady, Brian Johnson, Lisa Kempler, Roy Lurie, Ashley Mohamed, Joseph E. Pascente, David. R. Pickens, Edgardo Felipe Riveron, Michael Robinson, Loren Shure, Jack Sklanski, Sally Stowe, Craig Watson, and Greg Wolodkin. We also wish to acknowledge the organizations cited in the captions of many of the figures in the book for their permission to use that material.

Special thanks go to Tom Robbins, Rose Kernan, Alice Dworkin, Xiaohong Zhu, Bruce Kenselaar, and Jayne Conte at Prentice Hall for their commitment to excellence in all aspects of the production of the book. Their creativity, assistance, and patience are truly appreciated.

Rafael C. Gonzalez
Richard E. Woods
Steven L. Eddins

About the Authors

Rafael C. Gonzalez

R. C. Gonzalez received the B.S.E.E. degree from the University of Miami in 1965 and the M.E. and Ph.D. degrees in electrical engineering from the University of Florida, Gainesville, in 1967 and 1970, respectively. He joined the Electrical and Computer Engineering Department at the University of Tennessee, Knoxville (UTK) in 1970, where he became Associate Professor in 1973, Professor in 1978, and Distinguished Service Professor in 1984. He served as Chairman of the department from 1994 through 1997. He is currently a Professor Emeritus of Electrical and Computer Engineering at UTK.

He is the founder of the Image & Pattern Analysis Laboratory and the Robotics & Computer Vision Laboratory at the University of Tennessee. He also founded Perceptics Corporation in 1982 and was its president until 1992. The last three years of this period were spent under a full-time employment contract with Westinghouse Corporation, who acquired the company in 1989. Under his direction, Perceptics became highly successful in image processing, computer vision, and laser disk storage technologies. In its initial ten years, Perceptics introduced a series of innovative products, including: The world's first commercially-available computer vision system for automatically reading the license plate on moving vehicles; a series of large-scale image processing and archiving systems used by the U.S. Navy at six different manufacturing sites throughout the country to inspect the rocket motors of missiles in the Trident II Submarine Program; the market leading family of imaging boards for advanced Macintosh computers; and a line of trillion-byte laser disk products.

He is a frequent consultant to industry and government in the areas of pattern recognition, image processing, and machine learning. His academic honors for work in these fields include the 1977 UTK College of Engineering Faculty Achievement Award; the 1978 UTK Chancellor's Research Scholar Award; the 1980 Magnavox Engineering Professor Award; and the 1980 M. E. Brooks Distinguished Professor Award. In 1981 he became an IBM Professor at the University of Tennessee and in 1984 he was named a Distinguished Service Professor there. He was awarded a Distinguished Alumnus Award by the University of Miami in 1985, the Phi Kappa Phi Scholar Award in 1986, and the University of Tennessee's Nathan W. Dougherty Award for Excellence in Engineering in 1992. Honors for industrial accomplishment include the 1987 IEEE Outstanding Engineer Award for Commercial Development in Tennessee; the 1988 Albert Rose National Award for Excellence in Commercial Image Processing; the 1989 B. Otto Wheeley Award for Excellence in Technology Transfer; the 1989 Coopers and Lybrand Entrepreneur of the Year Award; the 1992 IEEE Region 3 Outstanding Engineer Award; and the 1993 Automated Imaging Association National Award for Technology Development.

Dr. Gonzalez is author or co-author of over 100 technical articles, two edited books, and five textbooks in the fields of pattern recognition, image processing, and robotics. His books are used in over 500 universities and research institutions throughout the world. He is listed in the prestigious Marquis *Who's Who in America*, Marquis *Who's Who in Engineering*, Marquis *Who's Who in the World*, and in 10 other national and international biographical citations. He is the co-holder of two U.S. Patents, and has been an associate editor of the *IEEE Transactions on*

Systems, Man and Cybernetics, and the *International Journal of Computer and Information Sciences*. He is a member of numerous professional and honorary societies, including Tau Beta Pi, Phi Kappa Phi, Eta Kappa Nu, and Sigma Xi. He is a Fellow of the IEEE.

Richard E. Woods

Richard E. Woods earned his B.S., M.S., and Ph.D. degrees in Electrical Engineering from the University of Tennessee, Knoxville. His professional experiences range from entrepreneurial to the more traditional academic, consulting, governmental, and industrial pursuits. Most recently, he founded MedData Interactive, a high technology company specializing in the development of handheld computer systems for medical applications. He was also a founder and Vice President of Perceptics Corporation, where he was responsible for the development of many of the company's quantitative image analysis and autonomous decision making products.

Prior to Perceptics and MedData, Dr. Woods was an Assistant Professor of Electrical Engineering and Computer Science at the University of Tennessee and prior to that, a computer applications engineer at Union Carbide Corporation. As a consultant, he has been involved in the development of a number of special-purpose digital processors for a variety of space and military agencies, including NASA, the Ballistic Missile Systems Command, and the Oak Ridge National Laboratory.

Dr. Woods has published numerous articles related to digital signal processing and is co-author of *Digital Image Processing*, the leading text in the field. He is a member of several professional societies, including Tau Beta Pi, Phi Kappa Phi, and the IEEE. In 1986, he was recognized as a Distinguished Engineering Alumnus of the University of Tennessee.

Steven L. Eddins

Steven L. Eddins is development manager of the image processing group at The MathWorks, Inc. He led the development of several versions of the company's Image Processing Toolbox. His professional interests include building software tools that are based on the latest research in image processing algorithms, and that have a broad range of scientific and engineering applications.

Prior to joining The MathWorks, Inc. in 1993, Dr. Eddins was on the faculty of the Electrical Engineering and Computer Science Department at the University of Illinois, Chicago. There he taught graduate and senior-level classes in digital image processing, computer vision, pattern recognition, and filter design, and he performed research in the area of image compression.

Dr. Eddins holds a B.E.E. (1986) and a Ph.D. (1990), both in electrical engineering from the Georgia Institute of Technology. He is a member of the IEEE.

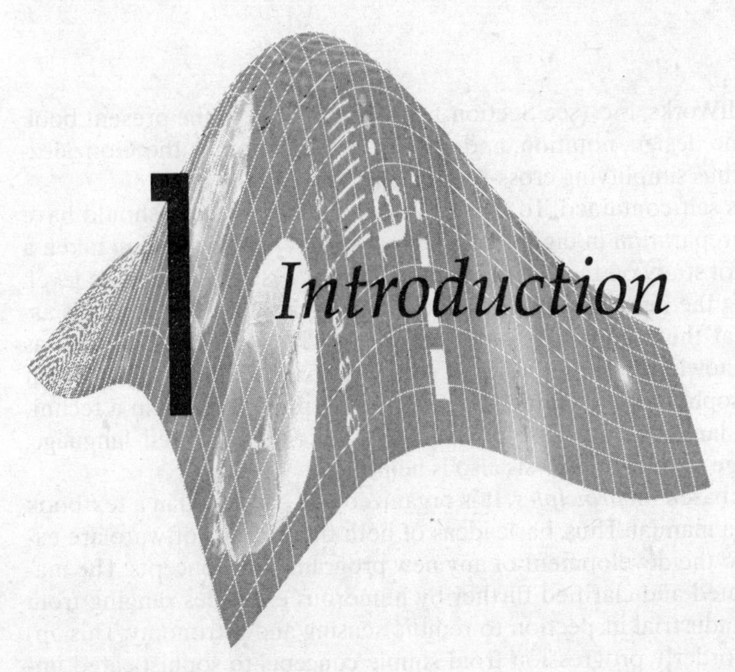

1 Introduction

Preview

Digital image processing is an area characterized by the need for extensive experimental work to establish the viability of proposed solutions to a given problem. In this chapter we outline how a theoretical base and state-of-the-art software can be integrated into a prototyping environment whose objective is to provide a set of well-supported tools for the solution of a broad class of problems in digital image processing.

1.1 Background

An important characteristic underlying the design of image processing systems is the significant level of testing and experimentation that normally is required before arriving at an acceptable solution. This characteristic implies that the ability to formulate approaches and quickly prototype candidate solutions generally plays a major role in reducing the cost and time required to arrive at a viable system implementation.

Little has been written in the way of instructional material to bridge the gap between theory and application in a well-supported software environment. The main objective of this book is to integrate under one cover a broad base of theoretical concepts with the knowledge required to implement those concepts using state-of-the-art image processing software tools. The theoretical underpinnings of the material in the following chapters are mainly from the leading textbook in the field: *Digital Image Processing*, by Gonzalez and Woods, published by Prentice Hall. The software code and supporting tools are based on the leading software package in the field: The *MATLAB Image Processing Toolbox*,†

---

†In the following discussion and in subsequent chapters we sometimes refer to *Digital Image Processing* by Gonzalez and Woods as "the Gonzalez-Woods book," and to the Image Processing Toolbox as "IPT" or simply as the "toolbox."

from The MathWorks, Inc. (see Section 1.3). The material in the present book shares the same design, notation, and style of presentation as the Gonzalez-Woods book, thus simplifying cross-referencing between the two.

The book is self-contained. To master its contents, the reader should have introductory preparation in digital image processing, either by having taken a formal course of study on the subject at the senior or first-year graduate level, or by acquiring the necessary background in a program of self-study. It is assumed also that the reader has some familiarity with MATLAB, as well as rudimentary knowledge of the basics of computer programming, such as that acquired in a sophomore- or junior-level course on programming in a technically oriented language. Because MATLAB is an array-oriented language, basic knowledge of matrix analysis also is helpful.

The book is based on *principles*. It is organized and presented in a textbook format, not as a manual. Thus, basic ideas of both theory and software are explained prior to the development of any new programming concepts. The material is illustrated and clarified further by numerous examples ranging from medicine and industrial inspection to remote sensing and astronomy. This approach allows orderly progression from simple concepts to sophisticated implementation of image processing algorithms. However, readers already familiar with MATLAB, IPT, and image processing fundamentals can proceed directly to specific applications of interest, in which case the functions in the book can be used as an extension of the family of IPT functions. All new functions developed in the book are fully documented, and the code for each is included either in a chapter or in Appendix C.

Over 60 new functions are developed in the chapters that follow. These functions complement and extend by 35% the set of about 175 functions in IPT. In addition to addressing specific applications, the new functions are clear examples of how to combine existing MATLAB and IPT functions with new code to develop prototypic solutions to a broad spectrum of problems in digital image processing. The toolbox functions, as well as the functions developed in the book, run under most operating systems. Consult the book Web site (see Section 1.5) for a complete list.

1.2 What Is Digital Image Processing?

An image may be defined as a two-dimensional function, $f(x, y)$, where x and y are *spatial coordinates*, and the amplitude of f at any pair of coordinates (x, y) is called the *intensity* or *gray level* of the image at that point. When x, y, and the amplitude values of f are all finite, discrete quantities, we call the image a *digital image*. The field of *digital image processing* refers to processing digital images by means of a digital computer. Note that a digital image is composed of a finite number of elements, each of which has a particular location and value. These elements are referred to as *picture elements*, *image elements*, *pels*, and *pixels*. *Pixel* is the term most widely used to denote the elements of a digital image. We consider these definitions formally in Chapter 2.

Vision is the most advanced of our senses, so it is not surprising that images play the single most important role in human perception. However, unlike humans, who are limited to the visual band of the electromagnetic (EM) spectrum, imaging machines cover almost the entire EM spectrum, ranging from gamma to radio waves. They can operate also on images generated by sources that humans are not accustomed to associating with images. These include ultrasound, electron microscopy, and computer-generated images. Thus, digital image processing encompasses a wide and varied field of applications.

There is no general agreement among authors regarding where image processing stops and other related areas, such as image analysis and computer vision, start. Sometimes a distinction is made by defining image processing as a discipline in which both the input and output of a process are images. We believe this to be a limiting and somewhat artificial boundary. For example, under this definition, even the trivial task of computing the average intensity of an image would not be considered an image processing operation. On the other hand, there are fields such as computer vision whose ultimate goal is to use computers to emulate human vision, including learning and being able to make inferences and take actions based on visual inputs. This area itself is a branch of artificial intelligence (AI), whose objective is to emulate human intelligence. The field of AI is in its earliest stages of infancy in terms of development, with progress having been much slower than originally anticipated. The area of image analysis (also called image understanding) is in between image processing and computer vision.

There are no clear-cut boundaries in the continuum from image processing at one end to computer vision at the other. However, one useful paradigm is to consider three types of computerized processes in this continuum: low-, mid-, and high-level processes. Low-level processes involve primitive operations such as image preprocessing to reduce noise, contrast enhancement, and image sharpening. A low-level process is characterized by the fact that both its inputs and outputs are images. Mid-level processes on images involve tasks such as segmentation (partitioning an image into regions or objects), description of those objects to reduce them to a form suitable for computer processing, and classification (recognition) of individual objects. A mid-level process is characterized by the fact that its inputs generally are images, but its outputs are attributes extracted from those images (e.g., edges, contours, and the identity of individual objects). Finally, higher-level processing involves "making sense" of an ensemble of recognized objects, as in image analysis, and, at the far end of the continuum, performing the cognitive functions normally associated with human vision.

Based on the preceding comments, we see that a logical place of overlap between image processing and image analysis is the area of recognition of individual regions or objects in an image. Thus, what we call in this book *digital image processing* encompasses processes whose inputs and outputs are images and, *in addition*, encompasses processes that extract attributes from images, up to and including the recognition of individual objects. As a simple illustration to clarify these concepts, consider the area of automated analysis of text. The processes of acquiring an image of the area containing the text, preprocessing that image, extracting (segmenting) the individual characters, describing the characters in a form suitable for computer processing, and recognizing those individual characters, are in the scope of what we call digital image processing in this book. Making sense of the content of the page may be viewed as being in the domain of image analysis and even computer vision, depending on the level of complexity implied by the statement "making sense." Digital image processing, as we have defined it, is used successfully in a broad range of areas of exceptional social and economic value.

1.3 Background on MATLAB and the Image Processing Toolbox

MATLAB is a high-performance language for technical computing. It integrates computation, visualization, and programming in an easy-to-use environment where problems and solutions are expressed in familiar mathematical notation. Typical uses include the following:

- Math and computation
- Algorithm development
- Data acquisition
- Modeling, simulation, and prototyping
- Data analysis, exploration, and visualization
- Scientific and engineering graphics
- Application development, including graphical user interface building

MATLAB is an interactive system whose basic data element is an array that does not require dimensioning. This allows formulating solutions to many technical computing problems, especially those involving matrix representations, in a fraction of the time it would take to write a program in a scalar non-interactive language such as C or Fortran.

The name MATLAB stands for *matrix laboratory*. MATLAB was written originally to provide easy access to matrix software developed by the LINPACK (Linear System Package) and EISPACK (Eigen System Package) projects. Today, MATLAB engines incorporate the LAPACK (Linear Algebra Package) and BLAS (Basic Linear Algebra Subprograms) libraries, constituting the state of the art in software for matrix computation.

In university environments, MATLAB is the standard computational tool for introductory and advanced courses in mathematics, engineering, and science. In industry, MATLAB is the computational tool of choice for research, development, and analysis. MATLAB is complemented by a family of application-specific solutions called *toolboxes*. The Image Processing Toolbox is a collection of MATLAB functions (called *M-functions* or *M-files*) that extend the capability of the MATLAB environment for the solution of digital image processing problems. Other toolboxes that sometimes are used to complement IPT are the Signal Processing, Neural Network, Fuzzy Logic, and Wavelet Toolboxes.

The MATLAB Student Version includes a full-featured version of MATLAB. The Student Version can be purchased at significant discounts at university bookstores and at the MathWorks' Web site (www.mathworks.com). Student versions of add-on products, including the Image Processing Toolbox, also are available.

1.4 Areas of Image Processing Covered in the Book

Every chapter in this book contains the pertinent MATLAB and IPT material needed to implement the image processing methods discussed. When a MATLAB or IPT function does not exist to implement a specific method, a new function is developed and documented. As noted earlier, a complete listing of every new function is included in the book. The remaining eleven chapters cover material in the following areas.

Chapter 2: Fundamentals. This chapter covers the fundamentals of MATLAB notation, indexing, and programming concepts. This material serves as foundation for the rest of the book.

Chapter 3: Intensity Transformations and Spatial Filtering. This chapter covers in detail how to use MATLAB and IPT to implement intensity transformation functions. Linear and nonlinear spatial filters are covered and illustrated in detail.

Chapter 4: Processing in the Frequency Domain. The material in this chapter shows how to use IPT functions for computing the forward and inverse fast Fourier transforms (FFTs), how to visualize the Fourier spectrum, and how to implement filtering in the frequency domain. Shown also is a method for generating frequency domain filters from specified spatial filters.

Chapter 5: Image Restoration. Traditional linear restoration methods, such as the Wiener filter, are covered in this chapter. Iterative, nonlinear methods, such as the Richardson-Lucy method and maximum-likelihood estimation for blind deconvolution, are discussed and illustrated. Geometric corrections and image registration also are covered.

Chapter 6: Color Image Processing. This chapter deals with pseudocolor and full-color image processing. Color models applicable to digital image processing are discussed, and IPT functionality in color processing is extended via implementation of additional color models. The chapter also covers applications of color to edge detection and region segmentation.

Chapter 7: Wavelets. In its current form, IPT does not have any wavelet transforms. A set of wavelet-related functions compatible with the Wavelet Toolbox is developed in this chapter that will allow the reader to implement all the wavelet-transform concepts discussed in the Gonzalez-Woods book.

Chapter 8: Image Compression. The toolbox does not have any data compression functions. In this chapter, we develop a set of functions that can be used for this purpose.

Chapter 9: Morphological Image Processing. The broad spectrum of functions available in IPT for morphological image processing are explained and illustrated in this chapter using both binary and gray-scale images.

Chapter 10: Image Segmentation. The set of IPT functions available for image segmentation are explained and illustrated in this chapter. New functions for Hough transform processing and region growing also are developed.

Chapter 11: Representation and Description. Several new functions for object representation and description, including chain-code and polygonal representations, are developed in this chapter. New functions are included also for object description, including Fourier descriptors, texture, and moment invariants. These functions complement an extensive set of region property functions available in IPT.

Chapter 12: Object Recognition. One of the important features of this chapter is the efficient implementation of functions for computing the Euclidean and Mahalanobis distances. These functions play a central role in pattern matching. The chapter also contains a comprehensive discussion on how to manipulate strings of symbols in MATLAB. String manipulation and matching are important in structural pattern recognition.

In addition to the preceding material, the book contains three appendices.

Appendix A: Contains a summary of all IPT and new image-processing functions developed in the book. Relevant MATLAB function also are included. This is a useful reference that provides a global overview of all functions in the toolbox and the book.

Appendix B: Contains a discussion on how to implement graphical user interfaces (GUIs) in MATLAB. GUIs are a useful complement to the material in the book because they simplify and make more intuitive the control of interactive functions.

Appendix C: New function listings are included in the body of a chapter when a new concept is explained. Otherwise the listing is included in Appendix C. This is true also for listings of functions that are lengthy. Deferring the listing of some functions to this appendix was done primarily to avoid breaking the flow of explanations in text material.

1.5 The Book Web Site

An important feature of this book is the support contained in the book Web site. The site address is

www.prenhall.com/gonzalezwoodseddins

This site provides support to the book in the following areas:

- Downloadable M-files, including all M-files in the book
- Tutorials

- Projects
- Teaching materials
- Links to databases, including all images in the book
- Book updates
- Background publications

The site is integrated with the Web site of the Gonzalez-Woods book:

www.prenhall.com/gonzalezwoods

which offers additional support on instructional and research topics.

1.6 Notation

Equations in the book are typeset using familiar italic and Greek symbols, as in $f(x, y) = A \sin(ux + vy)$ and $\phi(u, v) = \tan^{-1}[I(u, v)/R(u, v)]$. All MATLAB function names and symbols are typeset in monospace font, as in `fft2(f)`, `logical(A)`, and `roipoly(f, c, r)`.

The first occurrence of a MATLAB or IPT function is highlighted by use of the following icon on the page margin:

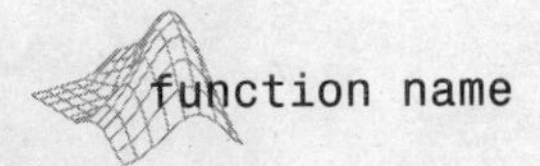

Similarly, the first occurrence of a new function developed in the book is highlighted by use of the following icon on the page margin:

`function name`

The symbol ▬ is used as a visual cue to denote the end of a function listing.

When referring to keyboard keys, we use bold letters, such as **Return** and **Tab**. We also use bold letters when referring to items on a computer screen or menu, such as **File** and **Edit**.

1.7 The MATLAB Working Environment

In this section we give a brief overview of some important operational aspects of using MATLAB.

1.7.1 The MATLAB Desktop

The MATLAB *desktop* is the main MATLAB application window. As Fig. 1.1 shows, the desktop contains five subwindows: the Command Window, the Workspace Browser, the Current Directory Window, the Command History Window, and one or more Figure Windows, which are shown only when the user displays a graphic.

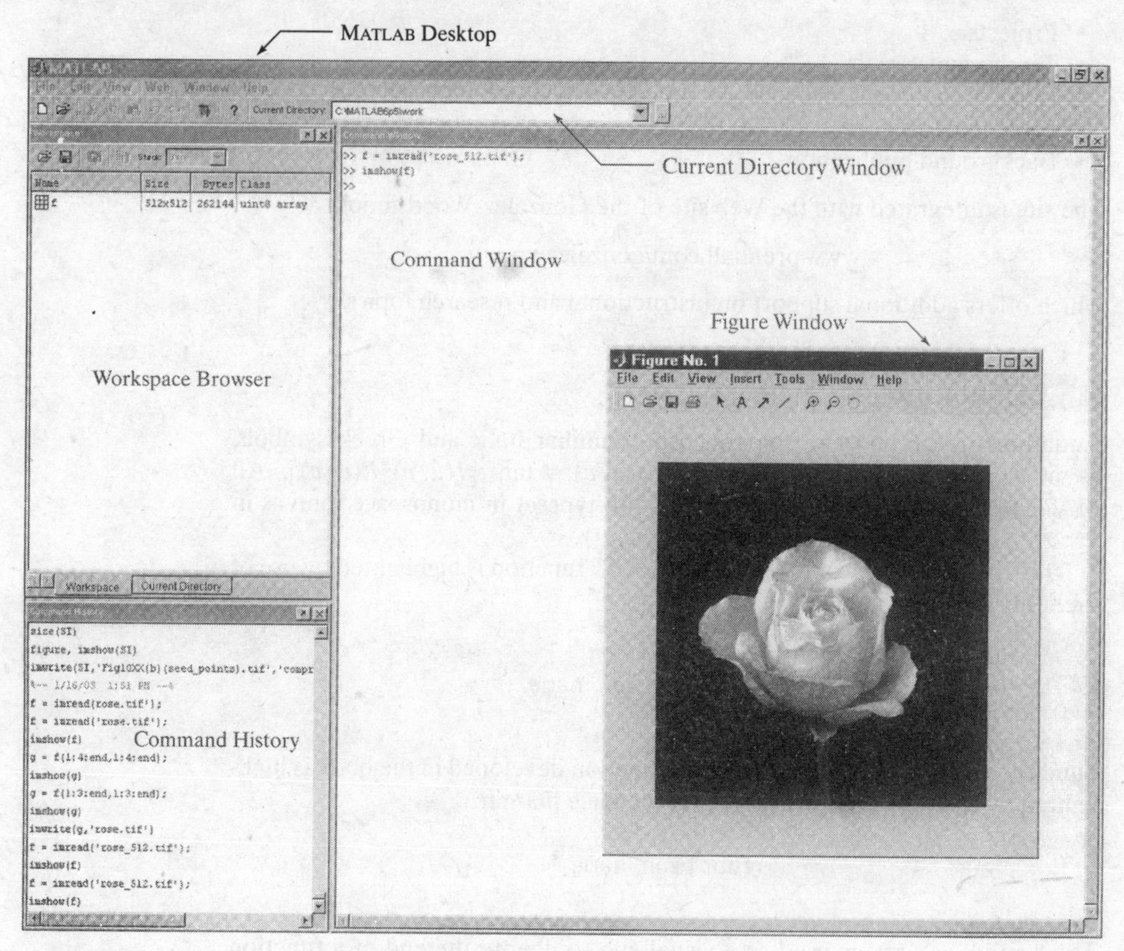

FIGURE 1.1 The MATLAB desktop and its principal components.

The *Command Window* is where the user types MATLAB commands and expressions at the prompt (>>) and where the outputs of those commands are displayed. MATLAB defines the *workspace* as the set of variables that the user creates in a work session. The *Workspace Browser* shows these variables and some information about them. Double-clicking on a variable in the Workspace Browser launches the *Array Editor*, which can be used to obtain information and in some instances edit certain properties of the variable.

The Current Directory tab above the Workspace tab shows the contents of the *current directory*, whose *path* is shown in the *Current Directory Window*. For example, in the Windows operating system the path might be as follows: C:\MATLAB\Work, indicating that directory "Work" is a subdirectory of the main directory "MATLAB," which is installed in drive C. Clicking on the arrow in the Current Directory Window shows a list of recently used paths. Clicking on the button to the right of the window allows the user to change the current directory.

MATLAB uses a *search path* to find M-files and other MATLAB-related files, which are organized in directories in the computer file system. Any file run in MATLAB must reside in the current directory or in a directory that is on the search path. By default, the files supplied with MATLAB and MathWorks toolboxes are included in the search path. The easiest way to see which directories are on the search path, or to add or modify a search path, is to select **Set Path** from the **File** menu on the desktop, and then use the **Set Path** dialog box. It is good practice to add any commonly used directories to the search path to avoid repeatedly having the change the current directory.

The *Command History Window* contains a record of the commands a user has entered in the Command Window, including both current and previous MATLAB sessions. Previously entered MATLAB commands can be selected and re-executed from the Command History Window by right-clicking on a command or sequence of commands. This action launches a menu from which to select various options in addition to executing the commands. This is a useful feature when experimenting with various commands in a work session.

1.7.2 Using the MATLAB Editor to Create M-Files

The MATLAB *editor* is both a text editor specialized for creating M-files and a graphical MATLAB debugger. The editor can appear in a window by itself, or it can be a subwindow in the desktop. M-files are denoted by the extension `.m`, as in `pixeldup.m`. The MATLAB editor window has numerous pull-down menus for tasks such as saving, viewing, and debugging files. Because it performs some simple checks and also uses color to differentiate between various elements of code, this text editor is recommended as the tool of choice for writing and editing M-functions. To open the editor, type `edit` at the prompt in the Command Window. Similarly, typing `edit filename` at the prompt opens the M-file `filename.m` in an editor window, ready for editing. As noted earlier, the file must be in the current directory, or in a directory in the search path.

1.7.3 Getting Help

The principal way to get help online[†] is to use the MATLAB *Help Browser*, opened as a separate window either by clicking on the question mark symbol (**?**) on the desktop toolbar, or by typing `helpbrowser` at the prompt in the Command Window. The Help Browser is a Web browser integrated into the MATLAB desktop that displays Hypertext Markup Language (HTML) documents. The Help Browser consists of two panes, the *help navigator pane*, used to find information, and the *display pane*, used to view the information. Self-explanatory tabs on the navigator pane are used to perform a search. For example, help on a specific function is obtained by selecting the **Search** tab, selecting **Function Name** as the **Search Type**, and then typing in the function name in the **Search for** field. It is good practice to open the Help Browser

---

[†]Use of the term *online* in this book refers to information, such as help files, available in a local computer system, not on the Internet.

at the beginning of a MATLAB session to have help readily available during code development or other MATLAB task.

Another way to obtain help for a specific function is by typing `doc` followed by the function name at the command prompt. For example, typing `doc format` displays documentation for the function called `format` in the display pane of the Help Browser. This command opens the browser if it is not already open.

M-functions have two types of information that can be displayed by the user. The first is called the *H1 line*, which contains the function name and a one-line description. The second is a block of explanation called the *Help text block* (these are discussed in detail in Section 2.10.1). Typing `help` at the prompt followed by a function name displays both the H1 line and the Help text for that function in the Command Window. Occasionally, this information can be more up to date than the information in the Help browser because it is extracted directly from the documentation of the M-function in question. Typing `lookfor` followed by a keyword displays all the H1 lines that contain that keyword. This function is useful when looking for a particular topic without knowing the names of applicable functions. For example, typing `lookfor edge` at the prompt displays all the H1 lines containing that keyword. Because the H1 line contains the function name, it then becomes possible to look at specific functions using the other help methods. Typing `lookfor edge -all` at the prompt displays the H1 line of all functions that contain the word `edge` in either the H1 line or the Help text block. Words that contain the characters `edge` also are detected. For example, the H1 line of a function containing the word `polyedge` in the H1 line or Help text would also be displayed.

It is common MATLAB terminology to use the term *help page* when referring to the information about an M-function displayed by any of the preceding approaches, excluding `lookfor`. It is highly recommended that the reader become familiar with all these methods for obtaining information because in the following chapters we often give only representative syntax forms for MATLAB and IPT functions. This is necessary either because of space limitations or to avoid deviating from a particular discussion more than is absolutely necessary. In these cases we simply introduce the syntax required to execute the function in the form required at that point. By being comfortable with online search methods, the reader can then explore a function of interest in more detail with little effort.

Finally, the MathWorks' Web site mentioned in Section 1.3 contains a large database of help material, contributed functions, and other resources that should be utilized when the online documentation contains insufficient information about a desired topic.

1.7.4 Saving and Retrieving a Work Session

There are several ways to save and load an entire work session (the contents of the Workspace Browser) or selected workspace variables in MATLAB. The simplest is as follows.

To save the entire workspace, simply right-click on any blank space in the Workspace Browser window and select **Save Workspace As** from the menu that appears. This opens a directory window that allows naming the file and selecting any folder in the system in which to save it. Then simply click **Save**. To save a selected variable from the Workspace, select the variable with a left click and then right-click on the highlighted area. Then select **Save Selection As** from the menu that appears. This again opens a window from which a folder can be selected to save the variable. To select multiple variables, use shift-click or control-click in the familiar manner, and then use the procedure just described for a single variable. All files are saved in double-precision, binary format with the extension `.mat`. These saved files commonly are referred to as *MAT-files*. For example, a session named, say, mywork_2003_02_10, would appear as the MAT-file mywork_2003_02_10.mat when saved. Similarly, a saved image called `final_image` (which is a single variable in the workspace) will appear when saved as final_image.mat.

To load saved workspaces and/or variables, left-click on the folder icon on the toolbar of the Workspace Browser window. This causes a window to open from which a folder containing the MAT-files of interest can be selected. Double-clicking on a selected MAT-file or selecting **Open** causes the contents of the file to be restored in the Workspace Browser window.

It is possible to achieve the same results described in the preceding paragraphs by typing `save` and `load` at the prompt, with the appropriate file names and path information. This approach is not as convenient, but it is used when formats other than those available in the menu method are required. As an exercise, the reader is encouraged to use the Help Browser to learn more about these two functions.

1.8 How References Are Organized in the Book

All references in the book are listed in the Bibliography by author and date, as in Soille [2003]. Most of the background references for the theoretical content of the book are from Gonzalez and Woods [2002]. In cases where this is not true, the appropriate new references are identified at the point in the discussion where they are needed. References that are applicable to all chapters, such as MATLAB manuals and other general MATLAB references, are so identified in the Bibliography.

Summary

In addition to a brief introduction to notation and basic MATLAB tools, the material in this chapter emphasizes the importance of a comprehensive prototyping environment in the solution of digital image processing problems. In the following chapter we begin to lay the foundation needed to understand IPT functions and introduce a set of fundamental programming concepts that are used throughout the book. The material in Chapters 3 through 12 spans a wide cross section of topics that are in the mainstream of digital image processing applications. However, although the topics covered are varied, the discussion in those chapters follows the same basic theme of demonstrating how combining MATLAB and IPT functions with new code can be used to solve a broad spectrum of image-processing problems.

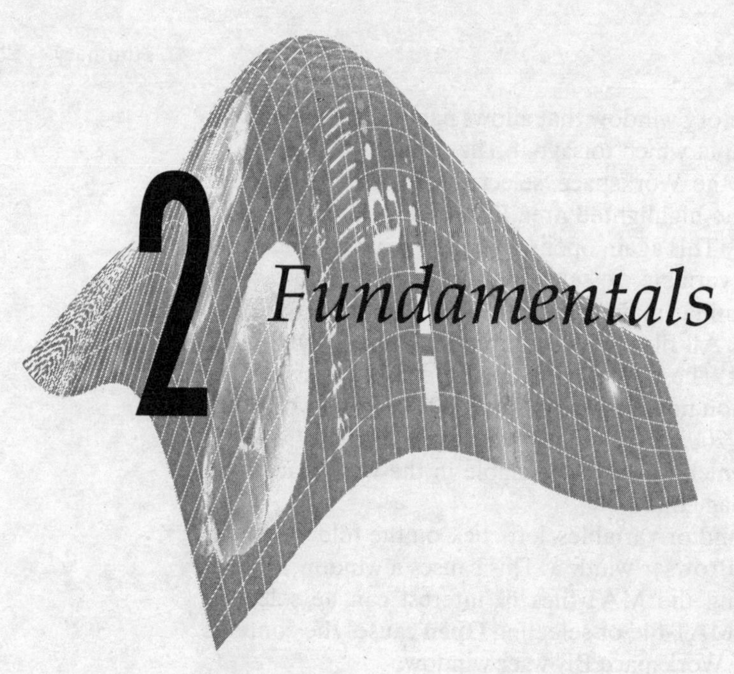

2 Fundamentals

Preview

As mentioned in the previous chapter, the power that MATLAB brings to digital image processing is an extensive set of functions for processing multidimensional arrays of which images (two-dimensional numerical arrays) are a special case. The Image Processing Toolbox (IPT) is a collection of functions that extend the capability of the MATLAB numeric computing environment. These functions, and the expressiveness of the MATLAB language, make many image-processing operations easy to write in a compact, clear manner, thus providing an ideal software prototyping environment for the solution of image processing problems. In this chapter we introduce the basics of MATLAB notation, discuss a number of fundamental IPT properties and functions, and introduce programming concepts that further enhance the power of IPT. Thus, the material in this chapter is the foundation for most of the material in the remainder of the book.

2.1 Digital Image Representation

An image may be defined as a two-dimensional function, $f(x, y)$, where x and y are *spatial* (plane) *coordinates*, and the amplitude of f at any pair of coordinates (x, y) is called the *intensity* of the image at that point. The term *gray level* is used often to refer to the intensity of monochrome images. Color images are formed by a combination of individual 2-D images. For example, in the RGB color system, a color image consists of three (red, green, and blue) individual component images. For this reason, many of the techniques developed for monochrome images can be extended to color images by processing the three component images individually. Color image processing is treated in detail in Chapter 6.

An image may be continuous with respect to the x- and y-coordinates, and also in amplitude. Converting such an image to digital form requires that the coordinates, as well as the amplitude, be digitized. Digitizing the coordinate values is called *sampling*; digitizing the amplitude values is called *quantization*. Thus, when x, y, and the amplitude values of f are all finite, discrete quantities, we call the image a *digital image*.

2.1.1 Coordinate Conventions

The result of sampling and quantization is a matrix of real numbers. We use two principal ways in this book to represent digital images. Assume that an image $f(x, y)$ is sampled so that the resulting image has M rows and N columns. We say that the image is of *size* $M \times N$. The values of the coordinates (x, y) are discrete quantities. For notational clarity and convenience, we use integer values for these discrete coordinates. In many image processing books, the image origin is defined to be at $(x, y) = (0, 0)$. The next coordinate values along the first row of the image are $(x, y) = (0, 1)$. It is important to keep in mind that the notation $(0, 1)$ is used to signify the second sample along the first row. It does not mean that these are the actual values of physical coordinates when the image was sampled. Figure 2.1(a) shows this coordinate convention. Note that x ranges from 0 to $M - 1$, and y from 0 to $N - 1$, in integer increments.

The coordinate convention used in the toolbox to denote arrays is different from the preceding paragraph in two minor ways. First, instead of using (x, y), the toolbox uses the notation (r, c) to indicate rows and columns. Note, however, that the order of coordinates is the same as the order discussed in the previous paragraph, in the sense that the first element of a coordinate tuple, (a, b), refers to a row and the second to a column. The other difference is that the origin of the coordinate system is at $(r, c) = (1, 1)$; thus, r ranges from 1 to M, and c from 1 to N, in integer increments. This coordinate convention is shown in Fig. 2.1(b).

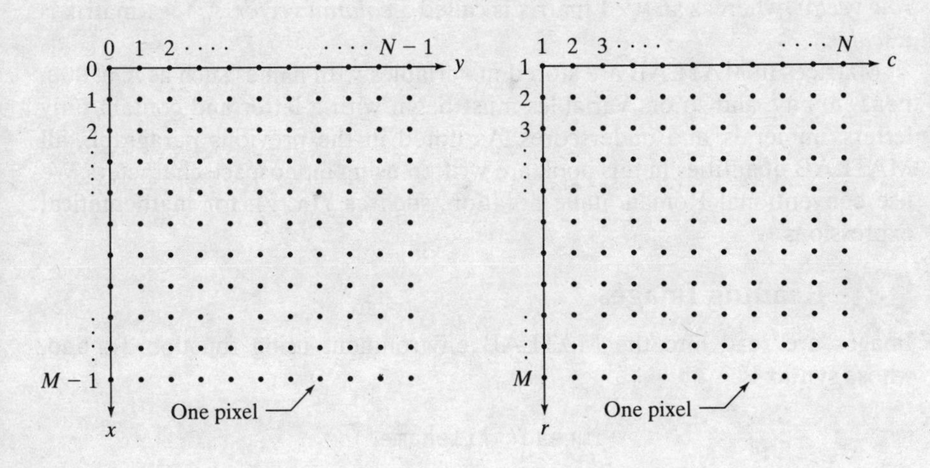

a b

FIGURE 2.1 Coordinate conventions used (a) in many image processing books, and (b) in the Image Processing Toolbox.

IPT documentation refers to the coordinates in Fig. 2.1(b) as *pixel coordinates*. Less frequently, the toolbox also employs another coordinate convention called *spatial coordinates*, which uses x to refer to columns and y to refers to rows. This is the opposite of our use of variables x and y. With very few exceptions, we do not use IPT's spatial coordinate convention in this book, but the reader will definitely encounter the terminology in IPT documentation.

2.1.2 Images as Matrices

The coordinate system in Fig. 2.1(a) and the preceding discussion lead to the following representation for a digitized image function:

$$f(x, y) = \begin{bmatrix} f(0,0) & f(0,1) & \cdots & f(0, N-1) \\ f(1,0) & f(1,1) & \cdots & f(1, N-1) \\ \vdots & \vdots & & \vdots \\ f(M-1,0) & f(M-1,1) & \cdots & f(M-1, N-1) \end{bmatrix}$$

The right side of this equation is a digital image by definition. Each element of this array is called an *image element*, *picture element*, *pixel*, or *pel*. The terms *image* and *pixel* are used throughout the rest of our discussions to denote a digital image and its elements.

MATLAB and IPT documentation use both the terms matrix *and* array, *mostly interchangeably. However, keep in mind that a matrix is two dimensional, whereas an array can have any finite dimension.*

A digital image can be represented naturally as a MATLAB matrix:

```
      ┌ f(1, 1)  f(1, 2)  ···  f(1, N) ┐
      │ f(2, 1)  f(2, 2)  ···  f(2, N) │
  f = │    ⋮        ⋮              ⋮    │
      └ f(M, 1)  f(M, 2)  ···  f(M, N) ┘
```

where `f(1, 1)` $= f(0,0)$ (note the use of a monospace font to denote MATLAB quantities). Clearly the two representations are identical, except for the shift in origin. The notation `f(p, q)` denotes the element located in row `p` and column `q`. For example, `f(6, 2)` is the element in the sixth row and second column of the matrix `f`. Typically we use the letters `M` and `N`, respectively, to denote the number of rows and columns in a matrix. A `1` × `N` matrix is called a *row vector*, whereas an `M` × `1` matrix is called a *column vector*. A `1` × `1` matrix is a *scalar*.

Matrices in MATLAB are stored in variables with names such as `A`, `a`, `RGB`, `real_array`, and so on. Variables must begin with a letter and contain only letters, numerals, and underscores. As noted in the previous paragraph, all MATLAB quantities in this book are written using monospace characters. We use conventional Roman, italic notation, such as $f(x, y)$, for mathematical expressions.

2.2 Reading Images

Images are read into the MATLAB environment using function `imread`, whose syntax is

```
imread('filename')
```

Format Name	Description	Recognized Extensions
TIFF	Tagged Image File Format	.tif, .tiff
JPEG	Joint Photographic Experts Group	.jpg, .jpeg
GIF	Graphics Interchange Format†	.gif
BMP	Windows Bitmap	.bmp
PNG	Portable Network Graphics	.png
XWD	X Window Dump	.xwd

† GIF is supported by `imread`, but not by `imwrite`.

TABLE 2.1 Some of the image/graphics formats supported by `imread` and `imwrite`, starting with MATLAB 6.5. Earlier versions support a subset of these formats. See online help for a complete list of supported formats.

Here, `filename` is a string containing the complete name of the image file (including any applicable extension). For example, the *command line*

```
>> f = imread('chestxray.jpg');
```

reads the JPEG (Table 2.1) image `chestxray` into image array `f`. Note the use of single quotes (') to delimit the string `filename`. The semicolon at the end of a command line is used by MATLAB for *suppressing* output. If a semicolon is not included, MATLAB displays the results of the operation(s) specified in that line. The prompt symbol (>>) designates the beginning of a command line, as it appears in the MATLAB Command Window (see Fig. 1.1).

semicolon (;)

prompt (>>)

When, as in the preceding command line, no path information is included in `filename`, `imread` reads the file from the *current directory* (see Section 1.7.1) and, if that fails, it tries to find the file in the MATLAB search path (see Section 1.7.1). The simplest way to read an image from a specified directory is to include a full or relative path to that directory in `filename`. For example,

In Windows, directories also are called folders.

```
>> f = imread('D:\myimages\chestxray.jpg');
```

reads the image from a folder called `myimages` on the `D:` drive, whereas

```
>> f = imread('.\myimages\chestxray.jpg');
```

reads the image from the `myimages` subdirectory of the current working directory. The Current Directory Window on the MATLAB desktop toolbar displays MATLAB's current working directory and provides a simple, manual way to change it. Table 2.1 lists some of the most popular image/graphics formats supported by `imread` and `imwrite` (`imwrite` is discussed in Section 2.4).

Function `size` gives the row and column dimensions of an image:

size

```
>> size(f)
ans =
   1024 1024
```

As in `size`*, many MATLAB and IPT functions can return more than one output argument. Multiple output arguments must be enclosed within square brackets,* [ ].

This function is particularly useful in programming when used in the following form to determine automatically the size of an image:

```
>> [M, N] = size(f);
```

This syntax returns the number of rows (`M`) and columns (`N`) in the image.

The `whos` function displays additional information about an array. For instance, the statement

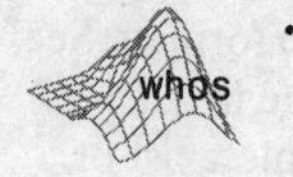


```
>> whos f
```

gives

```
  Name        Size         Bytes        Class
  f           1024x1024    1048576      uint8 array
  Grand total is 1048576 elements using 1048576 bytes
```

The `uint8` entry shown refers to one of several MATLAB data classes discussed in Section 2.5. A semicolon at the end of a `whos` line has no effect, so normally one is not used.

2.3 Displaying Images

Images are displayed on the MATLAB desktop using function `imshow`, which has the basic syntax:

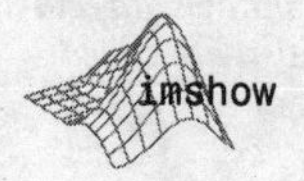


```
imshow(f, G)
```

where `f` is an image array, and `G` is the number of intensity levels used to display it. If `G` is omitted, it defaults to 256 levels. Using the syntax

```
imshow(f, [low high])
```

displays as black all values less than or equal to `low`, and as white all values greater than or equal to `high`. The values in between are displayed as intermediate intensity values using the default number of levels. Finally, the syntax

```
imshow(f, [ ])
```

sets variable `low` to the minimum value of array `f` and `high` to its maximum value. This form of `imshow` is useful for displaying images that have a low dynamic range or that have positive and negative values.

Function `pixval` is used frequently to display the intensity values of individual pixels interactively. This function displays a cursor overlaid on an image. As the cursor is moved over the image with the mouse, the coordinates of the cursor position and the corresponding intensity values are shown on a display that appears below the figure window. When working with color images, the coordinates as well as the red, green, and blue components are displayed. If the left button on the mouse is clicked and then held pressed, pixval displays the Euclidean distance between the initial and current cursor locations.

The syntax form of interest here is

```
pixval
```

pixval

which shows the cursor on the last image displayed. Clicking the X button on the cursor window turns it off.

EXAMPLE 2.1: Image reading and displaying.

■ (a) The following statements read from disk an image called `rose_512.tif`, extract basic information about the image, and display it using `imshow`:

```
>> f = imread('rose_512.tif');
>> whos f

   Name        Size             Bytes       Class
   f           512x512          262144      uint8 array
   Grand total is 262144 elements using 262144 bytes

>> imshow(f)
```

A semicolon at the end of an `imshow` line has no effect, so normally one is not used. Figure 2.2 shows what the output looks like on the screen. The figure number appears on the top, left of the window. Note the various pull-down menus and utility buttons. They are used for processes such as scaling, saving, and exporting the contents of the display window. In particular, the **Edit** menu has functions for editing and formatting results before they are printed or saved to disk.

FIGURE 2.2 Screen capture showing how an image appears on the MATLAB desktop. However, in most of the examples throughout this book, only the images themselves are shown. Note the figure number on the top, left part of the window.

If another image, g, is displayed using imshow, MATLAB replaces the image in the screen with the new image. To keep the first image and output a second image, we use function figure as follows:

```
>> figure, imshow(g)
```

Function figure *creates a figure window. When used without an argument, as shown here, it simply creates a new figure window. Typing* figure(n)*, forces figure number* n *to become visible.*

Using the statement

```
>> imshow(f), figure, imshow(g)
```

displays both images. Note that more than one command can be written on a line, as long as different commands are properly delimited by commas or semicolons. As mentioned earlier, a semicolon is used whenever it is desired to suppress screen outputs from a command line.

(b) Suppose that we have just read an image h and find that using imshow(h) produces the image in Fig. 2.3(a). It is clear that this image has a low dynamic range, which can be remedied for display purposes by using the statement

```
>> imshow(h, [ ])
```

Figure 2.3(b) shows the result. The improvement is apparent. ■

2.4 Writing Images

Images are written to disk using function imwrite, which has the following basic syntax:

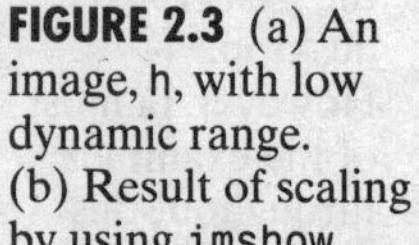

```
imwrite(f, 'filename')
```

With this syntax, the string contained in filename must include a recognized file format extension (see Table 2.1). Alternatively, the desired format can be specified explicitly with a third input argument. For example, the following command writes f to a TIFF file named patient10_run1:

```
>> imwrite(f, 'patient10_run1', 'tif')
```

or, alternatively,

```
>> imwrite(f, 'patient10_run1.tif')
```

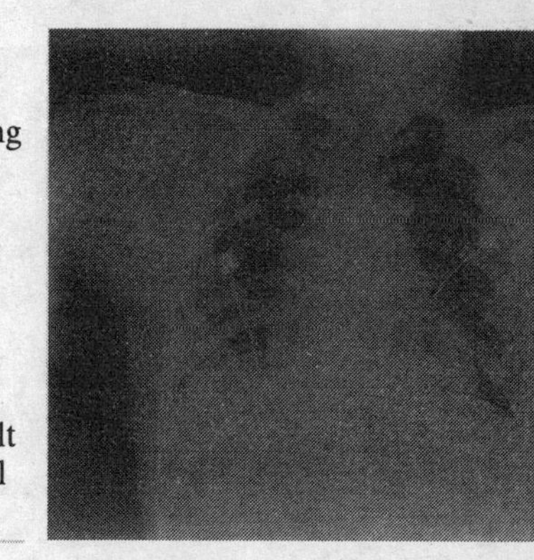

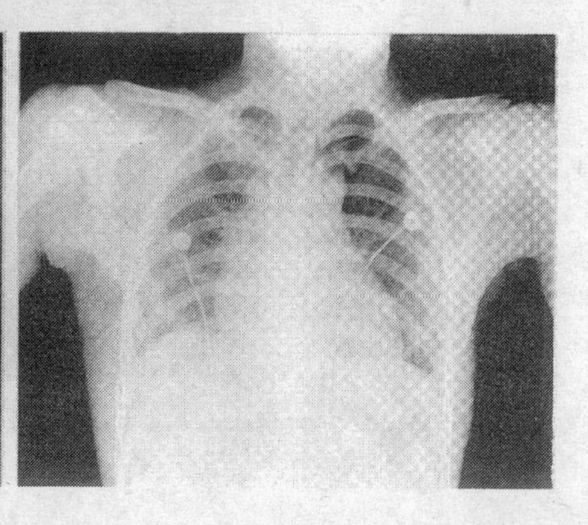

a b

FIGURE 2.3 (a) An image, h, with low dynamic range. (b) Result of scaling by using imshow (h,[]). (Original image courtesy of Dr. David R. Pickens, Dept. of Radiology & Radiological Sciences, Vanderbilt University Medical Center.)

If `filename` contains no path information, then `imwrite` saves the file in the current working directory.

The `imwrite` function can have other parameters, depending on the file format selected. Most of the work in the following chapters deals either with JPEG or TIFF images, so we focus attention here on these two formats.

A more general `imwrite` syntax applicable only to JPEG images is

```
imwrite(f, 'filename.jpg', 'quality', q)
```

where `q` is an integer between 0 and 100 (the lower the number the higher the degradation due to JPEG compression).

EXAMPLE 2.2: Writing an image and using function `imfinfo`.

■ Figure 2.4(a) shows an image, `f`, typical of sequences of images resulting from a given chemical process. It is desired to transmit these images on a routine basis to a central site for visual and/or automated inspection. In order to reduce storage and transmission time, it is important that the images be compressed as much as possible while not degrading their visual appearance beyond a reasonable level. In this case "reasonable" means no perceptible false contouring. Figures 2.4(b) through (f) show the results obtained by writing image `f` to disk (in JPEG format), with `q = 50, 25, 15, 5, and 0`, respectively. For example, for `q = 25` the applicable syntax is

```
>> imwrite(f, 'bubbles25.jpg', 'quality', 25)
```

The image for `q = 15` [Fig. 2.4(d)] has false contouring that is barely visible, but this effect becomes quite pronounced for `q = 5` and `q = 0`. Thus, an acceptable solution with some margin for error is to compress the images with `q = 25`. In order to get an idea of the compression achieved and to obtain other image file details, we can use function `imfinfo`, which has the syntax

```
imfinfo filename
```

`imfinfo`

where `filename` is the *complete* file name of the image stored in disk. For example,

```
>> imfinfo bubbles25.jpg
```

outputs the following information (note that some fields contain no information in this case):

```
         Filename:  'bubbles25.jpg'
      FileModDate:  '04-Jan-2003 12:31:26'
         FileSize:  13849
           Format:  'jpg'
    FormatVersion:  ''
            Width:  714
           Height:  682
         BitDepth:  8
        ColorType:  'grayscale'
  FormatSignature:  ''
          Comment:  {}
```

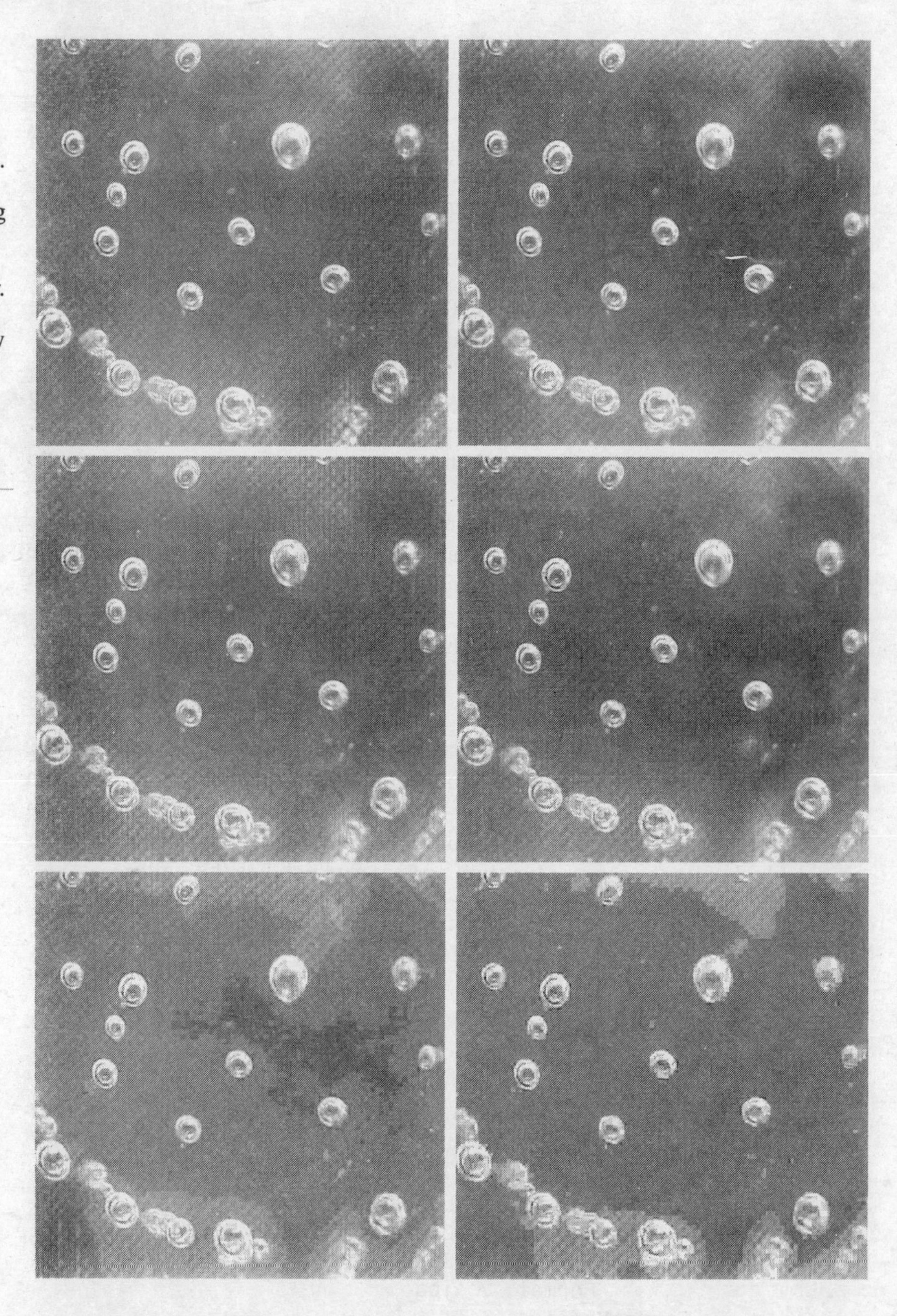

a b
c d
e f

FIGURE 2.4 (a) Original image. (b) through (f) Results of using `jpg` quality values `q = 50, 25, 15, 5,` and `0`, respectively. False contouring begins to be barely noticeable for `q = 15` [image (d)] but is quite visible for `q = 5` and `q = 0`.

See Example 2.11 for a function that creates all the images in Fig. 2.4 using a simple `for` *loop.*

where `FileSize` is in bytes. The number of bytes in the original image is computed simply by multiplying `Width` by `Height` by `BitDepth` and dividing the result by `8`. The result is 486948. Dividing this by `FileSize` gives the compression ratio: (486948/13849) = 35.16. This compression ratio was achieved while maintaining image quality consistent with the requirements of the application. In addition to the obvious advantages in storage space, this reduction allows the transmission of approximately 35 times the amount of uncompressed data per unit time.

The information fields displayed by `imfinfo` can be captured into a so-called *structure variable* that can be used for subsequent computations. Using the preceding image as an example, and assigning the name `K` to the structure variable, we use the syntax

Structures are discussed in Sections 2.10.6 and 11.1.1.

```
>> K = imfinfo('bubbles25.jpg');
```

to store into variable `K` all the information generated by command `imfinfo`. The information generated by `imfinfo` is appended to the structure variable by means of *fields*, separated from `K` by a dot. For example, the image height and width are now stored in structure fields `K.Height` and `K.Width`.

As an illustration, consider the following use of structure variable `K` to compute the compression ratio for `bubbles25.jpg`:

```
>> K = imfinfo('bubbles25.jpg');
>> image_bytes = K.Width*K.Height*K.BitDepth/8;
>> compressed_bytes = K.FileSize;
>> compression_ratio = image_bytes/compressed_bytes
compression_ratio =

    35.1612
```

Note that `imfinfo` was used in two different ways. The first was to type `imfinfo bubbles25.jpg` at the prompt, which resulted in the information being displayed on the screen. The second was to type `K = imfinfo('bubbles25.jpg')`, which resulted in the information generated by `imfinfo` being stored in `K`. These two different ways of calling `imfinfo` are an example of *command-function duality*, an important concept that is explained in more detail in the MATLAB online documentation. ■

To learn more about command function duality, consult the help page on this topic. See Section 1.7.3 regarding help pages.

A more general `imwrite` syntax applicable only to `tif` images has the form

```
imwrite(g, 'filename.tif', 'compression', 'parameter', ...
                           'resolution', [colres rowres])
```

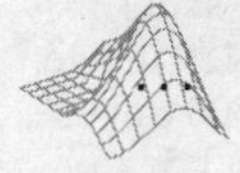

If a statement does not fit on one line, use an ellipsis (three periods), followed by **Return** *or* **Enter**, *to indicate that the statement continues on the next line. There are no spaces between the periods.*

where `'parameter'` can have one of the following principal values: `'none'` indicates no compression; `'packbits'` indicates packbits compression (the default for nonbinary images); and `'ccitt'` indicates ccitt compression (the default for binary images). The 1×2 array `[colres rowres]` contains two integers that give the column resolution and row resolution in dots-per-unit (the default values are [72 72]). For example, if the image dimensions are in inches, `colres` is the number of dots (pixels) per inch (dpi) in the vertical direction, and similarly for `rowres` in the horizontal direction. Specifying the resolution by a single scalar, `res`, is equivalent to writing `[res res]`.

EXAMPLE 2.3: Using `imwrite` parameters.

■ Figure 2.5(a) is an 8-bit X-ray image of a circuit board generated during quality inspection. It is in `jpg` format, at 200 dpi. The image is of size 450 × 450 pixels, so its dimensions are 2.25 × 2.25 inches. We want to store this image in `tif` format, with no compression, under the name `sf`. In addition, we want to reduce the size of the image to 1.5 × 1.5 inches while keeping the pixel count at 450 × 450. The following statement yields the desired result:

```
>> imwrite(f,'sf.tif','compression','none','resolution', ...
                                              [300 300])
```

The values of the vector `[colres rowres]` were determined by multiplying 200 dpi by the ratio 2.25/1.5, which gives 300 dpi. Rather than do the computation manually, we could write

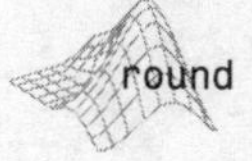


```
>> res = round(200*2.25/1.5);
>> imwrite(f, 'sf.tif', 'compression', 'none' ,'resolution', res)
```

where function `round` rounds its argument to the nearest integer. It is important to note that the number of pixels was not changed by these commands. Only the scale of the image changed. The original 450 × 450 image at 200 dpi is of size 2.25 × 2.25 inches. The new 300-dpi image is identical, except that its

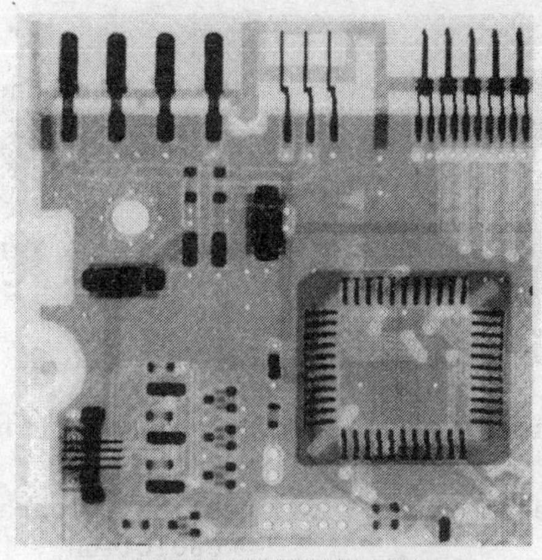

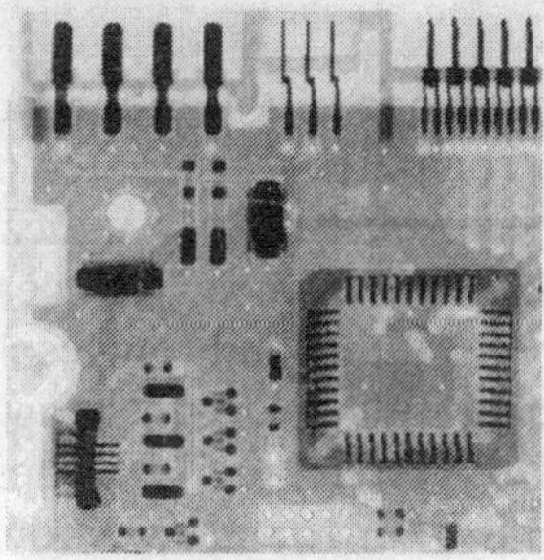

a
b

FIGURE 2.5 Effects of changing the dpi resolution while keeping the number of pixels constant. (a) A 450 × 450 image at 200 dpi (size = 2.25 × 2.25 inches). (b) The same 450 × 450 image, but at 300 dpi (size = 1.5 × 1.5 inches). (Original image courtesy of Lixi, Inc.)

450 × 450 pixels are distributed over a 1.5 × 1.5-inch area. Processes such as this are useful for controlling the size of an image in a printed document without sacrificing resolution. ■

Often, it is necessary to export images to disk the way they appear on the MATLAB desktop. This is especially true with plots, as shown in the next chapter. The *contents* of a figure window can be exported to disk in two ways. The first is to use the **File** pull-down menu in the figure window (see Fig. 2.2) and then choose **Export**. With this option, the user can select a location, file name, and format. More control over export parameters is obtained by using the `print` command:

```
print -fno -dfileformat -rresno filename
```

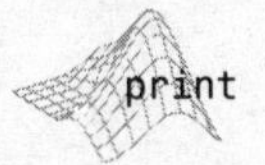

where *no* refers to the figure number in the figure window of interest, *fileformat* refers to one of the file formats in Table 2.1, *resno* is the resolution in dpi, and `filename` is the name we wish to assign the file. For example, to export the contents of the figure window in Fig. 2.2 as a `tif` file at 300 dpi, and under the name `hi_res_rose`, we would type

```
>> print -f1 -dtiff -r300 hi_res_rose
```

This command sends the file `hi_res_rose.tif` to the current directory.

If we simply type `print` at the prompt, MATLAB prints (to the default printer) the contents of the last figure window displayed. It is possible also to specify other options with `print`, such as a specific printing device.

2.5 Data Classes

Although we work with integer coordinates, the values of pixels themselves are not restricted to be integers in MATLAB. Table 2.2 lists the various *data classes*† supported by MATLAB and IPT for representing pixel values. The first eight entries in the table are referred to as *numeric* data classes. The ninth entry is the *char* class and, as shown, the last entry is referred to as the *logical* data class.

All *numeric* computations in MATLAB are done using `double` quantities, so this is also a frequent data class encountered in image processing applications. Class `uint8` also is encountered frequently, especially when reading data from storage devices, as 8-bit images are the most common representations found in practice. These two data classes, class `logical`, and, to a lesser degree, class `uint16`, constitute the primary data classes on which we focus in this book. Many IPT functions, however, support all the data classes listed in Table 2.2. Data class `double` requires 8 bytes to represent a number, `uint8` and `int8` require 1 byte each, `uint16` and `int16` require 2 bytes, and `uint32`,

†MATLAB documentation often uses the terms *data class* and *data type* interchangeably. In this book, we reserve use of the term *type* for images, as discussed in Section 2.6.

TABLE 2.2
Data classes. The first eight entries are referred to as *numeric* classes; the ninth entry is the *character* class, and the last entry is of class *logical*.

Name	Description
`double`	Double-precision, floating-point numbers in the approximate range -10^{308} to 10^{308} (8 bytes per element).
`uint8`	Unsigned 8-bit integers in the range [0, 255] (1 byte per element).
`uint16`	Unsigned 16-bit integers in the range [0, 65535] (2 bytes per element).
`uint32`	Unsigned 32-bit integers in the range [0, 4294967295] (4 bytes per element).
`int8`	Signed 8-bit integers in the range [−128, 127] (1 byte per element).
`int16`	Signed 16-bit integers in the range [−32768, 32767] (2 bytes per element).
`int32`	Signed 32-bit integers in the range [−2147483648, 2147483647] (4 bytes per element).
`single`	Single-precision floating-point numbers with values in the approximate range -10^{38} to 10^{38} (4 bytes per element).
`char`	Characters (2 bytes per element).
`logical`	Values are 0 or 1 (1 byte per element).

`int32`, and `single`, require 4 bytes each. The `char` data class holds characters in Unicode representation. A *character string* is merely a `1` × `n` array of characters. A `logical` array contains only the values 0 and 1, with each element being stored in memory using one byte per element. Logical arrays are created by using function `logical` (see Section 2.6.2) or by using relational operators (Section 2.10.2).

2.6 Image Types

The toolbox supports four types of images:

- Intensity images
- Binary images
- Indexed images
- RGB images

Most monochrome image processing operations are carried out using binary or intensity images, so our initial focus is on these two image types. Indexed and RGB color images are discussed in Chapter 6.

2.6.1 Intensity Images

An *intensity image* is a data matrix whose values have been scaled to represent intensities. When the elements of an intensity image are of class `uint8`, or class `uint16`, they have integer values in the range [0, 255] and [0, 65535], respectively. If the image is of class `double`, the values are floating-point numbers. Values of scaled, class `double` intensity images are in the range [0, 1] by convention.

2.6.2 Binary Images

Binary images have a very specific meaning in MATLAB. A *binary image* is a *logical* array of 0s and 1s. Thus, an array of 0s and 1s whose values are of data class, say, `uint8`, is not considered a binary image in MATLAB. A numeric array is converted to binary using function `logical`. Thus, if `A` is a numeric array consisting of 0s and 1s, we create a logical array `B` using the statement

```
B = logical(A)
```

If `A` contains elements other than 0s and 1s, use of the `logical` function converts all nonzero quantities to logical 1s and all entries with value 0 to logical 0s. Using relational and logical operators (see Section 2.10.2) also creates logical arrays.

To test if an array is logical we use the `islogical` function:

```
islogical(C)
```

See Table 2.9 for a list of other functions based on the `is*` *syntax.*

If `C` is a logical array, this function returns a 1. Otherwise it returns a 0. Logical arrays can be converted to numeric arrays using the data class conversion functions discussed in Section 2.7.1.

2.6.3 A Note on Terminology

Considerable care was taken in the previous two sections to clarify the use of the terms *data class* and *image type*. In general, we refer to an image as being a "`data_class` image_type image," where `data_class` is one of the entries from Table 2.2, and image_type is one of the image types defined at the beginning of this section. Thus, an image is characterized by *both* a class *and* a type. For instance, a statement discussing an "`unit8` intensity image" is simply referring to an intensity image whose pixels are of data class `unit8`. Some functions in the toolbox support all data classes, while others are very specific as to what constitutes a valid class. For example, the pixels in a binary image can only be of data class `logical`, as mentioned earlier.

2.7 Converting between Data Classes and Image Types

Converting between data classes and image types is a frequent operation in IPT applications. When converting between data classes, it is important to keep in mind the value ranges for each data class detailed in Table 2.2.

2.7.1 Converting between Data Classes

Converting between data classes is straightforward. The general syntax is

```
B = data_class_name(A)
```

where `data_class_name` is one of the names in the first column of Table 2.2. For example, suppose that `A` is an array of class `uint8`. A double-precision array, B, is generated by the command `B = double(A)`. This conversion is used routinely throughout the book because MATLAB expects operands in numerical computations to be double-precision, floating-point numbers. If `C` is an array of class `double` in which all values are in the range [0, 255] (but possibly containing fractional values), it can be converted to an `uint8` array with the command `D = uint8(C)`.

If an array of class `double` has any values outside the range [0, 255] and it is converted to class `uint8` in the manner just described, MATLAB converts to 0 all values that are less than 0, and converts to 255 all values that are greater than 255. Numbers in between are converted to integers by discarding their fractional parts. Thus, proper scaling of a `double` array so that its elements are in the range [0, 255] is necessary before converting it to `uint8`. As indicated in Section 2.6.2, converting any of the numeric data classes to `logical` results in an array with logical 1s in locations where the input array had nonzero values, and logical 0s in places where the input array contained 0s.

2.7.2 Converting between Image Classes and Types

M-function `change-class`, *discussed in Section 3.2.3, can be used for changing an input image to a specified class.*

The toolbox provides specific functions (Table 2.3) that perform the scaling necessary to convert between image classes and types. Function `im2uint8` detects the data class of the input and performs all the necessary scaling for the toolbox to recognize the data as valid image data. For example, consider the following 2×2 image `f` of class `double`, which could be the result of an intermediate computation:

```
f =

   -0.5    0.5
   0.75    1.5
```

Performing the conversion

```
>> g = im2uint8(f)
```

yields the result

```
g =

     0   128
   191   255
```

TABLE 2.3
Functions in IPT for converting between image classes and types. See Table 6.3 for conversions that apply specifically to color images.

Name	Converts Input to:	Valid Input Image Data Classes
`im2uint8`	`uint8`	`logical`, `uint8`, `uint16`, and `double`
`im2uint16`	`uint16`	`logical`, `uint8`, `uint16`, and `double`
`mat2gray`	`double` (in range [0, 1])	`double`
`im2double`	`double`	`logical`, `uint8`, `uint16`, and `double`
`im2bw`	`logical`	`uint8`, `uint16`, and `double`

from which we see that function im2uint8 sets to 0 all values in the input that are less than 0, sets to 255 all values in the input that are greater than 1, and multiplies all other values by 255. Rounding the results of the multiplication to the nearest integer completes the conversion. Note that the rounding behavior of im2uint8 is different from the data-class conversion function uint8 discussed in the previous section, which simply discards fractional parts.

Converting an arbitrary array of class double to an array of class double *scaled* to the range [0, 1] can be accomplished by using function mat2gray whose basic syntax is

```
g = mat2gray(A, [Amin, Amax])
```

where image g has values in the range 0 (black) to 1 (white). The specified parameters Amin and Amax are such that values less than Amin in A become 0 in g, and values greater than Amax in A correspond to 1 in g. Writing

```
>> g = mat2gray(A);
```

sets the values of Amin and Amax to the actual minimum and maximum values in A. The input is assumed to be of class double. The output also is of class double.

Function im2double converts an input to class double. If the input is of class uint8, uint16, or logical, function im2double converts it to class double with values in the range [0, 1]. If the input is already of class double, im2double returns an array that is equal to the input. For example, if an array of class double results from computations that yield values outside the range [0, 1], inputting this array into im2double will have no effect. As mentioned in the preceding paragraph, a double array having arbitrary values can be converted to a double array with values in the range [0, 1] by using function mat2gray.

As an illustration, consider the class uint8 image†

```
>> h = uint8([25 50; 128 200]);
```

Performing the conversion

```
>> g = im2double(h);
```

yields the result

```
g =

    0.0980    0.1961
    0.4706    0.7843
```

from which we infer that the conversion when the input is of class uint8 is done simply by dividing each value of the input array by 255. If the input is of class uint16 the division is by 65535.

---

†Section 2.8.2 explains the use of square brackets and semicolons to specify a matrix.

Finally, we consider conversion between binary and intensity image types. Function im2bw, which has the syntax

```
g = im2bw(f, T)
```

produces a binary image, g, from an intensity image, f, by thresholding. The output binary image g has values of 0 for all pixels in the input image with intensity values less than threshold T, and 1 for all other pixels. The value specified for T has to be in the range [0, 1], regardless of the class of the input. The output binary image is automatically declared as a logical array by im2bw. If we write g = im2bw(f), IPT uses a default value of 0.5 for T. If the input is an uint8 image, im2bw divides all its pixels by 255 and then applies either the default or a specified threshold. If the input is of class uint16, the division is by 65535. If the input is a double image, im2bw applies either the default or a specified threshold directly. If the input is a logical array, the output is identical to the input. A logical (binary) array can be converted to a numerical array by using any of the four functions in the first column of Table 2.3.

EXAMPLE 2.4: Converting between image classes and types.

■ (a) We wish to convert the following double image

```
>> f = [1 2; 3 4]
f =
     1     2
     3     4
```

to binary such that values 1 and 2 become 0 and the other two values become 1. First we convert it to the range [0, 1]:

```
>> g = mat2gray(f)
g =
          0    0.3333
     0.6667    1.0000
```

Then we convert it to binary using a threshold, say, of value 0.6:

```
>> gb = im2bw(g, 0.6)
gb =
     0     0
     1     1
```

As mentioned in Section 2.5, we can generate a binary array directly using relational operators (Section 2.10.2). Thus we get the same result by writing

```
>> gb = f > 2

gb =

     0     0
     1     1
```

We could store in a variable (say, gbv) the fact that gb is a logical array by using the islogical function, as follows:

```
>> gbv = islogical(gb)

gbv =

     1
```

(b) Suppose now that we want to convert gb to a numerical array of 0s and 1s of class double. This is done directly:

```
>> gbd = im2double(gb)

gbd =

     0     0
     1     1
```

If gb had been a numeric array of class uint8, applying im2double to it would have resulted in an array with values

```
       0          0
  0.0039     0.0039
```

because im2double would have divided all the elements by 255. This did not happen in the preceding conversion because im2double detected that the input was a logical array, whose only possible values are 0 and 1. If the input in fact had been an uint8 numeric array and we wanted to convert it to class double while keeping the 0 and 1 values, we would have converted the array by writing

```
>> gbd = double(gb)

gbd =

     0     0
     1     1
```

Finally, we point out that MATLAB supports nested statements, so we could have started with image f and arrived at the same result by using the one-line statement

```
>> gbd = im2double(im2bw(mat2gray(f), 0.6));
```

or by using partial groupings of these functions. Of course, the entire process could have been done in this case with a simpler command:

```
>> gbd = double(f > 2);
```

again demonstrating the compactness of the MATLAB language. ■

2.8 Array Indexing

MATLAB supports a number of powerful indexing schemes that simplify array manipulation and improve the efficiency of programs. In this section we discuss and illustrate basic indexing in one and two dimensions (i.e., vectors and matrices). More sophisticated techniques are introduced as needed in subsequent discussions.

2.8.1 Vector Indexing

As discussed in Section 2.1.2, an array of dimension $1 \times N$ is called a *row vector*. The elements of such a vector are accessed using one-dimensional indexing. Thus, v(1) is the first element of vector v, v(2) its second element, and so forth. The elements of vectors in MATLAB are enclosed by square brackets and are separated by spaces or by commas. For example,

```
>> v = [1 3 5 7 9]
v =
    1   3   5   7   9
>> v(2)
ans =
    3
```

A row vector is converted to a column vector using the *transpose operator* (.'):

Using a single quote without the period computes the conjugate transpose. When the data are real, both transposes can be used interchangeably. See Table 2.4.

```
>> w = v.'
w =
    1
    3
    5
    7
    9
```

To access *blocks* of elements, we use MATLAB's *colon* notation. For example, to access the first three elements of v we write

```
>> v(1:3)

ans =

     1     3     5
```

Similarly, we can access the second through the fourth elements

```
>> v(2:4)

ans =

     3     5     7
```

or all the elements from, say, the third through the last element:

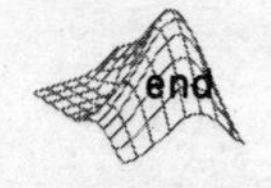


```
>> v(3:end)

ans =

     5     7     9
```

where end signifies the last element in the vector. If v is a vector, writing

```
>> v(:)
```

produces a column vector, whereas writing

```
>> v(1:end)
```

produces a row vector.

Indexing is not restricted to contiguous elements. For example,

```
>> v(1:2:end)

ans =
     1     5     9
```

The notation 1:2:end says to start at 1, count up by 2 and stop when the count reaches the last element. The steps can be negative:

```
>> v(end:-2:1)

ans =

     9     5     1
```

Here, the index count started at the last element, decreased by 2, and stopped when it reached the first element.

Function `linspace`, with syntax

```
x = linspace(a, b, n)
```

generates a row vector `x` of `n` elements linearly spaced between and including `a` and `b`. We use this function in several places in later chapters.

A vector can even be used as an index into another vector. For example, we can pick the first, fourth, and fifth elements of `v` using the command

```
>> v([1 4 5])

ans =

    1    7    9
```

As shown in the following section, the ability to use a vector as an index into another vector also plays a key role in matrix indexing.

2.8.2 Matrix Indexing

Matrices can be represented conveniently in MATLAB as a sequence of row vectors enclosed by square brackets and separated by semicolons. For example, typing

```
>> A = [1 2 3; 4 5 6; 7 8 9]
```

displays the 3×3 matrix

```
A =

    1    2    3
    4    5    6
    7    8    9
```

Note that the use of semicolons here is different from their use mentioned earlier to suppress output or to write multiple commands in a single line.

We select elements in a matrix just as we did for vectors, but now we need two indices: one to establish a row location and the other for the corresponding column. For example, to extract the element in the second row, third column, we write

```
>> A(2, 3)

ans =

    6
```

The colon operator is used in matrix indexing to select a two-dimensional block of elements out of a matrix. For example,

```
>> C3 = A(:, 3)
C3 =
     3
     6
     9
```

Here, use of the colon by itself is analogous to writing `A(1:3,3)`, which simply picks the third column of the matrix. Similarly, we extract the second row as follows:

```
>> R2 = A(2, :)
R2 =
     4     5     6
```

The following statement extracts the top two rows:

```
>> T2 = A(1:2, 1:3)
T2 =
     1     2     3
     4     5     6
```

To create a matrix `B` equal to `A` but with its last column set to 0s, we write

```
>> B = A;
>> B(:, 3) = 0
B =
     1     2     0
     4     5     0
     7     8     0
```

Operations using `end` are carried out in a manner similar to the examples given in the previous section for vector indexing. The following examples illustrate this.

```
>> A(end, end)
ans =
     9
```

```
>> A(end, end - 2)
ans =
    7
>> A(2:end, end:-2:1)
ans =
    6   4
    9   7
```

Using vectors to index into a matrix provides a powerful approach for element selection. For example,

```
>> E = A([1 3], [2 3])
E =
    2   3
    8   9
```

The notation `A([a b],[c d])` picks out the elements in `A` with coordinates (row `a`, column `c`), (row `a`, column `d`), (row `b`, column `c`), and (row `b`, column `d`). Thus, when we let `E = A([1 3], [2 3])` we are selecting the following elements in `A`: the element in row 1 column 2, the element in row 1 column 3, the element in row 3 column 2, and the element in row 3 column 3.

More complex schemes can be implemented using matrix addressing. A particularly useful addressing approach using matrices for indexing is of the form `A(D)`, where `D` is a logical array. For example, if

```
>> D = logical([1 0 0; 0 0 1; 0 0 0])
D =
    1   0   0
    0   0   1
    0   0   0
```

then

```
>> A(D)
ans =
    1
    6
```

Finally, we point out that use of a single colon as an index into a matrix selects all the elements of the array (on a column-by-column basis) and arranges them in the form of a column vector. For example, with reference to matrix `T2`,

```
>> v = T2(:)
v =

     1
     4
     2
     5
     3
     6
```

This use of the colon is helpful when, for example, we want to find the sum of all the elements of a matrix:

```
>> s = sum(A(:))
s =

    45
```

In general, `sum(v)` adds the values of all the elements of input vector `v`. If a matrix is input into `sum` [as in `sum(A)`], the output is a row vector containing the sums of each individual column of the input array (this behavior is typical of many MATLAB functions encountered in later chapters). By using a single colon in the manner just illustrated, we are in reality implementing the command

```
>> sum(sum(A));
```

because use of a single colon converts the matrix into a vector.

Using the colon notation is actually a form of *linear indexing* into a matrix or higher-dimensional array. In fact, MATLAB stores each array as a column of values regardless of the actual dimensions. This column consists of the array columns, appended end to end. For example, matrix `A` is stored in MATLAB as

```
     1
     4
     7
     2
     5
     8
     3
     6
     9
```

Accessing `A` with a single subscript indexes directly into this column. For example, `A(3)` accesses the third value in the column, the number `7`; `A(8)` accesses the eighth value, `6`, and so on. When we use the column notation, we are simply addressing all the elements, `A(1:end)`. This type of indexing is a basic staple in *vectorizing* loops for program optimization, as discussed in Section 2.10.4.

EXAMPLE 2.5: Some simple image operations using array indexing.

■ The image in Fig. 2.6(a) is a 1024 × 1024 intensity image, `f`, of class `uint8`. The image in Fig. 2.6(b) was flipped vertically using the statement

```
>> fp = f(end:-1:1, :);
```

The image shown in Fig. 2.6(c) is a section out of image (a), obtained using the command

```
>> fc = f(257:768, 257:768);
```

Similarly, Fig. 2.6(d) shows a subsampled image obtained using the statement

```
>> fs = f(1:2:end, 1:2:end);
```

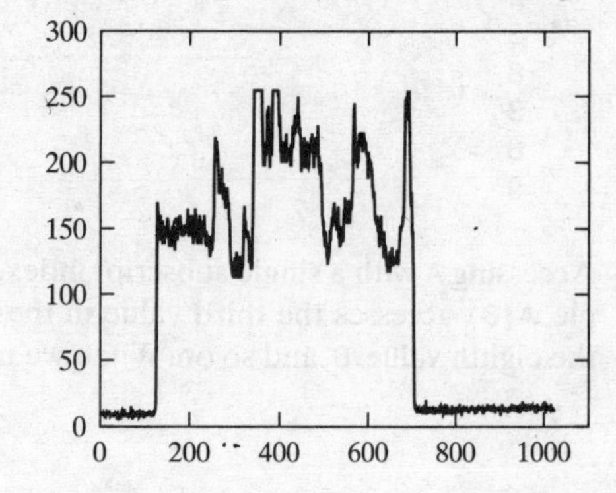

a b
c
d e

FIGURE 2.6 Results obtained using array indexing. (a) Original image. (b) Image flipped vertically. (c) Cropped image. (d) Subsampled image. (e) A horizontal scan line through the middle of the image in (a).

Finally, Fig. 2.6(e) shows a horizontal scan line through the middle of Fig. 2.6(a), obtained using the command

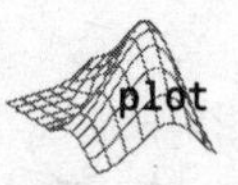


```
>> plot(f(512, :))
```

The `plot` function is discussed in detail in Section 3.3.1. ■

2.8.3 Selecting Array Dimensions

Operations of the form

```
operation(A, dim)
```

where `operation` denotes an applicable MATLAB operation, `A` is an array, and `dim` is a scalar, are used frequently in this book. For example, suppose that `A` is an array of size $M \times N$. The command

```
>> k = size(A, 1);
```

gives the size of `A` along its first dimension, which is defined by MATLAB as the vertical dimension. That is, this command gives the number of rows in `A`. Similarly, the second dimension of an array is in the horizontal direction, so the statement `size(A,2)` gives the number of columns in `A`. A *singleton dimension* is any dimension, `dim`, for which `size(A, dim) = 1`. Using these concepts, we could have written the last command in Example 2.5 as

```
>> plot(f(size(f, 1)/2, :))
```

MATLAB does not restrict the number of dimensions of an array, so being able to extract the components of an array in any dimension is an important feature. For the most part, we deal with 2-D arrays, but there are several instances (as when working with color or multispectral images) when it is necessary to be able to "stack" images along a third or higher dimension. We deal with this in Chapters 6, 11, and 12. Function `ndims`, with syntax

```
d = ndims(A)
```

gives the number of dimensions of array A. Function `ndims` never returns a value less than 2 because even scalars are considered two dimensional, in the sense that they are arrays of size 1×1.

2.9 Some Important Standard Arrays

Often, it is useful to be able to generate simple image arrays to try out ideas and to test the syntax of functions during development. In this section we introduce seven array-generating functions that are used in later chapters. If only one argument is included in any of the following functions, the result is a square array.

- `zeros(M, N)` generates an `M`×`N` matrix of 0s of class `double`.
- `ones(M, N)` generates an `M`×`N` matrix of 1s of class `double`.
- `true(M, N)` generates an `M`×`N` `logical` matrix of 1s.
- `false(M, N)` generates an `M`×`N` `logical` matrix of 0s.
- `magic(M)` generates an `M` × `M` "magic square." This is a square array in which the sum along any row, column, or main diagonal, is the same. Magic squares are useful arrays for testing purposes because they are easy to generate and their numbers are integers.
- `rand(M, N)` generates an `M`×`N` matrix whose entries are uniformly distributed random numbers in the interval [0, 1].
- `randn(M, N)` generates an `M`×`N` matrix whose numbers are normally distributed (i.e., Gaussian) random numbers with mean 0 and variance 1.

For example,

```
>> A = 5*ones(3, 3)
A =

     5     5     5
     5     5     5
     5     5     5
>> magic(3)
ans =
     8     1     6
     3     5     7
     4     9     2
>> B = rand(2, 4)
B =
    0.2311    0.4860    0.7621    0.0185
    0.6068    0.8913    0.4565    0.8214
```

2.10 Introduction to M-Function Programming

One of the most powerful features of the Image Processing Toolbox is its transparent access to the MATLAB programming environment. As will become evident shortly, MATLAB function programming is flexible and particularly easy to learn.

2.10.1 M-Files

So-called *M-files* in MATLAB can be scripts that simply execute a series of MATLAB statements, or they can be functions that can accept arguments and can produce one or more outputs. The focus of this section in on M-file functions. These functions extend the capabilities of both MATLAB and IPT to address specific, user-defined applications.

M-files are created using a text editor and are stored with a name of the form `filename.m`, such as `average.m` and `filter.m`. The components of a function M-file are

- The function definition line
- The H1 line
- Help text
- The function body
- Comments

The *function definition line* has the form

```
function [outputs] = name(inputs)
```

For example, a function to compute the sum and product (two different outputs) of two images would have the form

```
function [s, p] = sumprod(f, g)
```

where `f`, and `g` are the input images, `s` is the sum image, and `p` is the product image. The name `sumprod` is arbitrarily defined, but the word `function` always appears on the left, in the form shown. Note that the output arguments are enclosed by square brackets and the inputs are enclosed by parentheses. If the function has a single output argument, it is acceptable to list the argument without brackets. If the function has no output, only the word `function` is used, without brackets or equal sign. Function names must begin with a letter, and the remaining characters can be any combination of letters, numbers, and underscores. No spaces are allowed. MATLAB distinguishes function names up to 63 characters long. Additional characters are ignored.

Functions can be called at the command prompt; for example,

```
>> [s, p] = sumprod(f, g);
```

or they can be used as elements of other functions, in which case they become *subfunctions*. As noted in the previous paragraph, if the output has a single argument, it is acceptable to write it without the brackets, as in

```
>> y = sum(x);
```

The *H1 line* is the first text line. It is a single *comment* line that follows the function definition line. There can be no blank lines or leading spaces between the H1 line and the function definition line. An example of an H1 line is

```
% SUMPROD Computes the sum and product of two images.
```

As indicated in Section 1.7.3, the H1 line is the first text that appears when a user types

```
>> help function_name
```

at the MATLAB prompt. Also, as mentioned in that section, typing `lookfor keyword` displays all the H1 lines containing the string `keyword`. This line provides important summary information about the M-file, so it should be as descriptive as possible.

Help text is a text block that follows the H1 line, without any blank lines in between the two. Help text is used to provide comments and online help for the function. When a user types `help function_name` at the prompt, MATLAB displays all comment lines that appear between the function definition line and the first noncomment (executable or blank) line. The help system ignores any comment lines that appear after the Help text block.

The *function body* contains all the MATLAB code that performs computations and assigns values to output arguments. Several examples of MATLAB code are given later in this chapter.

All lines preceded by the symbol "`%`" that are not the H1 line or Help text are considered function *comment* lines and are not considered part of the Help text block. It is permissible to append comments to the end of a line of code.

M-files can be created and edited using any text editor and saved with the extension `.m` in a specified directory, typically in the MATLAB search path. Another way to create or edit an M-file is to use the `edit` function at the prompt. For example,

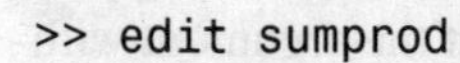

```
>> edit sumprod
```

opens for editing the file `sumprod.m` if the file exists in a directory that is in the MATLAB path or in the current directory. If the file cannot be found, MATLAB gives the user the option to create it. As noted in Section 1.7.2, the MATLAB editor window has numerous pull-down menus for tasks such as saving, viewing, and debugging files. Because it performs some simple checks and uses color to differentiate between various elements of code, this text editor is recommended as the tool of choice for writing and editing M-functions.

2.10.2 Operators

MATLAB operators are grouped into three main categories:

- Arithmetic operators that perform numeric computations
- Relational operators that compare operands quantitatively
- Logical operators that perform the functions AND, OR, and NOT

These are discussed in the remainder of this section.

Arithmetic Operators

dot notation

MATLAB has two different types of arithmetic operations. *Matrix arithmetic operations* are defined by the rules of linear algebra. *Array arithmetic operations* are carried out element by element and can be used with multidimensional arrays. The period (dot) character (`.`) distinguishes array operations from matrix operations. For example, `A*B` indicates matrix multiplication in the traditional sense, whereas `A.*B` indicates array multiplication, in the sense that the result is an array, the same size as `A` and `B`, in which each element is the product of corresponding elements of A and B. In other words, if C = A.*B, then C(I, J) = A(I, J)*B(I, J). Because matrix and array operations are the same for addition and subtraction, the character pairs .+ and .– are not used.

When writing an expression such as B = A, MATLAB makes a "note" that B is equal to A, but does not actually copy the data into B unless the contents of A change later in the program. This is an important point because using different variables to "store" the same information sometimes can enhance code clarity and readability. Thus, the fact that MATLAB does not duplicate information unless it is absolutely necessary is worth remembering when writing MATLAB code. Table 2.4 lists the MATLAB arithmetic operators, where A

TABLE 2.4 Array and matrix arithmetic operators. Computations involving these operators can be implemented using the operators themselves, as in A + B, or using the MATLAB functions shown, as in plus (A, B). The examples shown for arrays use matrices to simplify the notation, but they are easily extendable to higher dimensions.

Operator	Name	MATLAB Function	Comments and Examples
+	Array and matrix addition	plus(A, B)	a + b, A + B, or a + A.
–	Array and matrix subtraction	minus(A, B)	a – b, A – B, A – a, or a – A.
.*	Array multiplication	times(A, B)	C = A.*B, C(I, J) = A(I, J)*B(I, J).
*	Matrix multiplication	mtimes(A, B)	A*B, standard matrix multiplication, or a*A, multiplication of a scalar times all elements of A.
./	Array right division	rdivide(A, B)	C = A./B, C(I, J) = A(I, J)/B(I, J).
.\	Array left division	ldivide(A, B)	C = A.\B, C(I, J) = B(I, J)/A(I, J).
/	Matrix right division	mrdivide(A, B)	A/B is roughly the same as A*inv(B), depending on computational accuracy.
\	Matrix left division	mldivide(A, B)	A\B is roughly the same as inv(A)*B, depending on computational accuracy.
.^	Array power	power(A, B)	If C = A.^B, then C(I, J) = A(I, J)^B(I, J).
^	Matrix power	mpower(A, B)	See online help for a discussion of this operator.
.'	Vector and matrix transpose	transpose(A)	A.'. Standard vector and matrix transpose.
'	Vector and matrix complex conjugate transpose	ctranspose(A)	A'. Standard vector and matrix conjugate transpose. When A is real A.' = A'.
+	Unary plus	uplus (A)	+A is the same as 0 + A.
–	Unary minus	uminus (A)	–A is the same as 0 – A or –1*A.
:	Colon		Discussed in Section 2.8.

TABLE 2.5
The image arithmetic functions supported by IPT.

Function	Description
`imadd`	Adds two images; or adds a constant to an image.
`imsubtract`	Subtracts two images; or subtracts a constant from an image.
`immultiply`	Multiplies two images, where the multiplication is carried out between pairs of corresponding image elements; or multiplies a constant times an image.
`imdivide`	Divides two images, where the division is carried out between pairs of corresponding image elements; or divides an image by a constant.
`imabsdiff`	Computes the absolute difference between two images.
`imcomplement`	Complements an image. See Section 3.2.1.
`imlincomb`	Computes a linear combination of two or more images. See Section 5.3.1 for an example.

and `B` are matrices or arrays and `a` and `b` are scalars. All operands can be real or complex. The dot shown in the array operators is not necessary if the operands are scalars. Keep in mind that images are 2-D arrays, which are equivalent to matrices, so all the operators in the table are applicable to images.

The toolbox supports the *image* arithmetic functions listed in Table 2.5. Although these functions could be implemented using MATLAB arithmetic operators directly, the advantage of using the IPT functions is that they support the integer data classes whereas the equivalent MATLAB math operators require inputs of class `double`.

Example 2.6, to follow, uses functions `max` and `min`. The former function has the syntax forms

max
min

```
C = max(A)
C = max(A, B)
C = max(A, [ ], dim)
[C, I] = max(...)
```

In the first form, if `A` is a vector, `max(A)` returns its largest element; if `A` is a matrix, then `max(A)` treats the columns of `A` as vectors and returns a row vector containing the maximum element from each column. In the second form, `max(A, B)` returns an array the same size as `A` and `B` with the largest elements taken from `A` or `B`. In the third form, `max(A, [ ], dim)` returns the largest elements along the dimension of `A` specified by scalar `dim`. For example, `max(A, [ ], 1)` produces the maximum values along the first dimension (the rows) of `A`. Finally, `[C, I] = max(...)` also finds the indices of the maximum values of `A`, and returns them in output vector `I`. If there are several identical maximum values, the index of the first one found is returned. The dots indicate the syntax used on the right of any of the previous three forms. Function min has the same syntax forms just described.

EXAMPLE 2.6: Illustration of arithmetic operators and functions max and min.

■ Suppose that we want to write an M-function, call it fgprod, that multiplies two input images and outputs the product of the images, the maximum and minimum values of the product, and a normalized product image whose values are in the range [0, 1]. Using the text editor we write the desired function as follows:

```
function [p, pmax, pmin, pn] = improd(f, g)
%IMPROD Computes the product of two images.
%  [P, PMAX, PMIN, PN] = IMPROD(F, G)† outputs the element-by-
%  element product of two input images, F and G, the product
%  maximum and minimum values, and a normalized product array with
%  values in the range [0, 1]. The input images must be of the same
%  size.  They can be of class uint8, unit16, or double. The outputs
%  are of class double.

fd = double(f);
gd = double(g);
p = fd.*gd;
pmax = max(p(:));
pmin = min(p(:));
pn = mat2gray(p);
```

Note that the input images were converted to double using the function double instead of im2double because, if the inputs were of type uint8, im2double would convert them to the range [0, 1]. Presumably, we want p to contain the product of the original values. To obtain a normalized array, pn, in the range [0, 1] we used function mat2gray. Note also the use of single-colon indexing, as discussed in Section 2.8.

Suppose that f = [1 2; 3 4] and g = [1 2; 2 1]. Typing the preceding function at the prompt results in the following output:

```
>> [p, pmax, pmin, pn] = improd(f, g)
p =
     1   4
     6   4
pmax =
     6
pmin =
     1
```

†In MATLAB documentation, it is customary to use uppercase characters in the H1 line and in Help text when referring to function names and arguments. This is done to avoid confusion between program names/variables and normal explanatory text.

```
pn =

         0    0.6000

    1.0000    0.6000
```

Typing `help improd` at the prompt results in the following output:

```
>> help improd
IMPROD Computes the product of two images.
  [P, PMAX, PMIN, PN] = IMPROD(F, G) outputs the element-by-
  element product of two input images, F and G, the product
  maximum and minimum values, and a normalized product array with
  values in the range [0, 1]. The input images must be of the same
  size.  They can be of class uint8, unit16, or double. The outputs
  are of class double.
```

■

Relational Operators

MATLAB's relational operators are listed in Table 2.6. These operators compare corresponding elements of arrays of equal dimensions, on an element-by-element basis.

EXAMPLE 2.7: Relational operators.

■ Although the key use of relational operators is in flow control (e.g., in `if` statements), which is discussed in Section 2.10.3, we illustrate briefly how these operators can be used directly on arrays. Consider the following sequence of inputs and outputs:

```
>> A = [1 2 3; 4 5 6; 7 8 9]
A =
    1    2    3
    4    5    6
    7    8    9
>> B = [0 2 4; 3 5 6; 3 4 9]
B =
    0    2    4
    3    5    6
    3    4    9
>> A == B
```

TABLE 2.6 Relational operators.

Operator	Name
<	Less than
<=	Less than or equal to
>	Greater than
>=	Greater than or equal to
==	Equal to
~=	Not equal to

```
ans =

     0     1     0
     0     1     1
     0     0     1
```

Thus, we see that the operation A == B produces a logical array of the same dimensions as A and B, with 1s in locations where the corresponding elements of A and B match, and 0s elsewhere. As another illustration, the statement,

```
>> A >= B
ans =

     1     1     0
     1     1     1
     1     1     1
```

produces a logical array with 1s where the elements of A are greater than or equal to the corresponding elements of B and 0s elsewhere. ■

For vectors and rectangular arrays, both operands must have the same dimensions unless one operand is a scalar. In this case, MATLAB tests the scalar against every element of the other operand, yielding a logical array of the same size as the operand, with 1s in locations where the specified relation is satisfied and 0s elsewhere. If both operands are scalars, the result is a 1 if the specified relation is satisfied and 0 otherwise.

Logical Operators and Functions

Table 2.7 lists MATLAB's logical operators, and the following example illustrates some of their properties. Unlike most common interpretations of logical operators, the operators in Table 2.7 can operate on both logical *and* numeric data. MATLAB treats a logical 1 or nonzero numeric quantity as `true`, and a logical 0 or numeric 0 as `false` in all logical tests. For instance, the AND of two operands is 1 if both operands are logical 1s or nonzero numbers. The AND operation is 0 if either of its operands is logically or numerically 0, or if they both are logically or numerically 0.

TABLE 2.7 Logical operators.

Operator	Name
&	AND
|	OR
~	NOT

EXAMPLE 2.8:
Logical operators.

■ Consider the AND operation on the following numeric arrays:

```
>> A = [1 2 0; 0 4 5];
>> B = [1 -2 3; 0 1 1];
>> A & B

ans =

     1     1     0
     0     1     1
```

We see that the AND operator produces a logical array that is of the same size as the input arrays and has a 1 at locations where both operands are nonzero and 0s elsewhere. Note that all operations are done on pairs of corresponding elements of the arrays, as before.

The OR operator works in a similar manner. An OR expression is `true` if either operand is a logical 1 or nonzero numerical quantity, or if they both are logical 1s or nonzero numbers; otherwise it is `false`. The NOT operator works with a single input. Logically, if the operand is `true`, the NOT operator converts it to `false`. When using NOT with numeric data, any nonzero operand becomes 0, and any zero operand becomes 1. ■

MATLAB also supports the logical functions summarized in Table 2.8. The `all` and `any` functions are particularly useful in programming.

EXAMPLE 2.9:
Logical functions.

■ Consider the simple arrays `A = [1 2 3; 4 5 6]` and `B = [0 -1 1; 0 0 2]`. Substituting these arrays into the functions in Table 2.8 yield the following results:

```
>> xor(A, B)

ans =

     1     0     0
     1     1     0
```

TABLE 2.8
Logical functions.

Function	Comments
`xor` (exclusive OR)	The `xor` function returns a 1 only if both operands are logically different; otherwise `xor` returns a 0.
`all`	The `all` function returns a 1 if all the elements in a vector are nonzero; otherwise `all` returns a 0. This function operates columnwise on matrices.
`any`	The `any` function returns a 1 if any of the elements in a vector is nonzero; otherwise `any` returns a 0. This function operates columnwise on matrices.

```
>> all(A)
ans =
     1     1     1
>> any(A)
ans =
     1     1     1
>> all(B)
ans =
     0     0     1
>> any(B)
ans =
     0     1     1
```

Note how functions `all` and `any` operate on columns of `A` and `B`. For instance, the first two elements of the vector produced by `all(B)` are `0` because each of the first two columns of `B` contains at least one `0`; the last element is `1` because all elements in the last column of `B` are nonzero. ■

In addition to the functions listed in Table 2.8, MATLAB provides a number of other functions that test for the existence of specific conditions or values and return logical results. Some of these functions are listed in Table 2.9. A few of them deal with terms and concepts discussed earlier in this chapter (for example, see function `islogical` in Section 2.6.2); others are used in subsequent discussions. Keep in mind that the functions listed in Table 2.9 return a logical 1 when the condition being tested is true; otherwise they return a logical 0. When the argument is an array, some of the functions in Table 2.9 yield an array the same size as the argument containing logical 1s in the locations that satisfy the test performed by the function, and logical 0s elsewhere. For example, if `A = [1 2; 3 1/0]`, the function `isfinite(A)` returns the matrix `[1 1; 1 0]`, where the `0` (false) entry indicates that the last element of `A` is not finite.

Some Important Variables and Constants

The entries in Table 2.10 are used extensively in MATLAB programming. For example, `eps` typically is added to denominators in expressions to prevent overflow in the event that a denominator becomes zero.

TABLE 2.9
Some functions that return a logical 1 or a logical 0 depending on whether the value or condition in their arguments are `true` or `false`. See online help for a complete list.

Function	Description
`iscell(C)`	True if `C` is a cell array.
`iscellstr(s)`	True if `s` is a cell array of strings.
`ischar(s)`	True if `s` is a character string.
`isempty(A)`	True if `A` is the empty array, [ ].
`isequal(A, B)`	True if `A` and `B` have identical elements and dimensions.
`isfield(S, 'name')`	True if `'name'` is a field of structure `S`.
`isfinite(A)`	True in the locations of array `A` that are finite.
`isinf(A)`	True in the locations of array `A` that are infinite.
`isletter(A)`	True in the locations of `A` that are letters of the alphabet.
`islogical(A)`	True if `A` is a logical array.
`ismember(A, B)`	True in locations where elements of `A` are also in `B`.
`isnan(A)`	True in the locations of `A` that are `NaN`s (see Table 2.10 for a definition of `NaN`).
`isnumeric(A)`	True if `A` is a numeric array.
`isprime(A)`	True in locations of `A` that are prime numbers.
`isreal(A)`	True if the elements of `A` have no imaginary parts.
`isspace(A)`	True at locations where the elements of `A` are whitespace characters.
`issparse(A)`	True if `A` is a sparse matrix.
`isstruct(S)`	True if `S` is a structure.

TABLE 2.10
Some important variables and constants.

Function	Value Returned
`ans`	Most recent answer (variable). If no output variable is assigned to an expression, MATLAB automatically stores the result in `ans`.
`eps`	Floating-point relative accuracy. This is the distance between 1.0 and the next largest number representable using double-precision floating point.
`i` (or `j`)	Imaginary unit, as in `1 + 2i`.
`NaN` or `nan`	Stands for Not-a-Number (e.g., `0/0`).
`pi`	3.14159265358979
`realmax`	The largest floating-point number that your computer can represent.
`realmin`	The smallest floating-point number that your computer can represent.
`computer`	Your computer type.
`version`	MATLAB version string.

Number Representation

MATLAB uses conventional decimal notation, with an optional decimal point and leading plus or minus sign, for numbers. Scientific notation uses the letter `e` to specify a power-of-ten scale factor. Imaginary numbers use either `i` or `j` as a suffix. Some examples of valid number representations are

```
3            -99           0.0001
9.6397238    1.60210e-20   6.02252e23
1i           -3.14159j     3e5i
```

All numbers are stored internally using the long format specified by the Institute of Electrical and Electronics Engineers (IEEE) floating-point standard. Floating-point numbers have a finite precision of roughly 16 significant decimal digits and a finite range of approximately 10^{-308} to 10^{+308}.

2.10.3 Flow Control

The ability to control the flow of operations based on a set of predefined conditions is at the heart of all programming languages. In fact, conditional branching was one of two key developments that led to the formulation of general-purpose computers in the 1940s (the other development was the use of memory to hold stored programs and data). MATLAB provides the eight flow control statements summarized in Table 2.11. Keep in mind the observation made in the previous section that MATLAB treats a logical 1 or nonzero number as `true`, and a logical or numeric 0 as `false`.

TABLE 2.11 Flow control statements.

Statement	Description
`if`	`if`, together with `else` and `elseif`, executes a group of statements based on a specified logical condition.
`for`	Executes a group of statements a fixed (specified) number of times.
`while`	Executes a group of statements an indefinite number of times, based on a specified logical condition.
`break`	Terminates execution of a `for` or `while` loop.
`continue`	Passes control to the next iteration of a `for` or `while` loop, skipping any remaining statements in the body of the loop.
`switch`	`switch`, together with `case` and `otherwise`, executes different groups of statements, depending on a specified value or string.
`return`	Causes execution to return to the invoking function.
`try...catch`	Changes flow control if an error is detected during execution.

if, else, and elseif

Conditional statement `if` has the syntax

```
if expression
   statements
end
```

The *expression* is evaluated and, if the evaluation yields `true`, MATLAB executes one or more commands, denoted here as *statements*, between the `if` and `end` lines. If *expression* is `false`, MATLAB skips all the statements between the `if` and `end` lines and resumes execution at the line following the `end` line. When nesting `if`s, each `if` must be paired with a matching `end`.

The `else` and `elseif` statements further conditionalize the `if` statement. The general syntax is

```
if expression1
   statements1
elseif expression2
   statements2
else
   statements3
end
```

If *expression1* is `true`, *statements1* are executed and control is transferred to the `end` statement. If *expression1* evaluates to `false`, then *expression2* is evaluated. If this expression evaluates to `true`, then *statements2* are executed and control is transferred to the `end` statement. Otherwise (`else`) *statements3* are executed. Note that the `else` statement has no condition. The `else` and `elseif` statements can appear by themselves after an `if` statement; they do not need to appear in pairs, as shown in the preceding general syntax. It is acceptable to have multiple `elseif` statements.

EXAMPLE 2.10: Conditional branching and introduction of functions `error`, `length`, and `numel`.

■ Suppose that we want to write a function that computes the average intensity of an image. As discussed earlier, a two-dimensional array `f` can be converted to a column vector, `v`, by letting `v = f(:)`. Therefore, we want our function to be able to work with both vector and image inputs. The program should produce an error if the input is not a one- or two-dimensional array.

```
function av = average(A)
%AVERAGE Computes the average value of an array.
%  AV = AVERAGE(A) computes the average value of input
%  array, A, which must be a 1-D or 2-D array.

% Check the validity of the input. (Keep in mind that
% a 1-D array is a special case of a 2-D array.)
if ndims(A) > 2
   error('The dimensions of the input cannot exceed 2.')
end
```

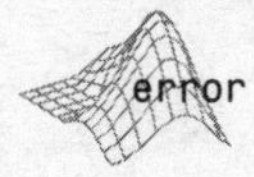

```
% Compute the average
av = sum(A(:))/length(A(:));
```

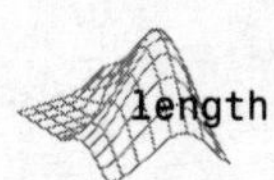

Note that the input is converted to a 1-D array by using `A(:)`. In general, `length(A)` returns the size of the longest dimension of an array, `A`. In this example, because `A(:)` is a vector, `length(A)` gives the number of elements of `A`. This eliminates the need to test whether the input is a vector or a 2-D array. Another way to obtain the number of elements in an array directly is to use function `numel`, whose syntax is

```
n = numel(A)
```

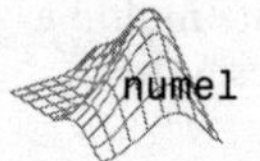

Thus, if `A` is an image, `numel(A)` gives its number of pixels. Using this function, the last executable line of the previous program becomes

```
av = sum(A(:))/numel(A);
```

Finally, note that the `error` function terminates execution of the program and outputs the message contained within the parentheses (the quotes shown are required). ■

for

As indicated in Table 2.11, a `for` loop executes a group of statements a specified number of times. The syntax is

```
for index = start:increment:end
   statements
end
```

It is possible to nest two or more `for` loops, as follows:

```
for index1 = start1:increment1:end
   statements1
   for index2 = start2:increment2:end
      statements2
   end
   additional loop1 statements
end
```

For example, the following loop executes 11 times:

```
count = 0;
for k = 0:0.1:1
   count = count + 1;
end
```

If the loop increment is omitted, it is taken to be 1. Loop increments also can be negative, as in `k = 0:-1:-10`. Note that no semicolon is necessary at the end of a `for` line. MATLAB automatically suppresses printing the values of a loop index. As discussed in detail in Section 2.10.4, considerable gains in program execution speed can be achieved by replacing `for` loops with so-called *vectorized code* whenever possible.

EXAMPLE 2.11: Using a `for` loop to write multiple images to file.

■ Example 2.2 compared several images using different JPEG quality values. Here, we show how to write those files to disk using a `for` loop. Suppose that we have an image, `f`, and we want to write it to a series of JPEG files with quality factors ranging from 0 to 100 in increments of 5. Further, suppose that we want to write the JPEG files with filenames of the form `series_xxx.jpg`, where `xxx` is the quality factor. We can accomplish this using the following `for` loop:

```
for q = 0:5:100
   filename = sprintf('series_%3d.jpg', q);
   imwrite(f, filename, 'quality', q);
end
```

Function `sprintf`, whose syntax in this case is

sprintf

```
s = sprintf('characters1%ndcharacters2', q)
```

See the help page for `sprintf` *for other syntax forms applicable to this function.*

writes formatted data as a string, `s`. In this syntax form, `characters1` and `characters2` are character strings, and `%nd` denotes a decimal number (specified by `q`) with `n` digits. In this example, `characters1` is `series_`, the value of `n` is 3, `characters2` is `.jpg`, and `q` has the values specified in the loop. ■

`while`

A `while` loop executes a group of statements for as long as the expression controlling the loop is `true`. The syntax is

```
while expression
   statements
end
```

As in the case of `for`, `while` loops can be nested:

```
while expression1
   statements1
   while expression2
      statements2
   end
   additional loop1 statements
end
```

For example, the following nested `while` loops terminate when both `a` and `b` have been reduced to 0:

```
a = 10;
b = 5;
while a
   a = a - 1;
   while b
      b = b - 1;
   end
end
```

Note that to control the loops we used MATLAB's convention of treating a numerical value in a logical context as `true` when it is nonzero and as `false` when it is 0. In other words, `while a` and `while b` evaluate to `true` as long as `a` and `b` are nonzero.

As in the case of `for` loops, considerable gains in program execution speed can be achieved by replacing `while` loops with vectorized code (Section 2.10.4) whenever possible.

break

As its name implies, `break` terminates the execution of a `for` or `while` loop. When a `break` statement is encountered, execution continues with the next statement outside the loop. In nested loops, `break` exits only from the innermost loop that contains it.

continue

The `continue` statement passes control to the next iteration of the `for` or `while` loop in which it appears, skipping any remaining statements in the body of the loop. In nested loops, `continue` passes control to the next iteration of the loop enclosing it.

switch

This is the statement of choice for controlling the flow of an M-function based on different types of inputs. The syntax is

```
switch switch_expression
   case case_expression
      statement(s)
   case {case_expression1, case_expression2,...}
      statement(s)
   otherwise
      statement(s)
end
```

The switch construct executes groups of statements based on the value of a variable or expression. The keywords case and otherwise delineate the groups. Only the first matching case is executed.† There must always be an end to match the switch statement. The curly braces are used when multiple expressions are included in the same case statement. As a simple example, suppose that we have an M-function that accepts an image f and converts it to a specified class, call it newclass. Only three image classes are acceptable for the conversion: uint8, uint16, and double. The following code fragment performs the desired conversion and outputs an error if the class of the input image is not one of the acceptable classes:

```
switch newclass
   case 'uint8'
      g = im2uint8(f);
   case 'uint16'
      g = im2uint16(f);
   case 'double'
      g = im2double(f);
   otherwise
      error('Unknown or improper image class.')
end
```

The switch construct is used extensively throughout the book.

EXAMPLE 2.12: Extracting a subimage from a given image.

■ In this example we write an M-function (based on for loops) to extract a rectangular subimage from an image. Although, as shown in the next section, we could do the extraction using a single MATLAB statement, we use the present example later to compare the speed between loops and vectorized code. The inputs to the function are an image, the size (number of rows and columns) of the subimage we want to extract, and the coordinates of the top, left corner of the subimage. Keep in mind that the image origin in MATLAB is at (1, 1), as discussed in Section 2.1.1.

```
function s = subim(f, m, n, rx, cy)
%SUBIM Extracts a subimage, s, from a given image, f.
%  The subimage is of size m-by-n, and the coordinates
%  of its top, left corner are (rx, cy).
s = zeros(m, n);
rowhigh = rx + m - 1;
colhigh = cy + n - 1;
xcount = 0;
for r = rx:rowhigh
   xcount = xcount + 1;
   ycount = 0;
```

†Unlike the C language switch construct, MATLAB's switch does not "fall through." That is, switch executes only the first matching case; subsequent matching cases do not execute. Therefore, break statements are not used.

```
   for c = cy:colhigh
      ycount = ycount + 1;
      s(xcount, ycount) = f(r, c);
   end
end
```

In the following section we give a significantly more efficient implementation of this code. As an exercise, the reader should implement the preceding program using `while` instead of `for` loops. ■

2.10.4 Code Optimization

As discussed in some detail in Section 1.3, MATLAB is a programming language specifically designed for array operations. Taking advantage of this fact whenever possible can result in significant increases in computational speed. In this section we discuss two important approaches for MATLAB code optimization: *vectorizing* loops and *preallocating* arrays.

Vectorizing Loops

Vectorizing simply means converting `for` and `while` loops to equivalent vector or matrix operations. As will become evident shortly, vectorization can result not only in significant gains in computational speed, but it also helps improve code readability. Although multidimensional vectorization can be difficult to formulate at times, the forms of vectorization used in image processing generally are straightforward.

We begin with a simple example. Suppose that we want to generate a 1-D function of the form

$$f(x) = A\sin(x/2\pi)$$

for $x = 0, 1, 2, \ldots, M - 1$. A `for` loop to implement this computation is

```
for x = 1:M   % Array indices in MATLAB cannot be 0.
   f(x) = A*sin((x - 1)/(2*pi));
end
```

However, this code can be made considerably more efficient by vectorizing it; that is, by taking advantage of MATLAB indexing, as follows:

```
x = 0:M - 1;
f = A*sin(x/(2*pi));
```

As this simple example illustrates, 1-D indexing generally is a simple process. When the functions to be evaluated have two variables, optimized indexing is slightly more subtle. MATLAB provides a direct way to implement 2-D function evaluations via function `meshgrid`, which has the syntax

```
[C, R] = meshgrid(c, r)
```

This function transforms the domain specified by *row* vectors `c` and `r` into arrays `C` and `R` that can be used for the evaluation of functions of two variables and 3-D surface plots (note that columns are listed first in both the input and output of `meshgrid`).

The rows of output array `C` are copies of the vector `c`, and the columns of the output array `R` are copies of the vector `r`. For example, suppose that we want to form a 2-D function whose elements are the sum of the squares of the values of coordinate variables `x` and `y` for `x = 0, 1, 2` and `y = 0, 1`. The vector `r` is formed from the row components of the coordinates: `r = [0 1 2]`. Similarly, `c` is formed from the column component of the coordinates: `c = [0 1]` (keep in mind that both `r` and `c` are row vectors here). Substituting these two vectors into `meshgrid` results in the following arrays:

```
>> [C, R]= meshgrid(c, r)
C =

     0     1
     0     1
     0     1

R =

     0     0
     1     1
     2     2
```

The function in which we are interested is implemented as

```
>> h = R.^2 + C.^2
```

which gives the following result:

```
h =

     0     1
     1     2
     4     5
```

Note that the dimensions of `h` are `length(r)` x `length(c)`. Also note, for example, that `h(1,1) = R(1,1)^2 + C(1,1)^2`. Thus, MATLAB automatically took care of indexing `h`. This is a potential source for confusion when 0s are involved in the coordinates because of the repeated warnings in this book and in manuals that MATLAB arrays cannot have 0 indices. As this simple illustration shows, when forming `h`, MATLAB used the *contents* of `R` and `C` for computations. The *indices* of `h`, `R`, and `C`, started at 1. The power of this indexing scheme is demonstrated in the following example.

EXAMPLE 2.13: An illustration of the computational advantages of vectorization, and intruduction of the timing functions `tic` and `toc`.

■ In this example we write an M-function to compare the implementation of the following two-dimensional image function using `for` loops and vectorization:

$$f(x, y) = A\sin(u_0x + v_0y)$$

for $x = 0, 1, 2, \ldots, M - 1$ and $y = 0, 1, 2, \ldots, N - 1$. We also introduce the timing functions `tic` and `toc`.

The function inputs are A, u_0, v_0, M and N. The desired outputs are the images generated by both methods (they should be identical), and the ratio of the time it takes to implement the function with `for` loops to the time it takes to implement it using vectorization. The solution is as follows:

```
function [rt, f, g] = twodsin(A, u0, v0, M, N)
%TWODSIN Compares for loops vs. vectorization.
%  The comparison is based on implementing the function
%  f(x, y) = Asin(u0x + v0y) for x = 0, 1, 2,..., M - 1 and
%  y = 0, 1, 2,..., N - 1. The inputs to the function are
%  M and N and the constants in the function.

% First implement using for loops.

tic   % Start timing.

for r = 1:M
   u0x = u0*(r - 1);
   for c = 1:N
      v0y = v0*(c - 1);
      f(r, c) = A*sin(u0x + v0y);
   end
end

t1 = toc;   % End timing.

% Now implement using vectorization.  Call the image g.

tic   % Start timing.

r = 0:M - 1;
c = 0:N - 1;
[C, R] = meshgrid(c, r);
g = A*sin(u0*R + v0*C);

t2 = toc;   % End timing.

% Compute the ratio of the two times.

rt = t1/(t2 + eps); % Use eps in case t2 is close to 0.
```

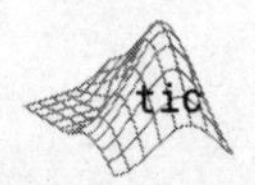

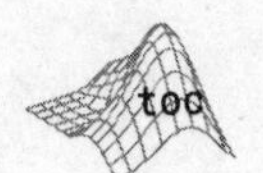

Running this function at the MATLAB prompt,

```
>> [rt, f, g] = twodsin(1, 1/(4*pi), 1/(4*pi), 512, 512);
```

FIGURE 2.7 Sinusoidal image generated in Example 2.13.

yielded the following value of `rt`:

```
>> rt
rt =
    34.2520
```

We convert the image generated (`f` and `g` are identical) to viewable form using function `mat2gray`:

```
>> g = mat2gray(g);
```

and display it using `imshow`,

```
>> imshow(g)
```

Figure 2.7 shows the result. ■

The vectorized code in Example 2.13 runs on the order of 30 times faster than the implementation based on `for` loops. This is a significant computational advantage that becomes increasingly meaningful as relative execution times become longer. For example, if `M` and `N` are large and the vectorized program takes 2 minutes to run, it would take over 1 hour to accomplish the same task using `for` loops. Numbers like these make it worthwhile to vectorize as much of a program as possible, especially if routine use of the program in envisioned.

The preceding discussion on vectorization is focused on computations involving the coordinates of an image. Often, we are interested in extracting and processing regions of an image. Vectorization of programs for extracting such regions is particularly simple if the region to be extracted is rectangular and encompasses all pixels within the rectangle, which generally is the case in this type of operation. The basic vectorized code to extract a region, `s`, of size `m` × `n` and with its top left corner at coordinates (`rx, cy`) is as follows:

```
rowhigh = rx + m - 1;
colhigh = cy + n - 1;
```

```
s = f(rx:rowhigh, cy:colhigh);
```

where `f` is the image from which the region is to be extracted. The `for` loops to accomplish the same thing were already worked out in Example 2.12. Implementing both methods and timing them as in Example 2.13 would show that the vectorized code runs on the order of 1000 times faster in this case than the code based on `for` loops.

Preallocating Arrays

Another simple way to improve code execution time is to preallocate the size of the arrays used in a program. When working with numeric or logical arrays, preallocation simply consists of creating arrays of 0s with the proper dimension. For example, if we are working with two images, `f` and `g`, of size 1024×1024 pixels, preallocation consists of the statements

```
>> f = zeros(1024); g = zeros(1024);
```

Preallocation also helps reduce memory fragmentation when working with large arrays. Memory can become fragmented due to dynamic memory allocation and deallocation. The net result is that there may be sufficient physical memory available during computation, but not enough contiguous memory to hold a large variable. Preallocation helps prevent this by allowing MATLAB to reserve sufficient memory for large data constructs at the beginning of a computation.

2.10.5 Interactive I/O

See Appendix B for details on constructing graphical user interfaces (GUIs).

Often, it is desired to write interactive M-functions that display information and instructions to users and accept inputs from the keyboard. In this section we establish a foundation for writing such functions.

Function `disp` is used to display information on the screen. Its syntax is

```
disp(argument)
```

If `argument` is an array, `disp` displays its contents. If `argument` is a text string, then `disp` displays the characters in the string. For example,

```
>> A = [1 2; 3 4];
>> disp(A)

   1   2
   3   4

>> sc = 'Digital Image Processing.';
>> disp(sc)

Digital Image Processing.

>> disp('This is another way to display text.')

This is another way to display text.
```

Note that only the contents of argument are displayed, without words like ans =, which we are accustomed to seeing on the screen when the value of a variable is displayed by omitting a semicolon at the end of a command line.

Function input is used for inputting data into an M-function. The basic syntax is

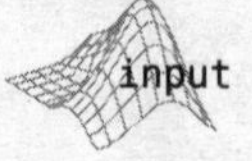


```
t = input('message')
```

This function outputs the words contained in message and waits for an input from the user, followed by a return, and stores the input in t. The input can be a single number, a character string (enclosed by single quotes), a vector (enclosed by square brackets and elements separated by spaces or commas), a matrix (enclosed by square brackets and rows separated by semicolons), or any other valid MATLAB data structure. The syntax

```
t = input('message', 's')
```

outputs the contents of message and accepts a *character* string whose elements can be separated by commas or spaces. This syntax is flexible because it allows multiple individual inputs. If the entries are intended to be numbers, the elements of the string (which are treated as characters) can be converted to numbers of class double by using the function str2num, which has the syntax

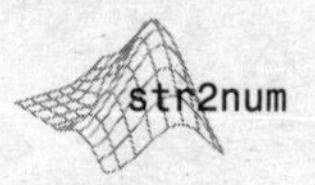


```
n = str2num(t)
```

See Section 12.4 for a detailed discussion of string operations.

For example,

```
>> t = input('Enter your data: ', 's')
Enter your data: 1, 2, 4
t =
   1 2 4
>> class(t)
ans =
   char
>> size(t)
ans =
   1   5
>> n = str2num(t)
n =
   1   2   4
```

```
>> size(n)
ans =
   1   3
>> class(n)
ans =
   double
```

Thus, we see that t is a 1 × 5 character array (the three numbers and the two spaces) and n is a 1 × 3 vector of numbers of class double.

If the entries are a mixture of characters and numbers, then we use one of MATLAB's string processing functions. Of particular interest in the present discussion is function strread, which has the syntax

```
[a, b, c, ...] = strread(cstr, 'format', 'param', 'value')
```

This function reads data from the character string cstr, using a specified format and param/value combinations. In this chapter the formats of interest are %f and %q, to denote floating-point numbers and character strings, respectively. For param we use delimiter to denote that the entities identified in format will be delimited by a character specified in value (typically a comma or space). For example, suppose that we have the string

See the help page for strread *for a list of the numerous syntax forms applicable to this function.*

```
>> t = '12.6, x2y, z';
```

To read the elements of this input into three variables a, b, and c, we write

```
>> [a, b, c] = strread(t, '%f%q%q', 'delimiter', ',')
a =
   12.6000
b =
   'x2y'
c =
   'z'
```

Output a is of class double; the quotes around outputs x2y and z indicate that b and c are cell arrays, which are discussed in the next section. We convert them to character arrays simply by letting

```
>> d = char(b)
d =
   x2y
```

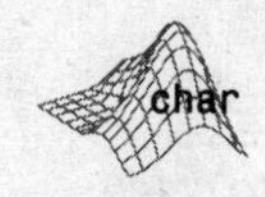

and similarly for `c`. The number (and order) of elements in the format string must match the number and type of expected output variables on the left. In this case we expect three inputs: one floating-point number followed by two character strings.

Function `strcmp (s1, s2)` *compares two strings,* `s1` *and* `s2`*, and returns a logical* `true (1)` *if the strings are equal; otherwise it returns a logical* `false (0)`.

Function `strcmp` is used to compare strings. For example, suppose that we have an M-function `g = imnorm(f, param)` that accepts an image, `f`, and a parameter `param` than can have one of two forms: `'norm1'`, and `'norm255'`. In the first instance, `f` is to be scaled to the range [0, 1]; in the second, it is to be scaled to the range [0, 255]. The output should be of class `double` in both cases. The following code fragment accomplishes the required normalization:

```
f = double(f);
f = f - min(f(:));
f = f./max(f(:));
if strcmp(param, 'norm1')
    g = f;
elseif strcmp(param, 'norm255')
    g = 255*f;
else
    error('Unknown value of param.')
end
```

An error would occur if the value specified in `param` is not `'norm1'` or `'norm255'`. Also, an error would be issued if other than all lowercase characters are used for either normalization factor. We can modify the function to accept either lower or uppercase characters by converting any input to lowercase using function `lower`, as follows:

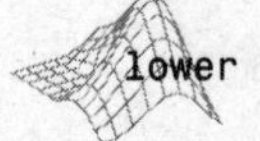


```
param = lower(param)
```

Similarly, if the code uses uppercase letters, we can convert any input character string to uppercase using function `upper`:

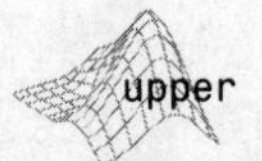


```
param = upper(param)
```

2.10.6 A Brief Introduction to Cell Arrays and Structures

Cell arrays and structures are discussed in detail in Section 11.1.1.

When dealing with mixed variables (e.g., characters and numbers), we can make use of cell arrays. A *cell array* in MATLAB is a multidimensional array whose elements are copies of other arrays. For example, the cell array

```
c = {'gauss', [1 0; 0 1], 3}
```

contains three elements: a character string, a 2×2 matrix, and a scalar (note the use of curly braces to enclose the arrays). To select the contents of a cell array we enclose an integer address in curly braces. In this case, we obtain the following results:

```
>> c{1}
ans =
   gauss
>> c{2}
ans =
   1  0
   0  1
>> c{3}
ans =
   3
```

An important property of cell arrays is that they contain *copies* of the arguments, not pointers to the arguments. For example, if we were working with cell array

```
c = {A, B}
```

in which A and B are matrices, and these matrices changed sometime later in a program, the contents of c would not change.

Structures are similar to cell arrays, in the sense that they allow grouping of a collection of dissimilar data into a single variable. However, unlike cell arrays where cells are addressed by numbers, the elements of structures are addressed by names called *fields*. Depending on the application, using fields adds clarity and readability to an M-function. For instance, letting S denote the structure variable and using the (arbitrary) field names char_string, matrix, and scalar, the data in the preceding example could be organized as a structure by letting

```
S.char_string = 'gauss';
S.matrix = [1 0; 0 1];
S.scalar = 3;
```

Note the use of a dot to append the various fields to the structure variable. Then, for example, typing S.matrix at the prompt, would produce

```
>> S.matrix
ans =
   1  0
   0  1
```

which agrees with the corresponding output for cell arrays. The clarity of using `S.matrix` as opposed to `c{2}` is evident in this case. This type of readability can be important if a function has numerous outputs that must be interpreted by a user.

Summary

The material in this chapter is the foundation for the discussions that follow. At this point, the reader should be able to retrieve an image from disk, process it via simple manipulations, display the result, and save it to disk. It is important to note that the key lesson from this chapter is how to combine MATLAB and IPT functions with programming constructs to generate solutions that expand the capabilities of those functions. In fact, this is the model of how material is presented in the following chapters. By combining standard functions with new code, we show prototypic solutions to a broad spectrum of problems of interest in digital image processing.

3 Intensity Transformations and Spatial Filtering

Preview

The term *spatial domain* refers to the image plane itself, and methods in this category are based on direct manipulation of pixels in an image. In this chapter we focus attention on two important categories of spatial domain processing: *intensity* (or *gray-level*) *transformations* and *spatial filtering*. The latter approach sometimes is referred to as *neighborhood processing*, or *spatial convolution*. In the following sections we develop and illustrate MATLAB formulations representative of processing techniques in these two categories. In order to carry a consistent theme, most of the examples in this chapter are related to image enhancement. This is a good way to introduce spatial processing because enhancement is highly intuitive and appealing, especially to beginners in the field. As will be seen throughout the book, however, these techniques are general in scope and have uses in numerous other branches of digital image processing.

3.1 Background

As noted in the preceding paragraph, spatial domain techniques operate directly on the pixels of an image. The spatial domain processes discussed in this chapter are denoted by the expression

$$g(x, y) = T[f(x, y)]$$

where $f(x, y)$ is the input image, $g(x, y)$ is the output (processed) image, and T is an operator on f, defined over a specified neighborhood about point (x, y). In addition, T can operate on a set of images, such as performing the addition of K images for noise reduction.

The principal approach for defining spatial neighborhoods about a point (x, y) is to use a square or rectangular region centered at (x, y), as Fig. 3.1 shows. The center of the region is moved from pixel to pixel starting, say, at the top, left

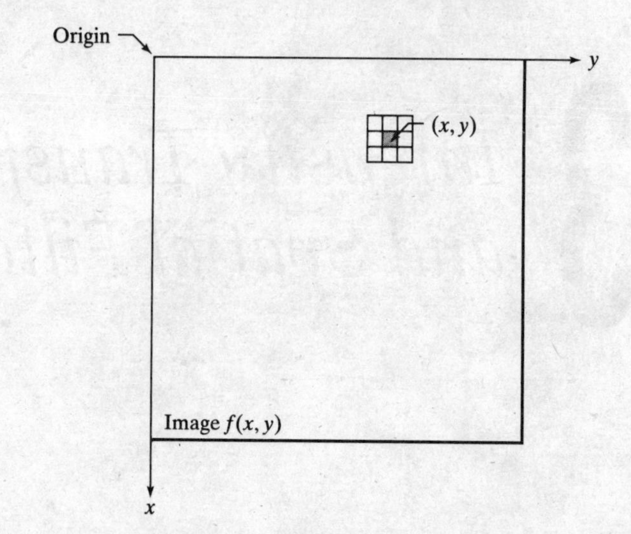

FIGURE 3.1 A neighborhood of size 3 × 3 about a point (x, y) in an image.

corner, and, as it moves, it encompasses different neighborhoods. Operator T is applied at each location (x, y) to yield the output, g, at that location. Only the pixels in the neighborhood are used in computing the value of g at (x, y).

The remainder of this chapter deals with various implementations of the preceding equation. Although this equation is simple conceptually, its computational implementation in MATLAB requires that careful attention be paid to data classes and value ranges.

3.2 Intensity Transformation Functions

The simplest form of the transformation T is when the neighborhood in Fig. 3.1 is of size 1×1 (a single pixel). In this case, the value of g at (x, y) depends only on the intensity of f at that point, and T becomes an *intensity* or *gray-level* transformation function. These two terms are used interchangeably, when dealing with monochrome (i.e., gray-scale) images. When dealing with color images, the term *intensity* is used to denote a color image component in certain color spaces, as described in Chapter 6.

Because they depend only on intensity values, and not explicitly on (x, y), intensity transformation functions frequently are written in simplified form as

$$s = T(r)$$

where r denotes the intensity of f and s the intensity of g, both at any corresponding point (x, y) in the images.

3.2.1 Function `imadjust`

Function `imadjust` is the basic IPT tool for intensity transformations of gray-scale images. It has the syntax

```
g = imadjust(f, [low_in high_in], [low_out high_out], gamma)
```

As illustrated in Fig. 3.2, this function maps the intensity values in image `f` to new values in `g`, such that values between `low_in` and `high_in` map to

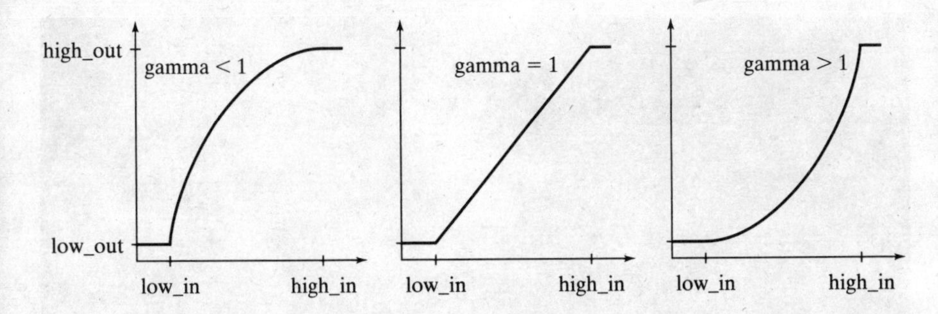

a b c

FIGURE 3.2 The various mappings available in function `imadjust`.

values between `low_out` and `high_out`. Values below `low_in` and above `high_in` are clipped; that is, values below `low_in` map to `low_out`, and those above `high_in` map to `high_out`. The input image can be of class `uint8`, `uint16`, or `double`, and the output image has the same class as the input. All inputs to function `imadjust`, other than `f`, are specified as values between 0 and 1, regardless of the class of `f`. If `f` is of class `uint8`, `imadjust` multiplies the values supplied by 255 to determine the actual values to use; if `f` is of class `uint16`, the values are multiplied by 65535. Using the empty matrix (`[ ]`) for `[low_in high_in]` or for `[low_out high_out]` results in the default values `[0 1]`. If `high_out` is less than `low_out`, the output intensity is reversed.

Parameter `gamma` specifies the shape of the curve that maps the intensity values in `f` to create `g`. If `gamma` is less than 1, the mapping is weighted toward higher (brighter) output values, as Fig. 3.2(a) shows. If `gamma` is greater than 1, the mapping is weighted toward lower (darker) output values. If it is omitted from the function argument, `gamma` defaults to 1 (linear mapping).

EXAMPLE 3.1: Using function `imadjust`.

■ Figure 3.3(a) is a digital mammogram image, `f`, showing a small lesion, and Fig. 3.3(b) is the negative image, obtained using the command

```
>> g1 = imadjust(f, [0 1], [1 0]);
```

This process, which is the digital equivalent of obtaining a photographic negative, is particularly useful for enhancing white or gray detail embedded in a large, predominantly dark region. Note, for example, how much easier it is to analyze the breast tissue in Fig. 3.3(b). The negative of an image can be obtained also with IPT function `imcomplement`:

`imcomplement`

```
g = imcomplement(f)
```

Figure 3.3(c) is the result of using the command

```
>> g2 = imadjust(f, [0.5 0.75], [0 1]);
```

which expands the gray scale region between 0.5 and 0.75 to the full [0, 1] range. This type of processing is useful for highlighting an intensity band of interest. Finally, using the command

```
>> g3 = imadjust(f, [ ], [ ], 2);
```

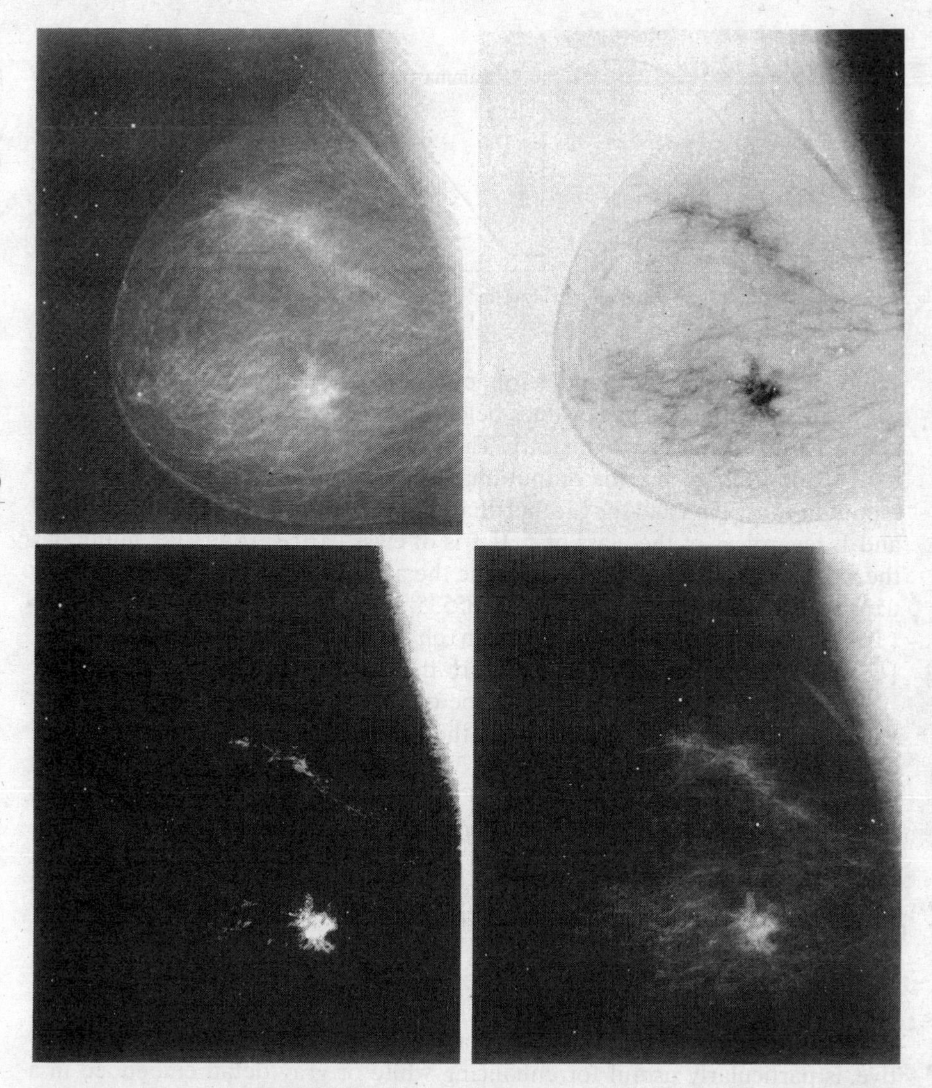

a b
c d

FIGURE 3.3 (a) Original digital mammogram. (b) Negative image. (c) Result of expanding the intensity range [0.5, 0.75]. (d) Result of enhancing the image with `gamma = 2`. (Original image courtesy of G. E. Medical Systems.)

produces a result similar to (but with more gray tones than) Fig. 3.3(c) by compressing the low end and expanding the high end of the gray scale [see Fig. 3.3(d)]. ■

3.2.2 Logarithmic and Contrast-Stretching Transformations

Logarithmic and contrast-stretching transformations are basic tools for dynamic range manipulation. Logarithm transformations are implemented using the expression

`log` *is the natural logarithm.* `log2` *and* `log10` *are the base* 2 *and base* 10 *logarithms, respectively.*

```
g = c*log(1 + double(f))
```

where `c` is a constant. The shape of this transformation is similar to the gamma curve shown in Fig. 3.2(a) with the low values set at 0 and the high values set to 1 on both scales. Note, however, that the shape of the gamma curve is variable, whereas the shape of the log function is fixed.

One of the principal uses of the log transformation is to compress dynamic range. For example, it is not unusual to have a Fourier spectrum (Chapter 4) with values in the range $[0, 10^6]$ or higher. When displayed on a monitor that is scaled linearly to 8 bits, the high values dominate the display, resulting in lost visual detail for the lower intensity values in the spectrum. By computing the `log`, a dynamic range on the order of, for example, 10^6, is reduced to approximately 14, which is much more manageable.

When performing a logarithmic transformation, it is often desirable to bring the resulting compressed values back to the full range of the display. For 8 bits, the easiest way to do this in MATLAB is with the statement

```
>> gs = im2uint8(mat2gray(g));
```

Use of `mat2gray` brings the values to the range [0, 1] and `im2uint8` brings them to the range [0, 255]. Later, in Section 3.2.3, we discuss a scaling function that automatically detects the class of the input and applies the appropriate conversion.

The function shown in Fig. 3.4(a) is called a *contrast-stretching* transformation function because it compresses the input levels lower than m into a narrow range of dark levels in the output image; similarly, it compresses the values above m into a narrow band of light levels in the output. The result is an image of higher contrast. In fact, in the limiting case shown in Fig. 3.4(b), the output is a binary image. This limiting function is called a *thresholding* function, which, as we discuss in Chapter 10, is a simple tool used for image segmentation. Using the notation introduced at the beginning of this section, the function in Fig. 3.4(a) has the form

$$s = T(r) = \frac{1}{1 + (m/r)^E}$$

where r represents the intensities of the input image, s the corresponding intensity values in the output image, and E controls the slope of the function. This equation is implemented in MATLAB for an entire image as

```
g = 1./(1 + (m./(double(f) + eps)).^E)
```

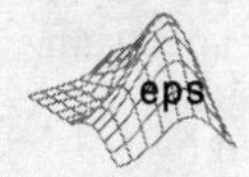

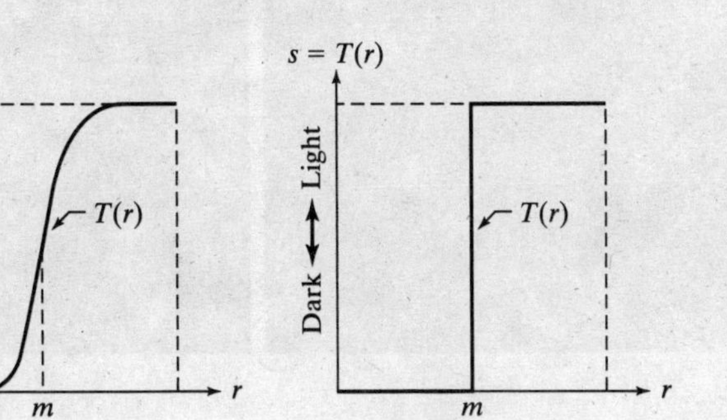

a b

FIGURE 3.4 (a) Contrast-stretching transformation. (b) Thresholding transformation.

Note the use of `eps` (see Table 2.10) to prevent overflow if `f` has any 0 values. Since the limiting value of $T(r)$ is 1, output values are scaled to the range [0, 1] when working with this type of transformation. The shape in Fig. 3.4(a) was obtained with `E = 20`.

EXAMPLE 3.2: Using a log transformation to reduce dynamic range.

■ Figure 3.5(a) is a Fourier spectrum with values in the range 0 to 1.5×10^6, displayed on a linearly scaled, 8-bit system. Figure 3.5(b) shows the result obtained using the commands

```
>> g = im2uint8(mat2gray(log(1 + double(f))));
>> imshow(g)
```

The visual improvement of `g` over the original image is quite evident. ■

3.2.3 Some Utility M-Functions for Intensity Transformations

In this section we develop two M-functions that incorporate various aspects of the intensity transformations introduced in the previous two sections. We show the details of the code for one of them to illustrate error checking, to introduce ways in which MATLAB functions can be formulated so that they can handle a variable number of inputs and/or outputs, and to show typical code formats used throughout the book. From this point on, detailed code of new M-functions is included in our discussions only when the purpose is to explain specific programming constructs, to illustrate the use of a new MATLAB or IPT function, or to review concepts introduced earlier. Otherwise, only the syntax of the function is explained, and its code is included in Appendix C. Also, in order to focus on the basic structure of the functions developed in the remainder of the book, this is the last section in which we show extensive use of error checking. The procedures that follow are typical of how error handling is programmed in MATLAB.

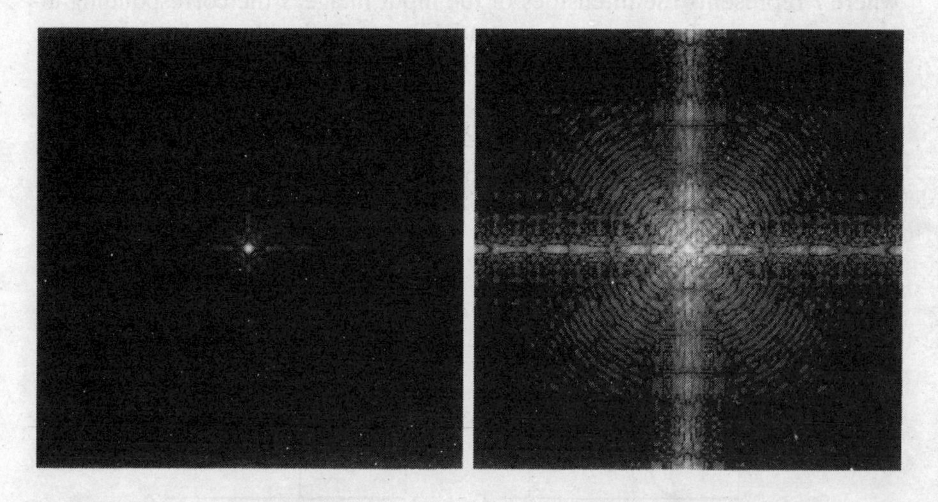

a b

FIGURE 3.5 (a) A Fourier spectrum. (b) Result obtained by performing a log transformation.

Handling a Variable Number of Inputs and/or Outputs

To check the number of arguments input into an M-function we use function `nargin`,

```
n = nargin
```

which returns the actual number of arguments input into the M-function. Similarly, function `nargout` is used in connection with the outputs of an M-function. The syntax is

```
n = nargout
```

For example, suppose that we execute the following M-function at the prompt:

```
>> T = testhv(4, 5);
```

Use of `nargin` within the body of this function would return a 2, while use of `nargout` would return a 1.

Function `nargchk` can be used in the body of an M-function to check if the correct number of arguments were passed. The syntax is

```
msg = nargchk(low, high, number)
```

This function returns the message `Not enough input parameters` if `number` is less than `low` or `Too many input parameters` if `number` is greater than `high`. If `number` is between `low` and `high` (inclusive), `nargchk` returns an empty matrix. A frequent use of function `nargchk` is to stop execution via the `error` function if the incorrect number of arguments is input. The number of actual input arguments is determined by the `nargin` function. For example, consider the following code fragment:

```
function G = testhv2(x, y, z)
  :
error(nargchk(2, 3, nargin));
  :
```

Typing

```
>> testhv2(6);
```

which only has one input argument would produce the error

```
Not enough input arguments.
```

and execution would terminate.

Often, it is useful to be able to write functions in which the number of input and/or output arguments is variable. For this, we use the variables `varargin` and `varargout`. In the declaration, `varargin` and `varargout` must be lowercase. For example,

`varargin`
`varargout`

```
function [m, n] = testhv3(varargin)
```

accepts a variable number of inputs into function `testhv3`, and

```
function [varargout] = testhv4(m, n, p)
```

returns a variable number of outputs from function `testhv4`. If function `testhv3` had, say, one fixed input argument, `x`, followed by a variable number of input arguments, then

```
function [m, n] = testhv3(x, varargin)
```

would cause `varargin` to start with the second input argument supplied by the user when the function is called. Similar comments apply to `varargout`. It is acceptable to have a function in which both the number of input and output arguments is variable.

When `varargin` is used as the input argument of a function, MATLAB sets it to a cell array (see Section 2.10.5) that accepts a variable number of inputs by the user. Because `varargin` is a cell array, an important aspect of this arrangement is that the call to the function can contain a mixed set of inputs. For example, assuming that the code of our hypothetical function `testhv3` is equipped to handle it, it would be perfectly acceptable to have a mixed set of inputs, such as

```
>> [m, n] = testhv3(f, [0  0.5  1.5], A, 'label');
```

where `f` is an image, the next argument is a row vector of length 3, `A` is a matrix, and `label'` is a character string. This is indeed a powerful feature that can be used to simplify the structure of functions requiring a variety of different inputs. Similar comments apply to `varargout`.

Another M-Function for Intensity Transformations

`changeclass` *is an undocumented IPT utility function. Its code is included in Appendix C.*

In this section we develop a function that computes the following transformation functions: negative, log, gamma and contrast stretching. These transformations were selected because we will need them later, and also to illustrate the mechanics involved in writing an M-function for intensity transformations. In writing this function we use function `changeclass`, which has the syntax

```
g = changeclass(newclass, f)
```

This function converts image `f` to the class specified in parameter `newclass` and outputs it as `g`. Valid values for `newclass` are `'uint8'`, `'uint16'`, and `'double'`.

Note in the following M-function, which we call `intrans`, how function options are formatted in the Help section of the code, how a variable number of inputs is handled, how error checking is interleaved in the code, and how the class of the output image is matched to the class of the input. Keep in mind when studying the following code that `varargin` is a cell array, so its elements are selected by using curly braces.

intrans

```
function g = intrans(f, varargin)
%INTRANS Performs intensity (gray-level) transformations.
%   G = INTRANS(F, 'neg') computes the negative of input image F.
%
%   G = INTRANS(F, 'log', C, CLASS) computes C*log(1 + F) and
%   multiplies the result by (positive) constant C. If the last two
%   parameters are omitted, C defaults to 1. Because the log is used
%   frequently to display Fourier spectra, parameter CLASS offers the
%   option to specify the class of the output as 'uint8' or
%   'uint16'. If parameter CLASS is omitted, the output is of the
%   same class as the input.
%
%   G = INTRANS(F, 'gamma', GAM) performs a gamma transformation on
%   the input image using parameter GAM (a required input).
%
%   G = INTRANS(F, 'stretch', M, E) computes a contrast-stretching
%   transformation using the expression 1./(1 + (M./(F +
%   eps)).^E). Parameter M must be in the range [0, 1]. The default
%   value for M is mean2(im2double(F)), and the default value for E
%   is 4.
%
%   For the 'neg', 'gamma', and 'stretch' transformations, double
%   input images whose maximum value is greater than 1 are scaled
%   first using MAT2GRAY. Other images are converted to double first
%   using IM2DOUBLE. For the 'log' transformation, double images are
%   transformed without being scaled; other images are converted to
%   double first using IM2DOUBLE.
%
%   The output is of the same class as the input, except if a
%   different class is specified for the 'log' option.

% Verify the correct number of inputs.
error(nargchk(2, 4, nargin))

% Store the class of the input for use later.
classin = class(f);
```

```
% If the input is of class double, and it is outside the range
% [0, 1], and the specified transformation is not 'log', convert the
% input to the range [0, 1].
if strcmp(class(f), 'double') & max(f(:)) > 1 & . . .
      ~strcmp(varargin{1}, 'log')
   f = mat2gray(f);
else % Convert to double, regardless of class(f).
   f = im2double(f);
end

% Determine the type of transformation specified.
method = varargin{1};

% Perform the intensity transformation specified.
switch method
case 'neg'
   g = imcomplement(f);

case 'log'
   if length(varargin) == 1
      c = 1;
   elseif length(varargin) == 2
      c = varargin{2};
   elseif length(varargin) == 3
      c = varargin{2};
      classin = varargin{3};
   else
      error('Incorrect number of inputs for the log option.')
   end
   g = c*(log(1 + double(f)));
case 'gamma'
   if length(varargin) < 2
      error('Not enough inputs for the gamma option.')
   end
   gam = varargin{2};
   g = imadjust(f, [ ], [ ], gam);
case 'stretch'
   if length(varargin) == 1
      % Use defaults.
      m = mean2(f);
      E = 4.0;
   elseif length(varargin) == 3
      m = varargin{2};
      E = varargin{3};
   else error('Incorrect number of inputs for the stretch option.')
   end
   g = 1./(1 + (m./(f + eps)).^E);
otherwise
   error('Unknown enhancement method.')
end

% Convert to the class of the input image.
g = changeclass(classin, g);
```

EXAMPLE 3.3: Illustration of function `intrans`.

■ As an illustration of function `intrans`, consider the image in Fig. 3.6(a), which is an ideal candidate for contrast stretching to enhance the skeletal structure. The result in Fig. 3.6(b) was obtained with the following call to `intrans`:

```
>> g = intrans(f, 'stretch', mean2(im2double(f)), 0.9);
>> figure, imshow(g)
```

`m = mean2(A)` *computes the mean (average) value of the elements of matrix* `A`.

Note how function `mean2` was used to compute the mean value of `f` directly inside the function call. The resulting value was used for `m`. Image `f` was converted to `double` using `im2double` in order to scale its values to the range [0, 1] so that the mean would also be in this range, as required for input `m`. The value of `E` was determined interactively. ■

An M-Function for Intensity Scaling

When working with images, results whose pixels span a wide negative to positive range of values are common. While this presents no problems during intermediate computations, it does become an issue when we want to use an 8-bit or 16-bit format for saving or viewing an image, in which case it often is desirable to scale the image to the full, maximum range, [0, 255] or [0, 65535]. The following M-function, which we call `gscale`, accomplishes this. In addition, the function can map the output levels to a specified range. The code for this function does not include any new concepts so we do not include it here. See Appendix C for the listing.

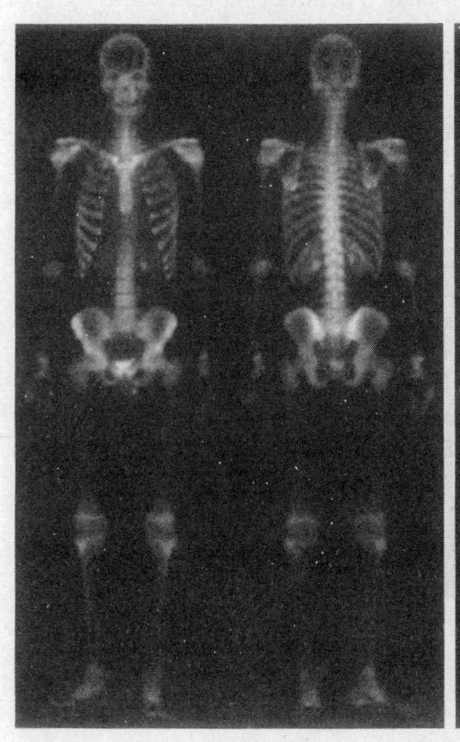
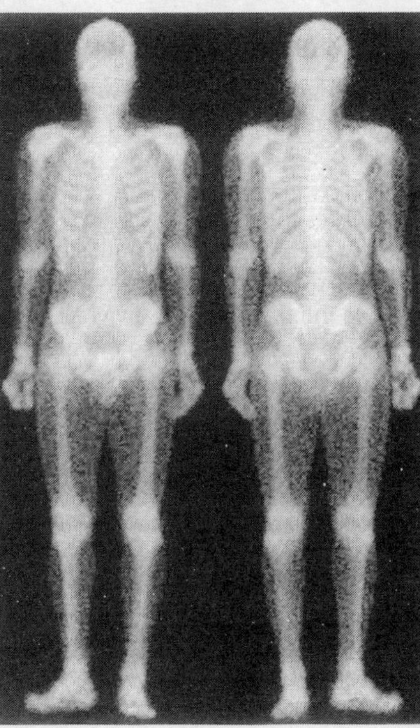

a b

FIGURE 3.6 (a) Bone scan image. (b) Image enhanced using a contrast-stretching transformation. (Original image courtesy of G. E. Medical Systems.)

The syntax of function `gscale` is

`gscale`

```
g = gscale(f, method, low, high)
```

where `f` is the image to be scaled. Valid values for `method` are `'full8'` (the default), which scales the output to the full range [0, 255], and `'full16'`, which scales the output to the full range [0, 65535]. If included, parameters `low` and `high` are ignored in these two conversions. A third valid value of `method` is `'minmax'`, in which case parameters `low` and `high`, both in the range [0, 1], must be provided. If `'minmax'` is selected, the levels are mapped to the range `[low, high]`. Although these values are specified in the range [0, 1], the program performs the proper scaling, depending on the class of the input, and then converts the output to the same class as the input. For example, if `f` is of class `uint8` and we specify `'minmax'` with the range [0, 0.5], the output also will be of class `uint8`, with values in the range [0, 128]. If `f` is of class `double` and its range of values is outside the range [0, 1], the program converts it to this range before proceeding. Function `gscale` is used in numerous places throughout the book.

3.3 Histogram Processing and Function Plotting

Intensity transformation functions based on information extracted from image intensity histograms play a basic role in image processing, in areas such as enhancement, compression, segmentation, and description. The focus of this section is on obtaining, plotting, and using histograms for image enhancement. Other applications of histograms are discussed in later chapters.

See Section 4.5.3 for a discussion of 2-D plotting techniques.

3.3.1 Generating and Plotting Image Histograms

The histogram of a digital image with L total possible intensity levels in the range $[0, G]$ is defined as the discrete function

$$h(r_k) = n_k$$

where r_k is the kth intensity level in the interval $[0, G]$ and n_k is the number of pixels in the image whose intensity level is r_k. The value of G is 255 for images of class `uint8`, 65535 for images of class `uint16`, and 1.0 for images of class `double`. Keep in mind that indices in MATLAB cannot be 0, so r_1 corresponds to intensity level 0, r_2 corresponds to intensity level 1, and so on, with r_L corresponding to level G. Note also that $G = L - 1$ for images of class `uint8` and `uint16`.

Often, it is useful to work with *normalized* histograms, obtained simply by dividing all elements of $h(r_k)$ by the total number of pixels in the image, which we denote by n:

$$p(r_k) = \frac{h(r_k)}{n} = \frac{n_k}{n}$$

for $k = 1, 2, \ldots, L$. From basic probability, we recognize $p(r_k)$ as an estimate of the probability of occurrence of intensity level r_k.

The core function in the toolbox for dealing with image histograms is `imhist`, which has the following basic syntax:

```
h = imhist(f, b)
```

where `f` is the input image, `h` is its histogram, $h(r_k)$, and `b` is the number of bins used in forming the histogram (if `b` is not included in the argument, `b = 256` is used by default). A bin is simply a subdivision of the intensity scale. For example, if we are working with `uint8` images and we let `b = 2`, then the intensity scale is subdivided into two ranges: 0 to 127 and 128 to 255. The resulting histogram will have two values: $h(1)$ equal to the number of pixels in the image with values in the interval $[0, 127]$, and $h(2)$ equal to the number of pixels with values in the interval $[128, 255]$. We obtain the normalized histogram simply by using the expression

```
p = imhist(f, b)/numel(f)
```

Recall from Section 2.10.3 that function `numel(f)` gives the number of elements in array `f` (i.e., the number of pixels in the image).

EXAMPLE 3.4: Computing and plotting image histograms.

■ Consider the image, `f`, from Fig. 3.3(a). The simplest way to plot its histogram is to use `imhist` with no output specified:

```
>> imhist(f);
```

Figure 3.7(a) shows the result. This is the histogram display default in the toolbox. However, there are many other ways to plot a histogram, and we take this opportunity to explain some of the plotting options in MATLAB that are representative of those used in image processing applications.

Histograms often are plotted using *bar* graphs. For this purpose we can use the function

```
bar(horz, v, width)
```

where `v` is a row vector containing the points to be plotted, `horz` is a vector of the same dimension as `v` that contains the increments of the horizontal scale, and `width` is a number between 0 and 1. If `horz` is omitted, the horizontal axis is divided in units from 0 to `length(v)`. When `width` is 1, the bars touch; when it is 0, the bars are simply vertical lines, as in Fig. 3.7(a). The default value is 0.8. When plotting a bar graph, it is customary to reduce the resolution of the horizontal axis by dividing it into bands. The following statements produce a bar graph, with the horizontal axis divided into groups of 10 levels:

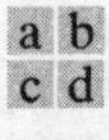

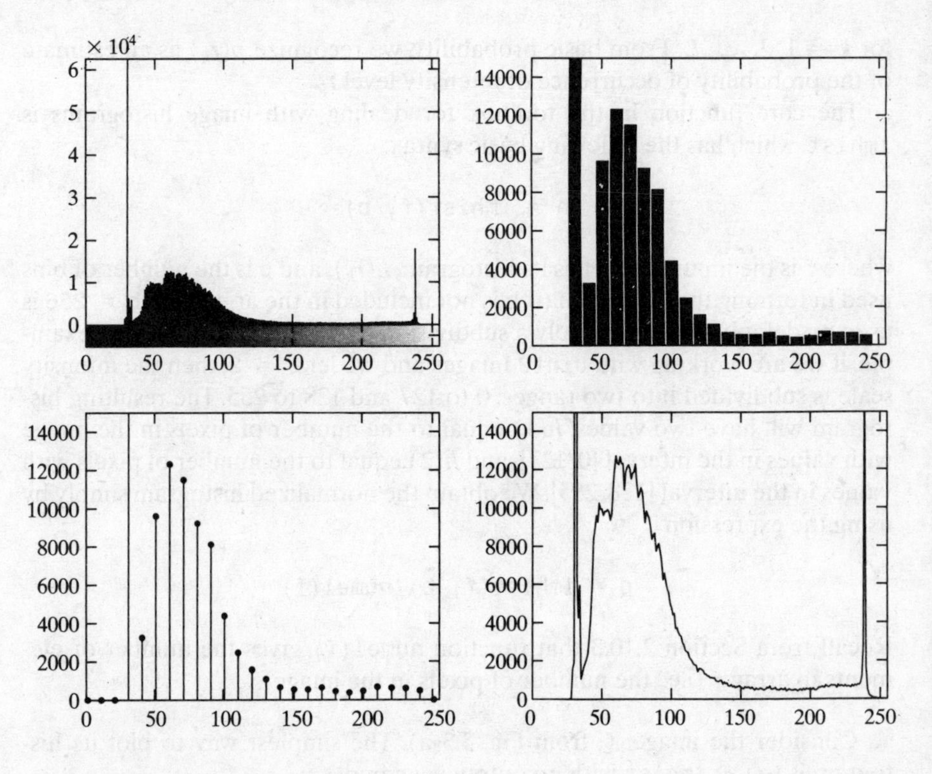

FIGURE 3.7 Various ways to plot an image histogram. (a) `imhist`, (b) `bar`, (c) `stem`, (d) `plot`.

```
>> h = imhist(f);
>> h1 = h(1:10:256);
>> horz = 1:10:256;
>> bar(horz, h1)
>> axis([0 255 0 15000])
>> set(gca, 'xtick', 0:50:255)
>> set(gca, 'ytick', 0:2000:15000)
```

set
gca
xtick
ytick

Figure 3.7(b) shows the result. The peak located at the high end of the intensity scale in Fig. 3.7(a) is missing in the bar graph as a result of the larger horizontal increments used in the plot.

The fifth statement in the preceding code was used to expand the lower range of the vertical axis for visual analysis, and to set the orizontal axis to the same range as in Fig. 3.7(a). The `axis` function has the syntax

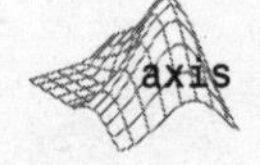


```
axis([horzmin  horzmax  vertmin  vertmax])
```

which sets the minimum and maximum values in the horizontal and vertical axes. In the last two statements, `gca` means "get current axis," (i.e., the axes of the figure last displayed) and `xtick` and `ytick` set the horizontal and vertical axes ticks in the intervals shown.

Axis labels can be added to the horizontal and vertical axes of a graph using the functions

```
xlabel('text string', 'fontsize', size)
ylabel('text string', 'fontsize', size)
```

where size is the font size in points. Text can be added to the body of the figure by using function text, as follows:

```
text(xloc, yloc, 'text string', 'fontsize', size)
```

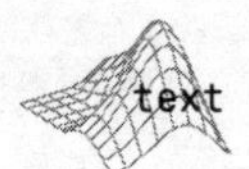

where xloc and yloc define the location where text starts. Use of these three functions is illustrated in Example 3.5. It is important to note that functions that set axis values and labels are used *after* the function has been plotted.

A title can be added to a plot using function title, whose basic syntax is

```
title('titlestring')
```

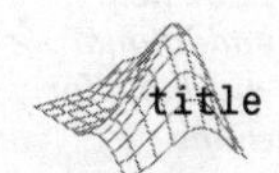

where titlestring is the string of characters that will appear on the title, centered above the plot.

A *stem* graph is similar to a bar graph. The syntax is

```
stem(horz, v, 'color_linestyle_marker', 'fill')
```

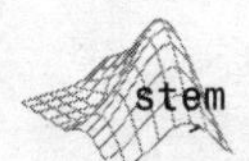

See the stem *help page for additional options available for this function.*

where v is row vector containing the points to be plotted, and horz is as described for bar. The argument,

```
color_linestyle_marker
```

is a triplet of values from Table 3.1. For example, stem(v, 'r--s') produces a stem plot where the lines and markers are red, the lines are dashed, and the markers are squares. If fill is used, and the marker is a circle, square, or diamond, the marker is filled with the color specified in color. The default color is black, the line default is solid, and the default marker is a circle. The stem graph in Fig. 3.7(c) was obtained using the statements

```
>> h = imhist(f);
>> h1 = h(1:10:256);
```

TABLE 3.1 Attributes for functions stem and plot. The none attribute is applicable only to function plot, and must be specified individually. See the syntax for function plot below.

Symbol	Color	Symbol	Line Style	Symbol	Marker
k	Black	-	Solid	+	Plus sign
w	White	--	Dashed	o	Circle
r	Red	:	Dotted	*	Asterisk
g	Green	-.	Dash-dot	.	Point
b	Blue	none	No line	x	Cross
c	Cyan			s	Square
y	Yellow			d	Diamond
m	Magenta			none	No marker

```
>> horz = 1:10:256;
>> stem(horz, h1, 'fill')
>> axis([0 255 0 15000])
>> set(gca, 'xtick', [0:50:255])
>> set(gca, 'ytick', [0:2000:15000])
```

Finally, we consider function `plot`, which plots a set of points by linking them with straight lines. The syntax is

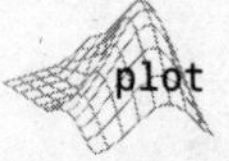


```
plot(horz, v, 'color_linestyle_marker')
```

See the `plot` *help page for additional options available for this function.*

where the arguments are as defined previously for stem plots. The values of `color`, `linestyle`, and `marker` are given in Table 3.1. As in `stem`, the attributes in `plot` can be specified as a triplet. When using `none` for `linestyle` or for `marker`, the attributes must be specified individually. For example, the command

```
>> plot(horz, v, 'color', 'g', 'linestyle', 'none', 'marker', 's')
```

plots green squares without connecting lines between them. The defaults for `plot` are solid black lines with no markers.

The plot in Fig. 3.7(d) was obtained using the following statements:

```
>> h = imhist(f);
>> plot(h)   % Use the default values.
>> axis([0 255 0 15000])
>> set(gca, 'xtick', [0:50:255])
>> set(gca, 'ytick', [0:2000:15000])
```

Function `plot` is used frequently to display transformation functions (see Example 3.5). ■

In the preceding discussion axis limits and tick marks were set manually. It is possible to set the limits and ticks automatically by using functions `ylim` and `xlim`, which, for our purposes here, have the syntax forms

```
ylim('auto')
xlim('auto')
```

Among other possible variations of the syntax for these two functions (see online help for details), there is a manual option, given by

```
ylim([ymin  ymax])
xlim([xmin  xmax])
```

which allows manual specification of the limits. If the limits are specified for only one axis, the limits on the other axis are set to `'auto'` by default. We use these functions in the following section.

Typing `hold on` at the prompt retains the current plot and certain axes properties so that subsequent graphing commands add to the existing graph. See Example 10.6 for an illustration.

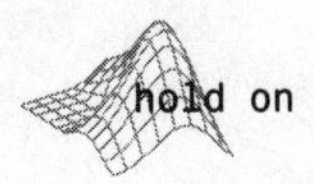

3.3.2 Histogram Equalization

Assume for a moment that intensity levels are continuous quantities normalized to the range [0, 1], and let $p_r(r)$ denote the probability density function (PDF) of the intensity levels in a given image, where the subscript is used for differentiating between the PDFs of the input and output images. Suppose that we perform the following transformation on the input levels to obtain output (processed) intensity levels, s,

$$s = T(r) = \int_0^r p_r(w)\,dw$$

where w is a dummy variable of integration. It can be shown (Gonzalez and Woods [2002]) that the probability density function of the output levels is *uniform*; that is,

$$p_s(s) = \begin{cases} 1 & \text{for } 0 \le s \le 1 \\ 0 & \text{otherwise} \end{cases}$$

In other words, the preceding transformation generates an image whose intensity levels are equally likely, and, in addition, cover the entire range [0, 1]. The net result of this intensity-level *equalization* process is an image with increased dynamic range, which will tend to have higher contrast. Note that the transformation function is really nothing more than the cumulative distribution function (CDF).

When dealing with discrete quantities we work with histograms and call the preceding technique *histogram equalization*, although, in general, the histogram of the processed image will not be uniform, due to the discrete nature of the variables. With reference to the discussion in Section 3.3.1, let $p_r(r_j)$, $j = 1, 2, \ldots, L$, denote the histogram associated with the intensity levels of a given image, and recall that the values in a normalized histogram are approximations to the probability of occurrence of each intensity level in the image. For discrete quantities we work with summations, and the equalization transformation becomes

$$\begin{aligned} s_k &= T(r_k) \\ &= \sum_{j=1}^{k} p_r(r_j) \\ &= \sum_{j=1}^{k} \frac{n_j}{n} \end{aligned}$$

for $k = 1, 2, \ldots, L$, where s_k is the intensity value in the output (processed) image corresponding to value r_k in the input image.

Histogram equalization is implemented in the toolbox by function `histeq`, which has the syntax

```
g = histeq(f, nlev)
```

where `f` is the input image and `nlev` is the number of intensity levels specified for the output image. If `nlev` is equal to L (the total number of *possible* levels in the input image), then `histeq` implements the transformation function, $T(r_k)$, directly. If `nlev` is less than L, then `histeq` attempts to distribute the levels so that they will approximate a flat histogram. Unlike `imhist`, the default value in `histeq` is `nlev = 64`. For the most part, we use the maximum possible number of levels (generally 256) for `nlev` because this produces a true implementation of the histogram-equalization method just described.

EXAMPLE 3.5: Histogram equalization.

■ Figure 3.8(a) is an electron microscope image of pollen, magnified approximately 700 times. In terms of needed enhancement, the most important features of this image are that it is dark and has a low dynamic range. This can be seen in the histogram in Fig. 3.8(b), in which the dark nature of the image is expected because the histogram is biased toward the dark end of the gray scale. The low dynamic range is evident from the fact that the "width" of the histogram is narrow with respect to the entire gray scale. Letting `f` denote the input image, the following sequence of steps produced Figs. 3.8(a) through (d):

```
>> imshow(f)
>> figure, imhist(f)
>> ylim('auto')
>> g = histeq(f, 256);
>> figure, imshow(g)
>> figure, imhist(g)
>> ylim('auto')
```

The images were saved to disk in tiff format at 300 dpi using `imwrite`, and the plots were similarly exported to disk using the `print` function discussed in Section 2.4.

The image in Fig. 3.8(c) is the histogram-equalized result. The improvements in average intensity and contrast are quite evident. These features also are evident in the histogram of this image, shown in Fig. 3.8(d). The increase in contrast is due to the considerable spread of the histogram over the entire intensity scale. The increase in overall intensity is due to the fact that the average intensity level in the histogram of the equalized image is higher (lighter) than the original. Although the histogram-equalization method just discussed does not produce a flat histogram, it has the desired characteristic of being able to increase the dynamic range of the intensity levels in an image.

If `A` *is a vector,* `B = cumsum(A)` *gives the sum of its elements. If* `A` *is a higher-dimensional array,* `B = cumsum(A, dim)` *given the sum along the dimension specified by* `dim`.

As noted earlier, the transformation function $T(r_k)$ is simply the cumulative sum of normalized histogram values. We can use function `cumsum` to obtain the transformation function, as follows:

cumsum

```
>> hnorm = imhist(f)./numel(f);
>> cdf = cumsum(hnorm);
```

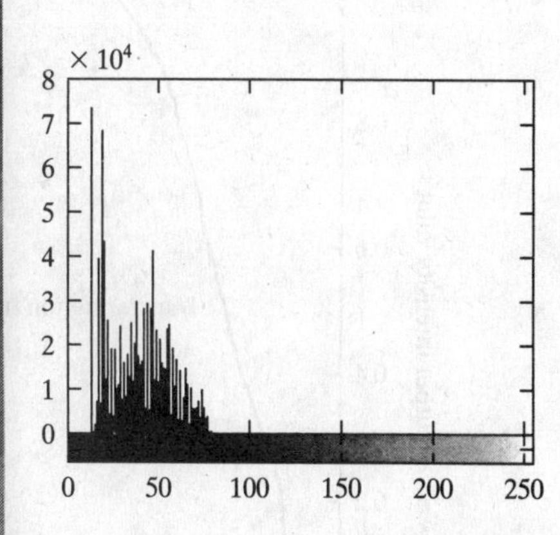

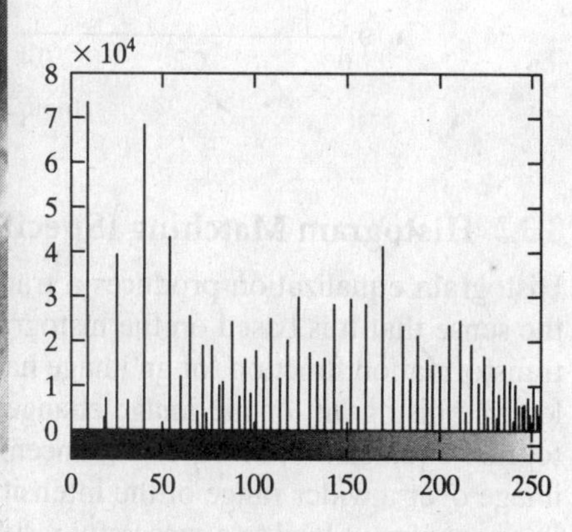

a b
c d

FIGURE 3.8 Illustration of histogram equalization. (a) Input image, and (b) its histogram. (c) Histogram-equalized image, and (d) its histogram. The improvement between (a) and (c) is quite visible. (Original image courtesy of Dr. Roger Heady, Research School of Biological Sciences, Australian National University, Canberra.)

A plot of `cdf`, shown in Fig. 3.9, was obtained using the following commands:

```
>> x = linspace(0, 1, 256);   % Intervals for [0, 1] horiz scale. Note
                              % the use of linspace from Sec. 2.8.1.
>> plot(x, cdf)               % Plot cdf vs. x.
>> axis([0 1 0 1])            % Scale, settings, and labels:
>> set(gca, 'xtick', 0:.2:1)
>> set(gca, 'ytick', 0:.2:1)
>> xlabel('Input intensity values', 'fontsize', 9)
>> ylabel('Output intensity values', 'fontsize', 9)
>> % Specify text in the body of the graph:
>> text(0.18, 0.5, 'Transformation function', 'fontsize', 9)
```

We can tell visually from this transformation function that a narrow range of input intensity levels is transformed into the full intensity scale in the output image. ■

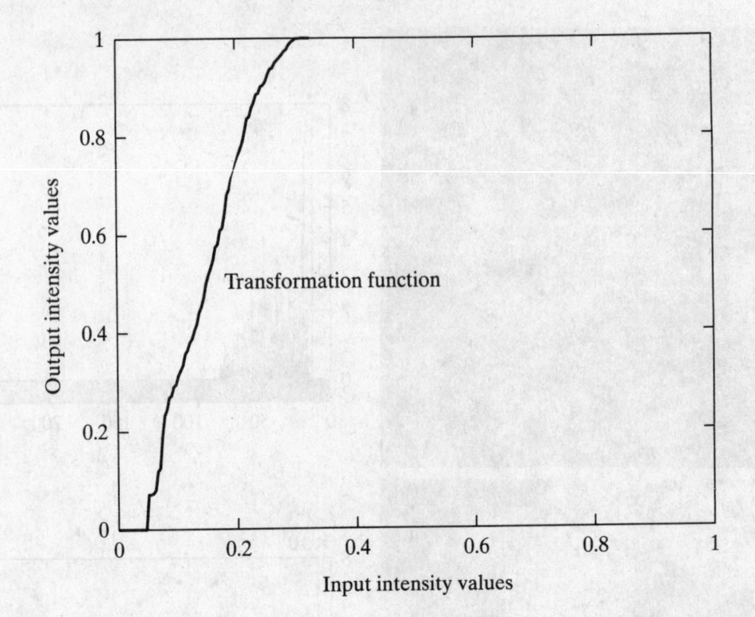

FIGURE 3.9 Transformation function used to map the intensity values from the input image in Fig. 3.8(a) to the values of the output image in Fig. 3.8(c).

3.3.3 Histogram Matching (Specification)

Histogram equalization produces a transformation function that is adaptive, in the sense that it is based on the histogram of a given image. However, once the transformation function for an image has been computed, it does not change unless the histogram of the image changes. As noted in the previous section, histogram equalization achieves enhancement by spreading the levels of the input image over a wider range of the intensity scale. We show in this section that this does not always lead to a successful result. In particular, it is useful in some applications to be able to specify the shape of the histogram that we wish the processed image to have. The method used to generate a processed image that has a specified histogram is called *histogram matching* or *histogram specification*.

The method is simple in principle. Consider for a moment continuous levels that are normalized to the interval [0, 1], and let r and z denote the intensity levels of the input and output images. The input levels have probability density function $p_r(r)$ and the output levels have the specified probability density function $p_z(z)$. We know from the discussion in the previous section that he transformation

$$s = T(r) = \int_0^r p_r(w)\, dw$$

results in intensity levels, s, that have a uniform probability density function, $p_s(s)$. Suppose now that we define a variable z with the property

$$H(z) = \int_0^z p_z(w)\, dw = s$$

Keep in mind that we are after an image with intensity levels z, which have the specified density $p_z(z)$. From the preceding two equations, it follows that

$$z = H^{-1}(s) = H^{-1}[T(r)]$$

We can find $T(r)$ from the input image (this is the histogram-equalization transformation discussed in the previous section), so it follows that we can use the preceding equation to find the transformed levels z whose PDF is the specified $p_z(z)$, as long as we can find H^{-1}. When working with discrete variables, we can guarantee that the inverse of H exists if $p_z(z)$ is a valid histogram (i.e., it has unit area and all its values are nonnegative), and none of its components is zero [i.e., no bin of $p_z(z)$ is empty]. As in histogram equalization, the discrete implementation of the preceding method only yields an approximation to the specified histogram.

The toolbox implements histogram matching using the following syntax in `histeq`:

```
g = histeq(f, hspec)
```

where `f` is the input image, `hspec` is the specified histogram (a row vector of specified values), and `g` is the output image, whose histogram approximates the specified histogram, `hspec`. This vector should contain integer counts corresponding to equally spaced bins. A property of `histeq` is that the histogram of `g` generally better matches `hspec` when `length(hspec)` is much smaller than the number of intensity levels in `f`.

EXAMPLE 3.6: Histogram matching.

■ Figure 3.10(a) shows an image, `f`, of the Mars moon, Phobos, and Fig. 3.10(b) shows its histogram, obtained using `imhist(f)`. The image is dominated by large, dark areas, resulting in a histogram characterized by a large concentration of pixels in the dark end of the gray scale. At first glance, one might conclude that histogram equalization would be a good approach to enhance this image, so that details in the dark areas become more visible. However, the result in Fig. 3.10(c), obtained using the command

```
>> f1 = histeq(f, 256);
```

shows that histogram equalization in fact did not produce a particularly good result in this case. The reason for this can be seen by studying the histogram of the equalized image, shown in Fig. 3.10(d). Here, we see that that the intensity levels have been shifted to the upper one-half of the gray scale, thus giving the image a washed-out appearance. The cause of the shift is the large concentration of dark components at or near 0 in the original histogram. In turn, the cumulative transformation function obtained from this histogram is steep, thus mapping the large concentration of pixels in the low end of the gray scale to the high end of the scale.

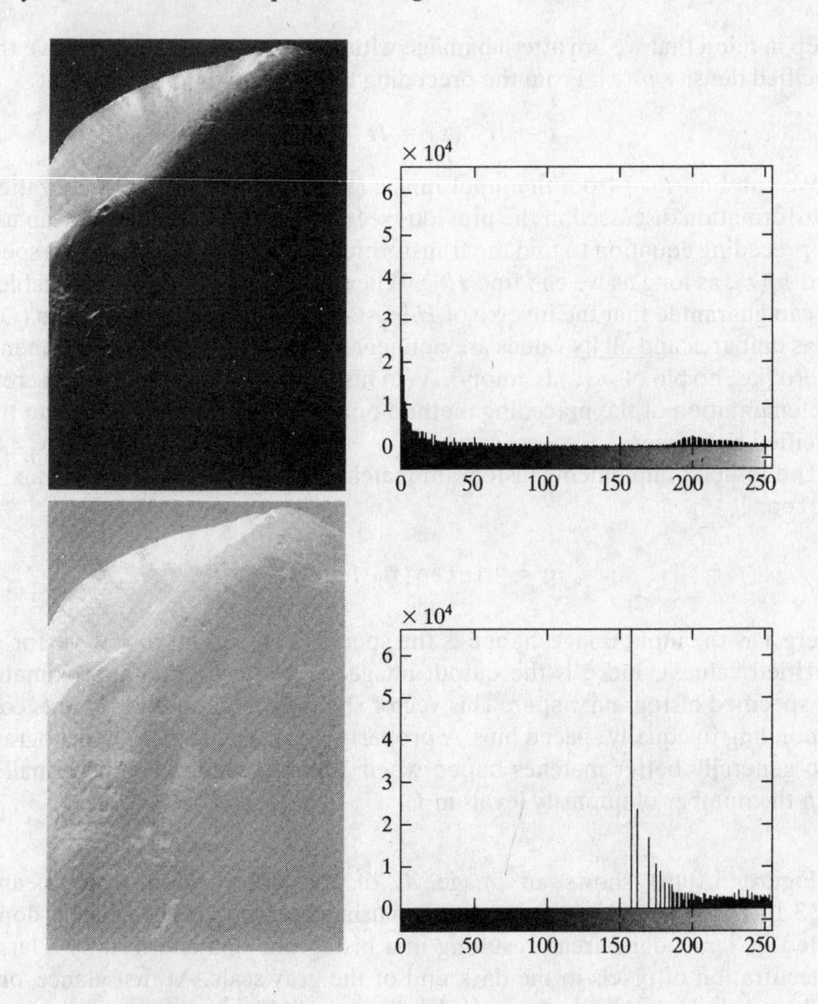

a b
c d

FIGURE 3.10
(a) Image of the Mars moon Phobos.
(b) Histogram.
(c) Histogram-equalized image.
(d) Histogram of (c).
(Original image courtesy of NASA).

One possibility for remedying this situation is to use histogram matching, with the desired histogram having a lesser concentration of components in the low end of the gray scale, and maintaining the general shape of the histogram of the original image. We note from Fig. 3.10(b) that the histogram is basically bimodal, with one large mode at the origin, and another, smaller, mode at the high end of the gray scale. These types of histograms can be modeled, for example, by using multimodal Gaussian functions. The following M-function computes a bimodal Gaussian function normalized to unit area, so it can be used as a specified histogram.

twomodegauss

```
function p = twomodegauss(m1, sig1, m2, sig2, A1, A2, k)
%TWOMODEGAUSS Generates a bimodal Gaussian function.
%  P = TWOMODEGAUSS(M1, SIG1, M2, SIG2, A1, A2, K) generates a bimodal,
%  Gaussian-like function in the interval [0, 1]. P is a 256-element
%  vector normalized so that SUM(P) equals 1. The mean and standard
%  deviation of the modes are (M1, SIG1) and (M2, SIG2), respectively.
%  A1 and A2 are the amplitude values of the two modes. Since the
```

```
%   output is normalized, only the relative values of A1 and A2 are
%   important. K is an offset value that raises the "floor" of the
%   function. A good set of values to try is M1 = 0.15, SIG1 = 0.05,
%   M2 = 0.75, SIG2 = 0.05, A1 = 1, A2 = 0.07, and K = 0.002.

c1 = A1 * (1 / ((2 * pi) ^ 0.5) * sig1);
k1 = 2 * (sig1 ^ 2);
c2 = A2 * (1 / ((2 * pi) ^ 0.5) * sig2);
k2 = 2 * (sig2 ^ 2);
z  = linspace(0, 1, 256);

p = k + c1 * exp(-((z - m1) .^ 2) ./ k1) + ...
    c2 * exp(-((z - m2) .^ 2) ./ k2);
p = p ./ sum(p(:));
```

The following interactive function accepts inputs from a keyboard and plots the resulting Gaussian function. Refer to Section 2.10.5 for an explanation of the functions `input` and `str2num`. Note how the limits of the plots are set.

manualhist

```
function p = manualhist
%MANUALHIST Generates a bimodal histogram interactively.
%   P = MANUALHIST generates a bimodal histogram using
%   TWOMODEGAUSS(m1, sig1, m2, sig2, A1, A2, k). m1 and m2 are the means
%   of the two modes and must be in the range [0, 1]. sig1 and sig2 are
%   the standard deviations of the two modes. A1 and A2 are
%   amplitude values, and k is an offset value that raises the
%   "floor" of histogram. The number of elements in the histogram
%   vector P is 256 and sum(P) is normalized to 1. MANUALHIST
%   repeatedly prompts for the parameters and plots the resulting
%   histogram until the user types an 'x' to quit, and then it returns the
%   last histogram computed.
%
%   A good set of starting values is: (0.15, 0.05, 0.75, 0.05, 1,
%   0.07, 0.002).

% Initialize.
repeats = true;
quitnow = 'x';

% Compute a default histogram in case the user quits before
% estimating at least one histogram.
p = twomodegauss(0.15, 0.05, 0.75, 0.05, 1, 0.07, 0.002);

% Cycle until an x is input.
while repeats
   s = input('Enter m1, sig1, m2, sig2, A1, A2, k OR x to quit:', 's');
   if s == quitnow
      break
   end

   % Convert the input string to a vector of numerical values and
   % verify the number of inputs.
   v = str2num(s);
   if numel(v) ~= 7
```

```
        disp('Incorrect number of inputs.')
        continue
    end
    p = twomodegauss(v(1), v(2), v(3), v(4), v(5), v(6), v(7));
    % Start a new figure and scale the axes. Specifying only xlim
    % leaves ylim on auto.
    figure, plot(p)
    xlim([0 255])
end
```

Since the problem with histogram equalization in this example is due primarily to a large concentration of pixels in the original image with levels near 0, a reasonable approach is to modify the histogram of that image so that it does not have this property. Figure 3.11(a) shows a plot of a function (obtained with program `manualhist`) that preserves the general shape of the original histogram, but has a smoother transition of levels in the dark region of the intensity scale. The output of the program, `p`, consists of 256 equally spaced points from this function and is the desired specified histogram. An image with the specified histogram was generated using the command

```
>> g = histeq(f, p);
```

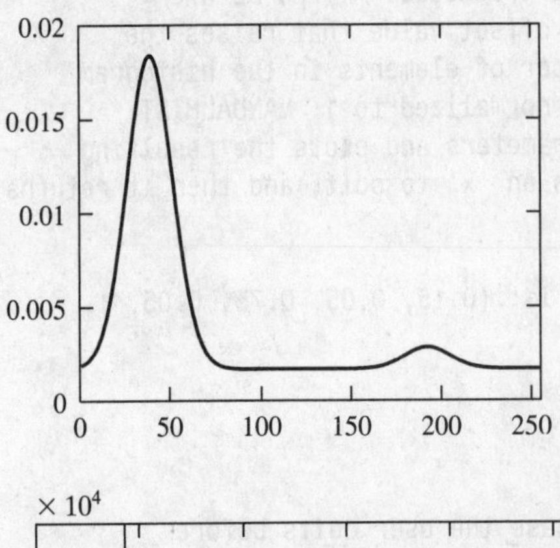

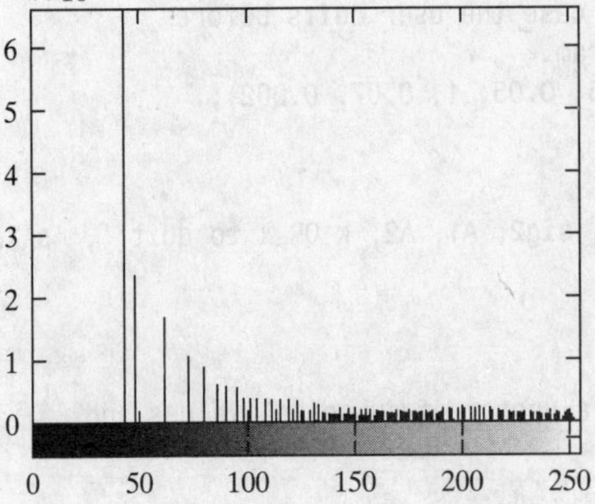

a b
c

FIGURE 3.11 (a) Specified histogram. (b) Result of enhancement by histogram matching. (c) Histogram of (b).

Figure 3.11(b) shows the result. The improvement over the histogram-equalized result in Fig. 3.10(c) is evident by comparing the two images. It is of interest to note that the specified histogram represents a rather modest change from the original histogram. This is all that was required to obtain a significant improvement in enhancement. The histogram of Fig. 3.11(b) is shown in Fig. 3.11(c). The most distinguishing feature of this histogram is how its low end has been moved closer to the lighter region of the gray scale, and thus closer to the specified shape. Note, however, that the shift to the right was not as extreme as the shift in the histogram shown in Fig. 3.10(d), which corresponds to the poorly enhanced image of Fig. 3.10(c). ■

3.4 Spatial Filtering

As mentioned in Section 3.1 and illustrated in Fig. 3.1, neighborhood processing consists of (1) defining a center point, (x, y); (2) performing an operation that involves only the pixels in a predefined neighborhood about that center point; (3) letting the result of that operation be the "response" of the process at *that* point; and (4) repeating the process for every point in the image. The process of moving the center point creates new neighborhoods, one for each pixel in the input image. The two principal terms used to identify this operation are *neighborhood processing* and *spatial filtering*, with the second term being more prevalent. As explained in the following section, if the computations performed on the pixels of the neighborhoods are linear, the operation is called *linear spatial filtering* (the term *spatial convolution* also used); otherwise it is called *nonlinear spatial filtering*.

3.4.1 Linear Spatial Filtering

The concept of *linear filtering* has its roots in the use of the Fourier transform for signal processing in the frequency domain, a topic discussed in detail in Chapter 4. In the present chapter, we are interested in filtering operations that are performed directly on the pixels of an image. Use of the term *linear spatial filtering* differentiates this type of process from *frequency domain filtering*.

The linear operations of interest in this chapter consist of multiplying each pixel in the neighborhood by a corresponding coefficient and summing the results to obtain the response at each point (x, y). If the neighborhood is of size $m \times n$, mn coefficients are required. The coefficients are arranged as a matrix, called a *filter*, *mask*, *filter mask*, *kernel*, *template*, or *window*, with the first three terms being the most prevalent. For reasons that will become obvious shortly, the terms *convolution filter*, *mask*, or *kernel*, also are used.

The mechanics of linear spatial filtering are illustrated in Fig. 3.12. The process consists simply of moving the center of the filter mask w from point to point in an image, f. At each point (x, y), the response of the filter at that point is the sum of products of the filter coefficients and the corresponding neighborhood pixels in the area spanned by the filter mask. For a mask of size $m \times n$, we assume typically that $m = 2a + 1$ and $n = 2b + 1$, where a and b

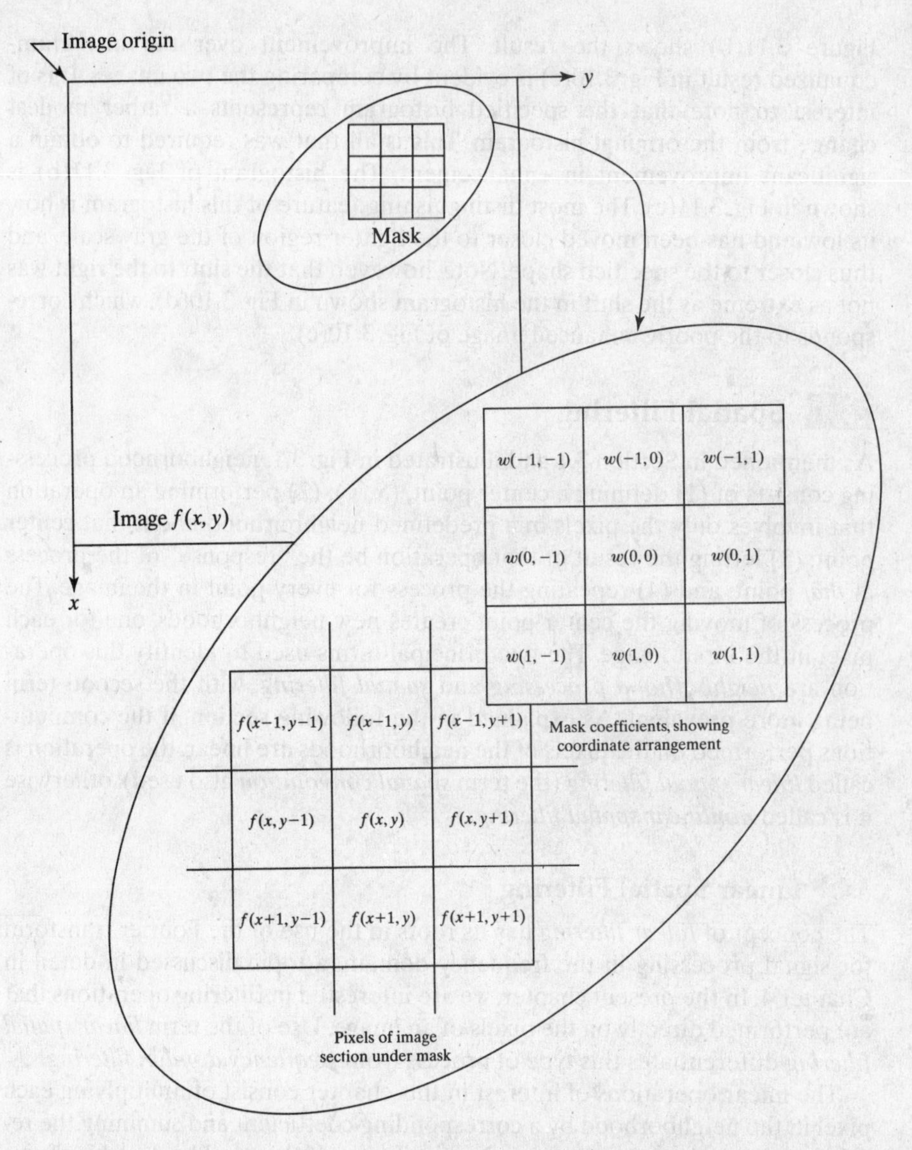

FIGURE 3.12 The mechanics of linear spatial filtering. The magnified drawing shows a 3×3 mask and the corresponding image neighborhood directly under it. The neighborhood is shown displaced out from under the mask for ease of readability.

are nonnegative integers. All this says is that our principal focus is on masks of odd sizes, with the smallest meaningful size being 3×3 (we exclude from our discussion the trivial case of a 1×1 mask). Although it certainly is not a requirement, working with odd-size masks is more intuitive because they have a unique center point.

There are two closely related concepts that must be understood clearly when performing linear spatial filtering. One is *correlation*; the other is *convolution*. Correlation is the process of passing the mask w by the image array f in the manner described in Fig. 3.12. Mechanically, convolution is the same process, except that w is rotated by 180° prior to passing it by f. These two concepts are best explained by some simple examples.

Figure 3.13(a) shows a one-dimensional function, f, and a mask, w. The origin of f is assumed to be its leftmost point. To perform the correlation of the two functions, we move w so that its rightmost point coincides with the origin of f, as shown in Fig. 3.13(b). Note that there are points between the two functions that do not overlap. The most common way to handle this problem is to pad f with as many 0s as are necessary to guarantee that there will always be corresponding points for the full excursion of w past f. This situation is shown in Fig. 3.13(c).

We are now ready to perform the correlation. The first value of correlation is the sum of products of the two functions in the position shown in Fig. 3.13(c). The sum of products is 0 in this case. Next, we move w one location to the right and repeat the process [Fig. 3.13(d)]. The sum of products again is 0. After four shifts [Fig. 3.13(e)], we encounter the first nonzero value of the correlation, which is $(2)(1) = 2$. If we proceed in this manner until w moves completely past f [the ending geometry is shown in Fig. 3.13(f)] we would get the result in Fig. 3.13(g). This set of values is the correlation of w and f. Note that, had we left w stationary and had moved f past w instead, the result would have been different, so the order matters.

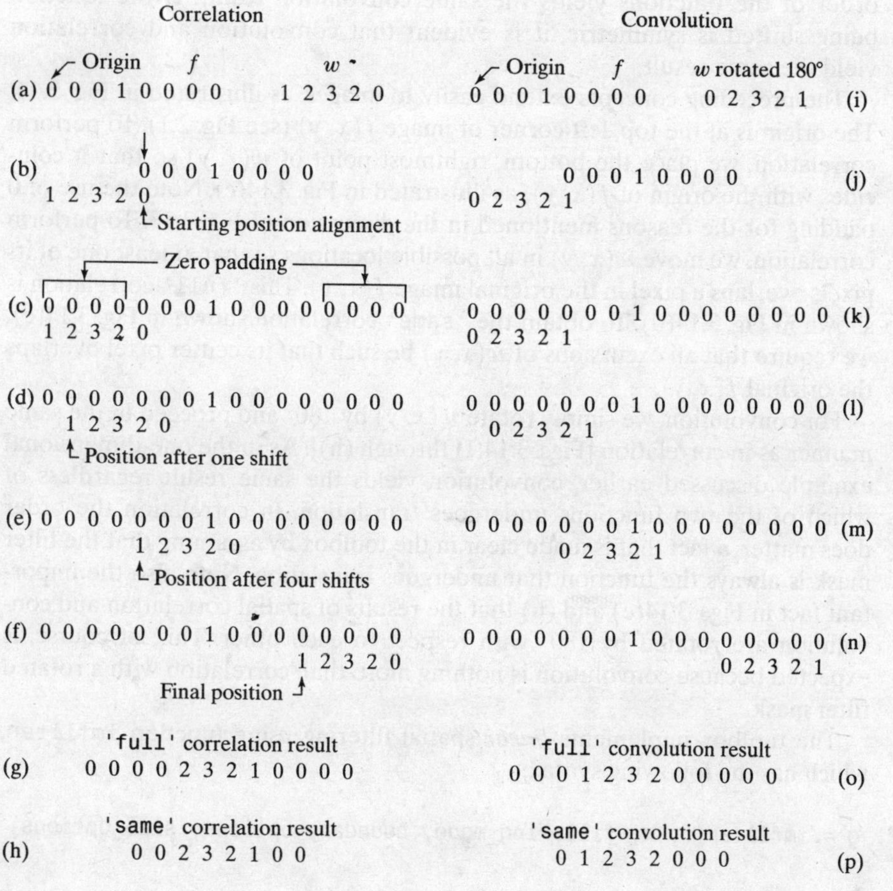

Correlation

Origin f w

(a) 0 0 0 1 0 0 0 0 1 2 3 2 0

(b) 0 0 0 1 0 0 0 0
1 2 3 2 0
Starting position alignment

Zero padding

(c) 0 0 0 0 0 0 0 1 0 0 0 0 0 0 0 0
1 2 3 2 0

(d) 0 0 0 0 0 0 0 1 0 0 0 0 0 0 0 0
1 2 3 2 0
Position after one shift

(e) 0 0 0 0 0 0 0 1 0 0 0 0 0 0 0 0
1 2 3 2 0
Position after four shifts

(f) 0 0 0 0 0 0 0 1 0 0 0 0 0 0 0 0
1 2 3 2 0
Final position

(g) 'full' correlation result
0 0 0 0 2 3 2 1 0 0 0 0

(h) 'same' correlation result
0 0 2 3 2 1 0 0

Convolution

Origin f w rotated 180°

0 0 0 1 0 0 0 0 0 2 3 2 1 (i)

0 0 0 1 0 0 0 0 (j)
0 2 3 2 1

0 0 0 0 0 0 0 1 0 0 0 0 0 0 0 0 (k)
0 2 3 2 1

0 0 0 0 0 0 0 1 0 0 0 0 0 0 0 0 (l)
0 2 3 2 1

0 0 0 0 0 0 0 1 0 0 0 0 0 0 0 0 (m)
0 2 3 2 1

0 0 0 0 0 0 0 1 0 0 0 0 0 0 0 0 (n)
0 2 3 2 1

'full' convolution result
0 0 0 1 2 3 2 0 0 0 0 0 (o)

'same' convolution result
0 1 2 3 2 0 0 0 (p)

FIGURE 3.13 Illustration of one-dimensional correlation and convolution.

The label `'full'` in the correlation shown in Fig. 3.13(g) is a flag (to be discussed later) used by the toolbox to indicate correlation using a padded image and computed in the manner just described. The toolbox provides another option, denoted by `'same'` [Fig. 3.13(h)] that produces a correlation that is the same size as f. This computation also uses zero padding, but the starting position is with the center point of the mask (the point labeled 3 in w) aligned with the origin of f. The last computation is with the center point of the mask aligned with the last point in f.

To perform convolution we rotate w by 180° and place its rightmost point at the origin of f, as shown in Fig. 3.13(j). We then repeat the sliding/computing process employed in correlation, as illustrated in Figs. 3.13(k) through (n). The `'full'` and `'same'` convolution results are shown in Figs. 3.13(o) and (p), respectively.

Function f in Fig. 3.13 is a discrete unit impulse function that is 1 at one location and 0 everywhere else. It is evident from the result in Figs. 3.13(o) or (p) that convolution basically just "copied" w at the location of the impulse. This simple copying property (called *sifting*) is a fundamental concept in linear system theory, and it is the reason why one of the functions is always rotated by 180° in convolution. Note that, unlike correlation, reversing the order of the functions yields the same convolution result. If the function being shifted is symmetric, it is evident that convolution and correlation yield the same result.

The preceding concepts extend easily to images, as illustrated in Fig. 3.14. The origin is at the top, left corner of image $f(x, y)$ (see Fig. 2.1). To perform correlation, we place the bottom, rightmost point of $w(x, y)$ so that it coincides with the origin of $f(x, y)$, as illustrated in Fig. 3.14(c). Note the use of 0 padding for the reasons mentioned in the discussion of Fig. 3.13. To perform correlation, we move $w(x, y)$ in all possible locations so that at least one of its pixels overlaps a pixel in the original image $f(x, y)$. This `'full'` correlation is shown in Fig. 3.14(d). To obtain the `'same'` correlation shown in Fig. 3.14(e), we require that all excursions of $w(x, y)$ be such that its center pixel overlaps the original $f(x, y)$.

For convolution, we simply rotate $w(x, y)$ by 180° and proceed in the same manner as in correlation [Figs. 3.14(f) through (h)]. As in the one-dimensional example discussed earlier, convolution yields the same result regardless of which of the two functions undergoes translation. In correlation the order does matter, a fact that is made clear in the toolbox by assuming that the filter mask is always the function that undergoes translation. Note also the important fact in Figs. 3.14(e) and (h) that the results of spatial correlation and convolution are rotated by 180° with respect to each other. This, of course, is expected because convolution is nothing more than correlation with a rotated filter mask.

The toolbox implements *linear* spatial filtering using function `imfilter`, which has the following syntax:

```
g = imfilter(f, w, filtering_mode, boundary_options, size_options)
```

```
                               Padded f
                               0 0 0 0 0 0 0 0 0
                               0 0 0 0 0 0 0 0 0
                               0 0 0 0 0 0 0 0 0
  ↙ Origin of f(x, y)          0 0 0 0 0 0 0 0 0
0 0 0 0 0                      0 0 0 0 1 0 0 0 0
0 0 0 0 0    w(x, y)           0 0 0 0 0 0 0 0 0
0 0 1 0 0     1 2 3            0 0 0 0 0 0 0 0 0
0 0 0 0 0     4 5 6            0 0 0 0 0 0 0 0 0
0 0 0 0 0     7 8 9            0 0 0 0 0 0 0 0 0
   (a)                               (b)

↘ Initial position for w   'full' correlation result   'same' correlation result
|1 2 3| 0 0 0 0 0 0        0 0 0 0 0 0 0 0 0            0 0 0 0 0
|4 5 6| 0 0 0 0 0 0        0 0 0 0 0 0 0 0 0            0 9 8 7 0
|7 8 9| 0 0 0 0 0 0        0 0 0 0 0 0 0 0 0            0 6 5 4 0
 0 0 0  0 0 0 0 0 0        0 0 0 9 8 7 0 0 0            0 3 2 1 0
 0 0 0  0 1 0 0 0 0        0 0 0 6 5 4 0 0 0            0 0 0 0 0
 0 0 0  0 0 0 0 0 0        0 0 0 3 2 1 0 0 0
 0 0 0  0 0 0 0 0 0        0 0 0 0 0 0 0 0 0
 0 0 0  0 0 0 0 0 0        0 0 0 0 0 0 0 0 0
 0 0 0  0 0 0 0 0 0        0 0 0 0 0 0 0 0 0
        (c)                       (d)                       (e)

↘ Rotated w                'full' convolution result   'same' convolution result
|9 8 7| 0 0 0 0 0 0        0 0 0 0 0 0 0 0 0            0 0 0 0 0
|6 5 4| 0 0 0 0 0 0        0 0 0 0 0 0 0 0 0            0 1 2 3 0
|3 2 1| 0 0 0 0 0 0        0 0 0 0 0 0 0 0 0            0 4 5 6 0
 0 0 0  0 0 0 0 0 0        0 0 0 1 2 3 0 0 0            0 7 8 9 0
 0 0 0  0 1 0 0 0 0        0 0 0 4 5 6 0 0 0            0 0 0 0 0
 0 0 0  0 0 0 0 0 0        0 0 0 7 8 9 0 0 0
 0 0 0  0 0 0 0 0 0        0 0 0 0 0 0 0 0 0
 0 0 0  0 0 0 0 0 0        0 0 0 0 0 0 0 0 0
 0 0 0  0 0 0 0 0 0        0 0 0 0 0 0 0 0 0
        (f)                       (g)                       (h)
```

FIGURE 3.14 Illustration of two-dimensional correlation and convolution. The 0s are shown in gray to simplify viewing.

where `f` is the input image, `w` is the filter mask, `g` is the filtered result, and the other parameters are summarized in Table 3.2. The `filtering_mode` specifies whether to filter using correlation (`'corr'`) or convolution (`'conv'`). The `boundary_options` deal with the border-padding issue, with the size of the border being determined by the size of the filter. These options are explained further in Example 3.7. The `size_options` are either `'same'` or `'full'`, as explained in Figs. 3.13 and 3.14.

The most common syntax for `imfilter` is

```
g = imfilter(f, w, 'replicate')
```

This syntax is used when implementing IPT standard linear spatial filters. These filters, which are discussed in Section 3.5.1, are prerotated by 180°, so we can use the correlation default in `imfilter`. From the discussion of Fig. 3.14, we know that performing correlation with a rotated filter is the same as performing convolution with the original filter. If the filter is symmetric about its center, then both options produce the same result.

TABLE 3.2 Options for function `imfilter`.

Options	Description
Filtering Mode	
`'corr'`	Filtering is done using correlation (see Figs. 3.13 and 3.14). This is the default.
`'conv'`	Filtering is done using convolution (see Figs. 3.13 and 3.14).
Boundary Options	
`P`	The boundaries of the input image are extended by padding with a value, `P` (written without quotes). This is the default, with value 0.
`'replicate'`	The size of the image is extended by replicating the values in its outer border.
`'symmetric'`	The size of the image is extended by mirror-reflecting it across its border.
`'circular'`	The size of the image is extended by treating the image as one period a 2-D periodic function.
Size Options	
`'full'`	The output is of the same size as the extended (padded) image (see Figs. 3.13 and 3.14).
`'same'`	The output is of the same size as the input. This is achieved by limiting the excursions of the center of the filter mask to points contained in the original image (see Figs. 3.13 and 3.14). This is the default.

When working with filters that are neither pre-rotated nor symmetric, and we wish to perform convolution, we have two options. One is to use the syntax

```
g = imfilter(f, w, 'conv', 'replicate')
```

`rot90(w, k)` *rotates* `w` *by* `k*90` *degrees, where* `k` *is an integer.*

The other approach is to preprocess w by using the function `rot90(w, 2)` to rotate it 180°, and then use `imfilter(f, w, 'replicate')`. Of course these two steps can be combined into one statement. The preceding syntax produces an image g that is of the same size as the input (i.e., the default in computation is the `'same'` mode discussed earlier).

Each element of the filtered image is computed using double-precision, floating-point arithmetic. However, `imfilter` converts the output image to the same class of the input. Therefore, if `f` is an integer array, then output elements that exceed the range of the integer type are truncated, and fractional values are rounded. If more precision is desired in the result, then `f` should be converted to class `double` by using `im2double` or `double` before using `imfilter`.

EXAMPLE 3.7: Using function `imfilter`.

■ Figure 3.15(a) is a class `double` image, `f`, of size 512 × 512 pixels. Consider the simple 31 × 31 filter

```
>> w = ones(31);
```

a b c
d e f

FIGURE 3.15 (a) Original image. (b) Result of using `imfilter` with default zero padding. (c) Result with the `'replicate'` option. (d) Result with the `'symmetric'` option. (e) Result with the `'circular'` option. (f) Result of converting the original image to class `uint8` and then filtering with the `'replicate'` option. A filter of size 31 × 31 with all 1s was used throughout.

which is proportional to an averaging filter. We did not divide the coefficients by $(31)^2$ to illustrate at the end of this example the scaling effects of using `imfilter` with an image of class `uint8`.

Convolving filter `w` with an image produces a blurred result. Because the filter is symmetric, we can use the correlation default in `imfilter`. Figure 3.15(b) shows the result of performing the following filtering operation:

```
>> gd = imfilter(f, w);
>> imshow(gd, [ ])
```

where we used the default boundary option, which pads the border of the image with 0's (black). As expected the edges between black and white in the filtered image are blurred, but so are the edges between the light parts of the image and the boundary. The reason, of course, is that the padded border is black. We can deal with this difficulty by using the `'replicate'` option

```
>> gr = imfilter(f, w, 'replicate');
>> figure, imshow(gr, [ ])
```

As Fig. 3.15(c) shows, the borders of the filtered image now appear as expected. In this case, equivalent results are obtained with the `'symmetric'` option

```
>> gs = imfilter(f, w, 'symmetric');
>> figure, imshow(gs, [ ])
```

Figure 3.15(d) shows the result. However, using the 'circular' option

```
>> gc = imfilter(f, w, 'circular');
>> figure, imshow(gc, [ ])
```

produced the result in Fig. 3.15(e), which shows the same problem as with zero padding. This is as expected because use of periodicity makes the black parts of the image adjacent to the light areas.

Finally, we illustrate how the fact that imfilter produces a result that is of the same class as the input can lead to difficulties if not handled properly:

```
>> f8 = im2uint8(f);
>> g8r = imfilter(f8, w, 'replicate');
>> figure, imshow(g8r, [ ])
```

Figure 3.15(f) shows the result of these operations. Here, when the output was converted to the class of the input (uint8) by imfilter, clipping caused some data loss. The reason is that the coefficients of the mask did not sum to the range [0, 1], resulting in filtered values outside the [0, 255] range. Thus, to avoid this difficulty, we have the option of normalizing the coefficients so that their sum is in the range [0, 1] (in the present case we would divide the coefficients by $(31)^2$, so the sum would be 1), or inputting the data in double format. Note, however, that even if the second option were used, the data usually would have to be normalized to a valid image format at some point (e.g., for storage) anyway. Either approach is valid; the key point is that data ranges have to be kept in mind to avoid unexpected results. ■

3.4.2 Nonlinear Spatial Filtering

Nonlinear spatial filtering is based on neighborhood operations also, and the mechanics of defining $m \times n$ neighborhoods by sliding the center point through an image are the same as discussed in the previous section. However, whereas linear spatial filtering is based on computing the sum of products (which is a linear operation), nonlinear spatial filtering is based, as the name implies, on nonlinear operations involving the pixels of a neighborhood. For example, letting the response at each center point be equal to the maximum pixel value in its neighborhood is a nonlinear filtering operation. Another basic difference is that the concept of a mask is not as prevalent in nonlinear processing. The idea of filtering carries over, but the "filter" should be visualized as a nonlinear function that operates on the pixels of a neighborhood, and whose response constitutes the response of the operation at the center pixel of the neighborhood.

The toolbox provides two functions for performing general nonlinear filtering: nlfilter and colfilt. The former performs operations directly in 2-D, while colfilt organizes the data in the form of columns. Although colfilt requires more memory, it generally executes significantly faster than nlfilter.

In most image processing applications speed is an overriding factor, so colfilt is preferred over nlfilt for implementing generalized nonlinear spatial filtering.

Given an input image, f, of size $M \times N$, and a neighborhood of size $m \times n$, function colfilt generates a matrix, call it A, of maximum size $mn \times MN$,[†] in which each column corresponds to the pixels encompassed by the neighborhood centered at a location in the image. For example, the first column corresponds to the pixels encompassed by the neighborhood when its center is located at the top, leftmost point in f. All required padding is handled transparently by colfilt (using zero padding).

The syntax of function colfilt is

```
g = colfilt(f, [m n], 'sliding', @fun, parameters)
```

where, as before, m and n are the dimensions of the filter region, 'sliding' indicates that the process is one of sliding the $m \times n$ region from pixel to pixel in the input image f, @fun references a function, which we denote arbitrarily as fun, and parameters indicates parameters (separated by commas) that may be required by function fun. The symbol @ is called a *function handle*, a MATLAB data type that contains information used in referencing a function. As will be demonstrated shortly, this is a particularly powerful concept.

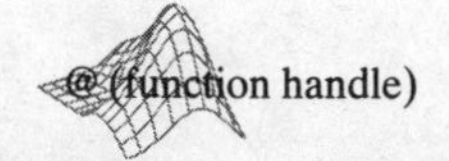

Because of the way in which matrix A is organized, function fun must operate on each of the columns of A individually and return a row vector, v, containing the results for all the columns. The kth element of v is the result of the operation performed by fun on the kth column of A. Since there can be up to MN columns in A, the maximum dimension of v is $1 \times MN$.

The linear filtering discussed in the previous section has provisions for padding to handle the border problems inherent in spatial filtering. When using colfilt, however, the input image must be padded explicitly before filtering. For this we use function padarray, which, for 2-D functions, has the syntax

```
fp = padarray(f, [r c], method, direction)
```

where f is the input image, fp is the padded image, [r c] gives the number of rows and columns by which to pad f, and method and direction are as explained in Table 3.3. For example, if f = [1 2; 3 4], the command

```
>> fp = padarray(f, [3 2], 'replicate', 'post')
```

[†] A *always* has *mn* rows, but the number of columns can vary, depending on the size of the input. Size selection is managed automatically by colfilt.

TABLE 3.3 Options for function padarray.

Options	Description
Method	
'symmetric'	The size of the image is extended by mirror-reflecting it across its border.
'replicate'	The size of the image is extended by replicating the values in its outer border.
'circular'	The size of the image is extended by treating the image as one period of a 2-D periodic function.
Direction	
'pre'	Pad before the first element of each dimension.
'post'	Pad after the last element of each dimension.
'both'	Pad before the first element and after the last element of each dimension. This is the default.

produces the result

```
fp =

     1     2     2     2
     3     4     4     4
     3     4     4     4
     3     4     4     4
     3     4     4     4
```

If direction is not included in the argument, the default is 'both'. If method is not included, the default padding is with 0's. If neither parameter is included in the argument, the default padding is 0 and the default direction is 'both'. At the end of computation, the image is cropped back to its original size.

EXAMPLE 3.8: Using function colfilt to implement a nonlinear spatial filter.

■ As an illustration of function colfilt, we implement a nonlinear filter whose response at any point is the geometric mean of the intensity values of the pixels in the neighborhood centered at that point. The geometric mean in a neighborhood of size $m \times n$ is the product of the intensity values in the neighborhood raised to the power $1/mn$. First we implement the nonlinear filter function, call it gmean:

```
function v = gmean(A)
mn = size(A, 1); % The length of the columns of A is always mn.
v = prod(A, 1).^(1/mn);
```

prod(A) *returns the product of the elements of* A. prod(A, dim) *returns the product of the elements of* A *along dimension* dim.

To reduce border effects, we pad the input image using, say, the 'replicate' option in function padarray:

```
>> f = padarray(f, [m n], 'replicate');
```

Finally, we call colfilt:

```
>> g = colfilt(f, [m n], 'sliding', @gmean);
```

There are several important points at play here. First, note that, although matrix A is part of the argument in function gmean, it is not included in the parameters in colfilt. This matrix is passed automatically to gmean by colfilt using the function handle. Also, because matrix A is managed automatically by colfilt, the number of columns in A is variable (but, as noted earlier, the number of rows, that is, the column length, is always mn). Therefore, the size of A must be computed each time the function in the argument is called by colfilt. The filtering process in this case consists of computing the product of all pixels in the neighborhood and then raising the result to the power $1/mn$. For any value of (x, y), the filtered result at that point is contained in the appropriate column in v. The function identified by the handle, @, can be any function callable from where the function handle was created. The key requirement is that the function operate on the columns of A and return a row vector containing the result for all individual columns. Function colfilt then takes those results and rearranges them to produce the output image, g. ■

Some commonly used nonlinear filters can be implemented in terms of other MATLAB and IPT functions such as imfilter and ordfilt2 (see Section 3.5.2). Function spfilt in Section 5.3, for example, implements the geometric mean filter in Example 3.8 in terms of imfilter and the MATLAB log and exp functions. When this is possible, performance usually is much faster, and memory usage is a fraction of the memory required by colfilt. Function colfilt, however, remains the best choice for nonlinear filtering operations that do not have such alternate implementations.

3.5 Image Processing Toolbox Standard Spatial Filters

In this section we discuss linear and nonlinear spatial filters supported by IPT. Additional nonlinear filters are implemented in Section 5.3.

3.5.1 Linear Spatial Filters

The toolbox supports a number of predefined 2-D linear spatial filters, obtained by using function fspecial, which generates a filter mask, w, using the syntax

```
w = fspecial('type', parameters)
```

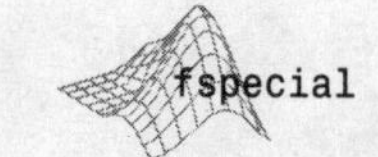

where 'type' specifies the filter type, and parameters further define the specified filter. The spatial filters supported by fspecial are summarized in Table 3.4, including applicable parameters for each filter.

TABLE 3.4
Spatial filters supported by function `fspecial`.

Type	Syntax and Parameters
`'average'`	`fspecial('average', [r c])`. A rectangular averaging filter of size `r` × `c`. The default is 3 × 3. A single number instead of `[r c]` specifies a square filter.
`'disk'`	`fspecial('disk', r)`. A circular averaging filter (within a square of size `2r + 1`) with radius `r`. The default radius is 5.
`'gaussian'`	`fspecial('gaussian', [r c], sig)`. A Gaussian lowpass filter of size `r` × `c` and standard deviation `sig` (positive). The defaults are 3 × 3 and 0.5. A single number instead of `[r c]` specifies a square filter.
`'laplacian'`	`fspecial('laplacian', alpha)`. A 3 × 3 Laplacian filter whose shape is specified by `alpha`, a number in the range [0, 1]. The default value for `alpha` is 0.5.
`'log'`	`fspecial('log', [r c], sig)`. Laplacian of a Gaussian (LoG) filter of size `r` × `c` and standard deviation `sig` (positive). The defaults are 5 × 5 and 0.5. A single number instead of `[r c]` specifies a square filter.
`'motion'`	`fspecial('motion', len, theta)`. Outputs a filter that, when convolved with an image, approximates linear motion (of a camera with respect to the image) of `len` pixels. The direction of motion is `theta`, measured in degrees, counterclockwise from the horizontal. The defaults are 9 and 0, which represents a motion of 9 pixels in the horizontal direction.
`'prewitt'`	`fspecial('prewitt')`. Outputs a 3 × 3 Prewitt mask, wv, that approximates a vertical gradient. A mask for the horizontal gradient is obtained by transposing the result: `wh = wv'`.
`'sobel'`	`fspecial('sobel')`. Outputs a 3 × 3 Sobel mask, sv, that approximates a vertical gradient. A mask for the horizontal gradient is obtained by transposing the result: `sh = sv'`.
`'unsharp'`	`fspecial('unsharp', alpha)`. Outputs a 3 × 3 unsharp filter. Parameter `alpha` controls the shape; it must be greater than 0 and less than or equal to 1.0; the default is 0.2.

EXAMPLE 3.9:
Using function `imfilter`.

■ We illustrate the use of `fspecial` and `imfilter` by enhancing an image with a Laplacian filter. The Laplacian of an image $f(x, y)$, denoted $\nabla^2 f(x, y)$, is defined as

$$\nabla^2 f(x, y) = \frac{\partial^2 f(x, y)}{\partial x^2} + \frac{\partial^2 f(x, y)}{\partial y^2}$$

Commonly used digital approximations of the second derivatives are

$$\frac{\partial^2 f}{\partial x^2} = f(x + 1, y) + f(x - 1, y) - 2f(x, y)$$

and

$$\frac{\partial^2 f}{\partial y^2} = f(x, y + 1) + f(x, y - 1) - 2f(x, y)$$

so that

$$\nabla^2 f = [f(x+1, y) + f(x-1, y) + f(x, y+1) + f(x, y-1)] - 4f(x, y)$$

This expression can be implemented at all points (x, y) in an image by convolving the image with the following spatial mask:

$$\begin{matrix} 0 & 1 & 0 \\ 1 & -4 & 1 \\ 0 & 1 & 0 \end{matrix}$$

An alternate definition of the digital second derivatives takes into account diagonal elements, and can be implemented using the mask

$$\begin{matrix} 1 & 1 & 1 \\ 1 & -8 & 1 \\ 1 & 1 & 1 \end{matrix}$$

Both derivatives sometimes are defined with the signs opposite to those shown here, resulting in masks that are the negatives of the preceding two masks.

Enhancement using the Laplacian is based on the equation

$$g(x, y) = f(x, y) + c[\nabla^2 f(x, y)]$$

where $f(x, y)$ is the input image, $g(x, y)$ is the enhanced image, and c is 1 if the center coefficient of the mask is positive, or -1 if it is negative (Gonzalez and Woods [2002]). Because the Laplacian is a derivative operator, it sharpens the image but drives constant areas to zero. Adding the original image back restores the gray-level tonality.

Function `fspecial('laplacian', alpha)` implements a more general Laplacian mask:

$$\begin{matrix} \dfrac{\alpha}{1+\alpha} & \dfrac{1-\alpha}{1+\alpha} & \dfrac{\alpha}{1+\alpha} \\ \dfrac{1-\alpha}{1+\alpha} & \dfrac{-4}{1+\alpha} & \dfrac{1-\alpha}{1+\alpha} \\ \dfrac{\alpha}{1+\alpha} & \dfrac{1-\alpha}{1+\alpha} & \dfrac{\alpha}{1+\alpha} \end{matrix}$$

which allows fine tuning of enhancement results. However, the predominant use of the Laplacian is based on the two masks just discussed.

We now proceed to enhance the image in Fig. 3.16(a) using the Laplacian. This image is a mildly blurred image of the North Pole of the moon. Enhancement in this case consists of sharpening the image, while preserving as much of its gray tonality as possible. First, we generate and display the Laplacian filter:

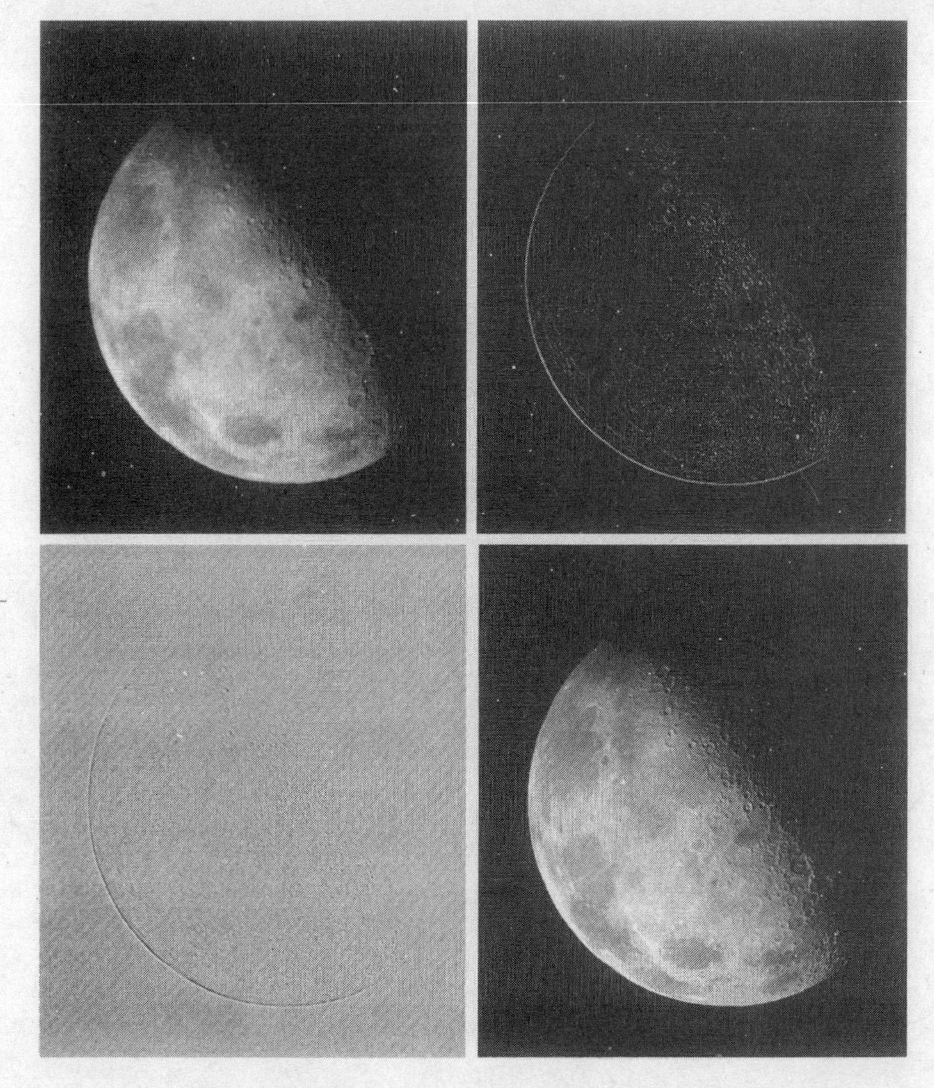

a b
c d

FIGURE 3.16 (a) Image of the North Pole of the moon. (b) Laplacian filtered image, using `uint8` formats. (c) Laplacian filtered image obtained using `double` formats. (d) Enhanced result, obtained by subtracting (c) from (a). (Original image courtesy of NASA.)

```
>> w = fspecial('laplacian', 0)
w =

    0.0000    1.0000    0.0000
    1.0000   -4.0000    1.0000
    0.0000    1.0000    0.0000
```

Note that the filter is of class `double`, and that its shape with `alpha = 0` is the Laplacian filter discussed previously. We could just as easily have specified this shape manually as

```
>> w = [0 1 0; 1 -4 1; 0 1 0];
```

Next we apply w to the input image, f, which is of class uint8:

```
>> g1 = imfilter(f, w, 'replicate');
>> imshow(g1, [ ])
```

Figure 3.16(b) shows the resulting image. This result looks reasonable, but has a problem: all its pixels are positive. Because of the negative center filter coefficient, we know that we can expect in general to have a Laplacian image with negative values. However, f in this case is of class uint8 and, as discussed in the previous section, filtering with imfilter gives an output that is of the same class as the input image, so negative values are truncated. We get around this difficulty by converting f to class double before filtering it:

```
>> f2 = im2double(f);
>> g2 = imfilter(f2, w, 'replicate');
>> imshow(g2, [ ])
```

The result, shown in Fig. 3.15(c), is more what a properly processed Laplacian image should look like.

Finally, we restore the gray tones lost by using the Laplacian by subtracting (because the center coefficient is negative) the Laplacian image from the original image:

```
>> g = f2 - g2;
>> imshow(g)
```

The result, shown in Fig. 3.16(d), is sharper than the original image. ■

EXAMPLE 3.10: Manually specifying filters and comparing enhancement techniques.

■ Enhancement problems often require the specification of filters beyond those available in the toolbox. The Laplacian is a good example. The toolbox supports a 3 × 3 Laplacian filter with a −4 in the center. Usually, sharper enhancement is obtained by using the 3 × 3 Laplacian filter that has a −8 in the center and is surrounded by 1s, as discussed earlier. The purpose of this example is to implement this filter manually, and also to compare the results obtained by using the two Laplacian formulations. The sequence of commands is as follows:

```
>>  f = imread('moon.tif');
>>  w4 = fspecial('laplacian', 0);  % Same as w in Example 3.9.
>>  w8 = [1 1 1; 1 -8 1; 1 1 1];
>>  f = im2double(f);
>>  g4 = f - imfilter(f, w4, 'replicate');
>>  g8 = f - imfilter(f, w8, 'replicate');
>>  imshow(f)
>>  figure, imshow(g4)
>>  figure, imshow(g8)
```

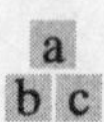

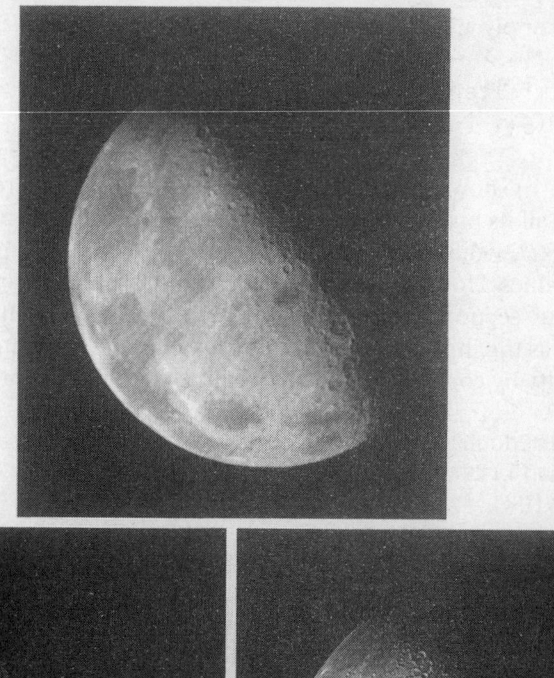

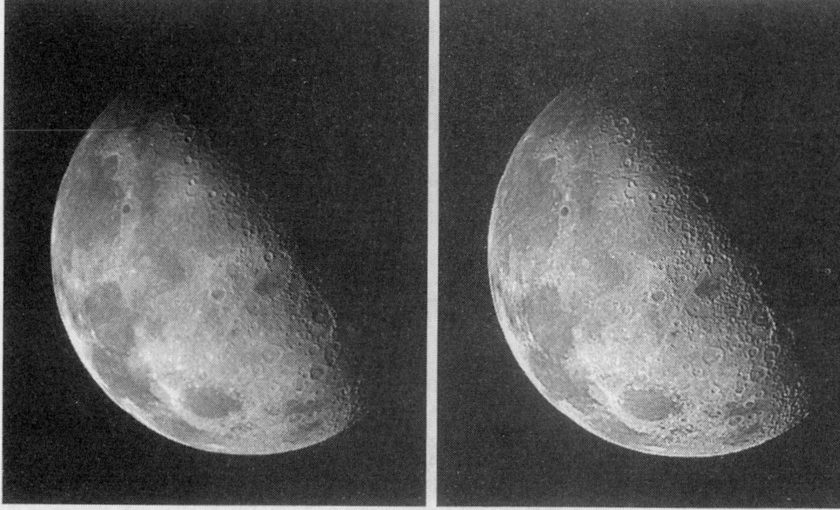

FIGURE 3.17 (a) Image of the North Pole of the moon. (b) Image enhanced using the Laplacian filter `'laplacian'`, which has a −4 in the center. (c) Image enhanced using a Laplacian filter with a −8 in the center.

Figure 3.17(a) shows the original moon image again for easy comparison. Fig. 3.17(b) is `g4`, which is the same as Fig. 3.16(d), and Fig. 3.17(c) shows `g8`. As expected, this result is significantly sharper than Fig. 3.17(b). ■

3.5.2 Nonlinear Spatial Filters

A commonly-used tool for generating nonlinear spatial filters in IPT is function `ordfilt2`, which generates *order-statistic filters* (also called *rank filters*). These are nonlinear spatial filters whose response is based on ordering (ranking) the pixels contained in an image neighborhood and then replacing the value of the center pixel in the neighborhood with the value determined by the ranking result. Attention is focused in this section on nonlinear filters generated by `ordfilt2`. A number of additional nonlinear filters are developed and implemented in Section 5.3.

The syntax of function `ordfilt2` is

```
g = ordfilt2(f, order, domain)
```

This function creates the output image `g` by replacing each element of `f` by the `order`-th element in the sorted set of neighbors specified by the nonzero elements in `domain`. Here, `domain` is an $m \times n$ matrix of 1s and 0s that specify the pixel locations in the neighborhood that are to be used in the computation. In this sense, `domain` acts like a mask. The pixels in the neighborhood that correspond to 0 in the `domain` matrix are not used in the computation. For example, to implement a *min filter* (order 1) of size $m \times n$ we use the syntax

```
g = ordfilt2(f, 1, ones(m, n))
```

In this formulation the 1 denotes the 1st sample in the ordered set of mn samples, and `ones(m, n)` creates an $m \times n$ matrix of 1s, indicating that all samples in the neighborhood are to be used in the computation.

In the terminology of statistics, a min filter (the first sample of an ordered set) is referred to as the 0th percentile. Similarly, the 100th percentile is the last sample in the ordered set, which is the mnth sample. This corresponds to a *max filter*, which is implemented using the syntax

```
g = ordfilt2(f, m*n, ones(m, n))
```

The best-known order-statistic filter in digital image processing is the *median*[†] *filter*, which corresponds to the 50th percentile. We can use MATLAB function `median` in `ordfilt2` to create a median filter:

```
g = ordfilt2(f, median(1:m*n), ones(m, n))
```

where `median(1:m*n)` simply computes the median of the ordered sequence $1, 2, \ldots, mn$. Function `median` has the general syntax

```
v = median(A, dim)
```

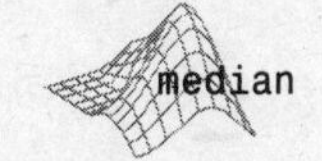

where `v` is vector whose elements are the median of `A` along dimension `dim`. For example, if `dim = 1`, each element of `v` is the median of the elements along the corresponding column of `A`.

[†]Recall that the median, ξ, of a set of values is such that half the values in the set are less than or equal to ξ, and half are greater than or equal to ξ.

Because of its practical importance, the toolbox provides a specialized implementation of the 2-D median filter:

```
g = medfilt2(f, [m n], padopt)
```

where the tuple `[m n]` defines a neighborhood of size `m` × `n` over which the median is computed, and `padopt` specifies one of three possible border padding options: `'zeros'` (the default), `'symmetric'` in which `f` is extended symmetrically by mirror-reflecting it across its border, and `'indexed'`, in which `f` is padded with 1s if it is of class `double` and with 0s otherwise. The default form of this function is

```
g = medfilt2(f)
```

which uses a 3×3 neighborhood to compute the median, and pads the border of the input with 0s.

EXAMPLE 3.11: Median filtering with function `medfilt2`.

■ Median filtering is a useful tool for reducing salt-and-pepper noise in an image. Although we discuss noise reduction in much more detail in Chapter 5, it will be instructive at this point to illustrate briefly the implementation of median filtering.

The image in Fig. 3.18(a) is an X-ray image, `f`, of an industrial circuit board taken during automated inspection of the board. Figure 3.18(b) is the same image corrupted by salt-and-pepper noise in which both the black and white points have a probability of occurrence of 0.2. This image was generated using function `imnoise`, which is discussed in detail in Section 5.2.1:

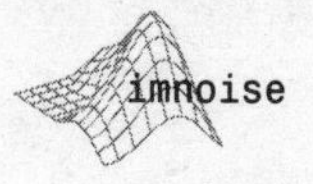


```
>> fn = imnoise(f, 'salt & pepper', 0.2);
```

Figure 3.18(c) is the result of median filtering this noisy image, using the statement:

```
>> gm = medfilt2(fn);
```

Considering the level of noise in Fig. 3.18(b), median filtering using the default settings did a good job of noise reduction. Note, however, the black specks around the border. These were caused by the black points surrounding the image (recall that the default pads the border with 0s). This type of effect can often be reduced by using the `'symmetric'` option:

```
>> gms = medfilt2(fn, 'symmetric');
```

The result, shown in Fig. 3.18(d), is close to the result in Fig. 3.18(c), except that the black border effect is not as pronounced. ■

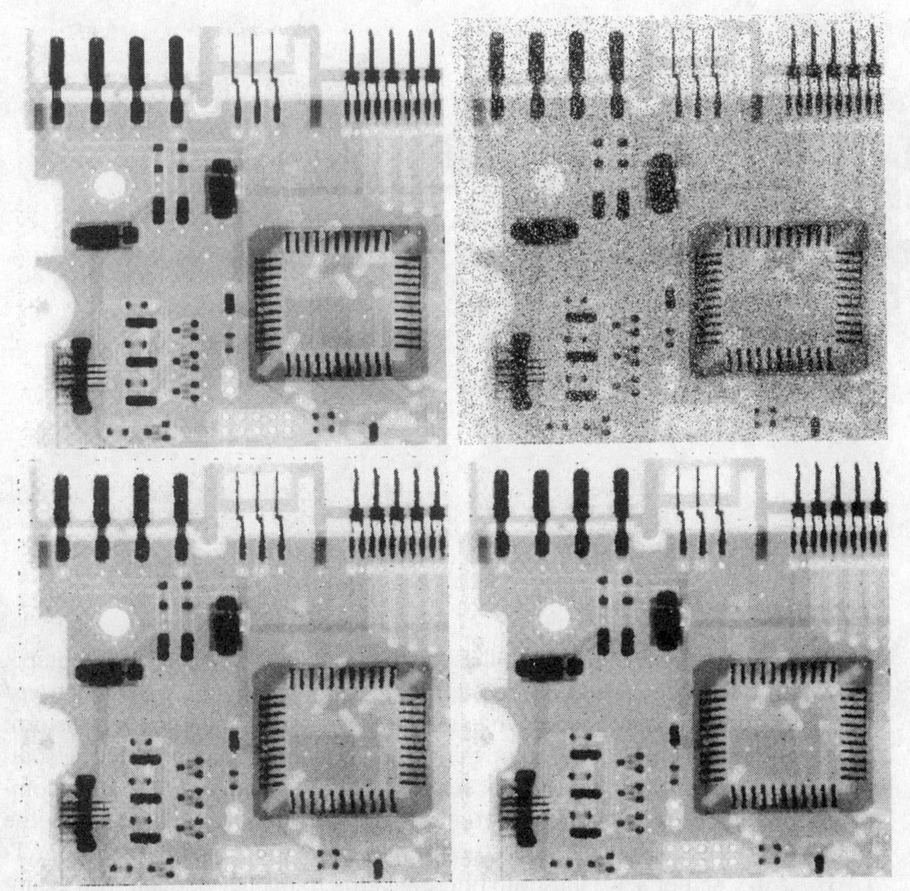

a b
c d

FIGURE 3.18 Median filtering, (a) X-ray image. (b) Image corrupted by salt-and-pepper noise. (c) Result of median filtering with `medfilt2` using the default settings. (d) Result of median filtering using the `'symmetric'` image extension option. Note the improvement in border behavior between (d) and (c). (Original image courtesy of Lixi, Inc.)

Summary

In addition to dealing with image enhancement, the material in this chapter is the foundation for numerous topics in subsequent chapters. For example, we will encounter spatial processing again in Chapter 5 in connection with image restoration, where we also take a closer look at noise reduction and noise-generating functions in MATLAB. Some of the spatial masks that were mentioned briefly here are used extensively in Chapter 10 for edge detection in segmentation applications. The concept of convolution and correlation is explained again in Chapter 4 from the perspective of the frequency domain. Conceptually, mask processing and the implementation of spatial filters will surface in various discussions throughout the book. In the process, we will extend the discussion begun here and introduce additional aspects of how spatial filters can be implemented efficiently in MATLAB.

4 Frequency Domain Processing

Preview

For the most part, this chapter parallels the filtering topics discussed in Chapter 3, but with all filtering carried out in the frequency domain via the Fourier transform. In addition to being a cornerstone of linear filtering, the Fourier transform offers considerable flexibility in the design and implementation of filtering solutions in areas such as image enhancement, image restoration, image data compression, and a host of other applications of practical interest. In this chapter, the focus is on the foundation of how to perform frequency domain filtering in MATLAB. As in Chapter 3, we illustrate filtering in the frequency domain with examples of image enhancement, including lowpass filtering, basic highpass filtering, and high-frequency emphasis filtering. We also show briefly how spatial and frequency domain processing can be used in combination to yield results that are superior to using either type of processing alone. The concepts and techniques developed in the following sections are quite general, as is amply illustrated by other applications of this material in Chapters 5, 8, and 11.

4.1 The 2-D Discrete Fourier Transform

Let $f(x, y)$, for $x = 0, 1, 2, \dots, M - 1$ and $y = 0, 1, 2, \dots, N - 1$, denote an $M \times N$ image. The 2-D, *discrete Fourier transform* (DFT) of f, denoted by $F(u, v)$, is given by the equation

$$F(u, v) = \sum_{x=0}^{M-1} \sum_{y=0}^{N-1} f(x, y) e^{-j2\pi(ux/M + vy/N)}$$

for $u = 0, 1, 2, \dots, M - 1$ and $v = 0, 1, 2, \dots, N - 1$. We could expand the exponential into sines and cosines with the variables u and v determining their frequencies (x and y are summed out). The *frequency domain* is simply the coordinate system spanned by $F(u, v)$ with u and v as (frequency) variables. This is analogous to the *spatial domain* studied in the previous chapter, which is the coordinate system spanned by $f(x, y)$, with x and y as (spatial) variables. The $M \times N$ rectangular region defined by $u = 0, 1, 2, \ldots, M - 1$ and $v = 0, 1, 2, \ldots, N - 1$ is often referred to as the *frequency rectangle*. Clearly, the frequency rectangle is of the same size as the input image.

The inverse, discrete Fourier transform is given by

$$f(x, y) = \frac{1}{MN} \sum_{u=0}^{M-1} \sum_{v=0}^{N-1} F(u, v) e^{j2\pi(ux/M + vy/N)}$$

for $x = 0, 1, 2, \ldots, M - 1$ and $y = 0, 1, 2, \ldots, N - 1$. Thus, given $F(u, v)$, we can obtain $f(x, y)$ back by means of the inverse DFT. The values of $F(u, v)$ in this equation sometimes are referred to as the *Fourier coefficients* of the expansion.

In some formulations of the DFT, the $1/MN$ term is placed in front of the transform and in others it is used in front of the inverse. To be consistent with MATLAB's implementation of the Fourier transform, we assume throughout the book that the term is in front of the inverse, as shown in the preceding equation. Because array indices in MATLAB start at 1, rather than 0, `F(1, 1)` and `f(1, 1)` in MATLAB correspond to the mathematical quantities $F(0, 0)$ and $f(0, 0)$ in the transform and its inverse.

The value of the transform at the origin of the frequency domain [i.e., $F(0, 0)$] is called the *dc* component of the Fourier transform. This terminology is from electrical engineering, where "dc" signifies direct current (current of zero frequency). It is not difficult to show that $F(0, 0)$ is equal to MN times the average value of $f(x, y)$.

Even if $f(x, y)$ is real, its transform in general is complex. The principal method of visually analyzing a transform is to compute its *spectrum* [i.e., the magnitude of $F(u, v)$] and display it as an image. Letting $R(u, v)$ and $I(u, v)$ represent the real and imaginary components of $F(u, v)$, the Fourier spectrum is defined as

$$|F(u, v)| = \left[R^2(u, v) + I^2(u, v)\right]^{1/2}$$

The *phase angle* of the transform is defined as

$$\phi(u, v) = \tan^{-1}\left[\frac{I(u, v)}{R(u, v)}\right]$$

The preceding two functions can be used to represent $F(u, v)$ in the familiar polar representation of a complex quantity:

$$F(u, v) = |F(u, v)| e^{-j\phi(u, v)}$$

The *power spectrum* is defined as the square of the magnitude:

$$\begin{aligned} P(u, v) &= |F(u, v)|^2 \\ &= R^2(u, v) + I^2(u, v) \end{aligned}$$

For purposes of visualization it typically is immaterial whether we view $|F(u, v)|$ or $P(u, v)$.

If $f(x, y)$ is real, its Fourier transform is conjugate symmetric about the origin; that is,

$$F(u, v) = F^*(-u, -v)$$

which implies that the Fourier spectrum also is symmetric about the origin:

$$|F(u, v)| = |F(-u, -v)|$$

It can be shown by direct substitution into the equation for $F(u, v)$ that

$$F(u, v) = F(u + M, v) = F(u, v + N) = F(u + M, v + N)$$

In other words, the DFT is infinitely periodic in both the u and v directions, with the periodicity determined by M and N. Periodicity is also a property of the inverse DFT:

$$f(x, y) = f(x + M, y) = f(x, y + N) = f(x + M, y + N)$$

That is, an image obtained by taking the inverse Fourier transform is also infinitely periodic. This is a frequent source of confusion because it is not at all intuitive that images resulting from taking the inverse Fourier transform should turn out to be periodic. It helps to remember that this is simply a mathematical property of the DFT and its inverse. Keep in mind also that DFT implementations compute only one period, so we work with arrays of size $M \times N$.

The periodicity issue becomes important when we consider *how* DFT data relate to the periods of the transform. For instance, Fig. 4.1(a) shows the spectrum of a one-dimensional transform, $F(u)$. In this case, the periodicity expression becomes $F(u) = F(u + M)$, from which it follows that $|F(u)| = |F(u + M)|$; also, because of symmetry, $|F(u)| = |F(-u)|$. The periodicity property indicates that $F(u)$ has a period of length M, and the symmetry property indicates that the magnitude of the transform is centered on the origin, as Fig. 4.1(a) shows. This figure and the preceding comments demonstrate that the magnitudes of the trans-

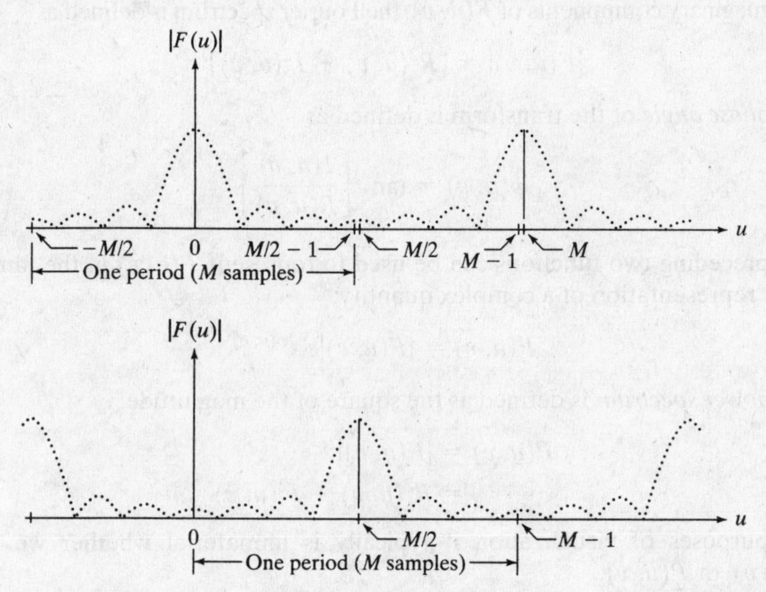

FIGURE 4.1 (a) Fourier spectrum showing back-to-back half periods in the interval $[0, M - 1]$. (b) Centered spectrum in the same interval, obtained by multiplying $f(x)$ by $(-1)^x$ prior to computing the Fourier transform.

form values from $M/2$ to $M - 1$ are repetitions of the values in the half period to the left of the origin. Because the 1-D DFT is implemented for only M points (i.e., for values of u in the interval $[0, M - 1]$), it follows that computing the 1-D transform yields two back-to-back half periods in this interval. We are interested in obtaining one full, *properly ordered* period in the interval $[0, M - 1]$. It is not difficult to show (Gonzalez and Woods [2002]) that the desired period is obtained by multiplying $f(x)$ by $(-1)^x$ prior to computing the transform. Basically, what this does is move the origin *of the transform* to the point $u = M/2$, as Fig. 4.1(b) shows. Now, the value of the spectrum at $u = 0$ in Fig. 4.1(b) corresponds to $|F(-M/2)|$ in Fig. 4.1(a). Similarly, the values at $|F(M/2)|$ and $|F(M - 1)|$ in Fig. 4.1(b) correspond to $|F(0)|$ and $|F(M/2 - 1)|$ in Fig. 4.1(a).

A similar situation exists with two-dimensional functions. Computing the 2-D DFT now yields transform points in the rectangular interval shown in Fig. 4.2(a), where the shaded area indicates values of $F(u, v)$ obtained by implementing the 2-D Fourier transform equation defined at the beginning of this section. The dashed rectangles are periodic repetitions, as in Fig. 4.1(a). The shaded region shows that the values of $F(u, v)$ now encompass four back-to-back quarter periods that meet at the point shown in Fig. 4.2(a). Visual analysis of the spectrum is simplified by moving the values at the origin of the transform to the center of the frequency rectangle. This can be accomplished by multiplying $f(x, y)$ by $(-1)^{x+y}$ prior to computing the 2-D Fourier transform. The periods then would align as shown in Fig. 4.2(b). As in the previous discussion for 1-D functions, the value of the spectrum at $(M/2, N/2)$ in Fig. 4.2(b) is the same as its value at $(0, 0)$ in Fig. 4.2(a), and the value at $(0, 0)$ in Fig. 4.2(b) is the same as the value at $(-M/2, -N/2)$ in Fig. 4.2(a). Similarly, the value at $(M - 1, N - 1)$ in Fig. 4.2(b) is the same as the value at $(M/2 - 1, N/2 - 1)$ in Fig. 4.2(a).

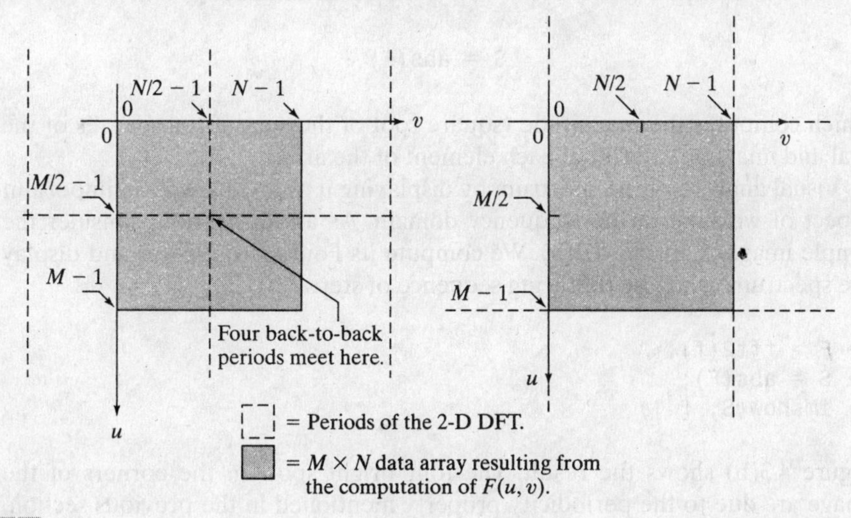

a b

FIGURE 4.2 (a) $M \times N$ Fourier spectrum (shaded), showing four back-to-back quarter periods contained in the spectrum data. (b) Spectrum obtained by multiplying $f(x, y)$ by $(-1)^{x+y}$ prior to computing the Fourier transform. Only one period is shown shaded because this is the data that would be obtained by an implementation of the equation for $F(u, v)$.

The preceding discussion for centering the transform by multiplying $f(x, y)$ by $(-1)^{x+y}$ is an important concept that is included here for completeness. When working in MATLAB, the approach is to compute the transform without multiplication by $(-1)^{x+y}$ and then to rearrange the data afterwards using function `fftshift`. This function and its use are discussed in the following section.

4.2 Computing and Visualizing the 2-D DFT in MATLAB

The DFT and its inverse are obtained in practice using a fast Fourier transform (FFT) algorithm. The FFT of an $M \times N$ image array `f` is obtained in the toolbox with function `fft2`, which has the simple syntax:

```
F = fft2(f)
```

This function returns a Fourier transform that is also of size $M \times N$, with the data arranged in the form shown in Fig. 4.2(a); that is, with the origin of the data at the top left, and with four quarter periods meeting at the center of the frequency rectangle.

As explained in Section 4.3.1, it is necessary to pad the input image with zeros when the Fourier transform is used for filtering. In this case, the syntax becomes

```
F = fft2(f, P, Q)
```

With this syntax, `fft2` pads the input with the required number of zeros so that the resulting function is of size $P \times Q$.

The Fourier spectrum is obtained by using function `abs`:

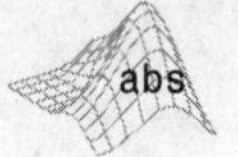


```
S = abs(F)
```

which computes the magnitude (square root of the sum of the squares of the real and imaginary parts) of each element of the array.

Visual analysis of the spectrum by displaying it as an image is an important aspect of working in the frequency domain. As an illustration, consider the simple image, `f`, in Fig. 4.3(a). We compute its Fourier transform and display the spectrum using the following sequence of steps:

```
>> F = fft2(f);
>> S = abs(F);
>> imshow(S, [ ])
```

Figure 4.3(b) shows the result. The four bright spots in the corners of the image are due to the periodicity property mentioned in the previous section.

IPT function `fftshift` can be used to move the origin of the transform to the center of the frequency rectangle. The syntax is

```
Fc = fftshift(F)
```

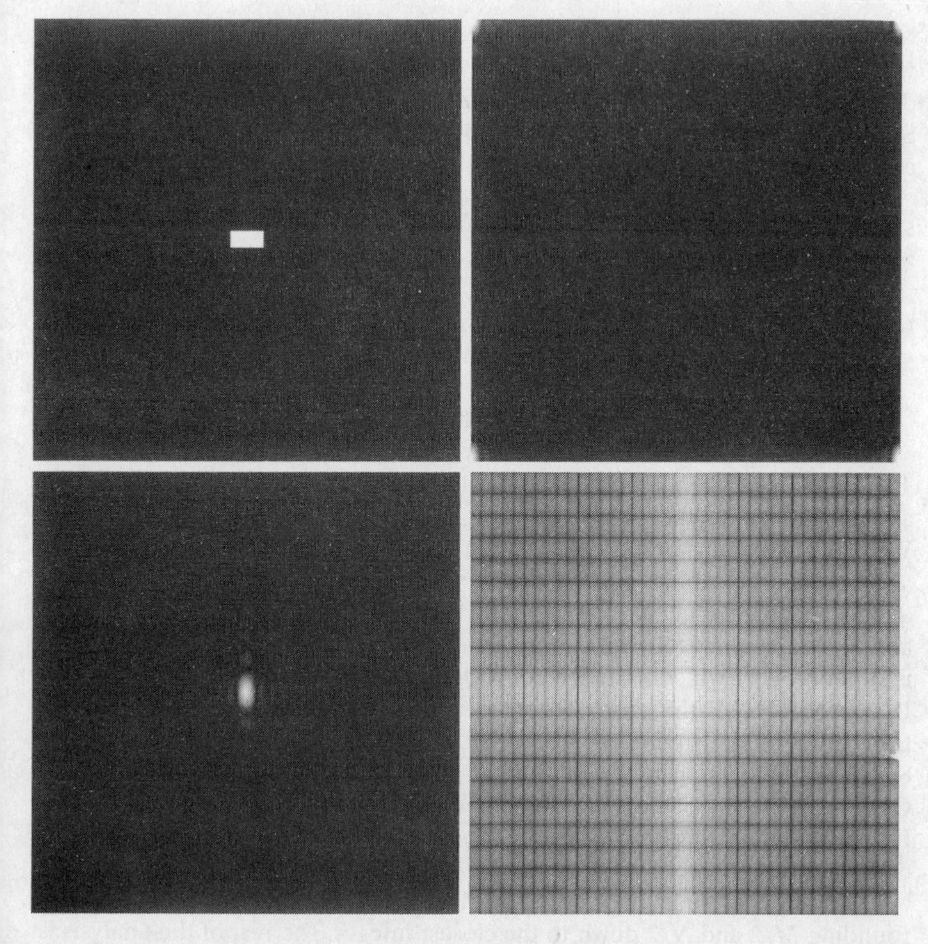

a b
c d

FIGURE 4.3
(a) A simple image.
(b) Fourier spectrum.
(c) Centered spectrum.
(d) Spectrum visually enhanced by a log transformation.

where `F` is the transform computed using `fft2` and `Fc` is the centered transform. Function `fftshift` operates by swapping quadrants of `F`. For example, if `a = [1 2; 3 4]`, `fftshift(a) = [4 3; 2 1]`. When applied to a transform after it has been computed, the net result of using `fftshift` is the same as if the input image had been multiplied by $(-1)^{x+y}$ prior to computing the transform. Note, however, that the two processes are *not* interchangeable. That is, letting $\Im[\cdot]$ denote the Fourier transform of the argument, we have that $\Im[(-1)^{x+y}f(x, y)]$ is equal to `fftshift(fft2(f))`, but this quantity is not equal to `fft2(fftshift(f))`.

In the present example, typing

```
>> Fc = fftshift(F);
>> imshow(abs(Fc), [ ])
```

yielded the image in Fig. 4.3(c). The result of centering is evident in this image.

Although the shift was accomplished as expected, the dynamic range of the values in this spectrum is so large (0 to 204000) compared to the 8 bits of the display that the bright values in the center dominate the result. As discussed in Section 3.2.2, this difficulty is handled via a log transformation. Thus, the commands

```
>> S2 = log(1 + abs(Fc));
>> imshow(S2, [ ])
```

resulted in Fig. 4.3(d). The increase in visual detail is evident in this image.

Function `ifftshift` reverses the centering. Its syntax is

```
F = ifftshift(Fc)
```

This function can be used also to convert a function that is initially centered on a rectangle to a function whose center is at the top, left corner of the rectangle. We make use of this property in Section 4.4.

While on the subject of centering, keep in mind that the center of the frequency rectangle is at $(M/2, N/2)$ if the variables u and v run from 0 to $M - 1$ and $N - 1$, respectively. For example, the center of an 8×8 frequency square is at point $(4, 4)$, which is the 5th point along each axis, counting up from $(0, 0)$. If, as in MATLAB, the variables run from 1 to M and 1 to N, respectively, then the center of the square is at $[(M/2) + 1, (N/2) + 1]$. In the case of our 8×8 example, the center would be at point $(5, 5)$, counting up from $(1, 1)$. Obviously, the two centers are the same point, but this can be a source of confusion when deciding how to specify the location of DFT centers in MATLAB computations.

If M and N are odd, the center for MATLAB computations is obtained by rounding $M/2$ and $N/2$ down to the closest integer. The rest of the analysis is as in the previous paragraph. For example, the center of a 7×7 region is at $(3, 3)$ if we count up from $(0, 0)$ and at $(4, 4)$ if we count up from $(1, 1)$. In either case, the center is the fourth point from the origin. If only one of the dimensions is odd, the center along that dimension is similarly obtained by rounding down in the manner just explained. Using MATLAB's function `floor`, and keeping in mind that the origin is at $(1, 1)$, the center of the frequency rectangle for MATLAB computations is at

`B = floor(A)` *rounds each element of* `A` *to the nearest integer less than or equal to its value. Function* `ceil` *rounds to the nearest integer greater than or equal to the value of each element of* `A`.

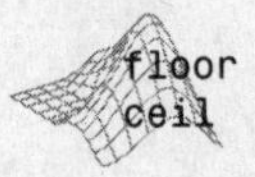


```
[floor(M/2) + 1, floor(N/2) + 1]
```

The center given by this expression is valid both for odd and even values of M and N.

Finally, we point out that the inverse Fourier transform is computed using function `ifft2`, which has the basic syntax

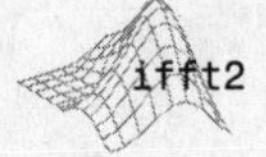


```
f = ifft2(F)
```

where `F` is the Fourier transform and `f` is the resulting image. If the input used to compute `F` is real, the inverse in theory should be real. In practice, however, the output of `ifft2` often has very small imaginary components resulting from round-off errors that are characteristic of floating point computations. Thus, it is good practice to extract the real part of the result after computing the inverse to obtain an image consisting only of real values. The two operations can be combined:

```
>> f = real(ifft2(F));
```

`real(arg)` *and* `imag(arg)` *extract the real and imaginary parts of* `arg`*, respectively.*

As in the forward case, this function has the alternate format `ifft2(F, P, Q)`, which pads `F` with zeros so that its size is $P \times Q$ before computing the inverse. This option is not used in the book.

4.3 Filtering in the Frequency Domain

Filtering in the frequency domain is quite simple conceptually. In this section we give a brief overview of the concepts involved in frequency domain filtering and its implementation in MATLAB.

4.3.1 Fundamental Concepts

The foundation for linear filtering in both the spatial and frequency domains is the convolution theorem, which may be written as†

$$f(x, y) * h(h, y) \Leftrightarrow H(u, v)F(u, v)$$

and, conversely,

$$f(x, y)h(h, y) \Leftrightarrow H(u, v) * G(u, v)$$

Here, the symbol "$*$" indicates convolution of the two functions, and the expressions on the sides of the double arrow constitute a Fourier transform pair. For example, the first expression indicates that convolution of two spatial functions can be obtained by computing the inverse Fourier transform of the product of the Fourier transforms of the two functions. Conversely, the forward Fourier transform of the convolution of two spatial functions gives the product of the transforms of the two functions. Similar comments apply to the second expression.

In terms of filtering, we are interested in the first of the two previous expressions. Filtering in the spatial domain consists of convolving an image $f(x, y)$ with a filter mask, $h(x, y)$. Linear spatial convolution is precisely as explained in Section 3.4.1. According to the convolution theorem, we can obtain the same result in the frequency domain by multiplying $F(u, v)$ by $H(u, v)$, the Fourier transform of the spatial filter. It is customary to refer to $H(u, v)$ as the *filter transfer function*.

Basically, the idea in frequency domain filtering is to select a filter transfer function that modifies $F(u, v)$ in a specified manner. For example, the filter in

---

†For digital images, these expressions are strictly valid only when $f(x, y)$ and $h(x, y)$ have been properly padded with zeros, as discussed later in this section.

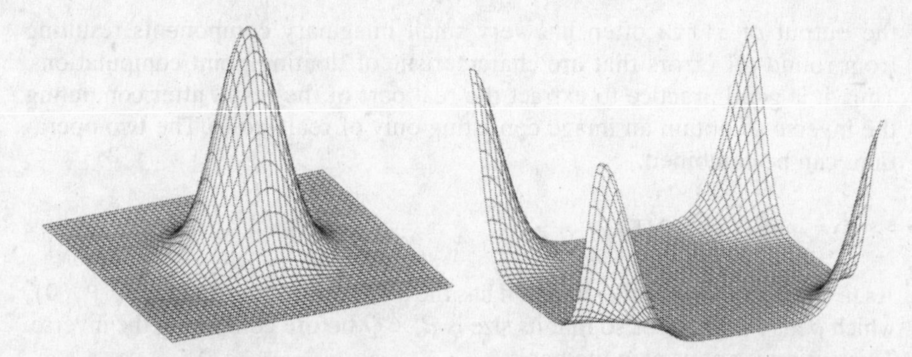

a b

FIGURE 4.4 Transfer functions of (a) a centered lowpass filter, and (b) the format used for DFT filtering. Note that these are frequency domain filters.

Fig. 4.4(a) has a transfer function that, when multiplied by a centered $F(u, v)$, attenuates the high-frequency components of $F(u, v)$, while leaving the low frequencies relatively unchanged. Filters with this characteristic are called *lowpass filters*. As discussed in Section 4.5.2, the net result of lowpass filtering is image blurring (smoothing). Figure 4.4(b) shows the same filter after it was processed with `fftshift`. This is the filter format used most frequently in the book when dealing with frequency domain filtering in which the Fourier transform of the input is not centered.

Based on the convolution theorem, we know that to obtain the corresponding filtered image in the spatial domain we simply compute the inverse Fourier transform of the product $H(u, v)F(u, v)$. It is important to keep in mind that the process just described is identical to what we would obtain by using convolution in the spatial domain, as long as the filter mask, $h(x, y)$, is the inverse Fourier transform of $H(u, v)$. In practice, spatial convolution generally is simplified by using small masks that attempt to capture the salient features of their frequency domain counterparts.

As noted in Section 4.1, images and their transforms are automatically considered periodic if we elect to work with DFTs to implement filtering. It is not difficult to visualize that convolving periodic functions can cause interference between adjacent periods if the periods are close with respect to the duration of the nonzero parts of the functions. This interference, called *wraparound error*, can be avoided by padding the functions with zeros, in the following manner.

Assume that functions $f(x, y)$ and $h(x, y)$ are of size $A \times B$ and $C \times D$, respectively. We form two *extended* (*padded*) functions, both of size $P \times Q$ by appending zeros to f and g. It can be shown that wraparound error is avoided by choosing

$$P \geq A + C - 1$$

and

$$Q \geq B + D - 1$$

Most of the work in this chapter deals with functions of the same size, $M \times N$, in which case we use the following padding values: $P \geq 2M - 1$ and $Q \geq 2N - 1$.

The following function, called `paddedsize`, computes the minimum even[†] values of P and Q required to satisfy the preceding equations. It also has an option to pad the inputs to form square images of size equal to the nearest integer power of 2. Execution time of FFT algorithms depends roughly on the number of prime factors in P and Q. These algorithms generally are faster when P and Q are powers of 2 than when P and Q are prime. In practice, it is advisable to work with square images and filters so that filtering is the same in both directions. Function `paddedsize` provides the flexibility to do this via the choice of the input parameters.

In function `paddedsize`, the vectors `AB`, `CD`, and `PQ` have elements `[A B]`, `[C D]`, and `[P Q]`, respectively, where these quantities are as defined above.

paddedsize

```
function PQ = paddedsize(AB, CD, PARAM)
%PADDEDSIZE Computes padded sizes useful for FFT-based filtering.
%   PQ = PADDEDSIZE(AB), where AB is a two-element size vector,
%   computes the two-element size vector PQ = 2*AB.
%
%   PQ = PADDEDSIZE(AB, 'PWR2') computes the vector PQ such that
%   PQ(1) = PQ(2) = 2^nextpow2(2*m), where m is MAX(AB).
%
%   PQ = PADDEDSIZE(AB, CD), where AB and CD are two-element size
%   vectors, computes the two-element size vector PQ. The elements
%   of PQ are the smallest even integers greater than or equal to
%   AB + CD - 1.
%
%   PQ = PADDEDSIZE(AB, CD, 'PWR2') computes the vector PQ such that
%   PQ(1) = PQ(2) = 2^nextpow2(2*m), where m is MAX([AB CD]).

if nargin == 1
   PQ = 2*AB;
elseif nargin == 2 & ~ischar(CD)
   PQ = AB + CD - 1;
   PQ = 2 * ceil(PQ / 2);
elseif nargin == 2
   m = max(AB); % Maximum dimension.

   % Find power-of-2 at least twice m.
   P = 2^nextpow2(2*m);
   PQ = [P, P];
elseif nargin == 3
   m = max([AB CD]); % Maximum dimension.
   P = 2^nextpow2(2*m);
   PQ = [P, P];
else
   error('Wrong number of inputs.')
end
```

`p = nextpow2(n)` *returns the smallest integer power of 2 that is greater than or equal to the absolute value of* `n`.

[†]It is customary to work with arrays of even dimensions to speed-up FFT computations.

With PQ thus computed using function paddedsize, we use the following syntax for fft2 to compute the FFT using zero padding:

```
F = fft2(f, PQ(1), PQ(2))
```

This syntax simply appends enough zeros to f such that the resulting image is of size PQ(1) × PQ(2), and then computes the FFT as previously described. Note that when using padding the filter function in the frequency domain must be of size PQ(1) × PQ(2) also.

EXAMPLE 4.1: Effects of filtering with and without padding.

■ The image, f, in Fig. 4.5(a) is used in this example to illustrate the difference between filtering with and without padding. In the following discussion we use function lpfilter to generate a Gaussian lowpass filters [similar to Fig. 4.4(b)] with a specified value of sigma (sig). This function is discussed in detail in Section 4.5.2, but the syntax is straightforward, so we use it here and defer further explanation of lpfilter to that section.

The following commands perform filtering without padding:

```
>> [M, N] = size(f);
>> F = fft2(f);
>> sig = 10;
>> H = lpfilter('gaussian', M, N, sig);
>> G = H.*F;
>> g = real(ifft2(G));
>> imshow(g, [ ])
```

Figure 4.5(b) shows image g. As expected, the image is blurred, but note that the vertical edges are not. The reason can be explained with the aid of Fig. 4.6(a), which shows graphically the implied periodicity in DFT computa-

a b c

FIGURE 4.5 (a) A simple image of size 256 × 256. (b) Image lowpass-filtered in the frequency domain without padding. (c) Image lowpass-filtered in the frequency domain with padding. Compare the light portion of the vertical edges in (b) and (c).

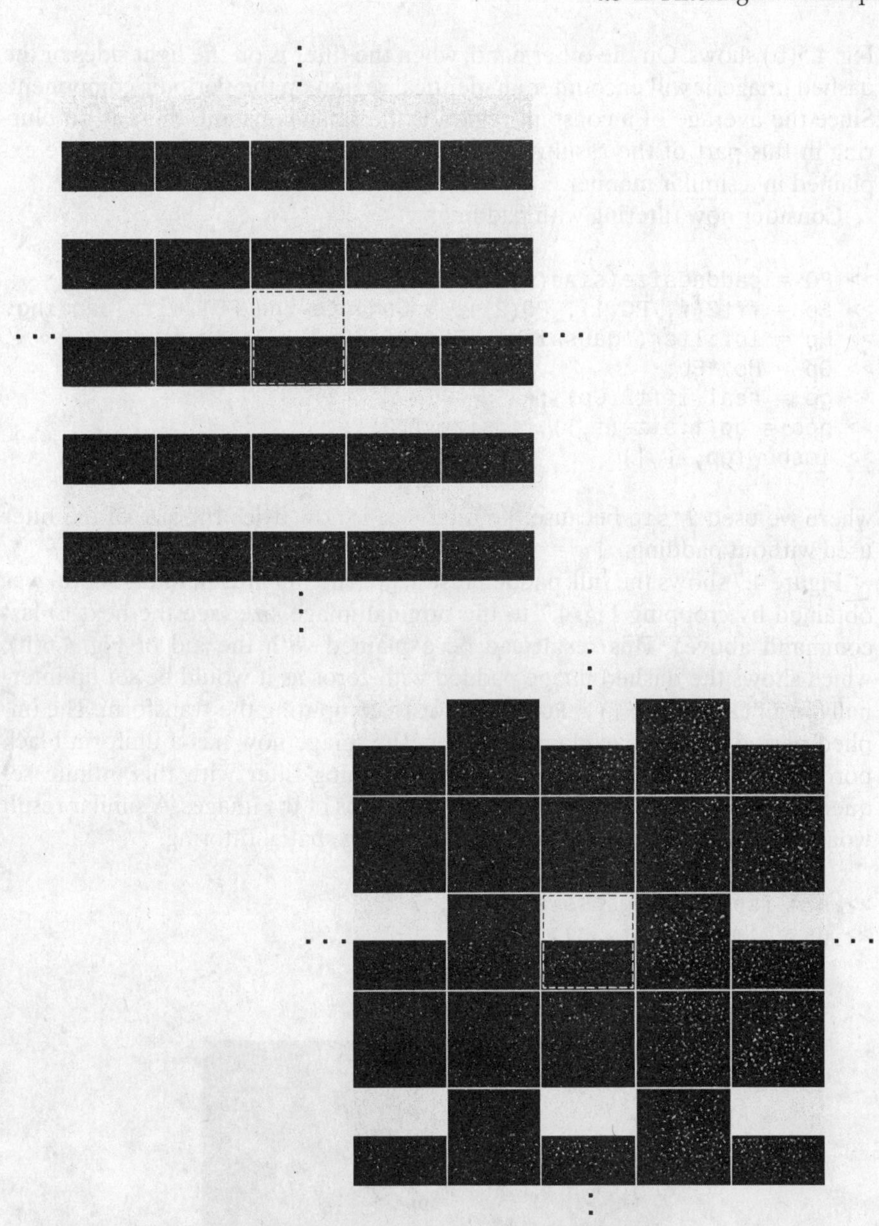

a
b

FIGURE 4.6 (a) Implied, infinite periodic sequence of the image in Fig. 4.5(a). The dashed region represents the data processed by `fft2`. (b) The same periodic sequence after padding with 0s. The thin white lines in both images are shown for convenience in viewing; they are not part of the data.

tions. The thin white lines between the images are included for convenience in viewing. They are not part of the data. The dashed lines are used to designate (arbitrarily) the $M \times N$ image processed by `fft2`. Imagine convolving a blurring filter with this infinite periodic sequence. It is clear that when the filter is passing through the top of the dashed image it will encompass part of the image itself and also the bottom part of the periodic component right above it. Thus, when a light and a dark region reside under the filter, the result will be a mid-gray, blurred output. This is precisely what the top of the image in

Fig. 4.5(b) shows. On the other hand, when the filter is on the light sides of the dashed image, it will encounter an identical region on the periodic component. Since the average of a constant region is the same constant, there is no blurring in this part of the result. Other parts of the image in Fig. 4.5(b) are explained in a similar manner.

Consider now filtering with padding:

```
>> PQ = paddedsize(size(f));
>> Fp = fft2(f, PQ(1), PQ(2)); % Compute the FFT with padding.
>> Hp = lpfilter('gaussian', PQ(1), PQ(2), 2*sig);
>> Gp = Hp.*Fp;
>> gp = real(ifft2(Gp));
>> gpc = gp(1:size(f,1), 1:size(f,2));
>> imshow(gp, [ ])
```

where we used `2*sig` because the filter size is now twice the size of the filter used without padding.

Figure 4.7 shows the full, padded result, gp. The final result in Fig. 4.5(c) was obtained by cropping Fig. 4.7 to the original image size (see the next-to-last command above). This result can be explained with the aid of Fig. 4.6(b), which shows the dashed image padded with zeros as it would be set up internally in `fft2(f, PQ(1), PQ(2))` prior to computing the transform. The implied periodicity is as explained earlier. The image now has a uniform black border all around it, so convolving a smoothing filter with this infinite sequence would show a gray blur in the light edges of the images. A similar result would be obtained by performing the following spatial filtering,

```
>> h = fspecial('gaussian', 15, 7);
>> gs = imfilter(f, h);
```

FIGURE 4.7 Full padded image resulting from `ifft2` after filtering. This image is of size 512×512 pixels.

Recall from Section 3.4.1 that this call to function `imfilter` pads the border of the image with 0s by default. ■

4.3.2 Basic Steps in DFT Filtering

The discussion in the previous section can be summarized in the following step-by-step procedure involving MATLAB functions, where `f` is the image to be filtered, `g` is the result, and it is assumed that the filter function $H(u, v)$ is of the same size as the padded image:

1. Obtain the padding parameters using function `paddedsize`:
```
PQ = paddedsize(size(f));
```
2. Obtain the Fourier transform with padding:
```
F = fft2(f, PQ(1), PQ(2));
```
3. Generate a filter function, `H`, of size `PQ(1)` × `PQ(2)` using any of the methods discussed in the remainder of this chapter. The filter must be in the format shown in Fig. 4.4(b). If it is centered instead, as in Fig. 4.4(a), let `H = fftshift(H)` before using the filter.
4. Multiply the transform by the filter:
```
G = H.*F;
```
5. Obtain the real part of the inverse FFT of `G`:
```
g = real(ifft2(G));
```
6. Crop the top, left rectangle to the original size:
```
g = g(1:size(f, 1), 1:size(f, 2));
```

This filtering procedure is summarized in Fig. 4.8. The preprocessing stage might encompass procedures such as determining image size, obtaining the padding parameters, and generating a filter. Postprocessing entails computing the real part of the result, cropping the image, and converting it to class `uint8` or `uint16` for storage.

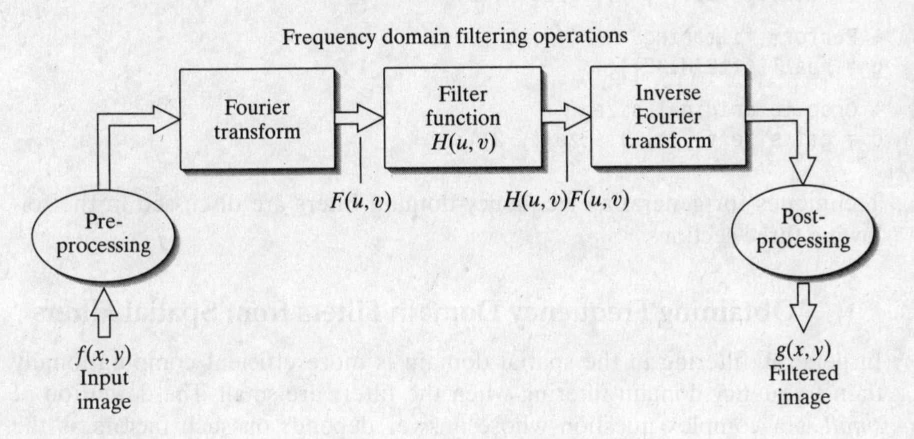

FIGURE 4.8 Basic steps for filtering in the frequency domain.

The filter function $H(u, v)$ in Fig. 4.8 multiplies both the real and imaginary parts of $F(u, v)$. If $H(u, v)$ is real, then the phase of the result is not changed, a fact that can be seen in the phase equation (Section 4.1) by noting that, if the multipliers of the real and imaginary parts are equal, they cancel out, leaving the phase angle unchanged. Filters that operate in this manner are called *zero-phase-shift filters*. These are the only types of linear filters considered in this chapter.

It is well known from linear system theory that, under certain mild conditions, inputting an impulse into a linear system completely characterizes the system. When working with finite, discrete data as we do in this book, the response of a linear system, including the response to an impulse, also is finite. If the linear system is just a spatial filter, then we can completely determine the filter simply by observing its response to an impulse. A filter determined in this manner is called a *finite-impulse-response* (FIR) *filter*. All the linear spatial filters in this book are FIR filters.

4.3.3 An M-function for Filtering in the Frequency Domain

The sequence of filtering steps described in the previous section is used throughout this chapter and parts of the next, so it will be convenient to have available an M-function that accepts as inputs an image and a filter function, handles all the filtering details, and outputs the filtered, cropped image. The following function does this.

dftfilt

```
function g = dftfilt(f, H)
%DFTFILT Performs frequency domain filtering.
%   G = DFTFILT(F, H) filters F in the frequency domain using the
%   filter transfer function H. The output, G, is the filtered
%   image, which has the same size as F. DFTFILT automatically pads
%   F to be the same size as H. Function PADDEDSIZE can be used
%   to determine an appropriate size for H.
%
%   DFTFILT assumes that F is real and that H is a real, uncentered,
%   circularly-symmetric filter function.

% Obtain the FFT of the padded input.
F = fft2(f, size(H, 1), size(H, 2));

% Perform filtering.
g = real(ifft2(H.*F));

% Crop to original size.
g = g(1:size(f, 1), 1:size(f, 2));
```

Techniques for generating frequency-domain filters are discussed in the following three sections.

4.4 Obtaining Frequency Domain Filters from Spatial Filters

In general, filtering in the spatial domain is more efficient computationally than frequency domain filtering when the filters are small. The definition of *small* is a complex question whose answer depends on such factors as the machine and algorithms used and on issues such the sizes of buffers, how well complex data are handled, and a host of other factors beyond the scope of this discussion. A comparison by Brigham [1988] using 1-D functions shows that filtering using an FFT algorithm can be faster than a spatial implementation when the functions have on the order of 32 points, so the numbers in question are not large. Thus, it is useful to know how to convert a spatial filter into an equivalent frequency domain filter in order to obtain meaningful comparisons between the two approaches.

One obvious approach for generating a frequency domain filter, `H`, that corresponds to a given spatial filter, `h`, is to let `H = fft2(h, PQ(1), PQ(2))`, where the values of vector `PQ` depend on the size of the image we want to filter, as discussed in the last section. However, we are interested in this section on two major topics: (1) how to convert spatial filters into equivalent frequency domain filters; and (2) how to compare the results between spatial domain filtering using function `imfilter`, and frequency domain filtering using the techniques discussed in the previous section. Because, as explained in detail in Section 3.4.1, `imfilter` uses correlation and the origin of the filter is considered at its center, a certain amount of data preprocessing is required to make the two approaches equivalent. The toolbox provides a function, `freqz2`, that does precisely this and outputs the corresponding filter in the frequency domain.

Function `freqz2` computes the frequency response of FIR filters, which, as mentioned at the end of Section 4.3.2, are the only linear filters considered in this book. The result is the desired filter in the frequency domain. The syntax of interest in the present discussion is

```
H = freqz2(h, R, C)
```

freqz2

where `h` is a 2-D spatial filter and `H` is the corresponding 2-D frequency domain filter. Here, `R` is the number of rows, and `C` the number of columns that we wish filter `H` to have. Generally, we let `R = PQ(1) and C = PQ(2)`, as explained in Section 4.3.1. If `freqz2` is written without an output argument, the absolute value of `H` is displayed on the MATLAB desktop as a 3-D perspective plot. The mechanics involved in using function `freqz2` are easily explained by an example.

EXAMPLE 4.2: A comparison of filtering in the spatial and frequency domains.

■ Consider the image, `f`, of size 600 × 600 pixels shown in Fig. 4.9(a). In what follows, we generate the frequency domain filter, `H`, corresponding to the Sobel spatial filter that enhances vertical edges (see Table 3.4). We then compare the result of filtering `f` in the spatial domain with the Sobel mask (using `imfilter`) against the result obtained by performing the equivalent process in the frequency domain. In practice, filtering with a small filter like a Sobel mask would be implemented directly in the spatial domain, as mentioned earlier. However, we selected this filter for demonstration purposes because its coefficients are simple and because the results of filtering are intuitive and straightforward to compare. Larger spatial filters are handled in exactly the same manner.

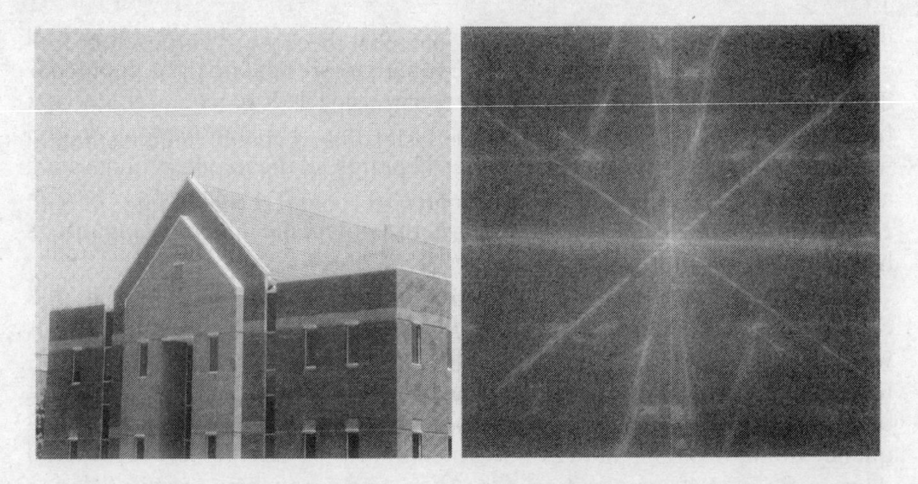

a b

FIGURE 4.9 (a) A gray-scale image. (b) Its Fourier spectrum.

Figure 4.9(b) is an image of the Fourier spectrum of `f`, obtained as follows:

```
>> F = fft2(f);
>> S = fftshift(log(1 + abs(F)));
>> S = gscale(S);
>> imshow(S)
```

Next, we generate the spatial filter using function `fspecial`:

```
h = fspecial('sobel')'
h =
     1     0    -1
     2     0    -2
     1     0    -1
```

To view a plot of the corresponding frequency domain filter we type

```
>> freqz2(h)
```

Figure 4.10(a) shows the result, with the axes suppressed (techniques for obtaining perspective plots are discussed in Section 4.5.3). The filter itself was obtained using the commands:

```
>> PQ = paddedsize(size(f));
>> H = freqz2(h, PQ(1), PQ(2));
>> H1 = ifftshift(H);
```

where, as noted earlier, `ifftshift` is needed to rearrange the data so that the origin is at the top, left of the frequency rectangle. Figure 4.10(b) shows a plot of `abs(H1)`. Figures 4.10(c) and (d) show the absolute values of `H` and `H1` in image form, displayed with the commands

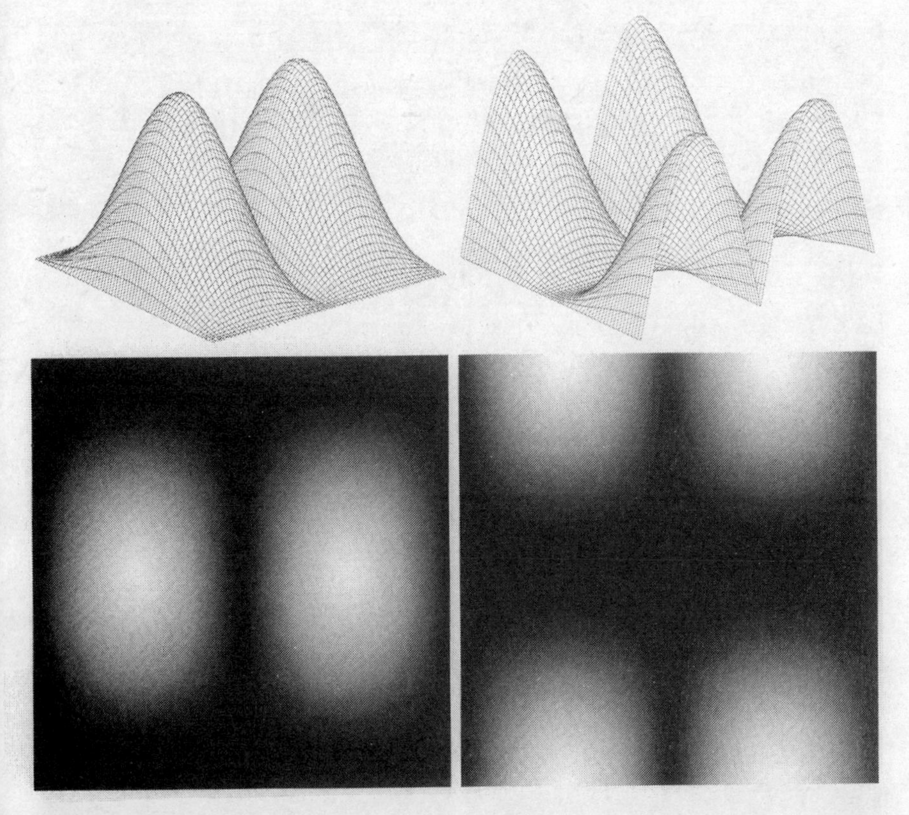

a b
c d

FIGURE 4.10 (a) Absolute value of the frequency domain filter corresponding to a vertical Sobel mask. (b) The same filter after processing with function `fftshift`. Figures (c) and (d) are the filters in (a) and (b) shown as images.

```
>> imshow(abs(H), [ ])
>> figure, imshow(abs(H1), [ ])
```

Next, we generate the filtered images. In the spatial domain we use

```
>> gs = imfilter(double(f), h);
```

We use `double(f)` *here so that* `imfilter` *will produce an output of class* `double`, *as explained in Section 3.4.1. The* `double` *format is required for some of the operations that follow.*

which pads the border of the image with 0s by default. The filtered image obtained by frequency domain processing is given by

```
gf = dftfilt(f, H1);
```

Figures 4.11(a) and (b) show the result of the commands:

```
>> imshow(gs, [ ])
>> figure, imshow(gf, [ ])
```

The gray tonality in the images is due to the fact that both `gs` and `gf` have negative values, which causes the average value of the images to be increased by the scaled `imshow` command. As discussed in Sections 6.6.1 and 10.1.3, the

a b
c d

FIGURE 4.11 (a) Result of filtering Fig. 4.9(a) in the spatial domain with a vertical Sobel mask. (b) Result obtained in the frequency domain using the filter shown in Fig. 4.10(b). Figures (c) and (d) are the absolute values of (a) and (b), respectively.

Sobel mask, h, generated above is used to detect vertical edges in an image using the absolute value of the response. Thus, it is more relevant to show the absolute values of the images just computed. Figures 4.11(c) and (d) show the images obtained using the commands

```
>> figure, imshow(abs(gs), [ ])
>> figure, imshow(abs(gf), [ ])
```

The edges can be seen more clearly by creating a thresholded binary image:

```
>> figure, imshow(abs(gs) > 0.2*abs(max(gs(:))))
>> figure, imshow(abs(gf) > 0.2*abs(max(gf(:))))
```

where the 0.2 multiplier was selected (arbitrarily) to show only the edges with strength greater than 20% of the maximum values of gs and gf. Figures 4.12(a) and (b) show the results.

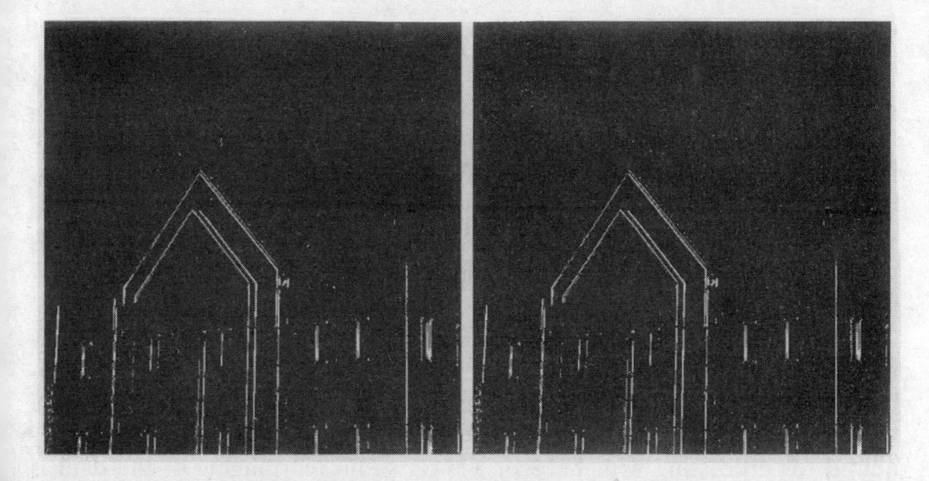

a b

FIGURE 4.12 Thresholded versions of Figs. 4.11(c) and (d), respectively, to show the principal edges more clearly.

The images obtained using spatial and frequency domain filtering are for all practical purposes identical, a fact that we confirm by computing their difference:

```
>> d = abs(gs - gf);
```

The maximum and minimum differences are

```
>> max(d(:))
ans =
   5.4015e-012
>> min(d(:))
ans =
   0
```

The approach just explained can be used to implement in the frequency domain the spatial filtering approach discussed in Sections 3.4.1 and 3.5.1, as well as any other FIR spatial filter of arbitrary size. ■

4.5 Generating Filters Directly in the Frequency Domain

In this section, we illustrate how to implement filter functions directly in the frequency domain. We focus on circularly symmetric filters that are specified as various functions of distance from the origin of the transform. The M-functions developed to implement these filters are a foundation that is easily extendable to other functions within the same framework. We begin by implementing several well-known smoothing (lowpass) filters. Then, we show how to use several of MATLAB's wireframe and surface plotting capabilities that aid in filter visualization. We conclude the section with a brief discussion of sharpening (highpass) filters.

4.5.1 Creating Meshgrid Arrays for Use in Implementing Filters in the Frequency Domain

Central to the M-functions in the following discussion is the need to compute distance functions from any point to a specified point in the frequency rectangle. Because FFT computations in MATLAB assume that the origin of the transform is at the top, left of the frequency rectangle, our distance computations are with respect to that point. The data can be rearranged for visualization purposes (so that the value at the origin is translated to the center of the frequency rectangle) by using function `fftshift`.

The following M-function, which we call `dftuv`, provides the necessary meshgrid array for use in distance computations and other similar applications. (See Section 2.10.4 for an explanation of function `meshgrid` used in the following code.). The meshgrid arrays generated by `dftuv` are in the order required for processing with `fft2` or `ifft2`, so no rearranging of the data is required.

`dftuv`

Function `find` *is discussed in Section 5.2.2.*

```
function [U, V] = dftuv(M, N)
%DFTUV Computes meshgrid frequency matrices.
%   [U, V] = DFTUV(M, N) computes meshgrid frequency matrices U and
%   V.  U and V are useful for computing frequency-domain filter
%   functions that can be used with DFTFILT.  U and V are both
%   M-by-N.

% Set up range of variables.
u = 0:(M - 1);
v = 0:(N - 1);

% Compute the indices for use in meshgrid.
idx = find(u > M/2);
u(idx) = u(idx) - M;
idy = find(v > N/2);
v(idy) = v(idy) - N;

% Compute the meshgrid arrays.
[V, U] = meshgrid(v, u);
```

EXAMPLE 4.3: Using function `dftuv`.

■ As an illustration, the following commands compute the distance squared from every point in a rectangle of size 8×5 to the origin of the rectangle:

```
>> [U, V] = dftuv(8, 5);
>> D = U.^2 + V.^2
D =
     0     1     4     4     1
     1     2     5     5     2
     4     5     8     8     5
     9    10    13    13    10
    16    17    20    20    17
     9    10    13    13    10
     4     5     8     8     5
     1     2     5     5     2
```

Note that the distance is 0 at the top, left, and the larger distances are in the center of the frequency rectangle, following the basic format explained in Fig. 4.2(a). We can use function `fftshift` to obtain the distances with respect to the center of the frequency rectangle,

```
>> fftshift(D)

ans =

    20    17    16    17    20
    13    10     9    10    13
     8     5     4     5     8
     5     2     1     2     5
     4     1     0     1     4
     5     2     1     2     5
     8     5     4     5     8
    13    10     9    10    13
```

The distance is now 0 at coordinates (5, 3), and the array is symmetric about this point. ■

4.5.2 Lowpass Frequency Domain Filters

An *ideal lowpass filter* (ILPF) has the transfer function

$$H(u, v) = \begin{cases} 1 & \text{if } D(u, v) \le D_0 \\ 0 & \text{if } D(u, v) > D_0 \end{cases}$$

where D_0 is a specified nonnegative number and $D(u, v)$ is the distance from point (u, v) to the center of the filter. The locus of points for which $D(u, v) = D_0$ is a circle. Keeping in mind that filter H multiplies the Fourier transform of an image, we see that an ideal filter "cuts off" (multiplies by 0) all components of F outside the circle and leaves unchanged (multiplies by 1) all components on, or inside, the circle. Although this filter is not realizable in analog form using electronic components, it certainly can be simulated in a computer using the preceding transfer function. The properties of ideal filters often are useful in explaining phenomena such as wraparound error.

A *Butterworth lowpass filter* (BLPF) of order n, with a cutoff frequency at a distance D_0 from the origin, has the transfer function

$$H(u, v) = \frac{1}{1 + [D(u, v)/D_0]^{2n}}$$

Unlike the ILPF, the BLPF transfer function does not have a sharp discontinuity at D_0. For filters with smooth transfer functions, it is customary to define a cutoff frequency locus at points for which $H(u, v)$ is down to a specified fraction of its maximum value. In the preceding equation, $H(u, v) = 0.5$ (down 50% from its maximum value of 1) when $D(u, v) = D_0$.

The transfer function of a *Gaussian lowpass filter* (GLPF) is given by

$$H(u, v) = e^{-D^2(u, v)/2\sigma^2}$$

where σ is the standard deviation. By letting $\sigma = D_0$, we obtain the following expression in terms of the cutoff parameter D_0:

$$H(u, v) = e^{-D^2(u, v)/2D_0^2}$$

When $D(u, v) = D_0$ the filter is down to 0.607 of its maximum value of 1.

EXAMPLE 4.4: Lowpass filtering.

■ As an illustration, we apply a Gaussian lowpass filter to the 500 × 500-pixel image, f, in Fig. 4.13(a). We use a value of D_0 equal to 5% of the padded image width. With reference to the filtering steps discussed in Section 4.3.2 we have

```
>> PQ = paddedsize(size(f));
>> [U, V] = dftuv(PQ(1), PQ(2));
>> D0 = 0.05*PQ(2);
>> F = fft2(f, PQ(1), PQ(2));
>> H = exp(-(U.^2 + V.^2)/(2*(D0^2)));
>> g = dftfilt(f, H);
```

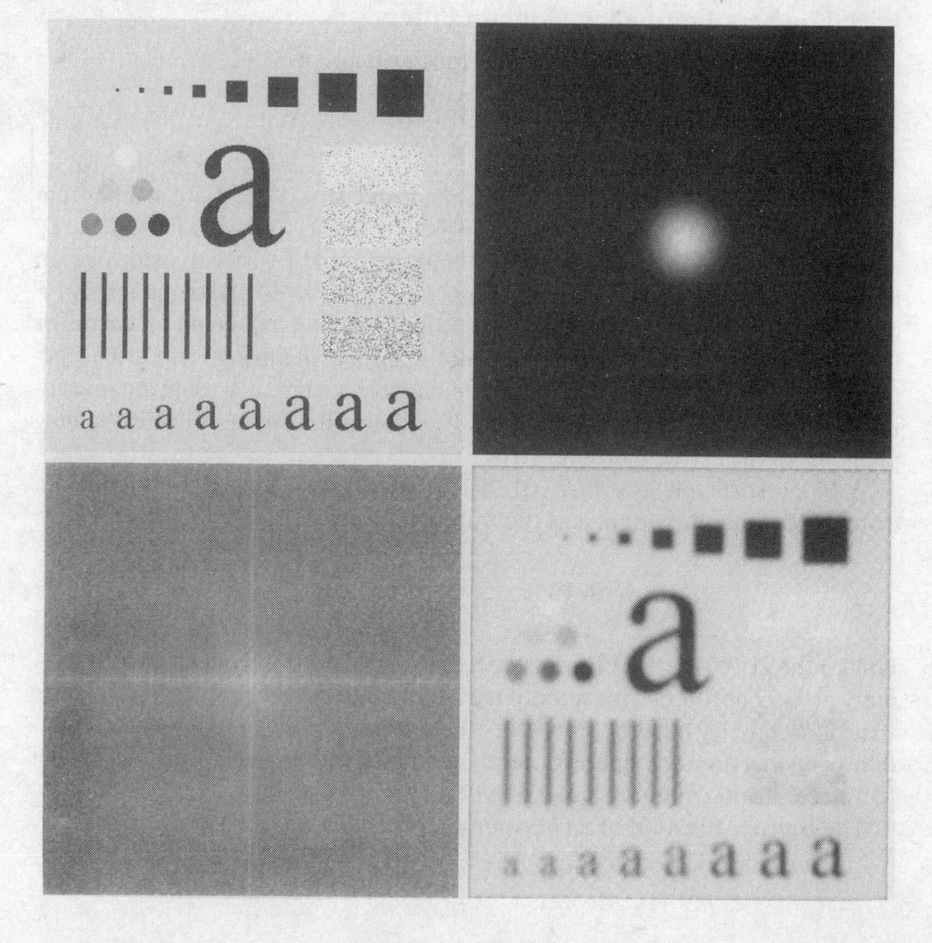

a b
c d

FIGURE 4.13 Lowpass filtering. (a) Original image. (b) Gaussian lowpass filter shown as an image. (c) Spectrum of (a). (d) Processed image.

We can view the filter as an image [Fig. 4.13(b)] by typing

```
>> figure, imshow(fftshift(H), [ ])
```

Similarly, the spectrum can be displayed as an image [Fig. 4.13(c)] by typing

```
>> figure, imshow(log(1 + abs(fftshift(F))), [ ])
```

Finally, Fig. 4.13(d) shows the output image, displayed using the command

```
>> figure, imshow(g, [ ])
```

As expected, this image is a blurred version of the original. ■

The following function generates the transfer functions of all the lowpass filters discussed in this section.

lpfilter

```
function [H, D] = lpfilter(type, M, N, D0, n)
%LPFILTER Computes frequency domain lowpass filters.
%   H = LPFILTER(TYPE, M, N, D0, n) creates the transfer function of
%   a lowpass filter, H, of the specified TYPE and size (M-by-N). To
%   view the filter as an image or mesh plot, it should be centered
%   using H = fftshift(H).
%
%   Valid values for TYPE, D0, and n are:
%
%   'ideal'     Ideal lowpass filter with cutoff frequency D0. n need
%               not be supplied. D0 must be positive.
%
%   'btw'       Butterworth lowpass filter of order n, and cutoff
%               D0. The default value for n is 1.0. D0 must be
%               positive.
%
%   'gaussian'  Gaussian lowpass filter with cutoff (standard
%               deviation) D0. n need not be supplied. D0 must be
%               positive.

% Use function dftuv to set up the meshgrid arrays needed for
% computing the required distances.
[U, V] = dftuv(M, N);

% Compute the distances D(U, V).
D = sqrt(U.^2 + V.^2);

% Begin filter computations.
switch type
case 'ideal'
   H = double(D <= D0);
case 'btw'
   if nargin == 4
      n = 1;
```

```
    end
    H = 1./(1 + (D./D0).^(2*n));
case 'gaussian'
    H = exp(-(D.^2)./(2*(D0^2)));
otherwise
    error('Unknown filter type.')
end
```

Function `lpfilter` is used again in Section 4.6 as the basis for generating highpass filters.

4.5.3 Wireframe and Surface Plotting

Plots of functions of one variable were introduced in Section 3.3.1. In the following discussion we introduce 3-D wireframe and surface plots, which are useful for visualizing the transfer functions of 2-D filters. The easiest way to draw a wireframe plot of a given 2-D function, `H`, is to use function `mesh`, which has the basic syntax

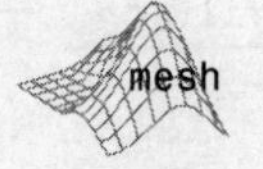


```
mesh(H)
```

This function draws a wireframe for `x = 1:M` and `y = 1:N`, where `[M, N] = size(H)`. Wireframe plots typically are unacceptably dense if `M` and `N` are large, in which case we plot every `k`th point using the syntax

```
mesh(H(1:k:end, 1:k:end))
```

As a rule of thumb, 40 to 60 subdivisions along each axis usually provide a good balance between resolution and appearance.

MATLAB plots mesh figures in color, by default. The command

```
colormap([0 0 0])
```

sets the wireframe to black (we discuss function `colormap` in Chapter 6). MATLAB also superimposes a grid and axes on a mesh plot. These can be turned off using the commands

```
grid off
axis off
```

They can be turned back on by replacing `off` with `on` in these two statements. Finally, the viewing point (location of the observer) is controlled by function `view`, which has the syntax

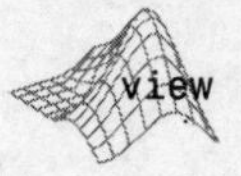


```
view(az, el)
```

As Fig. 4.14 shows, `az` and `el` represent azimuth and elevation angles (in degrees), respectively. The arrows indicate positive direction. The default values

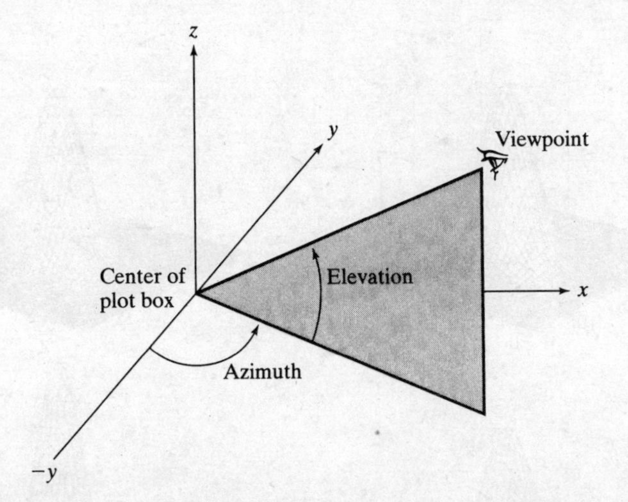

FIGURE 4.14 Geometry for function `view`.

are `az = -37.5` and `el = 30`, which place the viewer in the quadrant defined by the $-x$ and $-y$ axes, and looking into the quadrant defined by the positive x and y axes in Fig. 4.14.

To determine the current viewing geometry, we type

```
>> [az, el] = view;
```

To set the viewpoint to the default values, we type

```
>> view(3)
```

The viewpoint can be modified interactively by clicking on the **Rotate 3D** button in the figure window's toolbar and then clicking and dragging in the figure window.

As discussed in Chapter 6, it is possible to specify the viewer location in Cartesian coordinates, (x, y, z), which is ideal when working with RGB data. However, for general plot-viewing purposes, the method just discussed involves only two parameters and is more intuitive.

EXAMPLE 4.5: Wireframe plotting.

■ Consider a Gaussian lowpass filter similar to the one used in Example 4.4:

```
>> H = fftshift(lpfilter('gaussian', 500, 500, 50));
```

Figure 4.15(a) shows the wireframe plot produced by the commands

```
>> mesh(H(1:10:500, 1:10:500))
>> axis([0 50 0 50 0 1])
```

where the `axis` command is as described in Section 3.3.1, except that it contains a third range for the z axis.

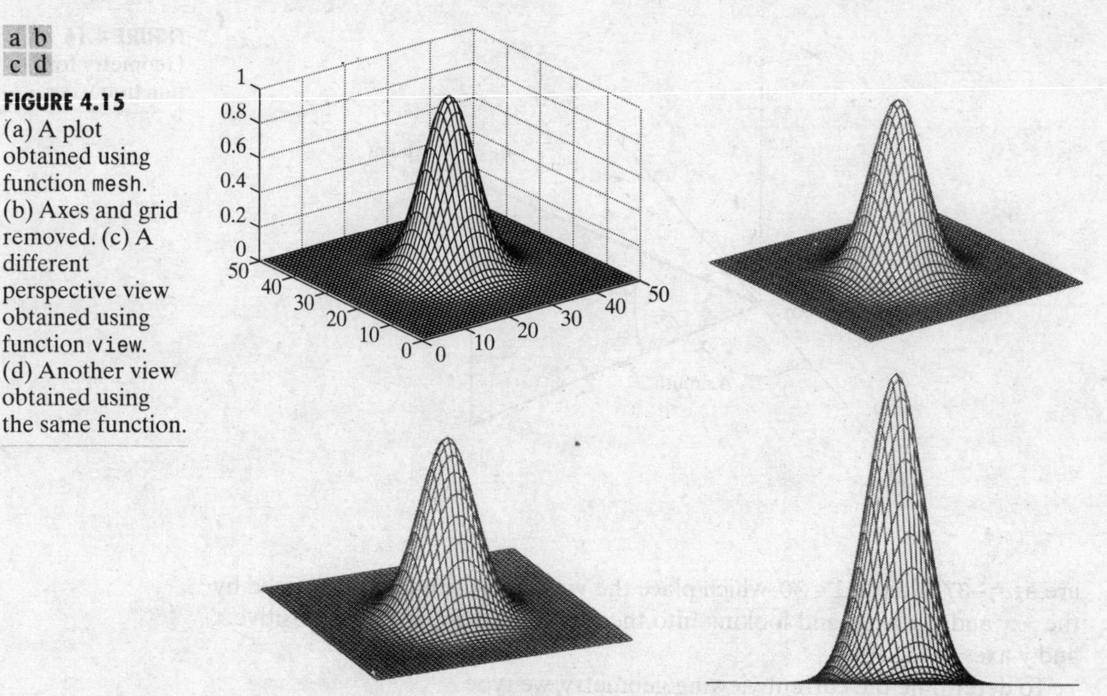

a b
c d

FIGURE 4.15 (a) A plot obtained using function `mesh`. (b) Axes and grid removed. (c) A different perspective view obtained using function `view`. (d) Another view obtained using the same function.

As noted earlier in this section, the wireframe is in color by default, transitioning from blue at the base to red at the top. We convert the plot lines to black and eliminate the axes and grid by typing

```
>> colormap([0 0 0])
>> axis off
>> grid off
```

Figure 4.15(b) shows the result. Figure 4.15(c) shows the result of the command

```
>> view(-25, 30)
```

which moved the observer slightly to the right, while leaving the elevation constant. Finally, Fig. 4.15(d) shows the result of leaving the azimuth at –25 and setting the elevation to 0:

```
>> view(-25, 0)
```

This example shows the significant plotting power of the simple function `mesh`. ■

Sometimes it is desirable to plot a function as a surface instead of as a wireframe. Function `surf` does this. Its basic syntax is

`surf`

```
surf(H)
```

This function produces a plot identical to `mesh`, with the exception that the quadrilaterals in the mesh are filled with colors (this is called *faceted shading*). To convert the colors to gray, we use the command

```
colormap(gray)
```

The `axis`, `grid`, and `view` functions work in the same way as described earlier for `mesh`. For example, Fig. 4.16(a) is the result of the following sequence of commands:

```
>> H = fftshift(lpfilter('gaussian', 500, 500, 50));
>> surf(H(1:10:500, 1:10:500))
>> axis([0 50 0 50 0 1])
>> colormap(gray)
>> grid off; axis off
```

The faceted shading can be smoothed and the mesh lines eliminated by interpolation using the command

```
shading interp
```

shading interp

Typing this command at the prompt produced Fig. 4.16(b).

When the objective is to plot an analytic function of two variables, we use `meshgrid` to generate the coordinate values and from these we generate the discrete (sampled) matrix to use in `mesh` or `surf`. For example, to plot the function

$$f(x, y) = xe^{(-x^2-y^2)}$$

from -2 to 2 in increments of 0.1 for both x and y, we write

```
>> [Y, X] = meshgrid(-2:0.1:2, -2:0.1:2);
>> Z = X.*exp(-X.^2 - Y.^2);
```

and then use `mesh(Z)` or `surf(Z)` as before. Recall from the discussion in Section 2.10.4 that that columns `(Y)` are listed first and rows `(X)` second in function `meshgrid`.

a b

FIGURE 4.16 (a) Plot obtained using function `surf`. (b) Result of using the command `shading interp`.

4.6 Sharpening Frequency Domain Filters

Just as lowpass filtering blurs an image, the opposite process, *highpass filtering*, sharpens the image by attenuating the low frequencies and leaving the high frequencies of the Fourier transform relatively unchanged. In this section we consider several approaches to highpass filtering.

4.6.1 Basic Highpass Filtering

Given the transfer function $H_{\text{lp}}(u, v)$ of a lowpass filter, we obtain the transfer function of the corresponding highpass filter by using the simple relation

$$H_{\text{hp}}(u, v) = 1 - H_{\text{lp}}(u, v).$$

Thus, function `lpfilter` developed in the previous section can be used as the basis for a highpass filter generator, as follows:

hpfilter

```
function H = hpfilter(type, M, N, D0, n)
%HPFILTER Computes frequency domain highpass filters.
%   H = HPFILTER(TYPE, M, N, D0, n) creates the transfer function of
%   a highpass filter, H, of the specified TYPE and size (M-by-N).
%   Valid values for TYPE, D0, and n are:
%
%   'ideal'     Ideal highpass filter with cutoff frequency D0. n
%               need not be supplied. D0 must be positive.
%
%   'btw'       Butterworth highpass filter of order n, and cutoff
%               D0. The default value for n is 1.0. D0 must be
%               positive.
%
%   'gaussian'  Gaussian highpass filter with cutoff (standard
%               deviation) D0. n need not be supplied. D0 must be
%               positive.

% The transfer function Hhp of a highpass filter is 1 - Hlp,
% where Hlp is the transfer function of the corresponding lowpass
% filter. Thus, we can use function lpfilter to generate highpass
% filters.
if nargin == 4
   n = 1; % Default value of n.
end

% Generate highpass filter.
Hlp = lpfilter(type, M, N, D0, n);
H = 1 - Hlp;
```

EXAMPLE 4.6: Highpass filters.

■ Figure 4.17 shows plots and images of ideal, Butterworth, and Gaussian highpass filters. The plot in Fig. 4.17(a) was generated using the commands

```
>> H = fftshift(hpfilter('ideal', 500, 500, 50));
>> mesh(H(1:10:500, 1:10:500));
>> axis([0 50 0 50 0 1])
```

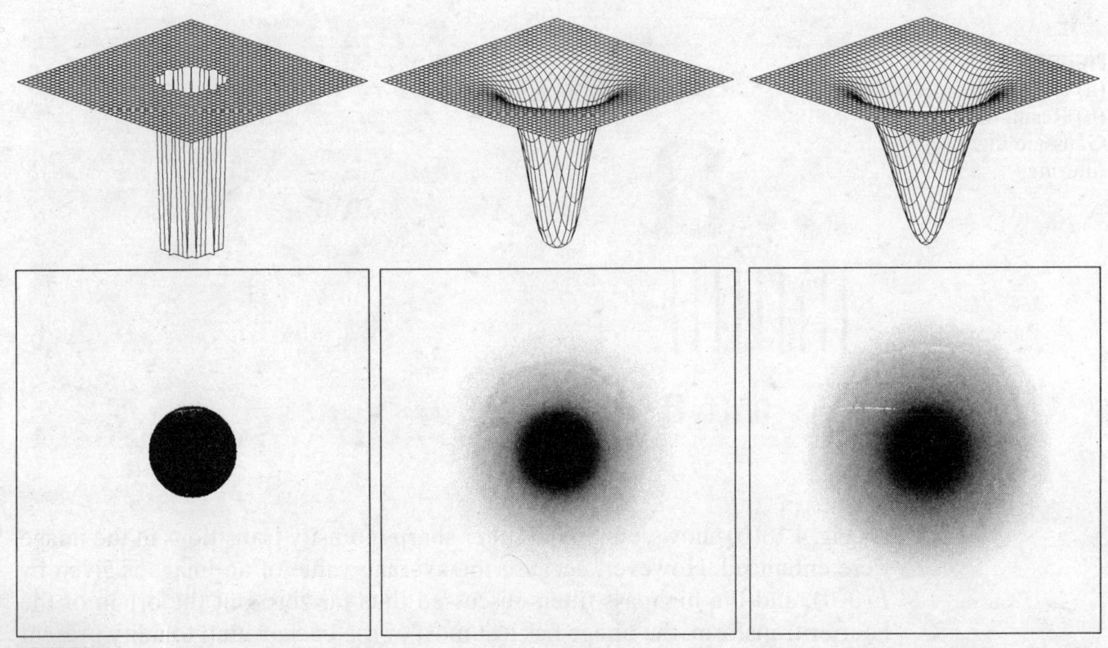

FIGURE 4.17 Top row: Perspective plots of ideal, Butterworth, and Gaussian highpass filters. Bottom row: Corresponding images.

```
>> colormap([0 0 0])
>> axis off
>> grid off
```

The corresponding image in Fig. 4.17(d) was generated using the command

```
>> figure, imshow(H, [ ])
```

where the thin black border is superimposed on the image to delineate its boundary. Similar commands yielded the rest of Fig. 4.17 (the Butterworth filter is of order 2). ■

EXAMPLE 4.7: Highpass filtering.

■ Figure 4.18(a) is the same test pattern, `f`, shown in Fig. 4.13(a). Figure 4.18(b), obtained using the following commands, shows the result of applying a Gaussian highpass filter to `f`:

```
>> PQ = paddedsize(size(f));
>> D0 = 0.05*PQ(1);
>> H = hpfilter('gaussian', PQ(1), PQ(2), D0);
>> g = dftfilt(f, H);
>> figure, imshow(g, [ ])
```

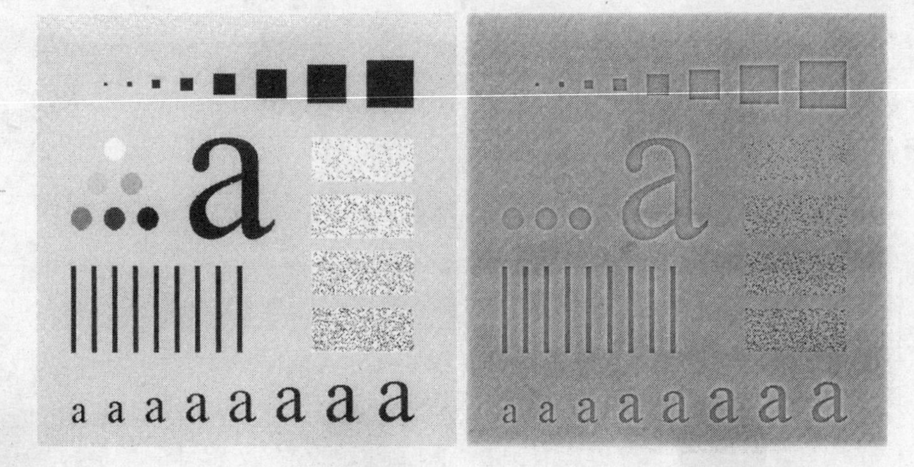

a b

FIGURE 4.18
(a) Original image.
(b) Result of Gaussian highpass filtering.

As Fig. 4.18(b) shows, edges and other sharp intensity transitions in the image were enhanced. However, because the average value of an image is given by $F(0, 0)$, and the highpass filters discussed thus far zero-out the origin of the Fourier transform, the image has lost most of the background tonality present in the original. This problem is addressed in the following section. ■

4.6.2 High-Frequency Emphasis Filtering

As mentioned in Example 4.7, highpass filters zero out the *dc* term, thus reducing the average value of an image to 0. An approach to compensate for this is to add an offset to a highpass filter. When an offset is combined with multiplying the filter by a constant greater than 1, the approach is called *high-frequency emphasis* filtering because the constant multiplier highlights the high frequencies. The multiplier increases the amplitude of the low frequencies also, but the low-frequency effects on enhancement are less than those due to high frequencies, as long as the offset is small compared to the multiplier. High-frequency emphasis has the transfer function

$$H_{\text{hfe}}(u, v) = a + bH_{\text{hp}}(u, v)$$

where a is the offset, b is the multiplier, and $H_{\text{hp}}(u, v)$ is the transfer function of a highpass filter.

EXAMPLE 4.8: Combining high-frequency emphasis and histogram equalization.

■ Figure 4.19(a) shows a chest X-ray image, `f`. X-ray imagers cannot be focused in the same manner as optical lenses, so the resulting images generally tend to be slightly blurred. The objective of this example is to sharpen Fig. 4.19(a). Because the gray levels in this particular image are biased toward the dark end of the gray scale, we also take this opportunity to give an example of how spatial domain processing can be used to complement frequency domain filtering.

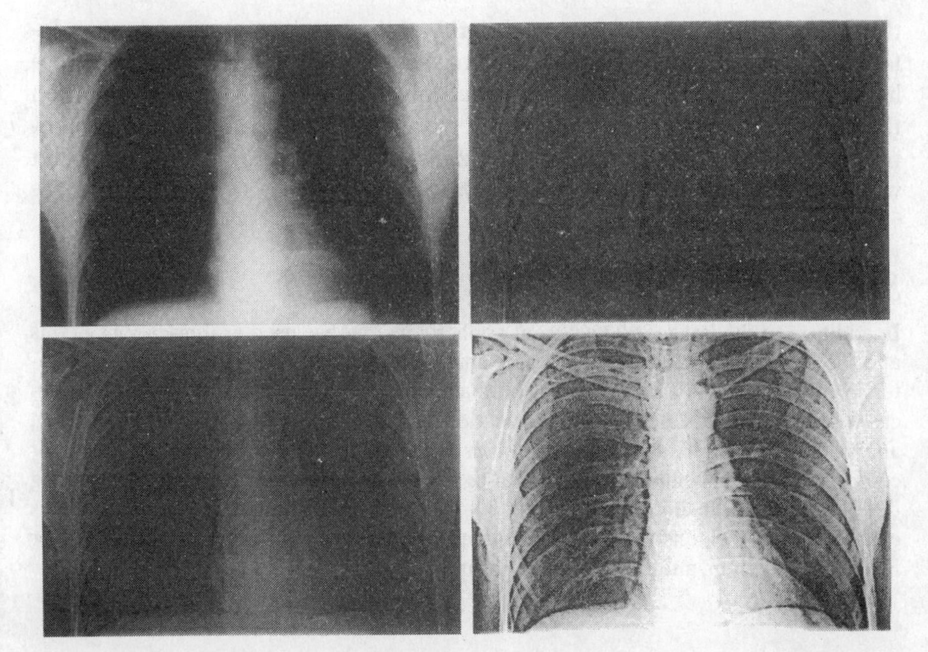

a b
c d

FIGURE 4.19 High-frequency emphasis filtering. (a) Original image. (b) Highpass filtering result. (c) High-frequency emphasis result. (d) Image (c) after histogram equalization. (Original image courtesy of Dr. Thomas R. Gest, Division of Anatomical Sciences, University of Michigan Medical School.)

Figure 4.19(b) shows the result of filtering Fig. 4.19(a) with a Butterworth highpass filter of order 2, and a value of D_0 equal to 5% of the vertical dimension of the padded image. Highpass filtering is not overly sensitive to the value of D_0, as long as the radius of the filter is not so small that frequencies near the origin of the transform are passed. As expected, the filtered result is rather featureless, but it shows faintly the principal edges in the image. The advantage of high-emphasis filtering (with $a = 0.5$ and $b = 2.0$ in this case) is shown in the image of Fig. 4.19(c), in which the gray-level tonality due to the low-frequency components was retained. The following sequence of commands was used to generate the processed images in Fig. 4.19, where `f` denotes the input image [the last command generated Fig. 4.19(d)]:

```
>> PQ = paddedsize(size(f));
>> D0 = 0.05*PQ(1);
>> HBW = hpfilter('btw', PQ(1), PQ(2), D0, 2);
>> H = 0.5 + 2*HBW;
>> gbw = dftfilt(f, HBW);
>> gbw = gscale(gbw);
>> ghf = dftfilt(f, H);
>> ghf = gscale(ghf);
>> ghe = histeq(ghf, 256);
```

As indicated in Section 3.3.2, an image characterized by gray levels in a narrow range of the gray scale is an ideal candidate for histogram equalization. As Fig. 4.19(d) shows, this indeed was an appropriate method to further enhance the image in this example. Note the clarity of the bone structure and other details that simply are not visible in any of the other three images. The final enhanced image appears a little noisy, but this is typical of X-ray images when their gray scale is expanded. The result obtained using a combination of high-frequency emphasis and histogram equalization is superior to the result that would be obtained by using either method alone. ■

Summary

In addition to the image enhancement applications that we used as illustrations in this and the preceding chapter, the concepts and techniques developed in these two chapters provide the basis for other areas of image processing addressed in subsequent discussions in the book. Intensity transformations are used frequently for intensity scaling, and spatial filtering is used extensively for image restoration in the next chapter, for color processing in Chapter 6, for image segmentation in Chapter 10, and for extracting descriptors from an image in Chapter 11. The Fourier techniques developed in this chapter are used extensively in the next chapter for image restoration, in Chapter 8 for image compression, and in Chapter 11 for image description.

5 Image Restoration

Preview

The objective of restoration is to improve a given image in some predefined sense. Although there are areas of overlap between image enhancement and image restoration, the former is largely a subjective process, while image restoration is for the most part an objective process. Restoration attempts to reconstruct or recover an image that has been degraded by using a priori knowledge of the degradation phenomenon. Thus, restoration techniques are oriented toward modeling the degradation and applying the inverse process in order to recover the original image.

This approach usually involves formulating a criterion of goodness that yields an optimal estimate of the desired result. By contrast, enhancement techniques basically are heuristic procedures designed to manipulate an image in order to take advantage of the psychophysical aspects of the human visual system. For example, contrast stretching is considered an enhancement technique because it is based primarily on the pleasing aspects it might present to the viewer, whereas removal of image blur by applying a deblurring function is considered a restoration technique.

In this chapter we explore how to use MATLAB and IPT capabilities to model degradation phenomena and to formulate restoration solutions. As in Chapters 3 and 4, some restoration techniques are best formulated in the spatial domain, while others are better suited for the frequency domain. Both methods are investigated in the sections that follow.

5.1 A Model of the Image Degradation/Restoration Process

As Fig. 5.1 shows, the degradation process is modeled in this chapter as a degradation function that, together with an additive noise term, operates on an input image $f(x, y)$ to produce a degraded image $g(x, y)$:

$$g(x, y) = H[f(x, y)] + \eta(x, y)$$

Given $g(x, y)$, some knowledge about the degradation function H, and some knowledge about the additive noise term $\eta(x, y)$, the objective of restoration is to obtain an estimate, $\hat{f}(x, y)$, of the original image. We want the estimate to be as close as possible to the original input image. In general, the more we know about H and η, the closer $\hat{f}(x, y)$ will be to $f(x, y)$.

If H is a *linear, spatially invariant* process, it can be shown that the degraded image is given in the *spatial domain* by

Following convention, we use an in-line asterisk in equations to denote convolution and a superscript asterisk to denote the complex conjugate. As required, we also use an asterisk in MATLAB expressions to denote multiplication. Care should be taken not to confuse these unrelated uses of the same symbol.

$$g(x, y) = h(x, y) * f(x, y) + \eta(x, y)$$

where $h(x, y)$ is the spatial representation of the degradation function and, as in Chapter 4, the symbol "$*$" indicates convolution. We know from the discussion in Section 4.3.1 that convolution in the spatial domain and multiplication in the frequency domain constitute a Fourier transform pair, so we may write the preceding model in an equivalent *frequency domain* representation:

$$G(u, v) = H(u, v)F(u, v) + N(u, v)$$

where the terms in capital letters are the Fourier transforms of the corresponding terms in the convolution equation. The degradation function $H(u, v)$ sometimes is called the *optical transfer function* (OTF), a term derived from the Fourier analysis of optical systems. In the spatial domain, $h(x, y)$ is referred to as the *point spread function* (PSF), a term that arises from letting $h(x, y)$ operate on a point of light to obtain the characteristics of the degradation for any type of input. The OTF and PSF are a Fourier transform pair, and the toolbox provides two functions, `otf2psf` and `psf2otf`, for converting between them.

Because the degradation due to a linear, space-invariant degradation function, H, can be modeled as convolution, sometimes the degradation process is referred to as "convolving the image with a PSF or OTF." Similarly, the restoration process is sometimes referred to as *deconvolution*.

In the following three sections, we assume that H is the identity operator, and we deal only with degradation due to noise. Beginning in Section 5.6 we look at several methods for image restoration in the presence of both H and η.

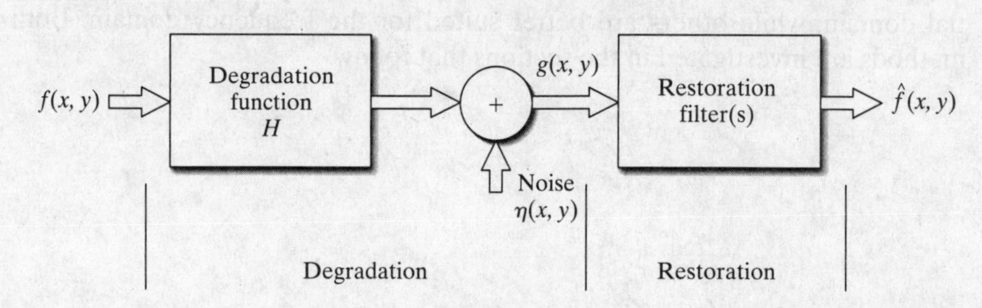

FIGURE 5.1 A model of the image degradation/restoration process.

5.2 Noise Models

The ability to simulate the behavior and effects of noise is central to image restoration. In this chapter, we are interested in two basic types of noise models: noise in the spatial domain (described by the noise probability density function), and noise in the frequency domain, described by various Fourier properties of the noise. With the exception of the material in Section 5.2.3, we assume in this chapter that noise is independent of image coordinates.

5.2.1 Adding Noise with Function `imnoise`

The toolbox uses function `imnoise` to corrupt an image with noise. This function has the basic syntax

```
g = imnoise(f, type, parameters)
```

where `f` is the input image, and `type` and `parameters` are as explained later. Function `imnoise` converts the input image to class `double` in the range [0, 1] before adding noise to it. This must be taken into account when specifying noise parameters. For example, to add Gaussian noise of mean 64 and variance 400 to an `uint8` image, we scale the mean to 64/255 and the variance to $400/(255)^2$ for input into `imnoise`. The syntax forms for this function are:

- `g = imnoise(f, 'gaussian', m, var)` adds Gaussian noise of mean `m` and variance `var` to image `f`. The default is zero mean noise with 0.01 variance.
- `g = imnoise(f, 'localvar', V)` adds zero-mean, Gaussian noise of local variance, `V`, to image `f`, where `V` is an array of the same size as `f` containing the desired variance values at each point.
- `g = imnoise(f, 'localvar', image_intensity, var)` adds zero-mean, Gaussian noise to image `f`, where the local variance of the noise, `var`, is a function of the image intensity values in `f`. The `image_intensity` and `var` arguments are vectors of the same size, and `plot(image_intensity, var)` plots the functional relationship between noise variance and image intensity. The `image_intensity` vector must contain normalized intensity values in the range [0, 1].
- `g = imnoise(f, 'salt & pepper', d)` corrupts image `f` with salt and pepper noise, where `d` is the noise density (i.e., the percent of the image area containing noise values). Thus, approximately `d*numel(f)` pixels are affected. The default is 0.05 noise density.
- `g = imnoise(f, 'speckle', var)` adds multiplicative noise to image `f`, using the equation `g = f + n*f`, where `n` is uniformly distributed random noise with mean 0 and variance `var`. The default value of `var` is 0.04.
- `g = imnoise(f, 'poisson')` generates Poisson noise from the data instead of adding artificial noise to the data. In order to comply with Poisson statistics, the intensities of `uint8` and `uint16` images must correspond to the number of photons (or any other quanta of information). Double-precision images are used when the number of photons per pixel is larger than 65535

(but less than 10^{12}). The intensity values vary between 0 and 1 and correspond to the number of photons divided by 10^{12}.

Several illustrations of `imnoise` are given in the following sections.

5.2.2 Generating Spatial Random Noise with a Specified Distribution

Often, it is necessary to be able to generate noise of types and parameters beyond those available in function `imnoise`. Spatial noise values are random numbers, characterized by a probability density function (PDF) or, equivalently, by the corresponding cumulative distribution function (CDF). Random number generation for the types of distributions in which we are interested follow some fairly simple rules from probability theory.

Numerous random number generators are based on expressing the generation problem in terms of random numbers with a uniform CDF in the interval (0, 1). In some instances, the base random number generator of choice is a generator of Gaussian random numbers with zero mean and unit variance. Although we can generate these two types of noise using `imnoise`, it is more meaningful in the present context to use MATLAB function `rand` for uniform random numbers and `randn` for normal (Gaussian) random numbers. These functions are explained later in this section.

The foundation of the approach described in this section is a well-known result from probability (Peebles [1993]) which states that if w is a uniformly distributed random variable in the interval (0, 1), then we can obtain a random variable z with a specified CDF, F_z, by solving the equation

$$z = F_z^{-1}(w)$$

This simple, yet powerful, result can be stated equivalently as finding a solution to the equation $F_z(z) = w$.

EXAMPLE 5.1: Using uniform random numbers to generate random numbers with a specified distribution.

■ Assume that we have a generator of uniform random numbers, w, in the interval (0, 1), and suppose that we want to use it to generate random numbers, z, with a Rayleigh CDF, which has the form

$$F_z(z) = \begin{cases} 1 - e^{-(z-a)^2/b} & \text{for } z \geq a \\ 0 & \text{for } z < a \end{cases}$$

To find z we solve the equation

$$1 - e^{-(z-a)^2/b} = w$$

or

$$z = a + \sqrt{b \ln(1 - w)}$$

Because the square root term is nonnegative, we are assured that no values of z less than a are generated. This is as required by the definition of the Rayleigh CDF. Thus, a uniform random number w from our generator can be used in the previous equation to generate a random variable z having a Rayleigh distribution with parameters a and b.

In MATLAB this result is easily generalized to an $M \times N$ array, R, of random numbers by using the expression

```
>> R = a + sqrt(b*log(1 - rand(M, N)));
```

where, as discussed in Section 3.2.2, log is the natural logarithm, and, as mentioned earlier, rand generates uniformly distributed random numbers in the interval $(0, 1)$. If we let M = N = 1, then the preceding MATLAB command line yields a single value from a random variable with a Rayleigh distribution characterized by parameters a and b. ■

The expression $z = a + \sqrt{b \ln(1 - w)}$ sometimes is called a *random number generator equation* because it establishes how to generate the desired random numbers. In this particular case, we were able to find a closed-form solution. As will be shown shortly, this is not always possible and the problem then becomes one of finding an applicable random number generator equation whose outputs will approximate random numbers with the specified CDF.

Table 5.1 lists the random variables of interest in the present discussion, along with their PDFs, CDFs, and random number generator equations. In some cases, as with the Rayleigh and exponential variables, it is possible to find a closed-form solution for the CDF and its inverse. This allows us to write an expression for the random number generator in terms of uniform random numbers, as illustrated in Example 5.1. In others, as in the case of the Gaussian and lognormal densities, closed-form solutions for the CDF do not exist, and it becomes necessary to find alternate ways to generate the desired random numbers. In the lognormal case, for instance, we make use of the knowledge that a lognormal random variable, z, is such that $\ln(z)$ has a Gaussian distribution and write the expression shown in Table 5.1 in terms of Gaussian random variables with zero mean and unit variance. Yet in other cases, it is advantageous to reformulate the problem to obtain an easier solution. For example, it can be shown that Erlang random numbers with parameters a and b can be obtained by adding b exponentially distributed random numbers that have parameter a (Leon-Garcia [1994]).

The random number generators available in imnoise and those shown in Table 5.1 play an important role in modeling the behavior of random noise in image-processing applications. We already saw the usefulness of the uniform distribution for generating random numbers with various CDFs. Gaussian noise is used as an approximation in cases such as imaging sensors operating at low light levels. Salt-and-pepper noise arises in faulty switching devices. The size of silver particles in a photographic emulsion is a random variable described by a lognormal distribution. Rayleigh noise arises in range imaging, while exponential and Erlang noise are useful in describing noise in laser imaging.

M-function imnoise2, listed later in this section, generates random numbers having the CDFs in Table 5.1. This function makes use of MATLAB function rand, which, for the purposes of this chapter, has the syntax

```
A = rand(M, N)
```

TABLE 5.1 Generation of random variables.

Name	PDF	Mean and Variance	CDF	Generator[†]
Uniform	$p_z(z) = \begin{cases} \dfrac{1}{b-a} & \text{if } a \le z \le b \\ 0 & \text{otherwise} \end{cases}$	$m = \dfrac{a+b}{2}, \quad \sigma^2 = \dfrac{(b-a)^2}{12}$	$F_z(z) = \begin{cases} 0 & z < a \\ \dfrac{z-a}{b-a} & a \le z \le b \\ 1 & z > b \end{cases}$	MATLAB function `rand`
Gaussian	$p_z(z) = \dfrac{1}{\sqrt{2\pi}b} e^{-(z-a)^2/2b^2}$ $-\infty < z < \infty$	$m = a, \quad \sigma^2 = b^2$	$F_z(z) = \int_{-\infty}^{z} p_z(v)\,dv$	MATLAB function `randn`
Salt & Pepper	$p_z(z) = \begin{cases} P_a & \text{for } z = a \\ P_b & \text{for } z = b \\ 0 & \text{otherwise} \end{cases}$ $b > a$	$m = aP_a + bP_b$ $\sigma^2 = (a-m)^2 P_a + (b-m)^2 P_b$	$F_z(z) = \begin{cases} 0 & \text{for } z < a \\ P_a & \text{for } a \le z < b \\ P_a + P_b & \text{for } b \le z \end{cases}$	MATLAB function `rand` with some additional logic
Lognormal	$p_z(z) = \dfrac{1}{\sqrt{2\pi}bz} e^{-[\ln(z)-a]^2/2b^2}$ $z > 0$	$m = e^{a+(b^2/2)}, \sigma^2 = [e^{b^2} - 1]e^{2a+b^2}$	$F_z(z) = \int_{0}^{z} p_z(v)\,dv$	$z = ae^{bN(0,1)}$
Rayleigh	$p_z(z) = \begin{cases} \dfrac{2}{b}(z-a)e^{-(z-a)^2/b} & z \ge a \\ 0 & z < a \end{cases}$	$m = a + \sqrt{\pi b/4}, \sigma^2 = \dfrac{b(4-\pi)}{4}$	$F_z(z) = \begin{cases} 1 - e^{-(z-a)^2/b} & z \ge a \\ 0 & z < a \end{cases}$	$z = a + \sqrt{b \ln[1 - U(0,1)]}$
Exponential	$p_z(z) = \begin{cases} ae^{-az} & z \ge 0 \\ 0 & z < 0 \end{cases}$	$m = \dfrac{1}{a}, \quad \sigma^2 = \dfrac{1}{a^2}$	$F_z(z) = \begin{cases} 1 - e^{-az} & z \ge 0 \\ 0 & z < 0 \end{cases}$	$z = -\dfrac{1}{a}\ln[1 - U(0,1)]$
Erlang	$p_z(z) = \dfrac{a^b z^{b-1}}{(b-1)!} e^{-az}$ $z \ge 0$	$m = \dfrac{b}{a}, \quad \sigma^2 = \dfrac{b}{a^2}$	$F_z(z) = \left[1 - e^{-az}\sum_{n=0}^{b-1}\dfrac{(az)^n}{n!}\right]$ $z \ge 0$	$z = E_1 + E_2 + \cdots + E_b$ (The E's are exponential random numbers with parameter a.)

[†] $N(0, 1)$ denotes normal (Gaussian) random numbers with mean 0 and a variance of 1. $U(0, 1)$ denotes uniform random numbers in the range (0, 1).

This function generates an array of size M × N whose entries are uniformly distributed numbers with values in the interval (0, 1). If N is omitted it defaults to M. If called without an argument, rand generates a single random number that changes each time the function is called. Similarly, the function

```
A = randn(M, N)
```

generates an M × N array whose elements are normal (Gaussian) numbers with zero mean and unit variance. If N is omitted it defaults to M. When called without an argument, randn generates a single random number.

Function imnoise2 also uses MATLAB function find, which has the following syntax forms:

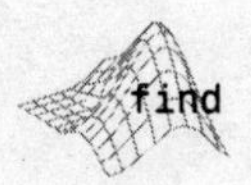


```
I = find(A)
[r, c] = find(A)
[r, c, v] = find(A)
```

The first form returns in I all the indices of array A that point to *nonzero* elements. If none is found, find returns an empty matrix. The second form returns the row and column indices of the nonzero entries in the matrix A. In addition to returning the row and column indices, the third form also returns the nonzero values of A as a column vector, v.

The first form treats the array A in the format A(:), so I is a column vector. This form is quite useful in image processing. For example, to find and set to 0 all pixels in an image whose values are less than 128 we write

```
>> I = find(A < 128);
>> A(I) = 0;
```

Recall that the logical statement A < 128 returns a 1 for the elements of A that satisfy the logical condition and 0 for those that do not. To set to 128 all pixels in the closed interval [64, 192] we write

```
>> I = find(A >= 64 & A <= 192);
>> A(I) = 128;
```

The first two forms of function find are used frequently in the remaining chapters of the book.

Unlike imnoise, the following M-function generates an M × N noise array, R, that is not scaled in any way. Another major difference is that imnoise outputs a noisy image, while imnoise2 produces the noise pattern itself. The user specifies the desired values for the noise parameters directly. Note that the noise array resulting from salt-and-pepper noise has three values: 0 corresponding to pepper noise, 1 corresponding to salt noise, and 0.5 corresponding to no noise.

This array needs to be processed further to make it useful. For example, to corrupt an image with this array, we find (using function `find`) all the coordinates in `R` that have value 0 and set the corresponding coordinates in the image to the smallest possible gray-level value (usually 0). Similarly, we find all the coordinates in `R` that have value 1 and set all the corresponding coordinates in the image to the highest possible value (usually 255 for an 8-bit image). This process simulates how salt-and-pepper noise affects an image in practice.

imnoise2

```
function R = imnoise2(type, M, N, a, b)
%IMNOISE2 Generates an array of random numbers with specified PDF.
%   R = IMNOISE2(TYPE, M, N, A, B) generates an array, R, of size
%   M-by-N, whose elements are random numbers of the specified TYPE
%   with parameters A and B. If only TYPE is included in the
%   input argument list, a single random number of the specified
%   TYPE and default parameters shown below is generated. If only
%   TYPE, M, and N are provided, the default parameters shown below
%   are used. If M = N = 1, IMNOISE2 generates a single random
%   number of the specified TYPE and parameters A and B.
%
%   Valid values for TYPE and parameters A and B are:
%
%   'uniform'         Uniform random numbers in the interval (A, B).
%                     The default values are (0, 1).
%   'gaussian'        Gaussian random numbers with mean A and standard
%                     deviation B. The default values are A = 0, B = 1.
%   'salt & pepper'   Salt and pepper numbers of amplitude 0 with
%                     probability Pa = A, and amplitude 1 with
%                     probability Pb = B. The default values are Pa =
%                     Pb = A = B = 0.05. Note that the noise has
%                     values 0 (with probability Pa = A) and 1 (with
%                     probability Pb = B), so scaling is necessary if
%                     values other than 0 and 1 are required. The noise
%                     matrix R is assigned three values. If R(x, y) =
%                     0, the noise at (x, y) is pepper (black). If
%                     R(x, y) = 1, the noise at (x, y) is salt
%                     (white). If R(x, y) = 0.5, there is no noise
%                     assigned to coordinates (x, y).
%   'lognormal'       Lognormal numbers with offset A and shape
%                     parameter B. The defaults are A = 1 and B =
%                     0.25.
%   'rayleigh'        Rayleigh noise with parameters A and B. The
%                     default values are A = 0 and B = 1.
%   'exponential'     Exponential random numbers with parameter A. The
%                     default is A = 1.
%   'erlang'          Erlang (gamma) random numbers with parameters A
%                     and B. B must be a positive integer. The
%                     defaults are A = 2 and B = 5. Erlang random
%                     numbers are approximated as the sum of B
%                     exponential random numbers.
```

```
% Set default values.
if nargin == 1
   a = 0; b = 1;
   M = 1; N = 1;
elseif nargin == 3
   a = 0; b = 1;
end

% Begin processing. Use lower(type) to protect against input
% being capitalized.
switch lower(type)
case 'uniform'
   R = a + (b - a)*rand(M, N);
case 'gaussian'
   R = a + b*randn(M, N);
case 'salt & pepper'
   if nargin <= 3
      a = 0.05; b = 0.05;
   end
   % Check to make sure that Pa + Pb is not > 1.
   if (a + b) > 1
      error('The sum Pa + Pb must not exceed 1.')
   end
   R(1:M, 1:N) = 0.5;
   % Generate an M-by-N array of uniformly-distributed random numbers
   % in the range (0, 1). Then, Pa*(M*N) of them will have values <=
   % a. The coordinates of these points we call 0 (pepper
   % noise). Similarly, Pb*(M*N) points will have values in the range
   % > a & <= (a + b). These we call 1 (salt noise).
   X = rand(M, N);
   c = find(X <= a);
   R(c) = 0;
   u = a + b;
   c = find(X > a & X <= u);
   R(c) = 1;
case 'lognormal'
   if nargin <= 3
      a = 1; b = 0.25;
   end
   R = a*exp(b*randn(M, N));
case 'rayleigh'
   R = a + (-b*log(1 - rand(M, N))).^0.5;
case 'exponential'
   if nargin <= 3
      a = 1;
   end
   if a <= 0
      error('Parameter a must be positive for exponential type.')
   end
   k = -1/a;
   R = k*log(1 - rand(M, N));
```

```
case 'erlang'
   if nargin <= 3
      a = 2; b = 5;
   end
   if (b ~= round(b) | b <= 0)
      error('Param b must be a positive integer for Erlang.')
   end
   k = -1/a;
   R = zeros(M, N);
   for j = 1:b
      R = R + k*log(1 - rand(M, N));
   end
otherwise
   error('Unknown distribution type.')
end
```

EXAMPLE 5.2: Histograms of data generated using the function `imnoise2`.

■ Figure 5.2 shows histograms of all the random number types in Table 5.1. The data for each plot were generated using function `imnoise2`. For example, the data for Fig. 5.2(a) were generated by the following command:

```
>> r = imnoise2('gaussian', 100000, 1, 0, 1);
```

This statement generated a column vector, `r`, with 100000 elements, each being a random number from a Gaussian distribution with mean 0 and standard deviation of 1. The histogram was then obtained using function `hist`, which has the syntax

hist

```
p = hist(r, bins)
```

where `bins` is the number of bins. We used `bins = 50` to generate the histograms in Fig. 5.2. The other histograms were generated in a similar manner. In each case, the parameters chosen were the default values listed in the explanation of function `imnoise2`. ■

5.2.3 Periodic Noise

Periodic noise in an image arises typically from electrical and/or electromechanical interference during image acquisition. This is the only type of spatially dependent noise that will be considered in this chapter. As discussed in Section 5.4, periodic noise is typically handled in an image by filtering in the frequency domain. Our model of periodic noise is a 2-D sinusoid with equation

$$r(x, y) = A\sin[2\pi u_0(x + B_x)/M + 2\pi v_0(y + B_y)/N]$$

where A is the amplitude, u_0 and v_0 determine the sinusoidal frequencies with respect to the x- and y-axis, respectively, and B_x and B_y are phase displacements with respect to the origin. The $M \times N$ DFT of this equation is

$$R(u, v) = j\frac{A}{2}[(e^{j2\pi u_0 B_x/M})\delta(u + u_0, v + v_0) - (e^{j2\pi v_0 B_y/N})\delta(u - u_0, v - v_0)]$$

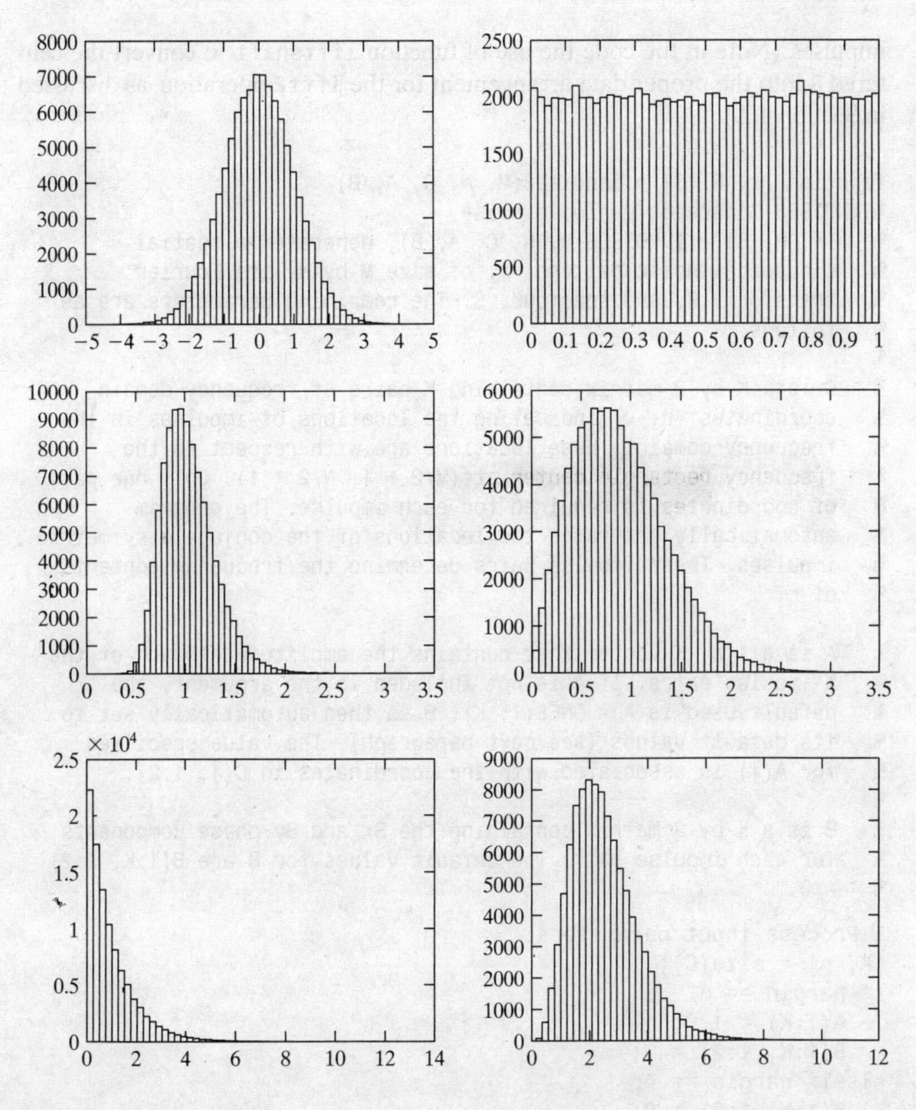

a b
c d
e f

FIGURE 5.2 Histograms of random numbers: (a) Gaussian, (b) uniform, (c) lognormal, (d) Rayleigh, (e) exponential, and (f) Erlang. In each case the default parameters listed in the explanation of function `imnoise2` were used.

which we see is a pair of complex conjugate impulses located at $(u + u_0, v + v_0)$ and $(u - u_0, v - v_0)$, respectively.

The following M-function accepts an arbitrary number of impulse locations (frequency coordinates), each with its own amplitude, frequencies, and phase displacement parameters, and computes $r(x, y)$ as the sum of sinusoids of the form described in the previous paragraph. The function also outputs the Fourier transform of the sum of sinusoids, $R(u, v)$, and the spectrum of $R(u, v)$. The sine waves are generated from the given impulse location information via the inverse DFT. This makes it more intuitive and simplifies visualization of frequency content in the spatial noise pattern. Only one pair of coordinates is required to define the location of an impulse. The program generates the conjugate symmetric impulses. (Note in the code the use of function `ifftshift` to convert the centered `R` into the proper data arrangement for the `ifft2` operation, as discussed in Section 4.2.)

imnoise3

```
function [r, R, S] = imnoise3(M, N, C, A, B)
%IMNOISE3 Generates periodic noise.
%   [r, R, S] = IMNOISE3(M, N, C, A, B), generates a spatial
%   sinusoidal noise pattern, r, of size M-by-N, its Fourier
%   transform, R, and spectrum, S. The remaining parameters are as
%   follows:
%
%   C is a K-by-2 matrix containing K pairs of frequency domain
%   coordinates (u, v) indicating the locations of impulses in the
%   frequency domain. These locations are with respect to the
%   frequency rectangle center at (M/2 + 1, N/2 + 1). Only one pair
%   of coordinates is required for each impulse. The program
%   automatically generates the locations of the conjugate symmetric
%   impulses. These impulse pairs determine the frequency content
%   of r.
%
%   A is a 1-by-K vector that contains the amplitude of each of the
%   K impulse pairs. If A is not included in the argument, the
%   default used is A = ONES(1, K). B is then automatically set to
%   its default values (see next paragraph). The value specified
%   for A(j) is associated with the coordinates in C(j, 1:2).
%
%   B is a K-by-2 matrix containing the Bx and By phase components
%   for each impulse pair. The default values for B are B(1:K, 1:2)
%   = 0.

% Process input parameters.
[K, n] = size(C);
if nargin == 3
   A(1:K) = 1.0;
   B(1:K, 1:2) = 0;
elseif nargin == 4
   B(1:K, 1:2) = 0;
end

% Generate R.
R = zeros(M, N);
for j = 1:K
   u1 = M/2 + 1 + C(j, 1); v1 = N/2 + 1 + C(j, 2);
   R(u1, v1) = i * (A(j)/2) * exp(i*2*pi*C(j, 1) * B(j, 1)/M);
   % Complex conjugate.
   u2 = M/2 + 1 - C(j, 1); v2 = N/2 + 1 - C(j, 2);
   R(u2, v2) = -i * (A(j)/2) * exp(i*2*pi*C(j, 2) * B(j, 2)/N);
end

% Compute spectrum and spatial sinusoidal pattern.
S = abs(R);
r = real(ifft2(ifftshift(R)));
```

EXAMPLE 5.3: Using function `imnoise3`.

■ Figures 5.3(a) and (b) show the spectrum and spatial sine noise pattern generated using the following commands:

```
>> C = [0 64; 0 128; 32 32; 64 0; 128 0; -32 32];
>> [r, R, S] = imnoise3(512, 512, C);
>> imshow(S, [ ])
>> figure, imshow(r, [ ])
```

Recall that the order of the coordinates is (u, v). These two values are specified with reference to the center of the frequency rectangle (see Section 4.2 for a definition of the coordinates of this center point). Figures 5.3(c) and (d) show the result obtained by repeating the previous commands, but with

```
>> C = [0 32; 0 64; 16 16; 32 0; 64 0; -16 16];
```

Similarly, Fig. 5.3(e) was obtained with

```
>> C = [6 32; -2 2];
```

Figure 5.3(f) was generated with the same `C`, but using a nondefault amplitude vector:

```
>> A = [1 5];
>> [r, R, S] = imnoise3(512, 512, C, A);
```

As Fig. 5.3(f) shows, the lower-frequency sine wave dominates the image. This is as expected because its amplitude is five times the amplitude of the higher-frequency component. ■

5.2.4 Estimating Noise Parameters

The parameters of periodic noise typically are estimated by analyzing the Fourier spectrum of the image. Periodic noise tends to produce frequency spikes that often can be detected even by visual inspection. Automated analysis is possible in situations in which the noise spikes are sufficiently pronounced, or when some knowledge about the frequency of the interference is available.

In the case of noise in the spatial domain, the parameters of the PDF may be known partially from sensor specifications, but it is often necessary to estimate them from sample images. The relationships between the mean, m, and variance, σ^2, of the noise, and the parameters a and b required to completely specify the noise PDFs of interest in this chapter are listed in Table 5.1. Thus, the problem becomes one of estimating the mean and variance from the sample image(s) and then using these estimates to solve for a and b.

Let z_i be a discrete random variable that denotes intensity levels in an image, and let $p(z_i)$, $i = 0, 1, 2, \ldots, L - 1$, be the corresponding normalized

a b
c d
e f

FIGURE 5.3
(a) Spectrum of specified impulses. (b) Corresponding sine noise pattern. (c) and (d) A similar sequence. (e) and (f) Two other noise patterns. The dots in (a) and (c) were enlarged to make them easier to see.

histogram, where L is the number of possible intensity values. A histogram component, $p(z_j)$, is an estimate of the probability of occurrence of intensity value z_j, and the histogram may be viewed as an approximation of the intensity PDF.

One of the principal approaches for describing the shape of a histogram is via its *central moments* (also called *moments about the mean*), which are defined as

$$\mu_n = \sum_{i=0}^{L-1} (z_i - m)^n p(z_i)$$

where n is the moment *order*, and m is the mean:

$$m = \sum_{i=0}^{L-1} z_i p(z_i)$$

Because the histogram is assumed to be normalized, the sum of all its components is 1, so, from the preceding equations, we see that $\mu_0 = 1$ and $\mu_1 = 0$. The second moment,

$$\mu_2 = \sum_{i=0}^{L-1} (z_i - m)^2 p(z_i)$$

is the variance. In this chapter, we are interested only in the mean and variance. Higher-order moments are discussed in Chapter 11.

Function `statmoments` computes the mean and central moments up to order n, and returns them in row vector `v`. Because the moment of order 0 is always 1, and the moment of order 1 is always 0, `statmoments` ignores these two moments and instead lets `v(1)` $= m$ and `v(k)` $= \mu_k$ for $k = 2, 3, \ldots, n$. The syntax is as follows (see Appendix C for the code):

```
[v, unv] = statmoments(p, n)
```

statmoments

where `p` is the histogram vector and `n` is the number of moments to compute. It is required that the number of components of `p` be equal to 2^8 for class `uint8` images, 2^{16} for class `uint16` images, and 2^8 or 2^{16} for images of class `double`. Output vector `v` contains the normalized moments based on values of the random variable that have been scaled to the range [0, 1], so all the moments are in this range also. Vector `unv` contains the same moments as `v`, but computed with the data in its original range of values. For example, if `length(p) = 256`, and `v(1) = 0.5`, then `unv(1)` would have the value `127.5`, which is half of the range `[0, 255]`.

Often, noise parameters must be estimated directly from a given noisy image or set of images. In this case, the approach is to select a region in an image with as featureless a background as possible, so that the variability of intensity values in the region will be due primarily to noise. To select a region of interest (ROI) in MATLAB we use function roipoly, which generates a polygonal ROI. This function has the basic syntax

roipoly

```
B = roipoly(f, c, r)
```

where f is the image of interest, and c and r are vectors of corresponding (sequential) column and row coordinates of the vertices of the polygon (note that columns are specified first). The output, B, is a binary image the same size as f with 0's outside the region of interest and 1's inside. Image B is used as a mask to limit operations to within the region of interest.

To specify a polygonal ROI interactively, we use the syntax

```
B = roipoly(f)
```

which displays the image f on the screen and lets the user specify the polygon using the mouse. If f is omitted, roipoly operates on the last image displayed. Using normal button clicks adds vertices to the polygon. Pressing **Backspace** or **Delete** removes the previously selected vertex. A shift-click, right-click, or double-click adds a final vertex to the selection and starts the fill of the polygonal region with 1s. Pressing **Return** finishes the selection without adding a vertex.

To obtain the binary image and a list of the polygon vertices, we use the construct

```
[B, c, r] = roipoly(. . .)
```

where roipoly(. . .) indicates any valid syntax for this function and, as before, c and r are the column and row coordinates of the vertices. This format is particularly useful when the ROI is specified interactively because it gives the coordinates of the polygon vertices for use in other programs or for later duplication of the same ROI.

The following function computes the histogram of an image within a polygonal region whose vertices are specified by vectors c and r, as in the preceding discussion. Note the use within the program of function roipoly to duplicate the polygonal region defined by c and r.

histroi

```
function [p, npix] = histroi(f, c, r)
%HISTROI Computes the histogram of an ROI in an image.
%   [P, NPIX] = HISTROI(F, C, R) computes the histogram, P, of a
%   polygonal region of interest (ROI) in image F. The polygonal
%   region is defined by the column and row coordinates of its
%   vertices, which are specified (sequentially) in vectors C and R,
%   respectively. All pixels of F must be >= 0. Parameter NPIX is the
%   number of pixels in the polygonal region.

% Generate the binary mask image.
B = roipoly(f, c, r);
```

```
% Compute the histogram of the pixels in the ROI.
p = imhist(f(B));

% Obtain the number of pixels in the ROI if requested in the output.
if nargout > 1
   npix = sum(B(:));
end
```

EXAMPLE 5.4: Estimating noise parameters.

■ Figure 5.4(a) shows a noisy image, denoted by f in the following discussion. The objective of this example is to estimate the noise type and its parameters using the techniques and tools developed thus far. Figure 5.4(b) shows the mask, B, generated interactively using the command:

```
>> [B, c, r] = roipoly(f);
```

Figure 5.4(c) was generated using the commands

```
>> [p, npix] = histroi(f, c, r);
>> figure, bar(p, 1)
```

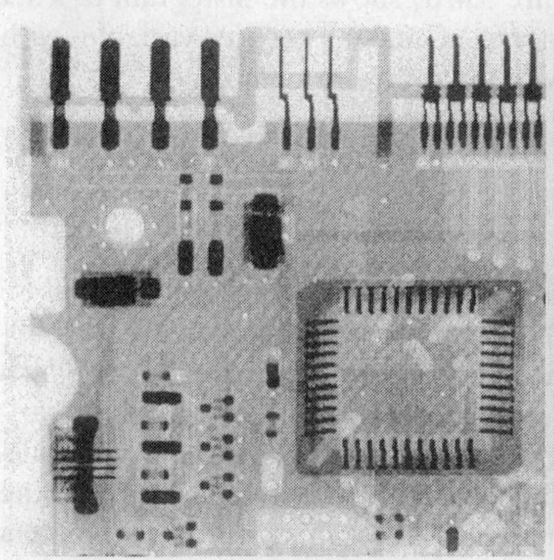

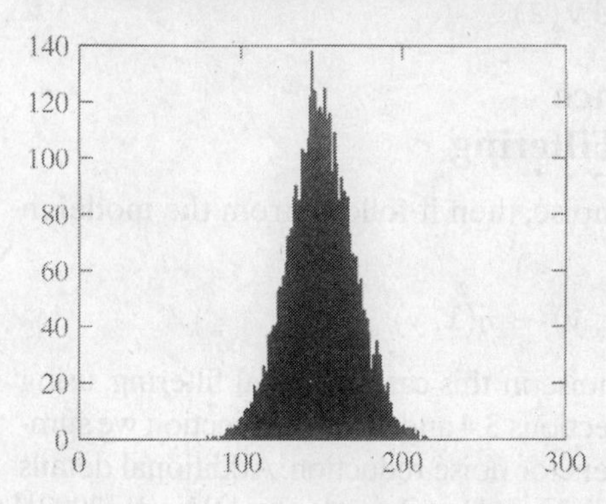

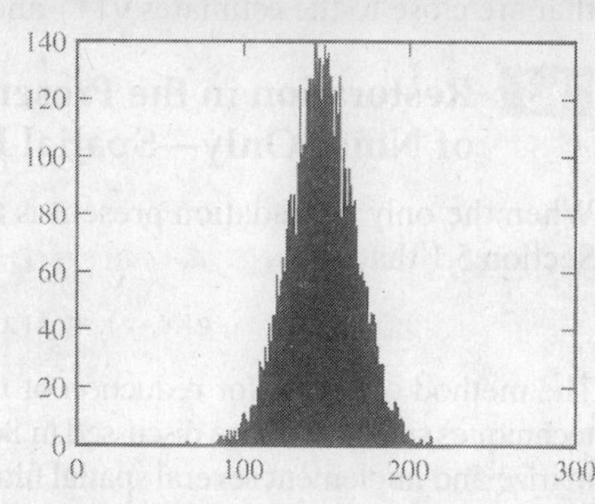

a b
c d

FIGURE 5.4 (a) Noisy image. (b) ROI generated interactively. (c) Histogram of ROI. (d) Histogram of Gaussian data generated using function imnoise2. (Original image courtesy of Lixi, Inc.)

The mean and variance of the region masked by B were obtained as follows:

```
>> [v, unv] = statmoments(h, 2);
>> v

v =

    0.5794    0.0063

>> unv
  147.7430   410.9313
```

It is evident from Fig. 5.4(c) that the noise is approximately Gaussian. In general, it is not possible to know the exact mean and variance of the noise because it is added to the gray levels of the image in region B. However, by selecting an area of nearly constant background level (as we did here), and because the noise appears Gaussian, we can estimate that the average gray level of the area B is reasonably close to the average gray level of the image without noise, indicating that the noise has zero mean. Also, the fact that the area has a nearly constant gray level tells us that the variability in the region defined by B is due primarily to the variance of the noise. (When feasible, another way to estimate the mean and variance of the noise is by imaging a target of constant, known gray level.) Figure 5.4(d) shows the histogram of a set of `npix` (this number is returned by `histroi`) Gaussian random variables with mean 147 and variance 400, obtained with the following commands:

```
>> X = imnoise2('gaussian', npix, 1, 147, 20);
>> figure, hist(X, 130)
>> axis([0 300 0 140])
```

where the number of bins in `hist` was selected so that the result would be compatible with the plot in Fig. 5.4(c). The histogram in this figure was obtained within function `histroi` using `imhist` (see the preceding code), which employs a different scaling than `hist`. We chose a set of `npix` random variables to generate `X`, so that the number of samples was the same in both histograms. The similarity between Figs. 5.4(c) and (d) clearly indicates that the noise is indeed well-approximated by a Gaussian distribution with parameters that are close to the estimates `v(1)` and `v(2)`. ■

5.3 Restoration in the Presence of Noise Only—Spatial Filtering

When the only degradation present is noise, then it follows from the model in Section 5.1 that

$$g(x, y) = f(x, y) + \eta(x, y)$$

The method of choice for reduction of noise in this case is spatial filtering, using techniques similar to those discussed in Sections 3.4 and 3.5. In this section we summarize and implement several spatial filters for noise reduction. Additional details on the characteristics of these filters are discussed by Gonzalez and Woods [2002].

5.3.1 Spatial Noise Filters

Table 5.2 lists the spatial filters of interest in this section, where S_{xy} denotes an $m \times n$ subimage (region) of the input noisy image, g. The subscripts on S indicate that the subimage is centered at coordinates (x, y), and $\hat{f}(x, y)$ (an estimate of f) denotes the filter response at those coordinates. The linear filters are implemented using function `imfilter` discussed in Section 3.4. The median, max, and min filters are nonlinear, order-statistic filters. The median filter can be implemented directly using IPT function `medfilt2`. The max and min filters are implemented using the more general order-filter function `ordfilt2` discussed in Section 3.5.2.

The following function, which we call `spfilt`, performs filtering in the spatial domain with any of the filters listed in Table 5.2. Note the use of function `imlincomb` (mentioned in Table 2.5) to compute the linear combination of the inputs. The syntax for this function is

```
B = imlincomb(c1, A1, c2, A2, . . ., ck, Ak)
```

which implements the equation

```
B = c1*A1 + c2*A2 + · · · + ck*Ak
```

where the `c`'s are real, `double` scalars, and the `A`'s are numeric arrays of the same class and size. Note also in subfunction `gmean` how function `warning` can be turned on and off. In this case, we are suppressing a warning that would be issued by MATLAB if the argument of the `log` function becomes 0. In general, `warning` can be used in any program. The basic syntax is

```
warning('message')
```

This function behaves exactly like function `disp`, except that it can be turned on and off with the commands `warning on` and `warning off`.

spfilt

```
function f = spfilt(g, type, m, n, parameter)
%SPFILT Performs linear and nonlinear spatial filtering.
%   F = SPFILT(G, TYPE, M, N, PARAMETER) performs spatial filtering
%   of image G using a TYPE filter of size M-by-N. Valid calls to
%   SPFILT are as follows:
%
%     F = SPFILT(G, 'amean', M, N)       Arithmetic mean filtering.
%     F = SPFILT(G, 'gmean', M, N)       Geometric mean filtering.
%     F = SPFILT(G, 'hmean', M, N)       Harmonic mean filtering.
%     F = SPFILT(G, 'chmean', M, N, Q)   Contraharmonic mean
%                                        filtering of order Q. The
%                                        default is Q = 1.5.
%     F = SPFILT(G, 'median', M, N)      Median filtering.
%     F = SPFILT(G, 'max', M, N)         Max filtering.
%     F = SPFILT(G, 'min', M, N)         Min filtering.
```

TABLE 5.2 Spatial filters. The variables m and n denote respectively the number of rows and columns of the filter neighborhood.

Filter Name	Equation	Comments
Arithmetic mean	$\hat{f}(x,y) = \frac{1}{mn} \sum_{(s,t)\in S_{xy}} g(s,t)$	Implemented using IPT functions `w = fspecial('average', [m, n])` and `f = imfilter(g, w)`.
Geometric mean	$\hat{f}(x,y) = \left[\prod_{(s,t)\in S_{xy}} g(s,t) \right]^{\frac{1}{mn}}$	This nonlinear filter is implemented using function `gmean` (see custom function `spfilt` in this section).
Harmonic mean	$\hat{f}(x,y) = \frac{mn}{\sum_{(s,t)\in S_{xy}} \frac{1}{g(s,t)}}$	This nonlinear filter is implemented using function `harmean` (see custom function `spfilt` in this section).
Contraharmonic mean	$\hat{f}(x,y) = \frac{\sum_{(s,t)\in S_{xy}} g(s,t)^{Q+1}}{\sum_{(s,t)\in S_{xy}} g(s,t)^{Q}}$	This nonlinear filter is implemented using function `charmean` (see custom function `spfilt` in this section).
Median	$\hat{f}(x,y) = \underset{(s,t)\in S_{xy}}{\text{median}} \{g(s,t)\}$	Implemented using IPT function `medfilt2`: `f = medfilt2(g, [m n])`.
Max	$\hat{f}(x,y) = \max_{(s,t)\in S_{xy}} \{g(s,t)\}$	Implemented using IPT function `ordfilt2`: `f = ordfilt2(g, m*n, ones(m, n))`.
Min	$\hat{f}(x,y) = \min_{(s,t)\in S_{xy}} \{g(s,t)\}$	Implemented using IPT function `ordfilt2`: `f = ordfilt2(g, 1, ones(m, n))`.
Midpoint	$\hat{f}(x,y) = \frac{1}{2}\left[\max_{(s,t)\in S_{xy}} \{g(s,t)\} + \min_{(s,t)\in S_{xy}} \{g(s,t)\} \right]$	Implemented as 0.5 times the sum of the max and min filtering operations.
Alpha-trimmed mean	$\hat{f}(x,y) = \frac{1}{mn-d} \sum_{(s,t)\in S_{xy}} g_r(s,t)$	The $d/2$ lowest and $d/2$ highest intensity levels of $g(s,t)$ in S_{xy} are deleted, $g_r(s,t)$ denotes the remaining $mn - d$ pixels in the neighborhood. Implemented using function `alphatrim` (see custom function `spfilt` in this section).

```
%      F = SPFILT(G, 'midpoint', M, N)    Midpoint filtering.
%      F = SPFILT(G, 'atrimmed', M, N, D) Alpha-trimmed mean filtering.
%                                         Parameter D must be a nonnegative
%                                         even integer; its default
%                                         value is D = 2.
%
%   The default values when only G and TYPE are input are M = N = 3,
%   Q = 1.5, and D = 2.

% Process inputs.
if nargin == 2
   m = 3; n = 3; Q = 1.5; d = 2;
elseif nargin == 5
   Q = parameter; d = parameter;
elseif nargin == 4
   Q = 1.5; d = 2;
else
   error('Wrong number of inputs.');
end

% Do the filtering.
switch type
case 'amean'
   w = fspecial('average', [m n]);
   f = imfilter(g, w, 'replicate');
case 'gmean'
   f = gmean(g, m, n);
case 'hmean'
   f = harmean(g, m, n);
case 'chmean'
   f = charmean(g, m, n, Q);
case 'median'
   f = medfilt2(g, [m n], 'symmetric');
case 'max'
   f = ordfilt2(g, m*n, ones(m, n), 'symmetric');
case 'min'
   f = ordfilt2(g, 1, ones(m, n), 'symmetric');
case 'midpoint'
   f1 = ordfilt2(g, 1, ones(m, n), 'symmetric');
   f2 = ordfilt2(g, m*n, ones(m, n), 'symmetric');
   f = imlincomb(0.5, f1, 0.5, f2);
case 'atrimmed'
   if (d < 0) | (d/2 ~= round(d/2))
      error('d must be a nonnegative, even integer.')
   end
   f = alphatrim(g, m, n, d);
otherwise
   error('Unknown filter type.')
end

%-------------------------------------------------------------------%
```

```
function f = gmean(g, m, n)
%  Implements a geometric mean filter.
inclass = class(g);
g = im2double(g);
% Disable log(0) warning.
warning off;
f = exp(imfilter(log(g), ones(m, n), 'replicate')).^(1 / m / n);
warning on;
f = changeclass(inclass, f);
%-------------------------------------------------------------------%
function f = harmean(g, m, n)
%  Implements a harmonic mean filter.
inclass = class(g);
g = im2double(g);
f = m * n ./ imfilter(1./(g + eps), ones(m, n), 'replicate');
f = changeclass(inclass, f);
%-------------------------------------------------------------------%
function f = charmean(g, m, n, q)
%  Implements a contraharmonic mean filter.
inclass = class(g);
g = im2double(g);
f = imfilter(g.^(q+1), ones(m, n), 'replicate');
f = f ./ (imfilter(g.^q, ones(m, n), 'replicate') + eps);
f = changeclass(inclass, f);
%-------------------------------------------------------------------%
function f = alphatrim(g, m, n, d)
%  Implements an alpha-trimmed mean filter.
inclass = class(g);
g = im2double(g);
f = imfilter(g, ones(m, n), 'symmetric');
for k = 1:d/2
   f = imsubtract(f, ordfilt2(g, k, ones(m, n), 'symmetric'));
end
for k = (m*n - (d/2) + 1):m*n
   f = imsubtract(f, ordfilt2(g, k, ones(m, n), 'symmetric'));
end
f = f / (m*n - d);
f = changeclass(inclass, f);
```

EXAMPLE 5.5: Using function `spfilt`.

■ The image in Fig. 5.5(a) is an `uint8` image corrupted by pepper noise only with probability 0.1. This image was generated using the following commands [`f` denotes the original image, which is Fig. 3.18(a)]:

```
>> [M, N] = size(f);
>> R = imnoise2('salt & pepper', M, N, 0.1, 0);
>> c = find(R == 0);
>> gp = f;
>> gp(c) = 0;
```

The image in Fig. 5.5(b), corrupted by salt noise only, was generated using the statements

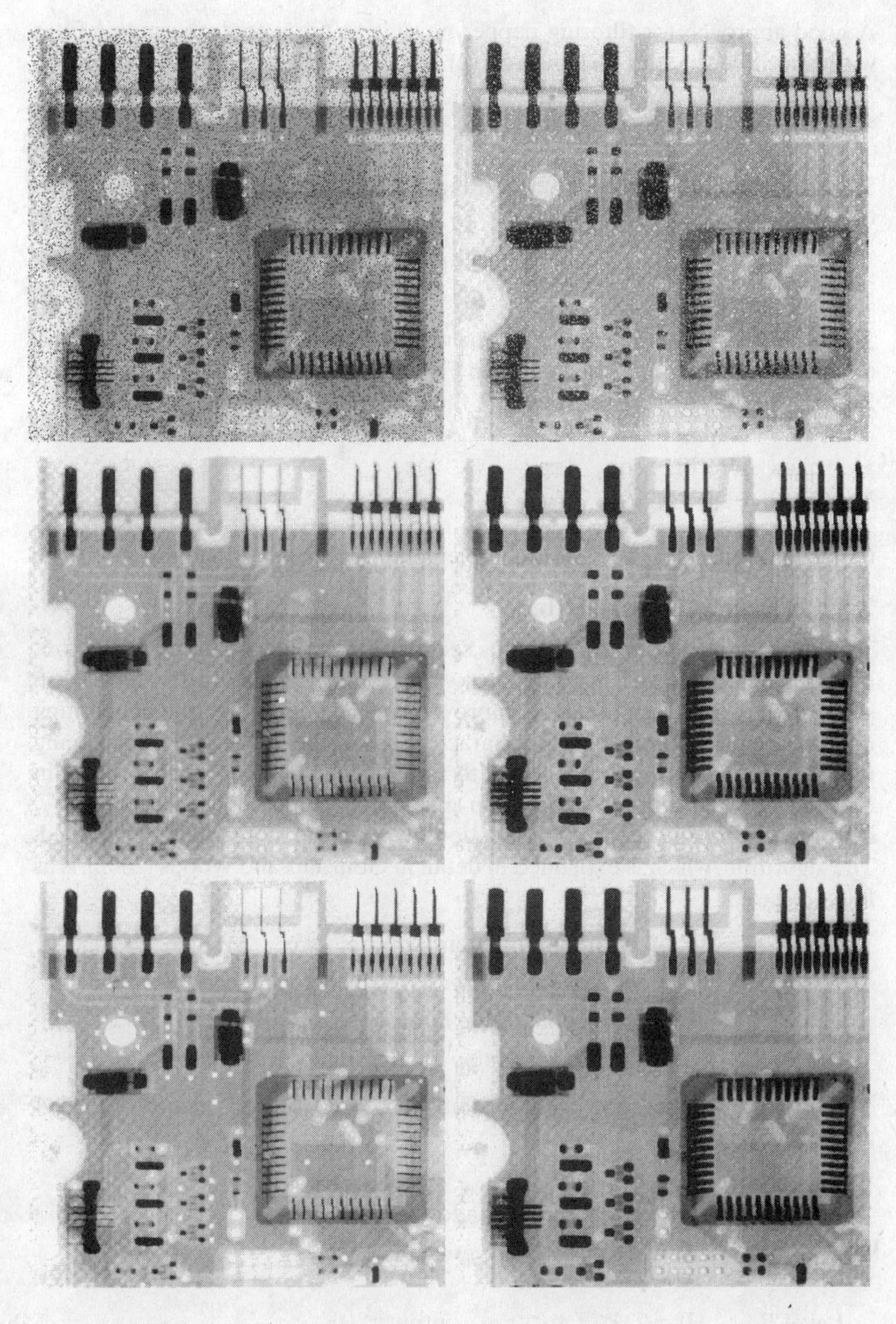

a b
c d
e f

FIGURE 5.5 (a) Image corrupted by pepper noise with probability 0.1. (b) Image corrupted by salt noise with the same probability. (c) Result of filtering (a) with a 3×3 contraharmonic filter of order $Q = 1.5$. (d) Result of filtering (b) with $Q = -1.5$. (e) Result of filtering (a) with a 3×3 max filter. (f) Result of filtering (b) with a 3×3 min filter.

```
>> R = imnoise2('salt & pepper', M, N, 0, 0.1);
>> c = find(R == 1);
>> gs = f;
>> gs(c) = 255;
```

A good approach for filtering pepper noise is to use a contraharmonic filter with a positive value of Q. Figure 5.5(c) was generated using the statement

```
>> fp = spfilt(gp, 'chmean', 3, 3, 1.5);
```

Similarly, salt noise can be filtered using a contraharmonic filter with a negative value of Q:

```
>> fs = spfilt(gs, 'chmean', 3, 3, -1.5);
```

Figure 5.5(d) shows the result. Similar results can be obtained using max and min filters. For example, the images in Figs. 5.5(e) and (f) were generated from Figs. 5.5(a) and (b), respectively, with the following commands:

```
>> fpmax = spfilt(gp, 'max', 3, 3);
>> fsmin = spfilt(gs, 'min', 3, 3);
```

Other solutions using `spfilt` are implemented in a similar manner. ■

5.3.2 Adaptive Spatial Filters

The filters discussed in the previous section are applied to an image without regard for how image characteristics vary from one location to another. In some applications, results can be improved by using filters capable of adapting their behavior depending on the characteristics of the image in the area being filtered. As an illustration of how to implement adaptive spatial filters in MATLAB, we consider in this section an adaptive median filter. As before, S_{xy} denotes a subimage centered at location (x, y) in the image being processed. The algorithm, which is explained in detail in Gonzalez and Woods [2002], is as follows: Let

$$\begin{aligned} z_{\min} &= \text{minimum intensity value in } S_{xy} \\ z_{\max} &= \text{maximum intensity value in } S_{xy} \\ z_{\text{med}} &= \text{median of the intensity values in } S_{xy} \\ z_{xy} &= \text{intensity value at coordinates } (x, y) \end{aligned}$$

The adaptive median filtering algorithm works in two levels, denoted level A and level B:

Level A: If $z_{\min} < z_{\text{med}} < z_{\max}$, go to level B
Else increase the window size
If window size $\leq S_{\max}$, repeat level A
Else output z_{med}

Level B: If $z_{\min} < z_{xy} < z_{\max}$, output z_{xy}
Else output z_{med}

where $S_{\max}$ denotes the maximum allowed size of the adaptive filter window. Another option in the last step in Level A is to output z_{xy} instead of the median. This produces a slightly less blurred result but can fail to detect salt (pepper)

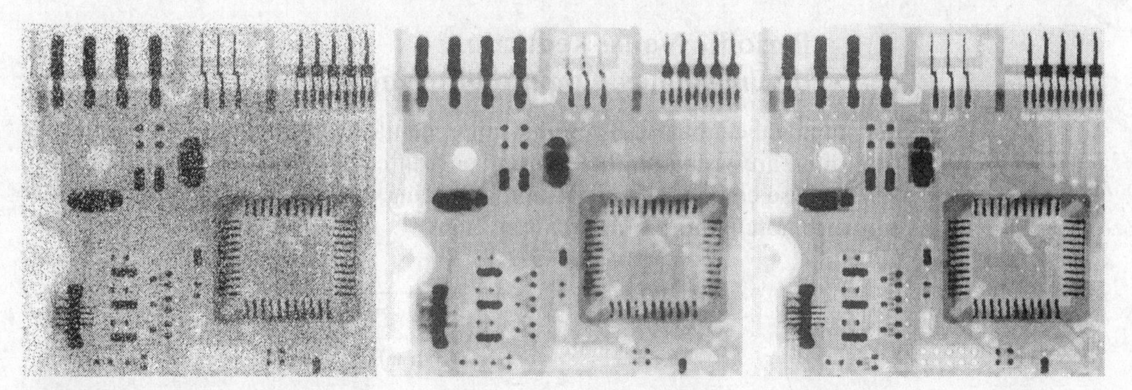

a b c

FIGURE 5.6 (a) Image corrupted by salt-and-pepper noise with density 0.25. (b) Result obtained using a median filter of size 7 × 7. (c) Result obtained using adaptive median filtering with $S_{max} = 7$.

noise embedded in a constant background having the same value as pepper (salt) noise.

An M-function that implements this algorithm, which we call `adpmedian`, is included in Appendix C. The syntax is

```
f = adpmedian(g, Smax)
```

`adpmedian`

where `g` is the image to be filtered, and, as defined above, `Smax` is the maximum allowed size of the adaptive filter window.

EXAMPLE 5.6: Adaptive median filtering.

■ Figure 5.6(a) shows the circuit board image, `f`, corrupted by salt-and-pepper noise, generated using the command

```
>> g = imnoise(f, 'salt & pepper', .25);
```

and Fig. 5.6(b) shows the result obtained using the command (see Section 3.5.2 regarding the use of `medfilt2`):

```
>> f1 = medfilt2(g, [7 7], 'symmetric');
```

This image is reasonably free of noise, but it is quite blurred and distorted (e.g., see the connector fingers in the top middle of the image). On the other hand, the command

```
>> f2 = adpmedian(g, 7);
```

yielded the image in Fig. 5.6(c), which is also reasonably free of noise, but is considerably less blurred and distorted than Fig. 5.6(b). ■

5.4 Periodic Noise Reduction by Frequency Domain Filtering

As noted in Section 5.2.3, periodic noise manifests itself as impulse-like bursts that often are visible in the Fourier spectrum. The principal approach for filtering these components is via notch filtering. The transfer function of a Butterworth notch filter of order n is given by

$$H(u, v) = \frac{1}{1 + \left[\frac{D_0^2}{D_1(u, v)D_2(u, v)}\right]^n}$$

where

$$D_1(u, v) = [(u - M/2 - u_0)^2 + (v - N/2 - v_0)^2]^{1/2}$$

and

$$D_2(u, v) = [(u - M/2 + u_0)^2 + (v - N/2 + v_0)^2]^{1/2}$$

where (u_0, v_0) (and by symmetry) $(-u_0, -v_0)$ are the locations of the "notches," and D_0 is a measure of their radius. Note that the filter is specified with respect to the center of the frequency rectangle, so it must be preprocessed with function `fftshift` prior to its use, as explained in Sections 4.2 and 4.3.

Writing an M-function for notch filtering follows the same principles used in Section 4.5. It is good practice to write the function so that multiple notches can be input, as in the approach used in Section 5.2.3 to generate multiple sinusoidal noise patterns. Once H has been obtained, filtering is done using function `dftfilt` explained in Section 4.3.3.

5.5 Modeling the Degradation Function

When equipment similar to the equipment that generated a degraded image is available, it is generally possible to determine the nature of the degradation by experimenting with various equipment settings. However, relevant imaging equipment availability is the exception, rather than the rule, in the solution of image restoration problems, and a typical approach is to experiment by generating PSFs and testing the results with various restoration algorithms. Another approach is to attempt to model the PSF mathematically. This approach is outside the mainstream of our discussion here; for an introduction to this topic see Gonzalez and Woods [2002]. Finally, when no information is available about the PSF, we can resort to "blind deconvolution" for inferring the PSF. This approach is discussed in Section 5.10. The focus of the remainder of the present section is on various techniques for modeling PSFs by using functions `imfilter` and `fspecial`, introduced in Sections 3.4 and 3.5, respectively, and the various noise-generating functions discussed earlier in this chapter.

One of the principal degradations encountered in image restoration problems is image blur. Blur that occurs with the scene and sensor at rest with respect to each other can be modeled by spatial or frequency domain lowpass filters. Another important degradation model is image blur due to uniform linear motion between the sensor and scene during image acquisition. Image blur can be modeled using IPT function `fspecial`:

```
PSF = fspecial('motion', len, theta)
```

This call to `fspecial` returns a PSF that approximates the effects of linear motion of a camera by `len` pixels. Parameter `theta` is in degrees, measured with respect to the positive horizontal axis in a counter-clockwise direction. The default value of `len` is 9 and the default `theta` is 0, which corresponds to motion of 9 pixels in the horizontal direction.

We use function `imfilter` to create a degraded image with a `PSF` that is either known or is computed by using the method just described:

```
>> g = imfilter(f, PSF, 'circular');
```

where `'circular'` (Table 3.2) is used to reduce border effects. We then complete the degraded image model by adding noise, as appropriate:

```
>> g = g + noise;
```

where `noise` is a random noise image of the same size as `g`, generated using one of the methods discussed in Section 5.2.

When comparing in a given situation the suitability of the various approaches discussed in this and the following sections, it is useful to use the same image or test pattern so that comparisons are meaningful. The test pattern generated by function `checkerboard` is particularly useful for this purpose because its size can be scaled without affecting its principal features. The syntax is

```
C = checkerboard(NP, M, N)
```

where `NP` is the number of pixels on the side of each square, `M` is the number of rows, and `N` is the number of columns. If `N` is omitted, it defaults to `M`. If both `M` and `N` are omitted, a square checkerboard with 8 squares on the side is generated. If, in addition, `NP` is omitted, it defaults to 10 pixels. The light squares on the left half of the checkerboard are white. The light squares on the right half of the checkerboard are gray. To generate a checkerboard in which all light squares are white we use the command

```
>> K = im2double(checkerboard(NP, M, N)) > 0.5;
```

Using the > operator produces a `logical` *result;* `im2double` *is used to produce an image of class* `double`, *which is consistent with the output format of function* `checkerboard`.

The images generated by function `checkerboard` are of class `double` with values in the range [0, 1].

Because some restoration algorithms are slow for large images, a good approach is to experiment with small images to reduce computation time and thus improve interactivity. In this case, it is useful for display purposes to be able to zoom an image by pixel replication. The following function does this (see Appendix C for the code):

pixeldup

```
B = pixeldup(A, m, n)
```

This function duplicates every pixel in A a total of m times in the vertical direction and n times in the horizontal direction. If n is omitted, it defaults to m.

EXAMPLE 5.7: Modeling a blurred, noisy image.

■ Figure 5.7(a) shows a checkerboard image generated by the command

```
>> f = checkerboard(8);
```

The degraded image in Fig. 5.7(b) was generated using the commands

```
>> PSF = fspecial('motion', 7, 45);
>> gb = imfilter(f, PSF, 'circular');
```

a b
c d

FIGURE 5.7 (a) Original image. (b) Image blurred using fspecial with len = 7, and theta = −45 degrees. (c) Noise image. (d) Sum of (b) and (c).

Note that the PSF is just a spatial filter. Its values are

```
>> PSF
PSF =
         0         0         0         0         0    0.0145         0
         0         0         0         0    0.0376    0.1283    0.0145
         0         0         0    0.0376    0.1283    0.0376         0
         0         0    0.0376    0.1283    0.0376         0         0
         0    0.0376    0.1283    0.0376         0         0         0
    0.0145    0.1283    0.0376         0         0         0         0
         0    0.0145         0         0         0         0         0
```

The noisy pattern in Fig. 5.7(c) was generated using the command

```
>> noise = imnoise(zeros(size(f)), 'gaussian', 0, 0.001);
```

Normally, we would have added noise to gb directly using `imnoise(gb, 'gaussian', 0, 0.001)`. However, the noise image is needed later in this chapter, so we computed it separately here.

The blurred noisy image in Fig. 5.7(d) was generated as

```
>> g = gb + noise;
```

The noise is not easily visible in this image because its maximum value is on the order of 0.15, whereas the maximum value of the image is 1. As shown in Sections 5.7 and 5.8, however, this level of noise is not insignificant when attempting to restore g. Finally, we point out that all images in Fig. 5.7 were zoomed to size 512×512 and displayed using a command of the form

```
>> imshow(pixeldup(f, 8), [ ])
```

The image in Fig. 5.7(d) is restored in Examples 5.8 and 5.9. ■

5.6 Direct Inverse Filtering

The simplest approach we can take to restoring a degraded image is to form an estimate of the form

$$\hat{F}(u, v) = \frac{G(u, v)}{H(u, v)}$$

and then obtain the corresponding estimate of the image by taking the inverse Fourier transform of $\hat{F}(u, v)$ [recall that $G(u, v)$ is the Fourier transform of the degraded image]. This approach is appropriately called *inverse filtering*. From the model discussed in Section 5.1, we can express our estimate as

$$\hat{F}(u, v) = F(u, v) + \frac{N(u, v)}{H(u, v)}$$

This deceptively simple expression tells us that, even if we knew $H(u, v)$ exactly, we could not recover $F(u, v)$ [and hence the original, undegraded image $f(x, y)$] because the noise component is a random function whose Fourier transform, $N(u, v)$, is not known. In addition, there usually is a problem in practice with function $H(u, v)$ having numerous zeros. Even if the term $N(u, v)$ were negligible, dividing it by vanishing values of $H(u, v)$ would dominate restoration estimates.

The typical approach when attempting inverse filtering is to form the ratio $\hat{F}(u, v) = G(u, v)/H(u, v)$ and then limit the frequency range for obtaining the inverse, to frequencies "near" the origin. The idea is that zeros in $H(u, v)$ are less likely to occur near the origin because the magnitude of the transform typically is at its highest value in that region. There are numerous variations of this basic theme, in which special treatment is given at values of (u, v) for which H is zero or near zero. This type of approach sometimes is called *pseudoinverse* filtering. In general, approaches based on inverse filtering of this type are seldom practical, as Example 5.8 in the next section shows.

5.7 Wiener Filtering

Wiener filtering (after N. Wiener, who first proposed the method in 1942) is one of the earliest and best known approaches to linear image restoration. A Wiener filter seeks an estimate $\hat{f}$ that minimizes the statistical error function

$$e^2 = E\{(f - \hat{f})^2\}$$

where E is the expected value operator and f is the undegraded image. The solution to this expression in the frequency domain is

$$\hat{F}(u, v) = \left[\frac{1}{H(u, v)} \frac{|H(u, v)|^2}{|H(u, v)|^2 + S_\eta(u, v)/S_f(u, v)}\right] G(u, v)$$

where

$H(u, v)$ = the degradation function

$|H(u, v)|^2 = H^*(u, v)H(u, v)$

$H^*(u, v)$ = the complex conjugate of $H(u, v)$

$S_\eta(u, v) = |N(u, v)|^2$ = the power spectrum of the noise

$S_f(u, v) = |F(u, v)|^2$ = the power spectrum of the undegraded image

The ratio $S_\eta(u, v)/S_f(u, v)$ is called the *noise-to-signal power ratio*. We see that if the noise power spectrum is zero for all relevant values of u and v, this ratio becomes zero and the Wiener filter reduces to the inverse filter discussed in the previous section.

Two related quantities of interest are the average noise power and the average image power, defined as

$$\eta_A = \frac{1}{MN} \sum_u \sum_v S_\eta(u, v)$$

and

$$f_A = \frac{1}{MN} \sum_u \sum_v S_f(u, v)$$

where M and N denote the vertical and horizontal sizes of the image and noise arrays, respectively. These quantities are scalar constants, and their ratio,

$$R = \frac{\eta_A}{f_A}$$

which is also a scalar, is used sometimes to generate a constant array in place of the function $S_\eta(u, v)/S_f(u, v)$. In this case, even if the actual ratio is not known, it becomes a simple matter to experiment interactively varying the constant and viewing the restored results. This, of course, is a crude approximation that assumes that the functions are constant. Replacing $S_\eta(u, v)/S_f(u, v)$ by a constant array in the preceding filter equation results in the so-called *parametric Wiener filter*. As illustrated in Example 5.8, the simple act of using a constant array can yield significant improvements over direct inverse filtering.

Wiener filtering is implemented in IPT using function `deconvwnr`, which has three possible syntax forms. In all these forms, `g` denotes the degraded image and `fr` is the restored image. The first syntax form,

```
fr = deconvwnr(g, PSF)
```

deconvwnr

assumes that the noise-to-signal ratio is zero. Thus, this form of the Wiener filter is the inverse filter mentioned in Section 5.6. The syntax

```
fr = deconvwnr(g, PSF, NSPR)
```

assumes that the noise-to-signal power ratio is known, either as a constant or as an array; the function accepts either one. This is the syntax used to implement the parametric Wiener filter, in which case `NSPR` would be an interactive scalar input. Finally, the syntax

```
fr = deconvwnr(g, PSF, NACORR, FACORR)
```

assumes that autocorrelation functions, `NACORR` and `FACORR`, of the noise and undegraded image are known. Note that this form of `deconvwnr` uses the autocorrelation of η and f instead of the power spectrum of these functions. From the correlation theorem we know that

$$|F(u, v)|^2 = \Im[f(x, y) \circ f(x, y)]$$

where " $\circ$ " denotes the correlation operation and $\Im$ denotes the Fourier transform. This expression indicates that we can obtain the autocorrelation function, $f(x, y) \circ f(x, y)$, for use in `deconvwnr` by computing the inverse Fourier transform of the power spectrum. Similar comments hold for the autocorrelation of the noise.

If the restored image exhibits ringing introduced by the discrete Fourier transform used in the algorithm, it sometimes helps to use function `edgetaper` prior to calling `deconvwnr`. The syntax is

edgetaper

```
J = edgetaper(I, PSF)
```

This function blurs the edges of the input image, `I`, using the point spread function, `PSF`. The output image, `J`, is the weighted sum of `I` and its blurred version. The weighting array, determined by the autocorrelation function of `PSF`, makes `J` equal to `I` in its central region, and equal to the blurred version of `I` near the edges.

EXAMPLE 5.8: Using function `deconvwnr` to restore a blurred, noisy image.

■ Figure 5.8(a) is the same as Fig. 5.7(d), and Fig. 5.8(b) was obtained using the command

```
>> fr1 = deconvwnr(g, PSF);
```

a b
c d

FIGURE 5.8 (a) Blurred, noisy image. (b) Result of inverse filtering. (c) Result of Wiener filtering using a constant ratio. (d) Result of Wiener filtering using autocorrelation functions.

where g is the corrupted image and PSF is the point spread function computed in Example 5.7. As noted earlier in this section, fr1 is the result of direct inverse filtering and, as expected, the result is dominated by the effects of noise. (As in Example 5.7, all displayed images were processed with pixeldup to zoom their size to 512 × 512 pixels.)

The ratio, R, discussed earlier in this section, was obtained using the original and noise images from Example 5.7:

```
>> Sn = abs(fft2(noise)).^2;                 % noise power spectrum
>> nA = sum(Sn(:))/prod(size(noise));  % noise average power
>> Sf = abs(fft2(f)).^2;                      % image power spectrum
>> fA = sum(Sf(:))/prod(size(f));         % image average power
>> R = nA/fA;
```

To restore the image using this ratio we write

```
>> fr2 = deconvwnr(g, PSF, R);
```

As Fig. 5.8(c) shows, this approach gives a significant improvement over direct inverse filtering.

Finally, we use the autocorrelation functions in the restoration (note the use of fftshift for centering):

```
>> NCORR = fftshift(real(ifft2(Sn)));
>> ICORR = fftshift(real(ifft2(Sf)));
>> fr3 = deconvwnr(g, PSF, NCORR, ICORR);
```

As Fig. 5.8(d) shows, the result is close to the original, although some noise is still evident. Because the original image and noise functions were known, we were able to estimate the correct parameters, and Fig. 5.8(d) is the best that can be accomplished with Wiener deconvolution in this case. The challenge in practice, when one (or more) of these quantities is not known, is the intelligent choice of functions used in experimenting, until an acceptable result is obtained. ■

5.8 Constrained Least Squares (Regularized) Filtering

Another well-established approach to linear restoration is *constrained least squares filtering*, called *regularized filtering* in IPT documentation. The definition of 2-D *discrete convolution* is

$$h(x, y)*f(x, y) = \frac{1}{MN}\sum_{m=0}^{M-1}\sum_{n=0}^{N-1} f(m, n)h(x - m, y - n)$$

Using this equation, we can express the linear degradation model discussed in Section 5.1, $g(x, y) = h(x, y)*f(x, y) + \eta(x, y)$, in vector-matrix form, as

$$\mathbf{g} = \mathbf{Hf} + \boldsymbol{\eta}$$

For example, suppose that $g(x, y)$ is of size $M \times N$. Then we can form the first N elements of the vector $\mathbf{g}$ by using the image elements in the first row of $g(x, y)$, the next N elements from the second row, and so on. The resulting vector will have dimensions $MN \times 1$. These also are the dimensions of $\mathbf{f}$ and $\boldsymbol{\eta}$, as these vectors are formed in the same manner. The matrix $\mathbf{H}$ then has dimensions $MN \times MN$. Its elements are given by the elements of the preceding convolution equation.

It would be reasonable to arrive at the conclusion that the restoration problem can now be reduced to simple matrix manipulations. Unfortunately, this is not the case. For instance, suppose that we are working with images of medium size; say $M = N = 512$. Then the vectors in the preceding matrix equation would be of dimension $262{,}144 \times 1$, and matrix $\mathbf{H}$ would be of dimensions $262{,}144 \times 262{,}144$. Manipulating vectors and matrices of these sizes is not a trivial task. The problem is complicated further by the fact that the inverse of $\mathbf{H}$ does not always exist due to zeros in the transfer function (see Section 5.6). However, formulating the restoration problem in matrix form does facilitate derivation of restoration techniques.

Although we do not derive the method of constrained least squares that we are about to present, central to this method is the issue of the sensitivity of the inverse of $\mathbf{H}$ mentioned in the previous paragraph. One way to deal with this issue is to base optimality of restoration on a measure of smoothness, such as the second derivative of an image (e.g., the Laplacian). To be meaningful, the restoration must be constrained by the parameters of the problem at hand. Thus, what is desired is to find the minimum of a criterion function, C, defined as

$$C = \sum_{x=0}^{M-1} \sum_{y=0}^{N-1} \left[\nabla^2 f(x, y)\right]^2$$

subject to the constraint

$$\|\mathbf{g} - \mathbf{H}\hat{\mathbf{f}}\|^2 = \|\boldsymbol{\eta}\|^2$$

where $\|\mathbf{w}\|^2 \triangleq \mathbf{w}^T\mathbf{w}$ is the Euclidean vector norm,† $\hat{\mathbf{f}}$ is the estimate of the undegraded image, and the Laplacian operator ∇^2 is as defined in Section 3.5.1.

The frequency domain solution to this optimization problem is given by the expression

$$\hat{F}(u, v) = \left[\frac{H^*(u, v)}{|H(u, v)|^2 + \gamma|P(u, v)|^2}\right] G(u, v)$$

where γ is a parameter that must be adjusted so that the constraint is satisfied (if γ is zero we have an inverse filter solution), and $P(u, v)$ is the Fourier transform of the function

---

†For a column vector $\mathbf{w}$ with n components, $\mathbf{w}^T\mathbf{w} = \sum_{k=1}^{n} w_k^2$, where w_k, is the kth component of $\mathbf{w}$.

$$p(x, y) = \begin{bmatrix} 0 & 1 & 0 \\ 1 & -4 & 1 \\ 0 & 1 & 0 \end{bmatrix}$$

We recognize this function as the Laplacian operator introduced in Section 3.5.1. The only unknowns in the preceding formulation are γ and $\|\boldsymbol{\eta}\|^2$. However, it can be shown that γ can be found iteratively if $\|\boldsymbol{\eta}\|^2$, which is proportional to the noise power (a scalar), is known.

Constrained least squares filtering is implemented in IPT by function `deconvreg`, which has the syntax

```
fr = deconvreg(g, PSF, NOISEPOWER, RANGE)
```

where `g` is the corrupted image, `fr` is the restored image, `NOISEPOWER` is proportional to $\|\boldsymbol{\eta}\|^2$, and `RANGE` is the range of values where the algorithm is limited to look for a solution for γ. The default range is $[10^{-9}, 10^{9}]$ (`[1e-10, 1e10]` in MATLAB notation). If the last two parameters are excluded from the argument, `deconvreg` produces an inverse filter solution. A good starting estimate for `NOISEPOWER` is $MN[\sigma_\eta^2 + m_\eta^2]$, where M and N are the dimensions of the image and the parameters inside the brackets are the noise variance and noise squared mean. This estimate is simply a starting point and, as the next example shows, the final value used can be quite different.

EXAMPLE 5.9: Using function `deconvreg` to restore a blurred, noisy image.

■ We now restore the image in Fig. 5.7(d) using `deconvreg`. The image is of size 64×64 and we know from Example 5.7 that the noise has a variance of 0.001 and zero mean. So, our initial estimate of `NOISEPOWER` is $(64)^2[0.001 - 0] \approx 4$. Figure 5.9(a) shows the result of using the command

```
>> fr = deconvreg(g, PSF, 4);
```

a b

FIGURE 5.9 (a) The image in Fig. 5.7(d) restored using a regularized filter with `NOISEPOWER` equal to 4. (b) The same image restored with `NOISEPOWER` equal to 0.4 and a `RANGE` of `[1e-7 1e7]`.

where g and PSF are from Example 5.7. The image was improved somewhat from the original, but obviously this is not a particularly good value for NOISEPOWER. After some experimenting with this parameter and parameter RANGE, we arrived at the result in Fig. 5.9(b), which was obtained using the command

```
>> fr = deconvreg(g, PSF, 0.4, [1e-7 1e7]);
```

Thus we see that we had to go down one order of magnitude on NOISEPOWER, and RANGE was tighter than the default. The Wiener filtering result in Fig. 5.8(d) is much better, but we obtained that result with full knowledge of the noise and image spectra. Without that information, the results obtainable by experimenting with the two filters often are comparable. ■

If the restored image exhibits ringing introduced by the discrete Fourier transform used in the algorithm, it usually helps to use function edgetaper (see Section 5.7) prior to calling deconvreg.

5.9 Iterative Nonlinear Restoration Using the Lucy-Richardson Algorithm

The image restoration methods discussed in the previous three sections are linear. They also are "direct" in the sense that, once the restoration filter is specified, the solution is obtained via one application of the filter. This simplicity of implementation, coupled with modest computational requirements and a well-established theoretical base, have made linear techniques a fundamental tool in image restoration for many years.

During the past two decades, nonlinear iterative techniques have been gaining acceptance as restoration tools that often yield results superior to those obtained with linear methods. The principal objections to nonlinear methods are that their behavior is not always predictable and that they generally require significant computational resources. The first objection often loses importance based on the fact that nonlinear methods have been shown to be superior to linear techniques in a broad spectrum of applications (Jansson [1997]). The second objection has become less of an issue due to the dramatic increase in inexpensive computing power over the last decade. The nonlinear method of choice in the toolbox is a technique developed by Richardson [1972] and by Lucy [1974], working independently. The toolbox refers to this method as the Lucy-Richardson (L-R) algorithm, but we also see it quoted in the literature as the Richardson-Lucy algorithm.

The L-R algorithm arises from a maximum-likelihood formulation (see Section 5.10) in which the image is modeled with Poisson statistics. Maximizing the likelihood function of the model yields an equation that is satisfied when the following iteration converges:

$$\hat{f}_{k+1}(x, y) = \hat{f}_k(x, y)\left[h(-x, -y) * \frac{g(x, y)}{h(x, y) * \hat{f}_k(x, y)}\right]$$

As before, "*" indicates convolution, $\hat{f}$ is the estimate of the undegraded image, and both g and h are as defined in Section 5.1. The iterative nature of the algorithm is evident. Its nonlinear nature arises from the division by $\hat{f}$ on the right side of the equation.

As with most nonlinear methods, the question of when to stop the L-R algorithm is difficult to answer in general. The approach often followed is to observe the output and stop the algorithm when a result acceptable in a given application has been obtained.

The L-R algorithm is implemented in IPT by function `deconvlucy`, which has the basic syntax

```
fr = deconvlucy(g, PSF, NUMIT, DAMPAR, WEIGHT)
```

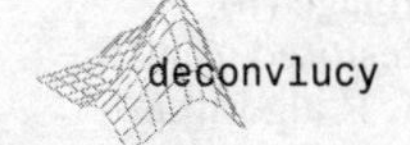

where `fr` is the restored image, `g` is the degraded image, `PSF` is the point spread function, `NUMIT` is the number of iterations (the default is 10), and `DAMPAR` and `WEIGHT` are defined as follows.

`DAMPAR` is a scalar that specifies the threshold deviation of the resulting image from image `g`. Iterations are suppressed for the pixels that deviate within the `DAMPAR` value from their original value. This suppresses noise generation in such pixels, preserving necessary image details. The default is 0 (no damping).

`WEIGHT` is an array of the same size as `g` that assigns a weight to each pixel to reflect its quality. For example, a bad pixel resulting from a defective imaging array can be excluded from the solution by assigning to it a zero weight value. Another useful application of this array is to let it adjust the weights of the pixels according to the amount of flat-field correction that may be necessary based on knowledge of the imaging array. When simulating blurring with a specified PSF (see Example 5.7), `WEIGHT` can be used to eliminate from computation pixels that are on the border of an image and thus are blurred differently by the PSF. If the PSF is of size $n \times n$, the border of zeros used in `WEIGHT` is of width `ceil(n/2)`. The default is a unit array of the same size as input image `g`.

If the restored image exhibits ringing introduced by the discrete Fourier transform used in the algorithm, it sometimes helps to use function `edgetaper` (see Section 5.7) prior to calling `deconvlucy`.

EXAMPLE 5.10: Using function `deconvlucy` to restore a blurred, noisy image.

■ Figure 5.10(a) shows an image generated using the command

```
>> f = checkerboard(8);
```

which produced a square image of size 64×64 pixels. As before, the size of the image was increased to size 512×512 for display purposes by using function `pixeldup`:

```
>> imshow(pixeldup(f, 8));
```

a b
c d
e f

FIGURE 5.10 (a) Original image. (b) Image blurred and corrupted by Gaussian noise. (c) through (f) Image (b) restored using the L-R algorithm with 5, 10, 20, and 100 iterations, respectively.

The following command generated a Gaussian PSF of size 7 × 7 with a standard deviation of 10:

```
>> PSF = fspecial('gaussian', 7, 10);
```

Next, we blurred image `f` using `PDF` and added to it Gaussian noise of zero mean and standard deviation of 0.01:

```
>> SD = 0.01;
>> g = imnoise(imfilter(f, PSF), 'gaussian', 0, SD^2);
```

Figure 5.10(b) shows the result.

The remainder of this example deals with restoring image `g` using function `deconvlucy`. For `DAMPAR` we specified a value equal to 10 times `SD`:

```
>> DAMPAR = 10*SD;
```

Array `WEIGHT` was created using the approach discussed in the preceding explanation of this parameter:

```
>> LIM = ceil(size(PSF, 1)/2);
>> WEIGHT = zeros(size(g));
>> WEIGHT(LIM + 1:end - LIM, LIM + 1:end - LIM) = 1;
```

Array `WEIGHT` is of size 64 × 64 with a border of 0s 4 pixels wide; the rest of the pixels are 1s.

The only variable left is `NUMIT`, the number of iterations. Figure 5.10(c) shows the result obtained using the commands

```
>> NUMIT = 5;
>> fr = deconvlucy(g, PSF, NUMIT, DAMPAR, WEIGHT);
>> imshow(pixeldup(fr, 8))
```

Although the image has improved somewhat, it is still blurry. Figures 5.10(d) and (e) show the results obtained using `NUMIT = 10` and `20`. The latter result is a reasonable restoration of the blurred, noisy image. In fact, further increases in the number of iterations did not produce dramatic improvements in the restored result. For example, Fig. 5.10(f) was obtained using 100 iterations. This image is only slightly sharper and brighter than the result obtained using 20 iterations. The thin black border seen in all results was caused by the 0s in array `WEIGHT`. ■

5.10 Blind Deconvolution

One of the most difficult problems in image restoration is obtaining a suitable estimate of the PSF to use in restoration algorithms such as those discussed in the preceding sections. As noted earlier, image restoration methods that are not based on specific knowledge of the PSF are called *blind deconvolution* algorithms.

An approach to blind deconvolution that has received significant attention over the past two decades is based on maximum-likelihood estimation (MLE), an optimization strategy used for obtaining estimates of quantities corrupted by random noise. Briefly, an interpretation of MLE is to think of image data as random quantities having a certain likelihood of being produced from a family of other possible random quantities. The likelihood function is expressed in terms of $g(x, y)$, $f(x, y)$, and $h(x, y)$ (see Section 5.1), and the problem then is to find the maximum of the likelihood function. In blind deconvolution the optimization problem is solved iteratively with specified constraints and, assuming convergence, the specific $f(x, y)$ and $h(x, y)$ that result in a maximum are the restored image *and* the PSF.

A derivation of MLE blind deconvolution is outside the scope of the present discussion, but the reader can gain a solid understanding of this area by consulting the following references: For background on maximum-likelihood estimation, see the classic book by Van Trees [1968]. For a review of some of the original image-processing work in this area see Dempster et al. [1977], and for some of its later extensions see Holmes [1992]. A good general reference book on deconvolution is Jansson [1997]. For detailed examples on the use of deconvolution in microscopy and in astronomy, see Holmes et al. [1995] and Hanisch et al. [1997], respectively.

The toolbox performs blind deconvolution via function `deconvblind`, which has the basic syntax

deconvblind

```
[fr, PSFe] = deconvblind(g, INITPSF)
```

where `g` is the degraded image, `INITPSF` is an initial estimate of the point spread function, `PSFe` is the final computed estimate of this function, and `fr` is the image restored using the estimated PSF. The algorithm used to obtain the restored image is the L-R iterative restoration algorithm explained in Section 5.9. The PSF estimation is affected strongly by the size of its initial guess, and less by its values (an array of 1s is a reasonable starting guess).

The number of iterations performed with the preceding syntax is 10 by default. Additional parameters may be included in the function to control the number of iterations and other features of the restoration, as in the following syntax:

```
[fr, PSFe] = deconvblind(g, INITPSF, NUMIT, DAMPAR, WEIGHT)
```

where `NUMIT`, `DAMPAR`, and `WEIGHT` are as described for the L-R algorithm in the previous section.

If the restored image exhibits ringing introduced by the discrete Fourier transform used in the algorithm, it sometimes helps to use function `edgetaper` (see Section 5.7) prior to calling `deconvblind`.

EXAMPLE 5.11: Using function `deconvblind` to estimate a PSF.

■ Figure 5.11(a) is the PSF used to generate the degraded image in Fig. 5.10(b):

```
>> PSF = fspecial('gaussian', 7, 10);
>> imshow(pixeldup(PSF, 73), [ ]);
```

a b
c d

FIGURE 5.11 (a) Original PSF. (b) through (d) Estimates of the PSF using 5, 10, and 20 iterations in function `deconvblind`.

As in Example 5.10, the degraded image in question was obtained with the commands

```
>> SD = 0.01;
>> g = imnoise(imfilter(f, PSF), 'gaussian', 0, SD^2);
```

In the present example we are interested in using function `deconvblind` to obtain an estimate of the PSF, given only the degraded image `g`. Figure 5.11(b) shows the PSF resulting from the following commands:

```
>> INITPSF = ones(size(PSF));
>> NUMIT = 5;
>> [fr, PSFe] = deconvblind(g, INITPSF, NUMIT, DAMPAR, WEIGHT);
>> imshow(pixeldup(PSFe, 73), [ ]);
```

where we used the same values as in Example 5.10 for `DAMPAR` and `WEIGHT`.

Figures 5.11(c) and (d), displayed in the same manner as `PSFe`, show the PSFs obtained with 10, and 20 iterations, respectively. The latter result is close to the true PSF in Fig. 5.11(a). ■

5.11 Geometric Transformations and Image Registration

We conclude this chapter with an introduction to geometric transformations for image restoration. Geometric transformations modify the spatial relationship between pixels in an image. They are often called *rubber-sheet transformations* because they may be viewed as printing an image on a sheet of rubber and then stretching this sheet according to a predefined set of rules.

Geometric transformations are used frequently to perform *image registration*, a process that takes two images of the same scene and aligns them so they can be merged for visualization, or for quantitative comparison. In the following sections, we discuss (1) spatial transformations and how to define and visualize them in MATLAB; (2) how to apply spatial transformations to images; and (3) how to determine spatial transformations for use in image registration.

5.11.1 Geometric Spatial Transformations

Suppose that an image, f, defined over a (w, z) coordinate system, undergoes geometric distortion to produce an image, g, defined over an (x, y) coordinate system. This transformation (of the coordinates) may be expressed as

$$(x, y) = T\{(w, z)\}$$

For example, if $(x, y) = T\{(w, v)\} = (w/2, z/2)$, the "distortion" is simply a shrinking of f by half in both spatial dimensions, as illustrated in Fig. 5.12.

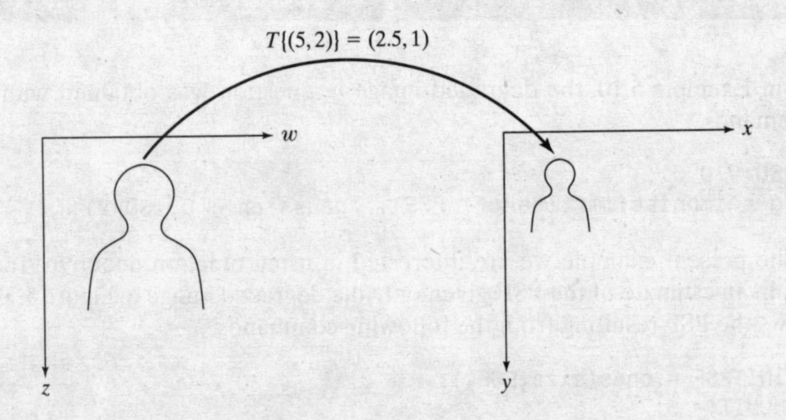

FIGURE 5.12 A simple spatial transformation. (Note that the xy-axes in this figure do not correspond to the image axis coordinate system defined in Section 2.1.1. As mentioned in that section, IPT on occasion uses the so-called spatial coordinate system in which y designates rows and x designates columns. This is the system used throughout this section in order to be consistent with IPT documentation on the topic of geometric transformations.)

One of the most commonly used forms of spatial transformations is the *affine transform* (Wolberg [1990]). The affine transform can be written in matrix form as

$$[x \quad y \quad 1] = [w \quad z \quad 1]\,\mathbf{T} = [w \quad z \quad 1]\begin{bmatrix} t_{11} & t_{12} & 0 \\ t_{21} & t_{22} & 0 \\ t_{31} & t_{32} & 1 \end{bmatrix}$$

This transformation can scale, rotate, translate, or shear a set of points, depending on the values chosen for the elements of **T**. Table 5.3 shows how to choose the values of the elements to achieve different transformations.

IPT represents spatial transformations using a so-called *tform structure*. One way to create such a structure is by using function `maketform`, whose calling syntax is

See Sections 2.10.6 and 11.1.1 for a discussion of structures.

```
tform = maketform(transform_type, transform_parameters)
```

TABLE 5.3 Types of affine transformations.

Type	Affine Matrix, T	Coordinate Equations	Diagram
Identity	$\begin{bmatrix} 1 & 0 & 0 \\ 0 & 1 & 0 \\ 0 & 0 & 1 \end{bmatrix}$	$x = w$ $y = z$	
Scaling	$\begin{bmatrix} s_x & 0 & 0 \\ 0 & s_y & 0 \\ 0 & 0 & 1 \end{bmatrix}$	$x = s_x w$ $y = s_y z$	
Rotation	$\begin{bmatrix} \cos\theta & \sin\theta & 0 \\ -\sin\theta & \cos\theta & 0 \\ 0 & 0 & 1 \end{bmatrix}$	$x = w\cos\theta - z\sin\theta$ $y = w\sin\theta + z\cos\theta$	
Shear (horizontal)	$\begin{bmatrix} 1 & 0 & 0 \\ \alpha & 1 & 0 \\ 0 & 0 & 1 \end{bmatrix}$	$x = w + \alpha z$ $y = z$	
Shear (vertical)	$\begin{bmatrix} 1 & \beta & 0 \\ 0 & 1 & 0 \\ 0 & 0 & 1 \end{bmatrix}$	$x = w$ $y = \beta w + z$	
Translation	$\begin{bmatrix} 1 & 0 & 0 \\ 0 & 1 & 0 \\ \delta_x & \delta_y & 1 \end{bmatrix}$	$x = w + \delta_x$ $y = z + \delta_y$	

The first input argument, `transform_type`, is one of these strings: `'affine'`, `'projective'`, `'box'`, `'composite'`, or `'custom'`. These transform types are described in Table 5.4, Section 5.11.3. Additional arguments depend on the transform type and are described in detail in the help page for `maketform`.

In this section our interest is on affine transforms. For example, one way to create an affine `tform` is to provide the **T** matrix directly, as in

```
>> T = [2 0 0; 0 3 0; 0 0 1];
>> tform = maketform('affine', T)
tform =
       ndims_in: 2
      ndims_out: 2
    forward_fcn: @fwd_affine
    inverse_fcn: @inv_affine
          tdata: [1 x 1 struct]
```

Although it is not necessary to use the fields of the `tform` structure directly to be able to apply it, information about **T**, as well as about $\mathbf{T}^{-1}$, is contained in the `tdata` field:

```
>> tform.tdata

ans =
       T: [3 x 3 double]
    Tinv: [3 x 3 double]

>> tform.tdata.T
ans =
    2    0    0
    0    3    0
    0    0    1

>> tform.tdata.Tinv
ans =
    0.5000         0         0
         0    0.3333         0
         0         0    1.0000
```

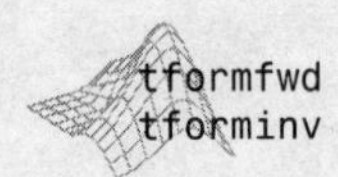

IPT provides two functions for applying a spatial transformation to points: `tformfwd` computes the forward transformation, $T\{(w, z)\}$, and `tforminv` computes the inverse transformation, $T^{-1}\{(x, y)\}$. The calling syntax for `tformfwd` is `XY = tformfwd(WZ, tform)`. Here, `WZ` is a $P \times 2$ matrix of points; each row of `WZ` contains the w and z coordinates of one point. Similarly, `XY` is a $P \times 2$ matrix of points; each row contains the x and y coordinates of a transformed point. For example, the following commands compute the forward transformation of a pair of points, followed by the inverse transform to verify that we get back the original data:

```
>> WZ = [1 1; 3 2];
>> XY = tformfwd(WZ, tform)
XY =
```

```
     2     3
     6     6

>> WZ2 = tforminv(XY, tform)
WZ2 =
     1     1
     3     2
```

To get a better feel for the effects of a particular spatial transformation, it is often useful to see how it transforms a set of points arranged on a grid. The following M-function, vistformfwd, constructs a grid of points, transforms the grid using tformfwd, and then plots the grid and the transformed grid side by side for comparison. Note the combined use of functions meshgrid (Section 2.10.4) and linspace (Section 2.8.1) for creating the grid. The following code also illustrates the use of some of the functions discussed thus far in this section.

vistformfwd

```
function vistformfwd(tform, wdata, zdata, N)
%VISTFORMFWD Visualize forward geometric transform.
%   VISTFORMFWD(TFORM, WRANGE, ZRANGE, N) shows two plots: an N-by-N
%   grid in the W-Z coordinate system, and the spatially transformed
%   grid in the X-Y coordinate system.  WRANGE and ZRANGE are
%   two-element vectors specifying the desired range for the grid. N
%   can be omitted, in which case the default value is 10.

if nargin < 4
   N = 10;
end

% Create the w-z grid and transform it.
[w, z] = meshgrid(linspace(wdata(1), zdata(2), N), ...
                  linspace(wdata(1), zdata(2), N));

wz = [w(:) z(:)];
xy = tformfwd([w(:) z(:)], tform);

% Calculate the minimum and maximum values of w and x,
% as well as z and y. These are used so the two plots can be
% displayed using the same scale.
x = reshape(xy(:, 1), size(w)); % reshape is discussed in Sec. 8.2.2.
y = reshape(xy(:, 2), size(z));
wx = [w(:); x(:)];
wxlimits = [min(wx) max(wx)];
zy = [z(:); y(:)];
zylimits = [min(zy) max(zy)];

% Create the w-z plot.
subplot(1,2,1) % See Section 7.2.1 for a discussion of this function.
plot(w, z, 'b'), axis equal, axis ij
hold on
plot(w', z', 'b')
hold off
xlim(wxlimits)
ylim(zylimits)
```

```
set(gca, 'XAxisLocation', 'top')
xlabel('w'), ylabel('z')

% Create the x-y plot.
subplot(1, 2, 2)
plot(x, y, 'b'), axis equal, axis ij
hold on
plot(x', y', 'b')
hold off
xlim(wxlimits)
ylim(zylimits)
set(gca, 'XAxisLocation', 'top')
xlabel('x'), ylabel('y')
```

EXAMPLE 5.12: Visualizing affine transforms using vistformfwd.

■ In this example we use `vistformfwd` to visualize the effect of several different affine transforms. We also explore an alternate way to create an affine `tform` using `maketform`. We start with an affine transform that scales horizontally by a factor of 3 and vertically by a factor of 2:

```
>> T1 = [3 0 0; 0 2 0; 0 0 1];
>> tform1 = maketform('affine', T1);
>> vistformfwd(tform1, [0 100], [0 100]);
```

Figures 5.13(a) and (b) show the result.

A shearing effect occurs when t_{21} or t_{12} is nonzero in the affine **T** matrix, such as

```
>> T2 = [1 0 0; .2 1 0; 0 0 1];
>> tform2 = maketform('affine', T2);
>> vistformfwd(tform2, [0 100], [0 100]);
```

Figures 5.13(c) and (d) show the effect of the shearing transform on a grid.

An interesting property of affine transforms is that the composition of several affine transforms is also an affine transform. Mathematically, affine transforms can be generated simply by using multiplication of the **T** matrices. The next block of code shows how to generate and visualize an affine transform that is a combination of scaling, rotation, and shear.

```
>> Tscale = [1.5 0 0; 0 2 0; 0 0 1];
>> Trotation = [cos(pi/4) sin(pi/4) 0
                -sin(pi/4) cos(pi/4) 0
                0 0 1];
>> Tshear = [1 0 0; .2 1 0; 0 0 1];
>> T3 = Tscale * Trotation * Tshear;
>> tform3 = maketform('affine', T3);
>> vistformfwd(tform3, [0 100], [0 100])
```

Figures 5.13(e) and (f) show the results. ■

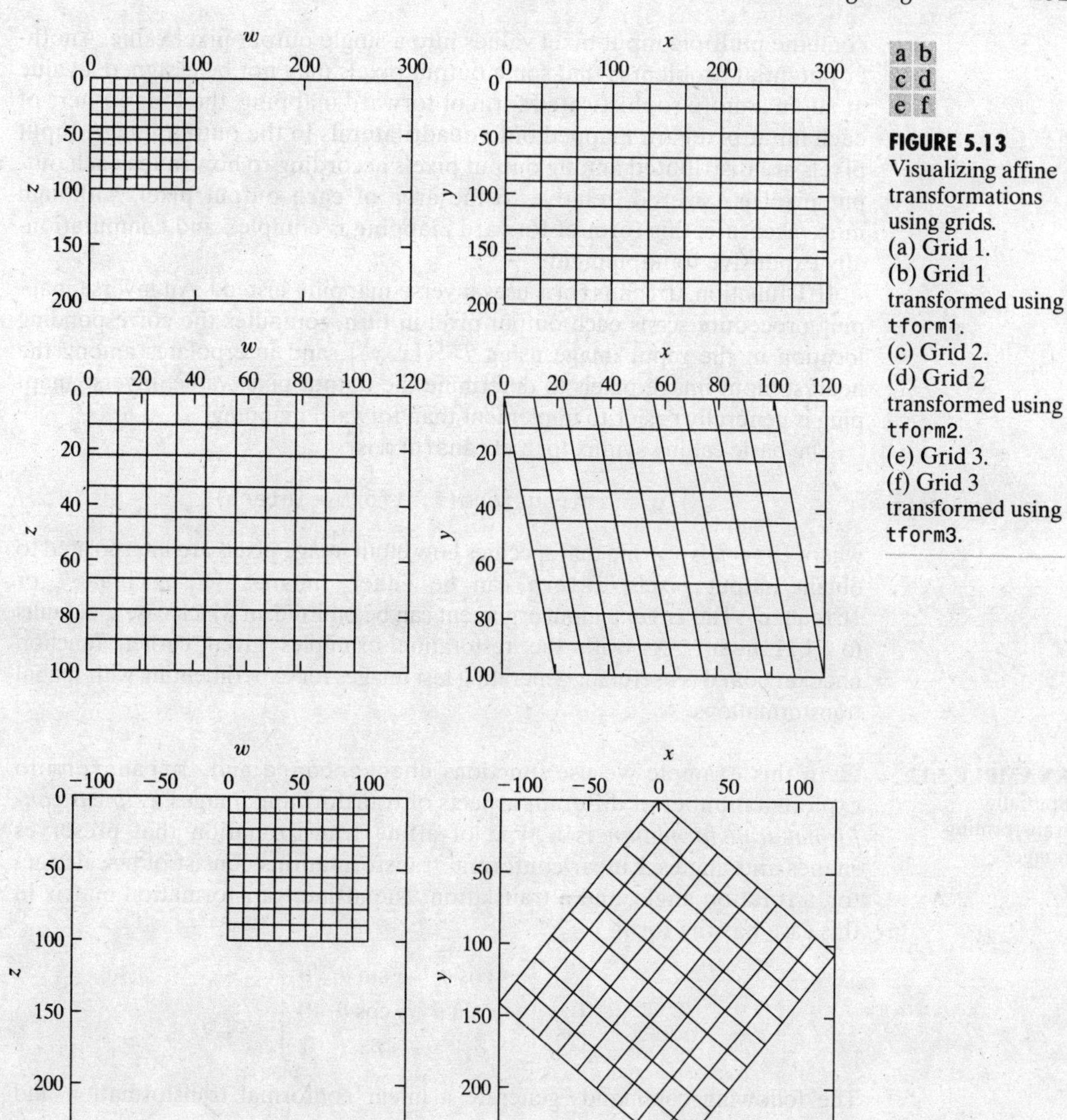

a b
c d
e f

FIGURE 5.13 Visualizing affine transformations using grids. (a) Grid 1. (b) Grid 1 transformed using `tform1`. (c) Grid 2. (d) Grid 2 transformed using `tform2`. (e) Grid 3. (f) Grid 3 transformed using `tform3`.

5.11.2 Applying Spatial Transformations to Images

Most computational methods for spatially transforming an image fall into one of two categories: methods that use *forward mapping*, and methods that use *inverse mapping*. Methods based on forward mapping scan each input pixel in turn, copying its value into the output image at the location determined by $T\{(w, z)\}$. One problem with the forward mapping procedure is that two or more different pixels in the input image could be transformed into the same pixel in the output image, raising the question of how to combine multiple input pixel values into a single output pixel value. Another potential problem is that some output pixels may not be assigned a value at all. In a more sophisticated form of forward mapping, the four corners of each input pixel are mapped onto quadrilaterals in the output image. Input pixels are distributed among output pixels according to how much each output pixel is covered, relative to the area of each output pixel. Although more accurate, this form of forward mapping is complex and computationally expensive to implement.

IPT function `imtransform` uses inverse mapping instead. An inverse mapping procedure scans each output pixel in turn, computes the corresponding location in the input image using $T^{-1}\{(x, y)\}$, and interpolates among the nearest input image pixels to determine the output pixel value. Inverse mapping is generally easier to implement than forward mapping.

The basic calling syntax for `imtransform` is

```
g = imtransform(f, tform, interp)
```

where `interp` is a string that specifies how input image pixels are interpolated to obtain output pixels; `interp` can be either `'nearest'`, `'bilinear'`, or `'bicubic'`. The `interp` input argument can be omitted, in which case it defaults to `'bilinear'`. As with the restoration examples given earlier, function `checkerboard` is useful for generating test images for experimenting with spatial transformations.

EXAMPLE 5.13: Spatially transforming images.

■ In this example we use functions `checkerboard` and `imtransform` to explore a number of different aspects of transforming images. A *linear conformal transformation* is a type of affine transformation that preserves shapes and angles. Linear conformal transformations consist of a scale factor, a rotation angle, and a translation. The affine transformation matrix in this case has the form

$$\mathbf{T} = \begin{bmatrix} s\cos\theta & s\sin\theta & 0 \\ -s\sin\theta & s\cos\theta & 0 \\ \delta_x & \delta_y & 1 \end{bmatrix}$$

The following commands generate a linear conformal transformation and apply it to a test image.

```
>> f = checkerboard(50);
>> s = 0.8;
>> theta = pi/6;
>> T = [s*cos(theta) s*sin(theta) 0
   -s*sin(theta) s*cos(theta) 0
     0  0   1];
>> tform = maketform('affine', T);
>> g = imtransform(f, tform);
```

Figures 5.14(a) and (b) show the original and transformed checkerboard images. The preceding call to `imtransform` used the default interpolation method,

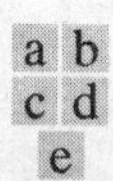

FIGURE 5.14 Affine transformations of the checkerboard image. (a) Original image. (b) Linear conformal transformation using the default interpolation (bilinear). (c) Using nearest neighbor interpolation. (d) Specifying an alternate fill value. (e) Controlling the output space location so that translation is visible.

`'bilinear'`. As mentioned earlier, we can select a different interpolation method, such as nearest neighbor, by specifying it explicitly in the call to `imtransform`:

```
>> g2 = imtransform(f, tform, 'nearest');
```

Figure 5.14(c) shows the result. Nearest neighbor interpolation is faster than bilinear interpolation, and it may be more appropriate in some situations, but it generally produces results inferior to those obtained with bilinear interpolation.

Function `imtransform` has several additional optional parameters that are useful at times. For example, passing it a `FillValue` parameter controls the color `imtransform` uses for pixels outside the domain of the input image:

```
>> g3 = imtransform(f, tform, 'FillValue', 0.5);
```

In Fig. 5.14(d) the pixels outside the original image are mid-gray instead of black.

Other extra parameters can help resolve a common source of confusion regarding translating images using `imtransform`. For example, the following commands perform a pure translation:

```
>> T2 = [1 0 0; 0 1 0; 50 50 1];
>> tform2 = maketform('affine', T2);
>> g4 = imtransform(f, tform2);
```

The result, however, would be identical to the original image in Fig. 5.14(a). This effect is caused by default behavior of `imtransform`. Specifically, `imtransform` determines the bounding box (see Section 11.4.1 for a definition of the term *bounding box*) of the output image in the output coordinate system, and by default it only performs inverse mapping over that bounding box. This effectively *undoes* the translation. By specifying the parameters `XData` and `YData`, we can tell `imtransform` exactly where in output space to compute the result. `XData` is a two-element vector that specifies the location of the left and right columns of the output image; `YData` is a two-element vector that specifies the location of the top and bottom rows of the output image. The following command computes the output image in the region between $(x, y) = (1, 1)$ and $(x, y) = (400, 400)$.

```
>> g5 = imtransform(f, tform2, 'XData', [1 400], 'YData', [1 400], ...
                    'FillValue', 0.5);
```

Figure 5.14(e) shows the result.

Other settings of `imtransform` and related IPT functions provide additional control over the result, particularly over how interpolation is performed. Most of the relevant toolbox documentation is in the help pages for functions `imtransform` and `makeresampler`. ■

5.11.3 Image Registration

Image registration methods seek to align two images of the same scene. For example, it may be of interest to align two or more images taken at roughly the same time, but using different instruments, such as an MRI (magnetic resonance imaging) scan and a PET (positron emission tomography) scan. Or, perhaps the images were taken at different times using the same instrument, such as satellite images of a given location taken several days, months, or even years apart. In either case, combining the images or performing quantitative analysis and comparisons requires compensating for geometric aberrations caused by differences in camera angle, distance, and orientation; sensor resolution; shift in subject position; and other factors.

The toolbox supports image registration based on the use of *control points*, also known as *tie points*, which are a subset of pixels whose locations in the two images are known or can be selected interactively. Figure 5.15 illustrates the idea of control points using a test pattern and a version of the test pattern that has undergone projective distortion. Once a sufficient number of control points have been chosen, IPT function `cp2tform` can be used to fit a specified type of spatial

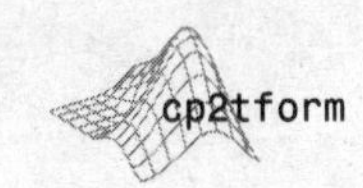

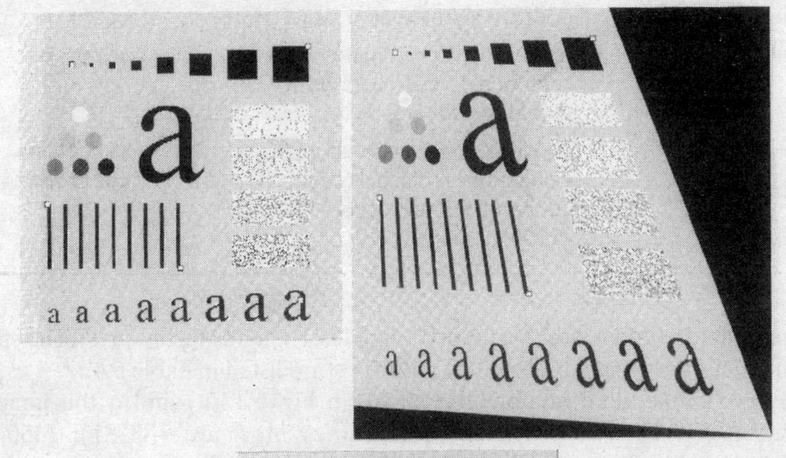

a b
c

FIGURE 5.15 Image registration based on control points. (a) Original image with control points (the small circles superimposed on the image). (b) Geometrically distorted image with control points. (c) Corrected image using a projective transformation inferred from the control points.

TABLE 5.4 Transformation types supported by `cp2tform` and `maketform`.

Transformation Type	Description	Functions
Affine	Combination of scaling, rotation, shearing, and translation. Straight lines remain straight and parallel lines remain parallel.	`maketform` `cp2tform`
Box	Independent scaling and translation along each dimension; a subset of affine.	`maketform`
Composite	A collection of spatial transformations that are applied sequentially.	`maketform`
Custom	User-defined spatial transform; user provides functions that define T and T^{-1}.	`maketform`
Linear conformal	Scaling (same in all dimensions), rotation, and translation; a subset of affine.	`cp2tform`
LWM	Local weighted mean; a locally-varying spatial transformation.	`cp2tform`
Piecewise linear	Locally varying spatial transformation.	`cp2tform`
Polynomial	Input spatial coordinates are a polynomial function of output spatial coordinates.	`cp2tform`
Projective	As with the affine transformation, straight lines remain straight, but parallel lines converge toward vanishing points.	`maketform` `cp2tform`

transformation to the control points (using least squares techniques). The spatial transformation types supported by `cp2tform` are listed in Table 5.4.

For example, let `f` denote the image in Fig. 5.15(a) and `g` the image in Fig. 5.15(b). The control point coordinates in `f` are (83, 81), (450, 56), (43, 293), (249, 392), and (436, 442). The corresponding control point locations in `g` are (68, 66), (375, 47), (42, 286), (275, 434), and (523, 532). Then the commands needed to align image `g` to image `f` are as follows:

```
>> basepoints = [83 81; 450 56; 43 293; 249 392; 436 442];
>> inputpoints = [68 66; 375 47; 42 286; 275 434; 523 532];
>> tform = cp2tform(inputpoints, basepoints, 'projective');
>> gp = imtransform(g, tform, 'XData', [1 502], 'YData', [1 502]);
```

Figure 5.15(c) shows the transformed image.

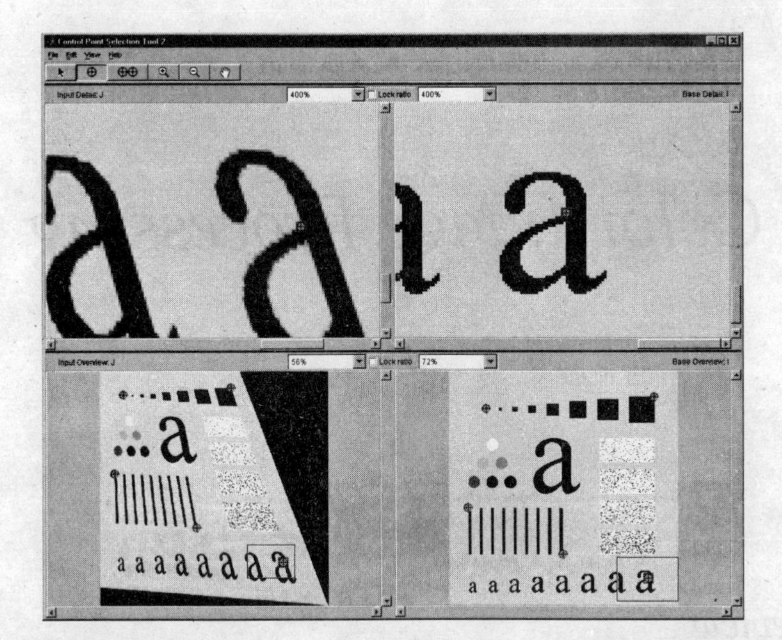

FIGURE 5.16 Interactive tool for choosing control points.

The toolbox includes a graphical user interface designed for the interactive selection of control points on a pair of images. Figure 5.16 shows a screen capture of this tool, which is invoked by the command `cpselect`.

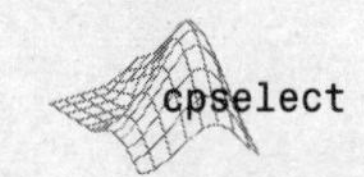

Summary

The material in this chapter is a good overview of how MATLAB and IPT functions can be used for image restoration, and how they can be used as the basis for generating models that help explain the degradation to which an image has been subjected. The capabilities of IPT for noise generation were enhanced significantly by the development in this chapter of functions `imnoise2` and `imnoise3`. Similarly, the spatial filters available in function `spfilt`, especially the nonlinear filters, are a significant extension of IPT's capabilities in this area. These functions are perfect examples of how relatively simple it is to incorporate MATLAB and IPT functions into new code to create applications that enhance the capabilities of an already large set of existing tools.

6 Color Image Processing

Preview

In this chapter we discuss fundamentals of color image processing using the Image Processing Toolbox and extend some of its functionality by developing additional color generation and transformation functions. The discussion in this chapter assumes familiarity on the part of the reader with the principles and terminology of color image processing at an introductory level.

6.1 Color Image Representation in MATLAB

As noted in Section 2.6, the Image Processing Toolbox handles color images either as indexed images or RGB (red, green, blue) images. In this section we discuss these two image types in some detail.

6.1.1 RGB Images

An RGB *color image* is an $M \times N \times 3$ array of *color pixels*, where each color pixel is a triplet corresponding to the red, green, and blue components of an RGB image at a specific spatial location (see Fig. 6.1). An RGB image may be viewed as a "stack" of three gray-scale images that, when fed into the red, green, and blue inputs of a color monitor, produce a color image on the screen. By convention, the three images forming an RGB color image are referred to as the red, green, and blue *component images*. The data class of the component images determines their range of values. If an RGB image is of class `double`, the range of values is $[0, 1]$. Similarly, the range of values is $[0, 255]$ or $[0, 65535]$ for RGB images of class `uint8` or `uint16`, respectively. The number of bits used to represent the pixel values of the component images determines the *bit depth* of an RGB image. For example, if each component image is an 8-bit image, the corresponding RGB image is said to be 24 bits deep. Generally, the number of bits in all component images is the same. In this case, the number of

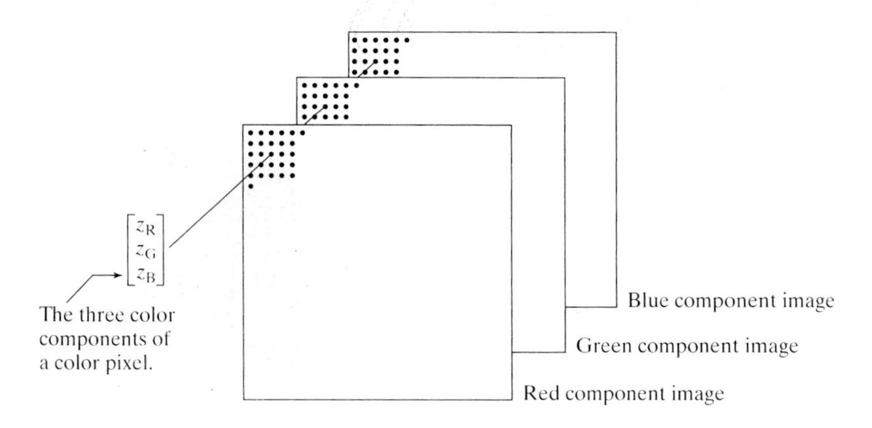

FIGURE 6.1 Schematic showing how pixels of an RGB color image are formed from the corresponding pixels of the three component images.

possible colors in an RGB image is $(2^b)^3$, where b is the number of bits in each component image. For the 8-bit case, the number is 16,777,216 colors.

Let `fR`, `fG`, and `fB` represent three RGB component images. An RGB image is formed from these images by using the `cat` (concatenate) operator to stack the images:

```
rgb_image = cat(3, fR, fG, fB)
```

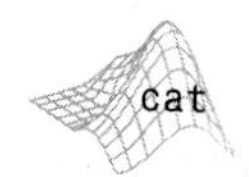

The order in which the images are placed in the operand matters. In general, `cat(dim, A1, A2, . . .)` concatenates the arrays along the dimension specified by `dim`. For example, if `dim = 1`, the arrays are arranged vertically, if `dim = 2`, they are arranged horizontally, and, if `dim = 3`, they are stacked in the third dimension, as in Fig. 6.1.

If all component images are identical, the result is a gray-scale image. Let `rgb_image` denote an RGB image. The following commands extract the three component images:

```
>> fR = rgb_image(:, :, 1);
>> fG = rgb_image(:, :, 2);
>> fB = rgb_image(:, :, 3);
```

The RGB *color space* usually is shown graphically as an RGB color cube, as depicted in Fig. 6.2. The vertices of the cube are the *primary* (red, green, and blue) and *secondary* (cyan, magenta, and yellow) colors of light.

Often, it is useful to be able to view the color cube from any perspective. Function `rgbcube` is used for this purpose. The syntax is

```
rgbcube(vx, vy, vz)
```

rgbcube

Typing `rgbcube(vx, vy, vz)` at the prompt produces an RGB cube on the MATLAB desktop, viewed from point `(vx, vy, vz)`. The resulting image can be saved to disk using function `print`, discussed in Section 2.4. The code for this function follows. It is self-explanatory.

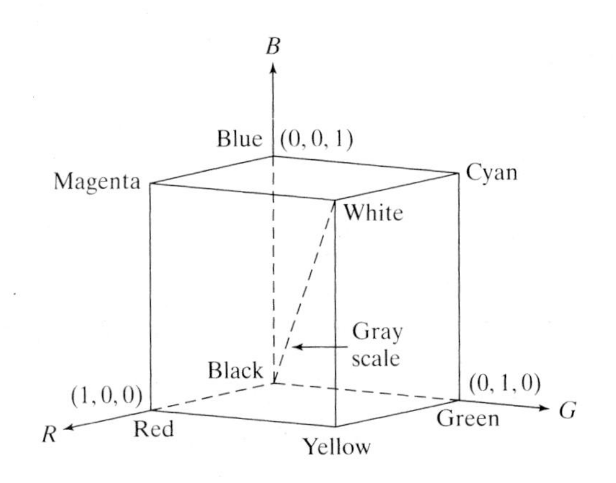

a b

FIGURE 6.2 (a) Schematic of the RGB color cube showing the primary and secondary colors of light at the vertices. Points along the main diagonal have gray values from black at the origin to white at point (1, 1, 1). (b) The RGB color cube.

```
function rgbcube(vx, vy, vz)
%RGBCUBE Displays an RGB cube on the MATLAB desktop.
%   RGBCUBE(VX, VY, VZ) displays an RGB color cube, viewed from point
%   (VX, VY, VZ). With no input arguments, RGBCUBE uses (10, 10, 4)
%   as the default viewing coordinates. To view individual color
%   planes, use the following viewing coordinates, where the first
%   color in the sequence is the closest to the viewing axis, and the
%   other colors are as seen from that axis, proceeding to the right
%   right (or above), and then moving clockwise.
%
%       ---------------------------------------------------
%            COLOR PLANE                    ( vx,  vy,  vz)
%       ---------------------------------------------------
%        Blue-Magenta-White-Cyan            (  0,   0,  10)
%        Red-Yellow-White-Magenta           ( 10,   0,   0)
%        Green-Cyan-White-Yellow            (  0,  10,   0)
%        Black-Red-Magenta-Blue             (  0, -10,   0)
%        Black-Blue-Cyan-Green              (-10,   0,   0)
%        Black-Red-Yellow-Green             (  0,   0, -10)

% Set up parameters for function patch.
vertices_matrix = [0 0 0;0 0 1;0 1 0;0 1 1;1 0 0;1 0 1;1 1 0;1 1 1];
faces_matrix = [1 5 6 2;1 3 7 5;1 2 4 3;2 4 8 6;3 7 8 4;5 6 8 7];
colors = vertices_matrix;
% The order of the cube vertices was selected to be the same as
% the  order of the (R,G,B) colors (e.g., (0,0,0) corresponds to
% black, (1,1,1) corresponds to white, and so on.)

% Generate RGB cube using function patch.
patch('Vertices', vertices_matrix, 'Faces', faces_matrix, ...
      'FaceVertexCData', colors, 'FaceColor', 'interp', ...
      'EdgeAlpha', 0)

% Set up viewing point.
if nargin == 0
   vx = 10; vy = 10; vz = 4;
```

Function `patch` *creates filled, 2-D polygons based on specified property/value pairs. For more information about* `patch`, *see the MATLAB help page for this function.*

patch

```
elseif nargin ~= 3
   error('Wrong number of inputs.')
end
axis off
view([vx, vy, vz])
axis square
```

6.1.2 Indexed Images

An *indexed* image has two components: a *data matrix* of integers, X, and a *colormap matrix*, map. Matrix map is an $m \times 3$ array of class double containing floating-point values in the range [0, 1]. The length, m, of the map is equal to the number of colors it defines. Each row of map specifies the red, green, and blue components of a single color. An indexed image uses "direct mapping" of pixel intensity values to colormap values. The color of each pixel is determined by using the corresponding value of integer matrix X as a pointer into map. If X is of class double, then all of its components with values less than or equal to 1 point to the first row in map, all components with value 2 point to the second row, and so on. If X is of class uint8 or uint16, then all components with value 0 point to the first row in map, all components with value 1 point to the second row, and so on. These concepts are illustrated in Fig. 6.3.

If three columns of map *are equal, then the colormap becomes a* grayscale map.

To display an indexed image we write

```
>> imshow(X, map)
```

or, alternatively,

```
>> image(X)
>> colormap(map)
```

A colormap is stored with an indexed image and is automatically loaded with the image when function imread is used to load the image.

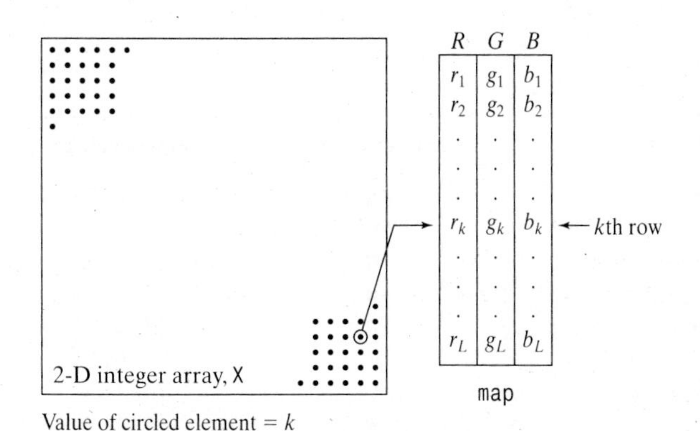

FIGURE 6.3 Elements of an indexed image. Note that the value of an element of integer array X determines the row number in the colormap. Each row contains an RGB triplet, and L is the total number of rows.

Sometimes it is necessary to approximate an indexed image by one with fewer colors. For this we use function `imapprox`, whose syntax is

```
[Y, newmap] = imapprox(X, map, n)
```

This function returns an array `Y` with colormap `newmap`, which has at most `n` colors. The input array `X` can be of class `uint8`, `uint16`, or `double`. The output `Y` is of class `uint8` if `n` is less than or equal to 256. If `n` is greater than 256, `Y` is of class `double`.

When the number of rows in `map` is less than the number of distinct integer values in `X`, multiple values in `X` are displayed using the same color in `map`. For example, suppose that `X` consists of four vertical bands of equal width, with values 1, 64, 128, and 256. If we specify the colormap `map = [0 0 0; 1 1 1]`, then all the elements in `X` with value 1 would point to the first row (black) of the map and all the other elements would point to the second row (white). Thus, the command `imshow(X, map)` would display an image with a black band followed by three white bands. In fact, this would be true until the length of the map became 65, at which time the display would be a black band, followed by a gray band, followed by two white bands. Nonsensical image displays can result if the length of the map exceeds the allowed range of values of the elements of `X`.

There are several ways to specify a color map. One approach is to use the statement

```
>> map(k, :) = [r(k) g(k) b(k)]
```

where `[r(k) g(k) b(k)]` are RGB values that specify one row of a colormap. The map is filled out by varying `k`.

Table 6.1 lists the RGB values for some basic colors. Any of the three formats shown in the table can be used to specify colors. For example, the background color of a figure can be changed to green by using any of the following three statements:

```
>> whitebg('g')
>> whitebg('green')
>> whitebg([0 1 0])
```

TABLE 6.1 RGB values of some basic colors. The long or short names (enclosed by quotes) can be used instead of the numerical triplet to specify an RGB color.

Long name	Short name	RGB values
Black	k	[0 0 0]
Blue	b	[0 0 1]
Green	g	[0 1 0]
Cyan	c	[0 1 1]
Red	r	[1 0 0]
Magenta	m	[1 0 1]
Yellow	y	[1 1 0]
White	w	[1 1 1]

Other colors in addition to the ones shown in Table 6.1 involve fractional values. For instance, `[.5 .5 .5]` is gray, `[.5 0 0]` is dark red, and `[.49 1 .83]` is aquamarine.

MATLAB provides several predefined color maps, accessed using the command

```
>> colormap(map_name)
```

colormap

which sets the colormap to the matrix `map_name`; an example is

```
>> colormap(copper)
```

where `copper` is one of the prespecified MATLAB colormaps. The colors in this map vary smoothly from black to bright copper. If the last image displayed was an indexed image, this command changes its colormap to `copper`. Alternatively, the image can be displayed directly with the desired colormap:

```
>> imshow(X, copper)
```

Table 6.2 lists some of the colormaps available in MATLAB. The length (number of colors) of these colormaps can be specified by enclosing the number in parentheses. For example, `gray(16)` generates a colormap with 16 shades of gray.

6.1.3 IPT Functions for Manipulating RGB and Indexed Images

Table 6.3 lists the IPT functions suitable for converting between RGB, indexed, and gray-scale images. For clarity of notation in this section, we use `rgb_image` to denote RGB images, `gray_image` to denote gray-scale images, `bw` to denote black and white images, and `X`, to denote the data matrix component of indexed images. Recall that an indexed image is composed of an integer data matrix and a colormap matrix.

Function `dither` is applicable both to gray-scale and color images. Dithering is a process used mostly in the printing and publishing industry to give the visual impression of shade variations on a printed page that consists of dots. In the case of gray-scale images, dithering attempts to capture shades of gray by producing a binary image of black dots on a white background (or vice versa). The sizes of the dots vary, from small dots in light areas to increasingly larger dots for dark areas. The key issue in implementing a dithering algorithm is a tradeoff between "accuracy" of visual perception and computational complexity. The dithering approach used in IPT is based on the Floyd-Steinberg algorithm (see Floyd and Steinberg [1975], and Ulichney [1987]). The syntax used by function `dither` for gray-scale images is

```
bw = dither(gray_image)
```

dither

where, as noted earlier, `gray_image` is a gray-scale image and `bw` is the dithered result (a binary image).

TABLE 6.2 Some of the MATLAB predefined colormaps.

Name	Description
`autumn`	Varies smoothly from red, through orange, to yellow.
`bone`	A gray-scale colormap with a higher value for the blue component. This colormap is useful for adding an "electronic" look to gray-scale images.
`colorcube`	Contains as many regularly spaced colors in RGB color space as possible, while attempting to provide more steps of gray, pure red, pure green, and pure blue.
`cool`	Consists of colors that are shades of cyan and magenta. It varies smoothly from cyan to magenta.
`copper`	Varies smoothly from black to bright copper.
`flag`	Consists of the colors red, white, blue, and black. This colormap completely changes color with each index increment.
`gray`	Returns a linear gray-scale colormap.
`hot`	Varies smoothly from black, through shades of red, orange, and yellow, to white.
`hsv`	Varies the hue component of the hue-saturation-value color model. The colors begin with red, pass through yellow, green, cyan, blue, magenta, and return to red. The colormap is particularly appropriate for displaying periodic functions.
`jet`	Ranges from blue to red, and passes through the colors cyan, yellow, and orange.
`lines`	Produces a colormap of colors specified by the `ColorOrder` property and a shade of gray. Consult online help regarding function `ColorOrder`.
`pink`	Contains pastel shades of pink. The pink colormap provides sepia tone colorization of grayscale photographs.
`prism`	Repeats the six colors red, orange, yellow, green, blue, and violet.
`spring`	Consists of colors that are shades of magenta and yellow.
`summer`	Consists of colors that are shades of green and yellow.
`white`	This is an all white monochrome colormap.
`winter`	Consists of colors that are shades of blue and green.

TABLE 6.3 IPT functions for converting between RGB, indexed, and gray-scale intensity images.

Function	Purpose
`dither`	Creates an indexed image from an RGB image by dithering.
`grayslice`	Creates an indexed image from a gray-scale intensity image by multilevel thresholding.
`gray2ind`	Creates an indexed image from a gray-scale intensity image.
`ind2gray`	Creates a gray-scale intensity image from an indexed image.
`rgb2ind`	Creates an indexed image from an RGB image.
`ind2rgb`	Creates an RGB image from an indexed image.
`rgb2gray`	Creates a gray-scale image from an RGB image.

When working with color images, dithering is used principally in conjunction with function rgb2ind to reduce the number of colors in an image. This function is discussed later in this section.

Function grayslice has the syntax

```
X = grayslice(gray_image, n)
```

grayslice

This function produces an indexed image by thresholding gray_image with threshold values

$$\frac{1}{n}, \frac{2}{n}, \ldots, \frac{n-1}{n}$$

As noted earlier, the resulting indexed image can be viewed with the command imshow(X, map) using a map of appropriate length [e.g., jet(16)]. An alternate syntax is

```
X = grayslice(gray_image, v)
```

where v is a vector whose values are used to threshold gray_image. When used in conjunction with a colormap, grayslice is a basic tool for pseudocolor image processing, where specified gray intensity bands are assigned different colors. The input image can be of class uint8, uint16, or double. The threshold values in v must between 0 and 1, even if the input image is of class uint8 or uint16. The function performs the necessary scaling.

Function gray2ind, with syntax

```
[X, map] = gray2ind(gray_image, n)
```

gray2ind

scales, then rounds image gray_image to produce an indexed image X with colormap gray(n). If n is omitted, it defaults to 64. The input image can be of class uint8, uint16, or double. The class of the output image X is uint8 if n is less than or equal to 256, or of class uint16 if n is greater than 256.

Function ind2gray, with the syntax

```
gray_image = ind2gray(X, map)
```

ind2gray

converts an indexed image, composed of X and map, to a gray-scale image. Array X can be of class uint8, uint16, or double. The output image is of class double.

The syntax of interest in this chapter for function rgb2ind has the form

```
[X, map] = rgb2ind(rgb_image, n, dither_option)
```

rgb2ind

where n determines the length (number of colors) of map, and dither_option can have one of two values: 'dither' (the default) dithers, if necessary, to achieve better color resolution at the expense of spatial resolution; conversely, 'nodither' maps each color in the original image to the closest color in the new map (depending on the value of n). No dithering is performed. The input image can be of class uint8, uint16, or double. The output array, X, is of class uint8 if n is less than or equal to 256; otherwise it is of class uint16. Example 6.1 shows the effect that dithering has on color reduction.

Function ind2rgb, with syntax

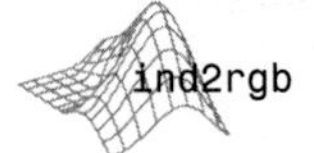


```
rgb_image = ind2rgb(X, map)
```

converts the matrix X and corresponding colormap map to RGB format; X can be of class uint8, uint16, or double. The output RGB image is an $M \times N \times 3$ array of class double.

Finally, function rgb2gray, with syntax

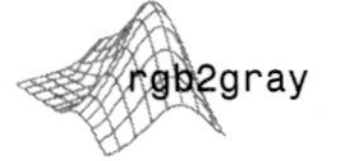


```
gray_image = rgb2gray(rgb_image)
```

converts an RGB image to a gray-scale image. The input RGB image can be of class uint8, uint16, or double; the output image is of the same class as the input.

EXAMPLE 6.1: Illustration of some of the functions in Table 6.3.

■ Function rgb2ind is quite useful for reducing the number of colors in an RGB image. As an illustration of this function, and of the advantages of using the dithering option, consider Fig. 6.4(a), which is a 24-bit RGB image, f. Figures 6.4(b) and (c) show the results of using the commands

```
>> [X1, map1] = rgb2ind(f, 8, 'nodither');
>> imshow(X1, map1)
```

and

```
>> [X2, map2] = rgb2ind(f, 8, 'dither');
>> figure, imshow(X2, map2)
```

Both images have only 8 colors, which is a significant reduction in the number of possible colors in f, which, for a 24-bit RGB image exceeds 16 million, as mentioned earlier. Figure 6.4(b) has noticeable false contouring, especially in the center of the large flower. The dithered image shows better tonality, and considerably less false contouring, a result of the "randomness" introduced by dithering. The image is a little blurred, but it certainly is visually superior to Fig. 6.4(b).

The effects of dithering are usually better illustrated with gray-scale images. Figures 6.4(d) and (e) were obtained using the commands

```
>> g = rgb2gray(f);
>> g1 = dither(g);
>> figure, imshow(g); figure, imshow(g1)
```

a
b c
d e

FIGURE 6.4
(a) RGB image.
(b) Number of colors reduced to 8 without dithering.
(c) Number of colors reduced to 8 with dithering.
(d) Gray-scale version of (a) obtained using function `rgb2gray`.
(e) Dithered gray-scale image (this is a binary image).

The image in Fig. 6.4(e) is a binary image, which again represents a significant degree of data reduction. By looking at Figs. 6.4(c) and (e), it is clear why dithering is such a staple in the printing and publishing industry, especially in situations (such as in newspapers) where paper quality and printing resolution are low. ■

6.2 Converting to Other Color Spaces

As explained in the previous section, the toolbox represents colors as RGB values, directly in an RGB image, or indirectly in an indexed image, where the colormap is stored in RGB format. However, there are other color spaces (also called *color models*) whose use in some applications may be more convenient and/or appropriate. These include the NTSC, YCbCr, HSV, CMY, CMYK, and HSI color spaces. The toolbox provides conversion functions from RGB to the NTSC, YCbCr, HSV and CMY color spaces, and back. Functions for converting to and from the HSI color space are developed later in this section.

6.2.1 NTSC Color Space

The NTSC color system is used in television in the United States. One of the main advantages of this format is that gray-scale information is separate from color data, so the same signal can be used for both color and monochrome television sets. In the NTSC format, image data consists of three components: *luminance* (Y), *hue* (I), and *saturation* (Q), where the choice of the letters YIQ is conventional. The luminance component represents gray-scale information, and the other two components carry the color information of a TV signal. The YIQ components are obtained from the RGB components of an image using the transformation

$$\begin{bmatrix} Y \\ I \\ Q \end{bmatrix} = \begin{bmatrix} 0.299 & 0.587 & 0.114 \\ 0.596 & -0.274 & -0.322 \\ 0.211 & -0.523 & 0.312 \end{bmatrix} \begin{bmatrix} R \\ G \\ B \end{bmatrix}$$

Note that the elements of the first row sum to 1 and the elements of the next two rows sum to 0. This is as expected because for a gray-scale image all the RGB components are equal, so the I and Q components should be 0 for such an image. Function `rgb2ntsc` performs the transformation:

```
yiq_image = rgb2ntsc(rgb_image)
```

where the input RGB image can be of class `uint8`, `uint16`, or `double`. The output image is an $M \times N \times 3$ array of class `double`. Component image `yiq_image(:, :, 1)` is the luminance, `yiq_image(:, :, 2)` is the hue, and `yiq_image(:, :, 3)` is the saturation image.

Similarly, the RGB components are obtained from the YIQ components using the transformation:

$$\begin{bmatrix} R \\ G \\ B \end{bmatrix} = \begin{bmatrix} 1.000 & 0.956 & 0.621 \\ 1.000 & -0.272 & -0.647 \\ 1.000 & -1.106 & 1.703 \end{bmatrix} \begin{bmatrix} Y \\ I \\ Q \end{bmatrix}$$

IPT function `ntsc2rgb` implements this equation:

```
rgb_image = ntsc2rgb(yiq_image)
```

ntsc2rgb

Both the input and output images are of class `double`.

6.2.2 The YCbCr Color Space

The YCbCr color space is used widely in digital video. In this format, luminance information is represented by a single component, Y, and color information is stored as two color-difference components, Cb and Cr. Component Cb is the difference between the blue component and a reference value, and component Cr is the difference between the red component and a reference value (Poynton [1996]). The transformation used by IPT to convert from RGB to YCbCr is

$$\begin{bmatrix} Y \\ Cb \\ Cr \end{bmatrix} = \begin{bmatrix} 16 \\ 128 \\ 128 \end{bmatrix} + \begin{bmatrix} 65.481 & 128.553 & 24.966 \\ -37.797 & -74.203 & 112.000 \\ 112.000 & -93.786 & -18.214 \end{bmatrix} \begin{bmatrix} R \\ G \\ B \end{bmatrix}$$

The conversion function is

```
ycbcr_image = rgb2ycbcr(rgb_image)
```

rgb2ycbcr

The input RGB image can be of class `uint8`, `uint16`, or `double`. The output image is of the same class as the input. A similar transformation converts from YCbCr back to RGB:

```
rgb_image = ycbcr2rgb(ycbcr_image)
```

ycbcr2rgb

The input YCbCr image can be of class `uint8`, `uint16`, or `double`. The output image is of the same class as the input.

To see the transformation matrix used to convert from YCbCr to RGB, type the following command at the prompt:
`>> edit ycbcr2rgb`.

6.2.3 The HSV Color Space

HSV (hue, saturation, value) is one of several color systems used by people to select colors (e.g., of paints or inks) from a color wheel or palette. This color system is considerably closer than the RGB system to the way in which humans experience and describe color sensations. In artist's terminology, hue, saturation, and value refer approximately to tint, shade, and tone.

The HSV color space is formulated by looking at the RGB color cube along its gray axis (the axis joining the black and white vertices), which results in the hexagonally shaped color palette shown in Fig. 6.5(a). As we move along the vertical (gray) axis in Fig. 6.5(b), the size of the hexagonal plane that is perpendicular to the axis changes, yielding the volume depicted in the figure. Hue is

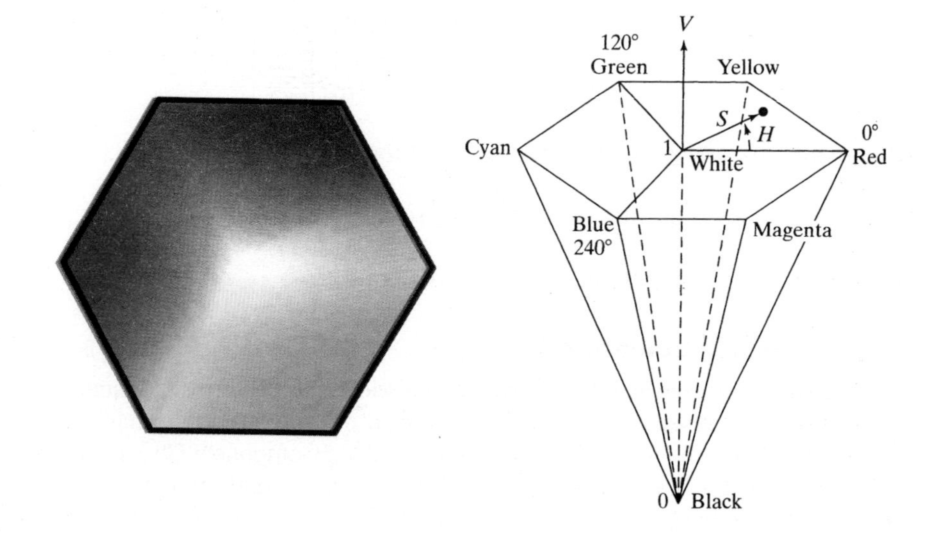

a b

FIGURE 6.5
(a) The HSV color hexagon.
(b) The HSV hexagonal cone.

expressed as an angle around a color hexagon, typically using the red axis as the 0° axis. Value is measured along the axis of the cone. The $V = 0$ end of the axis is black. The $V = 1$ end of the axis is white, which lies in the center of the full color hexagon in Fig. 6.5(a). Thus, this axis represents all shades of gray. Saturation (purity of the color) is measured as the distance from the V axis.

The HSV color system is based on cylindrical coordinates. Converting from RGB to HSV is simply a matter of developing the equations to map RGB values (which are in Cartesian coordinates) to cylindrical coordinates. This topic is treated in detail in most texts on computer graphics (e.g., see Rogers [1997]) so we do not develop the equations here.

The MATLAB function for converting from RGB to HSV is `rgb2hsv`, whose syntax is

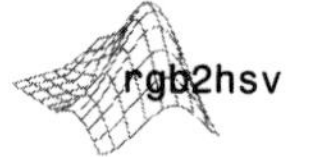


```
hsv_image = rgb2hsv(rgb_image)
```

The input RGB image can be of class `uint8`, `uint16`, or `double`; the output image is of class `double`. The function for converting from HSV back to RGB is `hsv2rgb`:

```
rgb_image = hsv2rgb(hsv_image)
```

The input image must be of class `double`. The output also is of class `double`.

6.2.4 The CMY and CMYK Color Spaces

Cyan, magenta, and yellow are the secondary colors of light or, alternatively, the primary colors of pigments. For example, when a surface coated with cyan pigment is illuminated with white light, no red light is reflected from the surface. That is, the cyan pigment subtracts red light from reflected white light, which itself is composed of equal amounts of red, green, and blue light.

Most devices that deposit colored pigments on paper, such as color printers and copiers, require CMY data input or perform an RGB to CMY conversion internally. This conversion is performed using the simple equation

$$\begin{bmatrix} C \\ M \\ Y \end{bmatrix} = \begin{bmatrix} 1 \\ 1 \\ 1 \end{bmatrix} - \begin{bmatrix} R \\ G \\ B \end{bmatrix}$$

where the assumption is that all color values have been normalized to the range [0, 1]. This equation demonstrates that light reflected from a surface coated with pure cyan does not contain red (that is, $C = 1 - R$ in the equation). Similarly, pure magenta does not reflect green, and pure yellow does not reflect blue. The preceding equation also shows that RGB values can be obtained easily from a set of CMY values by subtracting the individual CMY values from 1.

In theory, equal amounts of the pigment primaries, cyan, magenta, and yellow should produce black. In practice, combining these colors for printing produces a muddy-looking black. So, in order to produce true black (which is the predominant color in printing), a fourth color, *black*, is added, giving rise to the CMYK color model. Thus, when publishers talk about "four-color printing," they are referring to the three-colors of the CMY color model plus black.

Function `imcomplement` introduced in Section 3.2.1 can be used to convert from RGB to CMY:

```
cmy_image = imcomplement(rgb_image)
```

We use this function also to convert a CMY image to RGB:

```
rgb_image = imcomplement(cmy_image)
```

6.2.5 The HSI Color Space

With the exception of HSV, the color spaces discussed thus far are not well suited for *describing* colors in terms that are practical for human interpretation. For example, one does not refer to the color of an automobile by giving the percentage of each of the pigment primaries composing its color.

When humans view a color object, we tend to describe it by its hue, saturation, and brightness. *Hue* is an attribute that describes a pure color (e.g., pure yellow, orange, or red), whereas *saturation* gives a measure of the degree to which a pure color is diluted by white light. *Brightness* is a subjective descriptor that is practically impossible to measure. It embodies the achromatic notion of *intensity* and is a key factor in describing color sensation. We do know that intensity (gray level) is a most useful descriptor of monochromatic images. This quantity definitely is measurable and easily interpretable.

The color space we are about to present, called the *HSI* (hue, saturation, intensity) *color space*, decouples the intensity component from the color-carrying information (hue and saturation) in a color image. As a result, the HSI model is an ideal tool for developing image-processing algorithms based on color descriptions that are natural and intuitive to humans who, after all, are the developers and users of these algorithms. The HSV color space is somewhat similar, but its focus is on presenting colors that are meaningful when interpreted in terms of a color artist's palette.

As discussed in Section 6.1.1, an RGB color image is composed of three monochrome intensity images, so it should come as no surprise that we should be able to extract intensity from an RGB image. This becomes quite clear if we take the color cube from Fig. 6.2 and stand it on the black, (0, 0, 0), vertex, with the white vertex, (1, 1, 1), directly above it, as Fig. 6.6(a) shows. As noted in connection with Fig. 6.2, the intensity is along the line joining these two vertices. In the arrangement shown in Fig. 6.6, the line (intensity axis) joining the black and white vertices is vertical. Thus, if we wanted to determine the intensity component of any color point in Fig. 6.6, we would simply pass a plane *perpendicular* to the intensity axis and containing the color point. The intersection of the plane with the intensity axis would give us an intensity value in the range [0, 1]. We also note with a little thought that the saturation (purity) of a color increases as a function of distance from the intensity axis. In fact, the saturation of points on the intensity axis is zero, as evidenced by the fact that all points along this axis are gray.

In order to see how hue can be determined from a given RGB point, consider Fig. 6.6(b), which shows a plane defined by three points, (black, white, and cyan). The fact that the black and white points are contained in the plane tells us that the intensity axis also is contained in the plane. Furthermore, we see that *all* points contained in the plane segment defined by the intensity axis and the boundaries of the cube have the *same* hue (cyan in this case). This is because the colors inside a color triangle are various combinations or mixtures of the three vertex colors. If two of those vertices are black and white, and the third is a color point, all points on the triangle must have the same hue since the black and white components do not contribute to changes in hue (of course, the intensity and saturation of points in this triangle do change). By rotating the shaded plane about the vertical intensity axis, we would obtain different hues. From these concepts we arrive at the conclusion that the hue, saturation, and intensity values required to form the HSI space can be obtained from the RGB color cube. That is, we can convert any RGB point to a corresponding point is the HSI color model by working out the geometrical formulas describing the reasoning just outlined in the preceding discussion.

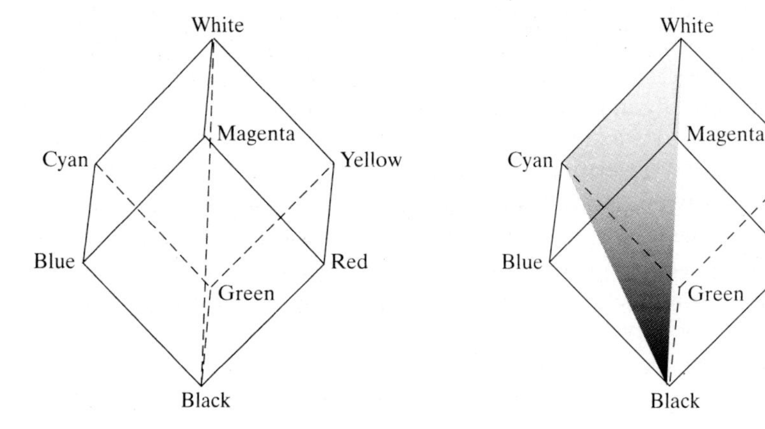

a b

FIGURE 6.6 Relationship between the RGB and HSI color models.

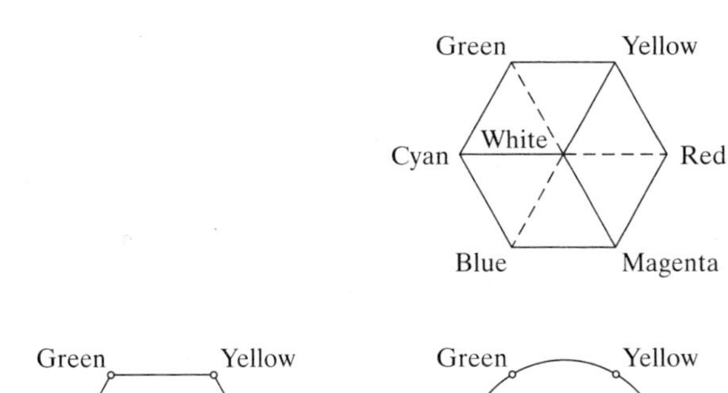

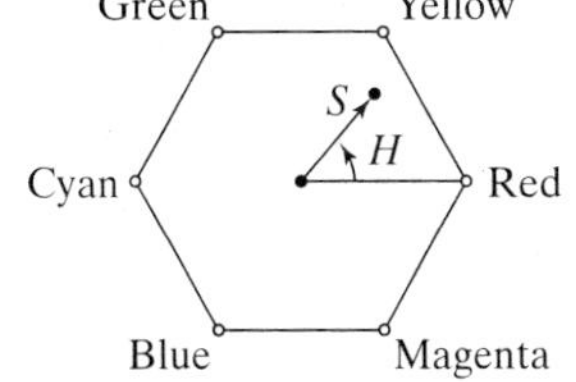

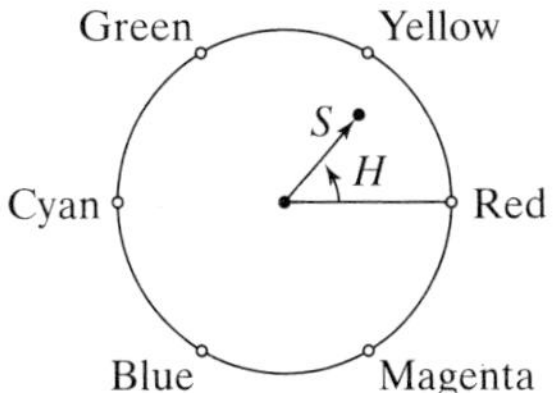

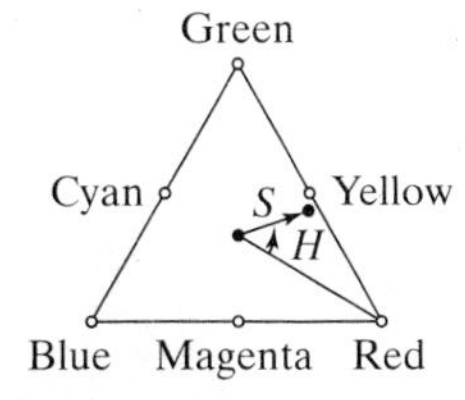

a
b c d

FIGURE 6.7 Hue and saturation in the HSI color model. The dot is an arbitrary color point. The angle from the red axis gives the hue, and the length of the vector is the saturation. The intensity of all colors in any of these planes is given by the position of the plane on the vertical intensity axis.

Based on the preceding discussion, we see that the HSI space consists of a vertical intensity axis and the locus of color points that lie on a plane perpendicular to this axis. As the plane moves up and down the intensity axis, the boundaries defined by the intersection of the plane with the faces of the cube have either a triangular or hexagonal shape. This can be visualized more readily by looking at the cube down its gray-scale axis, as shown in Fig. 6.7(a). In this plane we see that the primary colors are separated by 120°. The secondary colors are 60° from the primaries, which means that the angle between secondary colors also is 120°.

Figure 6.7(b) shows the hexagonal shape and an arbitrary color point (shown as a dot). The hue of the point is determined by an angle from some reference point. Usually (but not always) an angle of 0° from the red axis designates 0 hue, and the hue increases counterclockwise from there. The saturation (distance from the vertical axis) is the length of the vector from the origin to the point. Note that the origin is defined by the intersection of the color plane with the vertical intensity axis. The important components of the HSI color space are the vertical intensity axis, the length of the vector to a color point, and the angle this vector makes with the red axis. Therefore, it is not unusual to see the HSI plane defined is terms of the hexagon just discussed, a triangle, or even a circle, as Figs. 6.7(c) and (d) show. The shape chosen is not important because any one of these shapes can be warped into one of the other two by a geometric transformation. Figure 6.8 shows the HSI model based on color triangles and also on circles.

Converting Colors from RGB to HSI

In the following discussion we give the RGB to HSI conversion equations without derivation. See the book Web site (the address is listed in Section 1.5) for a detailed derivation of these equations. Given an image in RGB color format, the H component of each RGB pixel is obtained using the equation

$$H = \begin{cases} \theta & \text{if } B \leq G \\ 360 - \theta & \text{if } B > G \end{cases}$$

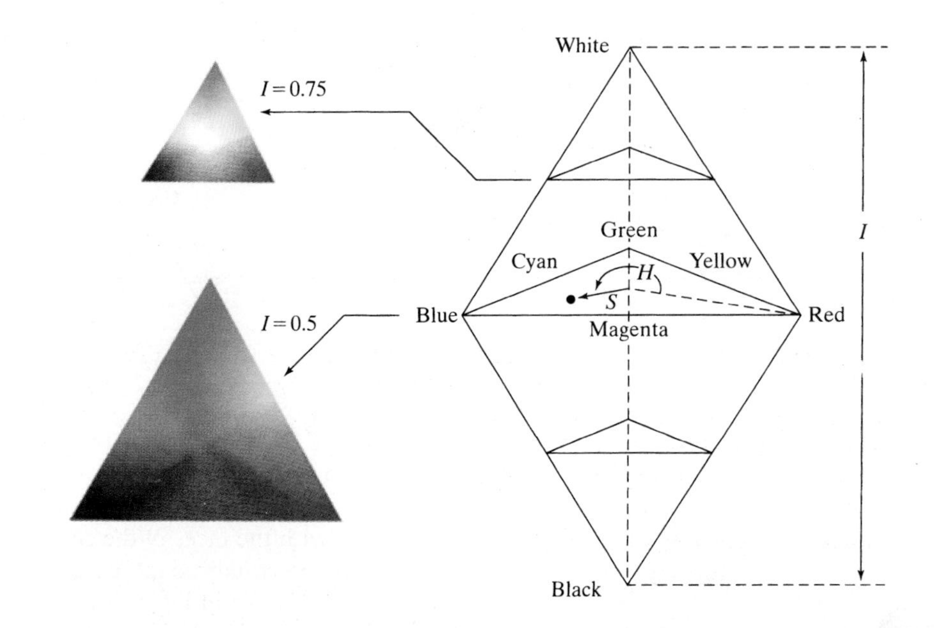

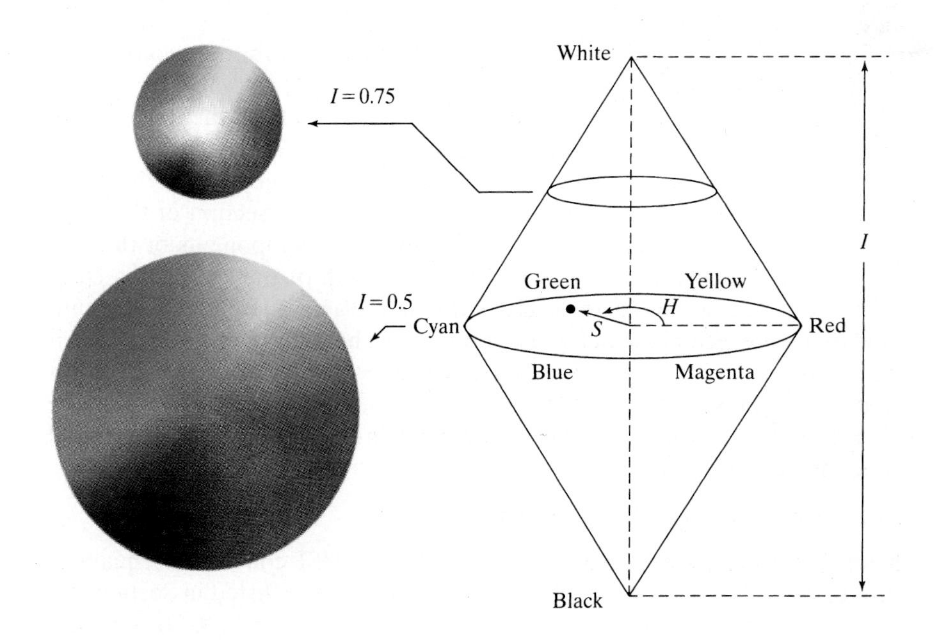

a
b

FIGURE 6.8 The HSI color model based on (a) triangular and (b) circular color planes. The triangles and circles are perpendicular to the vertical intensity axis.

with

$$\theta = \cos^{-1}\left\{\frac{\frac{1}{2}[(R - G) + (R - B)]}{[(R - G)^2 + (R - B)(G - B)]^{1/2}}\right\}$$

The saturation component is given by

$$S = 1 - \frac{3}{(R + G + B)}[\min(R, G, B)]$$

Finally, the intensity component is given by

$$I = \frac{1}{3}(R + G + B)$$

It is assumed that the RGB values have been normalized to the range [0, 1], and that angle θ is measured with respect to the red axis of the HSI space, as indicated in Fig. 6.7. Hue can be normalized to the range [0, 1] by dividing by 360° all values resulting from the equation for H. The other two HSI components already are in this range if the given RGB values are in the interval [0, 1].

Converting Colors from HSI to RGB

Given values of HSI in the interval [0, 1], we now find the corresponding RGB values in the same range. The applicable equations depend on the values of H. There are three sectors of interest, corresponding to the 120° intervals in the separation of primaries (see Fig. 6.7). We begin by multiplying H by 360°, which returns the hue to its original range of [0°, 360°].

RG sector ($0° \leq H < 120°$): When H is in this sector, the RGB components are given by the equations

$$B = I(1 - S)$$

$$R = I\left[1 + \frac{S \cos H}{\cos(60° - H)}\right]$$

and

$$G = 3I - (R + B)$$

GB sector ($120° \leq H < 240°$): If the given value of H is in this sector, we first subtract 120° from it:

$$H = H - 120°$$

Then the RGB components are

$$R = I(1 - S)$$

$$G = I\left[1 + \frac{S \cos H}{\cos(60° - H)}\right]$$

and

$$B = 3I - (R + G)$$

BR sector ($240° \le H \le 360°$): Finally, if H is in this range, we subtract 240° from it:

$$H = H - 240°$$

Then the RGB components are

$$G = I(1 - S)$$

$$B = I\left[1 + \frac{S\cos H}{\cos(60° - H)}\right]$$

and

$$R = 3I - (G + B)$$

Use of these equations for image processing is discussed later in this chapter.

An M-function for Converting from RGB to HSI

The following function,

rgb2hsi

```
hsi = rgb2hsi(rgb)
```

implements the equations just discussed for converting from RGB to HSI. To simplify the notation, we use `rgb` and `hsi` to denote RGB and HSI images, respectively. The documentation in the code details the use of this function.

```
function hsi = rgb2hsi(rgb)
%RGB2HSI Converts an RGB image to HSI.
%   HSI = RGB2HSI(RGB) converts an RGB image to HSI. The input image
%   is assumed to be of size M-by-N-by-3, where the third dimension
%   accounts for three image planes: red, green, and blue, in that
%   order. If all RGB component images are equal, the HSI conversion
%   is undefined. The input image can be of class double (with values
%   in the range [0, 1]), uint8, or uint16.
%
%   The output image, HSI, is of class double, where:
%      hsi(:, :, 1) = hue image normalized to the range [0, 1] by
%                     dividing all angle values by 2*pi.
%      hsi(:, :, 2) = saturation image, in the range [0, 1].
%      hsi(:, :, 3) = intensity image, in the range [0, 1].

% Extract the individual component immages.
rgb = im2double(rgb);
r = rgb(:, :, 1);
g = rgb(:, :, 2);
b = rgb(:, :, 3);

% Implement the conversion equations.
num = 0.5*((r - g) + (r - b));
```

```
den = sqrt((r - g).^2 + (r - b).*(g - b));
theta = acos(num./(den + eps));

H = theta;
H(b > g) = 2*pi - H(b > g);
H = H/(2*pi);

num = min(min(r, g), b);
den = r + g + b;
den(den == 0) = eps;
S = 1 - 3.* num./den;

H(S == 0) = 0;

I = (r + g + b)/3;

% Combine all three results into an hsi image.
hsi = cat(3, H, S, I);
```

An M-function for Converting from HSI to RGB

The following function,

```
rgb = hsi2rgb(hsi)
```

hsi2rgb

implements the equations for converting from HSI to RGB. The documentation in the code details the use of this function.

```
function rgb = hsi2rgb(hsi)
%HSI2RGB Converts an HSI image to RGB.
%   RGB = HSI2RGB(HSI) converts an HSI image to RGB, where HSI
%   is assumed to be of class double with:
%      hsi(:, :, 1) = hue image, assumed to be in the range
%                     [0, 1] by having been divided by 2*pi.
%      hsi(:, :, 2) = saturation image, in the range [0, 1].
%      hsi(:, :, 3) = intensity image, in the range [0, 1].
%
%   The components of the output image are:
%      rgb(:, :, 1) = red.
%      rgb(:, :, 2) = green.
%      rgb(:, :, 3) = blue.

% Extract the individual HSI component images.
H = hsi(:, :, 1) * 2 * pi;
S = hsi(:, :, 2);
I = hsi(:, :, 3);

% Implement the conversion equations.
R = zeros(size(hsi, 1), size(hsi, 2));
G = zeros(size(hsi, 1), size(hsi, 2));
B = zeros(size(hsi, 1), size(hsi, 2));

% RG sector (0 <= H < 2*pi/3).
idx = find( (0 <= H) & (H < 2*pi/3));
B(idx) = I(idx) .* (1 - S(idx));
```

```
R(idx) = I(idx) .* (1 + S(idx) .* cos(H(idx)) ./ ...
                                        cos(pi/3 - H(idx)));
G(idx) = 3*I(idx) - (R(idx) + B(idx));

% BG sector (2*pi/3 <= H < 4*pi/3).
idx = find( (2*pi/3 <= H) & (H < 4*pi/3) );
R(idx) = I(idx) .* (1 - S(idx));
G(idx) = I(idx) .* (1 + S(idx) .* cos(H(idx) - 2*pi/3) ./ ...
                    cos (pi - H(idx)));
B(idx) = 3*I(idx) - (R(idx) + G(idx));

% BR sector.
idx = find( (4*pi/3 <= H) & (H <= 2*pi));
G(idx) = I(idx) .* (1 - S(idx));
B(idx) = I(idx) .* (1 + S(idx) .* cos(H(idx) - 4*pi/3) ./ ...
                                        cos(5*pi/3 - H(idx)));
R(idx) = 3*I(idx) - (G(idx) + B(idx));

% Combine all three results into an RGB image. Clip to [0, 1] to
% compensate for floating-point arithmetic rounding effects.
rgb = cat(3, R, G, B);
rgb = max(min(rgb, 1), 0);
```

EXAMPLE 6.2: Converting from RGB to HSI.

■ Figure 6.9 shows the hue, saturation, and intensity components of an image of an RGB cube on a white background, similar to the image in Fig. 6.2(b). Figure 6.9(a) is the hue image. Its most distinguishing feature is the discontinuity in value along a 45° line in the front (red) plane of the cube. To understand the reason for this discontinuity, refer to Fig. 6.2(b), draw a line from the red to the white vertices of the cube, and select a point in the middle of this line. Starting at that point, draw a path to the right, following the cube around until you return to the starting point. The major colors encountered on this path are yellow, green, cyan, blue, magenta, and back to red. According to Fig. 6.7, the value of hue along this path should increase

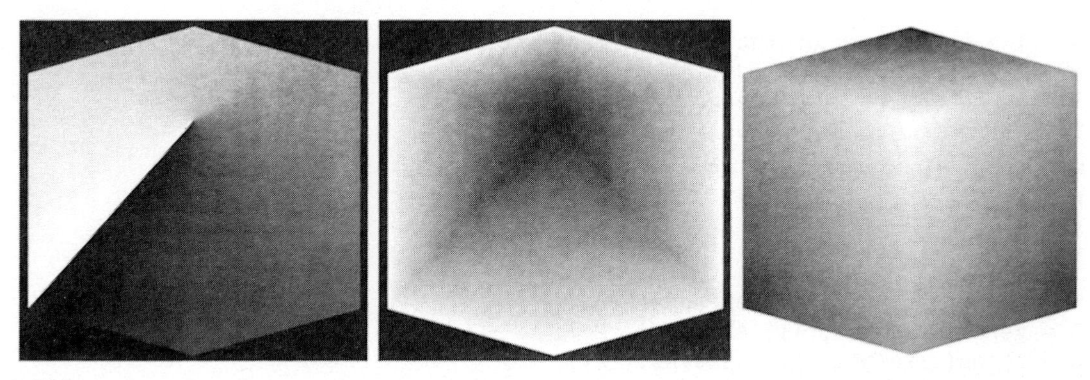

a b c

FIGURE 6.9 HSI component images of an image of an RGB color cube. (a) Hue, (b) saturation, and (c) intensity images.

from 0° to 360° (i.e., from the lowest to highest possible values of hue). This is precisely what Fig. 6.9(a) shows because the lowest value is represented as black and the highest value as white in the figure.

The saturation image in Fig. 6.9(b) shows progressively darker values toward the white vertex of the RGB cube, indicating that colors become less and less saturated as they approach white. Finally, every pixel in the intensity image shown in Fig. 6.9(c) is the average of the RGB values at the corresponding pixel in Fig. 6.2(b). Note that the background in this image is white because the intensity of the background in the color image is white. It is black in the other two images because the hue and saturation of white are zero. ■

6.3 The Basics of Color Image Processing

In this section we begin the study of processing techniques applicable to color images. Although they are far from being exhaustive, the techniques developed in the sections that follow are illustrative of how color images are handled for a variety of image-processing tasks. For the purposes of the following discussion we subdivide color image processing into three principal areas: (1) *color transformations* (also called *color mappings*); (2) *spatial processing* of individual color planes; and (3) *color vector processing*. The first category deals with processing the pixels of each color plane based strictly on their values and not on their spatial coordinates. This category is analogous to the material in Section 3.2 dealing with intensity transformations. The second category deals with spatial (neighborhood) filtering of *individual* color planes and is analogous to the discussion in Sections 3.4 and 3.5 on spatial filtering.

The third category deals with techniques based on processing all components of a color image simultaneously. Because full-color images have at least three components, color pixels really are vectors. For example, in the RGB system, each color point can be interpreted as a vector extending from the origin to that point in the RGB coordinate system (see Fig. 6.2).

Let $\mathbf{c}$ represent an arbitrary vector in RGB color space:

$$\mathbf{c} = \begin{bmatrix} c_R \\ c_G \\ c_B \end{bmatrix} = \begin{bmatrix} R \\ G \\ B \end{bmatrix}$$

This equation indicates that the components of $\mathbf{c}$ are simply the RGB components of a color image at a point. We take into account the fact that the color components are a function of coordinates (x, y) by using the notation

$$\mathbf{c}(x, y) = \begin{bmatrix} c_R(x, y) \\ c_G(x, y) \\ c_B(x, y) \end{bmatrix} = \begin{bmatrix} R(x, y) \\ G(x, y) \\ B(x, y) \end{bmatrix}$$

For an image of size $M \times N$, there are MN such vectors, $\mathbf{c}(x, y)$, for $x = 0, 1, 2, \ldots, M - 1$ and $y = 0, 1, 2, \ldots, N - 1$.

In some cases, equivalent results are obtained whether color images are processed one plane at a time or as vector quantities. However, as explained in

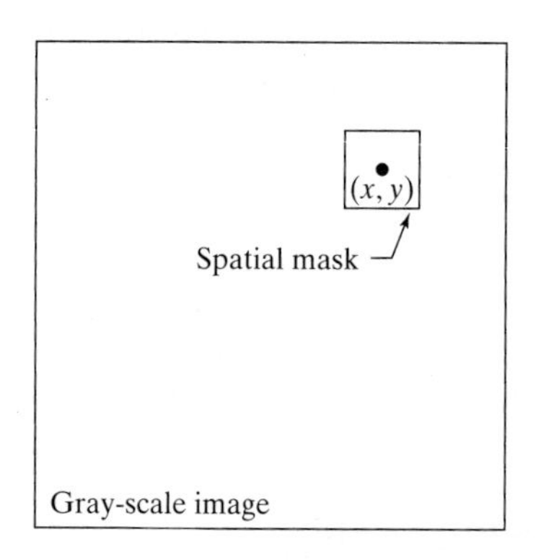

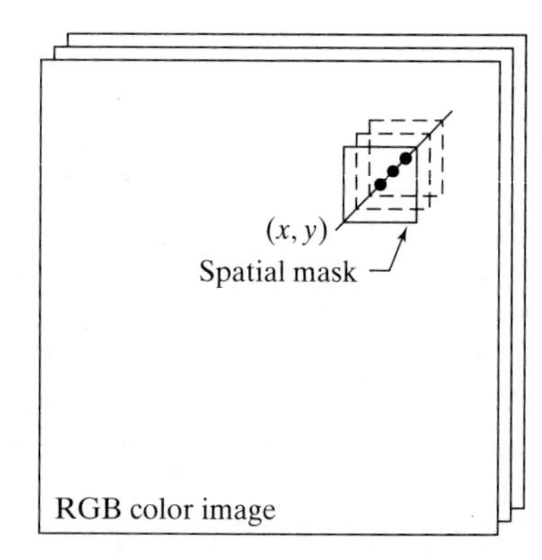

a b

FIGURE 6.10 Spatial masks for gray-scale and RGB color images.

more detail in Section 6.6, this is not always the case. In order for independent color component and vector-based processing to be equivalent, two conditions have to be satisfied: First, the process has to be applicable to both vectors and scalars. Second, the operation on each component of a vector must be independent of the other components. As an illustration, Fig. 6.10 shows spatial neighborhood processing of gray-scale and full-color images. Suppose that the process is neighborhood averaging. In Fig. 6.10(a), averaging would be accomplished by summing the gray levels of all the pixels in the neighborhood and dividing by the total number of pixels in the neighborhood. In Fig. 6.10(b) averaging would be done by summing all the vectors in the neighborhood and dividing each component by the total number of vectors in the neighborhood. But each component of the average vector is the sum of the pixels in the image corresponding to that component, which is the same as the result that would be obtained if the averaging were done on the neighborhood of each component image individually, and then the color vector were formed.

6.4 Color Transformations

The techniques described in this section are based on processing the color components of a color image or intensity component of a monochrome image within the context of a single color model. For color images, we restrict attention to transformations of the form

$$s_i = T_i(r_i), \quad i = 1, 2, \dots, n$$

where r_i and s_i are the color components of the input and output images, n is the dimension of (or number of color components in) the color space of r_i, and the T_i are referred to as *full-color transformation* (or *mapping*) functions.

If the input images are monochrome, then we write an equation of the form

$$s_i = T_i(r), \quad i = 1, 2, \dots, n$$

where r denotes gray-level values, s_i and T_i are as above, and n is the number of color components in s_i. This equation describes the mapping of gray levels into arbitrary colors, a process frequently referred to as a *pseudocolor transformation* or *pseudocolor mapping*. Note that the first equation can be used to process monochrome images in RGB space if we let $r_1 = r_2 = r_3 = r$. In either case, the equations given here are straightforward extensions of the intensity transformation equation introduced in Section 3.2. As is true of the transformations in that section, all n pseudo- or full-color transformation functions $\{T_1, T_2, \ldots, T_n\}$ are independent of the spatial image coordinates (x, y).

Some of the gray-scale transformations introduced in Chapter 3, like `imcomplement`, which computes the negative of an image, are independent of the gray-level content of the image being transformed. Others, like `histeq`, which depends on gray-level distribution, are adaptive, but the transformation is fixed once the necessary parameters have been estimated. And still others, like `imadjust`, which requires the user to select appropriate curve shape parameters, are often best specified interactively. A similar situation exists when working with pseudo- and full-color mappings—particularly when human viewing and interpretation (e.g., for color balancing) are involved. In such applications, the selection of appropriate mapping functions is best accomplished by directly manipulating graphical representations of candidate functions and viewing their combined effect (in real time) on the images being processed.

Figure 6.11 illustrates a simple but powerful way to specify mapping functions graphically. Figure 6.11(a) shows a transformation that is formed by linearly interpolating three *control points* (the circled coordinates in the figure); Fig. 6.11(b) shows the transformation that results from a cubic spline interpolation of the same three points; and Figs. 6.11(c) and (d) provide more complex linear and cubic spline interpolations, respectively. Both types of interpolation are supported in MATLAB. Linear interpolation is implemented by using

```
z = interp1q(x, y, xi)
```

`interp1q`

which returns a column vector containing the values of the linearly interpolated 1-D function `z` at points `xi`. Column vectors `x` and `y` specify the horizontal and vertical coordinate pairs of the underlying control points. The elements of `x` must increase monotonically. The length of `z` is equal to the length of `xi`. Thus, for example,

```
>> z = interp1q([0 255]', [0 255]', [0: 255]')
```

produces a 256-element one-to-one mapping connecting control points (0, 0) and (255, 255)—that is, `z = [0 1 2 . . . 255]'`.

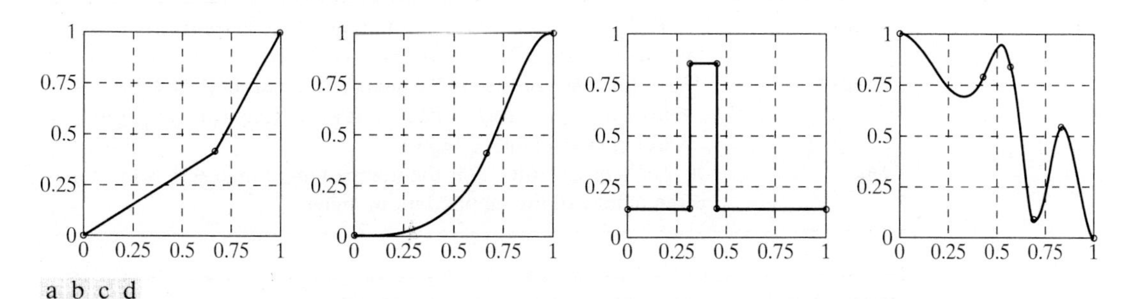

FIGURE 6.11 Specifying mapping functions using control points: (a) and (c) linear interpolation, and (b) and (d) cubic spline interpolation.

In a similar manner, cubic spline interpolation is implemented using the `spline` function,

spline

```
z = spline(x, y, xi)
```

where variables `z`, `x`, `y`, and `xi` are as described in the previous paragraph for `interp1q`. However, the `xi` must be distinct for use in function `spline`. Moreover, if `y` contains two more elements than `x`, its first and last entries are assumed to be the end slopes of the cubic spline. The function depicted in Fig. 6.11(b), for example, was generated using zero-valued end slopes.

The specification of transformation functions can be made interactive by graphically manipulating the control points that are input to functions `interp1q` and `spline` and displaying in real time the results of the transformation functions on the images being processed. The `ice` (interactive color editing) function does precisely this. Its syntax is

ice

```
g = ice('Property Name', 'Property Value', . . .)
```

The development of function `ice`, *given in Appendix B, is a comprehensive illustration of how to design a graphical user interface (GUI) in MATLAB.*

where `'Property Name'` and `'Property Value'` must appear in pairs, and the dots indicate repetitions of the pattern consisting of corresponding input pairs. Table 6.4 lists the valid pairs for use in function `ice`. Some examples are given later in this section.

With reference to the `'wait'` parameter, when the `'on'` option is selected either explicitly or by default, the output `g` is the processed image. In this case, `ice` takes control of the process, including the cursor, so nothing can be typed on the command window until the function is closed, at which time the final result is image `g`. When `'off'` is selected, `g` is the *handle*† of the processed image, and control is returned immediately to the command window; therefore, new commands can be typed with the `ice` function still active. To obtain the properties of an image with handle `g` we use the `get` function

get

```
h = get(g)
```

This function returns all properties and applicable current values of the graphics object identified by the handle `g`. The properties are stored in structure `h`,

TABLE 6.4 Valid inputs for function `ice`.

Property Name	Property Value
`'image'`	An RGB or monochrome input image, `f`, to be transformed by interactively specified mappings.
`'space'`	The color space of the components to be modified. Possible values are `'rgb'`, `'cmy'`, `'hsi'`, `'hsv'`, `'ntsc'` (or `'yiq'`), and `'ycbcr'`. The default is `'rgb'`.
`'wait'`	If `'on'` (the default), `g` is the mapped input image. If `'off'`, `g` is the handle of the mapped input image.

†Whenever MATLAB creates a graphics object, it assigns an identifier (called a *handle*) to the object, used to access the object's properties. Graphics handles are useful when modifying the appearance of graphs or creating custom plotting commands by writing M-files that create and manipulate objects directly.

so typing `h` at the prompt lists all the properties of the processed image (see Sections 2.10.6 and 11.1.1 for an explanation of structures). To extract a particular property, we type `h.PropertyName`.

Letting `f` denote an RGB or monochrome image, the following are examples of the syntax of function `ice`:

```
>> ice                                    % Only the ice graphical
                                          % interface is displayed.
>> g = ice('image', f);                   % Shows and returns the mapped
                                          % image g.
>> g = ice('image', f, 'wait', 'off');    % Shows g and returns
                                          % the handle.
>> g = ice('image', f, 'space', 'hsi');   % Maps RGB image f in HSI space.
```

Note that when a color space other than RGB is specified, the input image (whether monochrome or RGB) is transformed to the specified space before any mapping is performed. The mapped image is then converted to RGB for output. The output of `ice` is always RGB; its input is always monochrome or RGB. If we type `g = ice('image', f)`, an image and graphical user interface (GUI) like that shown in Fig. 6.12 appear on the MATLAB desktop. Initially,

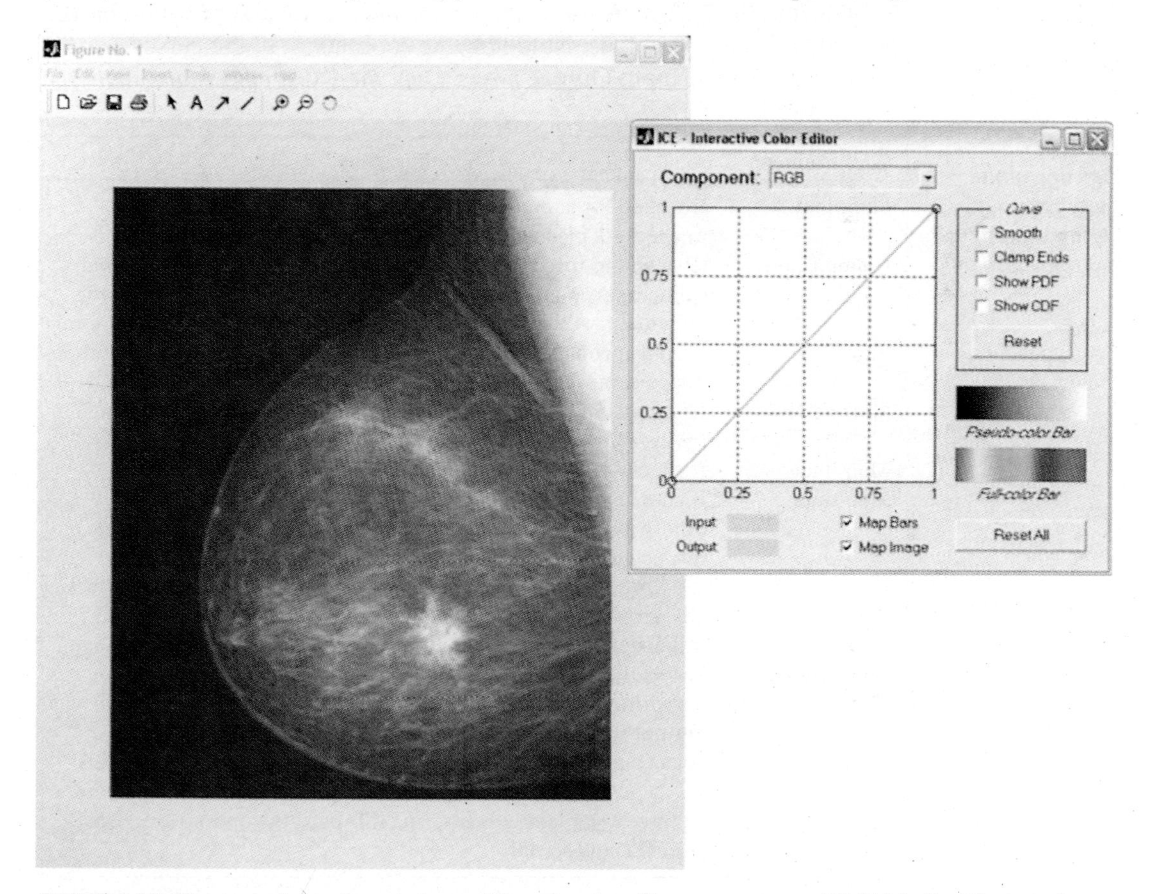

FIGURE 6.12 The typical opening windows of function `ice`. (Image courtesy of G. E. Medical Systems.)

TABLE 6.5
Manipulating control points with the mouse.

Mouse Action[†]	Result
Left Button	Move control point by pressing and dragging.
Left Button + Shift Key	Add control point. The location of the control point can be changed by dragging (while still pressing the Shift Key).
Left Button + Control Key	Delete control point.

[†] For three button mice, the left, middle, and right buttons correspond to the move, add, and delete operations in the table.

the transformation curve is a straight line with a control point at each end. Control points are manipulated with the mouse, as summarized in Table 6.5. Table 6.6 lists the function of the other GUI components. The following examples show typical applications of function `ice`.

EXAMPLE 6.3:
Inverse mappings: monochrome negatives and color complements.

■ Figure 6.13(a) shows the `ice` interface after the default RGB curve of Fig. 6.12 is modified to produce an inverse or negative mapping function. To create the new mapping function, control point $(0, 0)$ is moved (by clicking and dragging it to the upper-left corner) to $(0, 1)$, and control point $(1, 1)$ is moved similarly to coordinate $(1, 0)$. Note how the coordinates of the cursor are displayed in red in the Input/Output boxes. Only the RGB map is modified; the

TABLE 6.6
Function of the checkboxes and pushbuttons in the `ice` GUI.

GUI Element	Function
Smooth	Checked for cubic spline (smooth curve) interpolation. If unchecked, piecewise linear interpolation is used.
Clamp Ends	Checked to force the starting and ending curve slopes in cubic spline interpolation to 0. Piecewise linear interpolation is not affected.
Show PDF	Display probability density function(s) [i.e., histogram(s)] of the image components affected by the mapping function.
Show CDF	Display cumulative distribution function(s) instead of PDFs. (Note: PDFs and CDFs cannot be displayed simultaneously.)
Map Image	If checked, image mapping is enabled; otherwise it is not.
Map Bars	If checked, pseudo- and full-color bar mapping is enabled; otherwise the unmapped bars (a gray wedge and hue wedge, respectively) are displayed.
Reset	Initialize the currently displayed mapping function and uncheck all curve parameters.
Reset All	Initialize all mapping functions.
Input/Output	Shows the coordinates of a *selected* control point on the transformation curve. Input refers to the horizontal axis, and Output to the vertical axis.
Component	Select a mapping function for interactive manipulation. In RGB space, possible selections include R, G, B, and RGB (which maps all three color components). In HSI space, the options are H, S, I, and HSI, and so on.

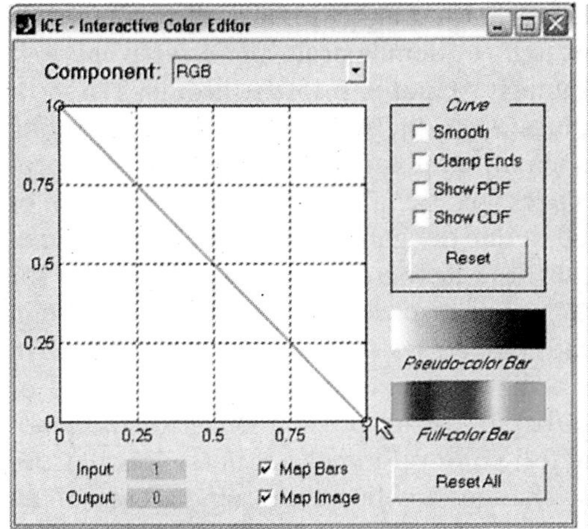

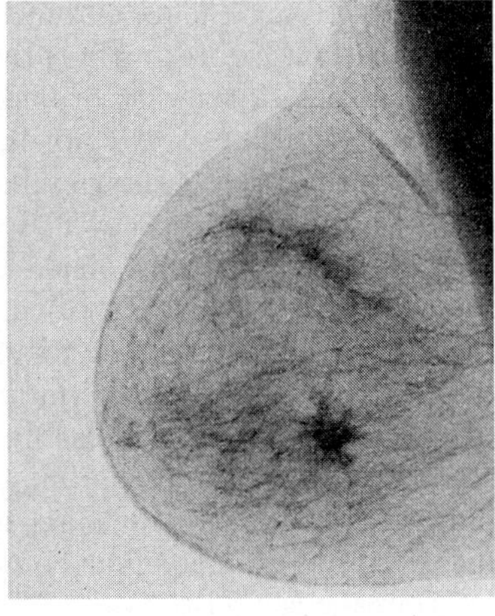

a b

FIGURE 6.13 (a) A negative mapping function, and (b) its effect on the monochrome image of Fig. 6.12.

individual R, G, and B maps are left in their 1 : 1 default states (see the Component entry in Table 6.6). For monochrome inputs, this guarantees monochrome outputs. Figure 6.13(b) shows the monochrome negative that results from the inverse mapping. Note that it is identical to Fig. 3.3(b), which was obtained using the `imcomplement` function. The pseudocolor bar in Fig. 6.13(a) is the "photographic negative" of the original gray-scale bar in Fig. 6.12.

Default (i.e., 1:1) mappings are not shown in most examples.

Inverse or negative mapping functions also are useful in color processing. As can be seen in Figs. 6.14(a) and (b), the result of the mapping is reminiscent of conventional color film negatives. For instance, the red stick of chalk in the bottom row of Fig. 6.14(a) is transformed to cyan in Fig. 6.14(b)—the *color complement* of red. The complement of a primary color is the mixture of the other two primaries (e.g., cyan is blue plus green). As in the gray-scale case, color complements are useful for enhancing detail that is embedded in dark regions of color—particularly when the regions are dominant in size. Note that the *Full-color Bar* in Fig. 6.13(a) contains the complements of the hues in the *Full-color Bar* of Fig. 6.12. ■

a b

FIGURE 6.14 (a) A full color image, and (b) its negative (color complement).

EXAMPLE 6.4: Monochrome and color contrast enhancement.

■ Consider next the use of function `ice` for monochrome and color contrast manipulation. Figures 6.15(a) through (c) demonstrate the effectiveness of `ice` in processing monochrome images. Figures 6.15(d) through (f) show similar effectiveness for color inputs. As in the previous example, mapping functions that are not shown remain in their default or 1:1 state. In both processing sequences, the Show PDF checkbox is enabled. Thus, the histogram of the aerial photo in (a) is displayed under the gamma-shaped mapping function (see Section 3.2.1) in (c); and three histograms are provided in (f) for the color image in (d)—one for each of its three color components. Although the S-shaped mapping function in (f) increases the contrast of the image in (d) [compare it to (e)], it also has a slight effect on hue. The small change of color is virtually imperceptible in (e), but is an obvious result of the mapping, as can be seen in the mapped full-color reference bar in (f). Recall from the previous example that equal changes to the three components of an RGB image can have a dramatic effect on color (see the color complement mapping in Fig. 6.14). ■

a b c
d e f

FIGURE 6.15 Using function `ice` for monochrome and full color contrast enhancement: (a) and (d) are the input images, both of which have a "washed-out" appearance; (b) and (e) show the processed results; (c) and (f) are the `ice` displays. (Original monochrome image for this example courtesy of NASA.)

The red, green, and blue components of the input images in Examples 6.3 and 6.4 are mapped identically—that is, using the same transformation function. To avoid the specification of three identical functions, function `ice` provides an "all components" function (the RGB curve when operating in the RGB color space) that is used to map all input components. The remaining examples demonstrate transformations in which the three components are processed differently.

EXAMPLE 6.5: Pseudocolor mappings.

■ As noted earlier, when a monochrome image is represented in the RGB color space and the resulting components are mapped independently, the transformed result is a pseudocolor image in which input image gray levels have been replaced by arbitrary colors. Transformations that do this are useful because the human eye can distinguish between millions of colors—but relatively few shades of gray. Thus, pseudocolor mappings are used frequently to make small changes in gray level visible to the human eye or to highlight important gray-scale regions. In fact, the principal use of pseudocolor is human visualization—the interpretation of gray-scale events in an image or sequence of images via gray-to-color assignments.

Figure 6.16(a) is an X-ray image of a weld (the horizontal dark region) containing several cracks and porosities (the bright white streaks running through the middle of the image). A pseudocolor version of the image in shown in

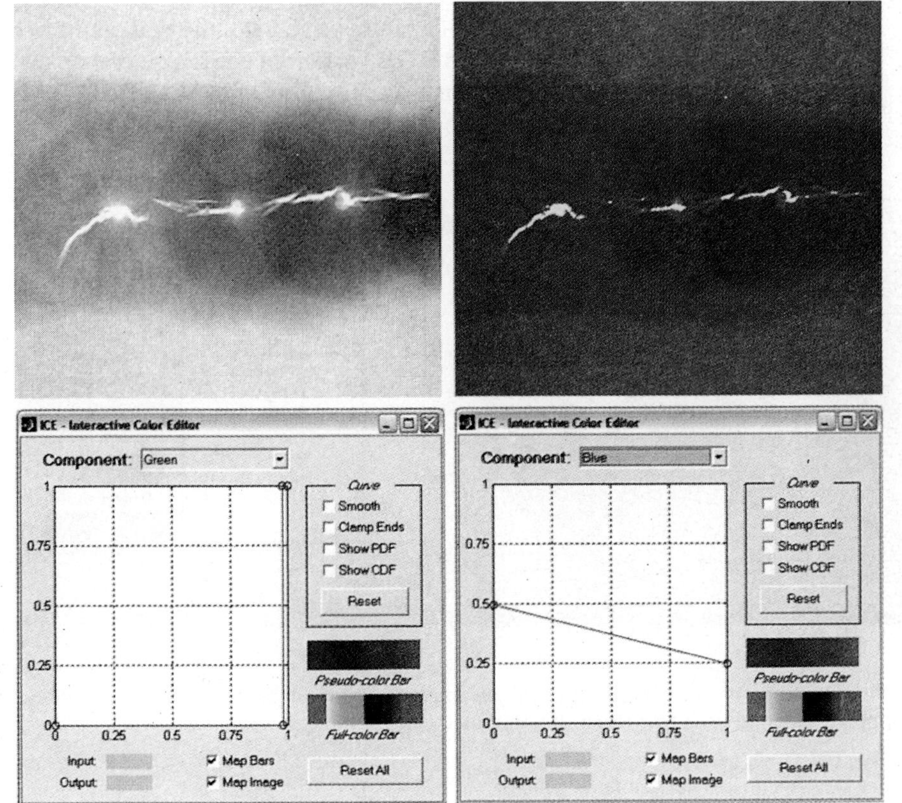

a b
c d

FIGURE 6.16 (a) X-ray of a defective weld; (b) a pseudocolor version of the weld; (c) and (d) mapping functions for the green and blue components. (Original image courtesy of X-TEK Systems, Ltd.)

Fig. 6.16(b); it was generated by mapping the green and blue components of the RGB-converted input using the mapping functions in Figs. 6.16(c) and (d). Note the dramatic visual difference that the pseudocolor mapping makes. The GUI pseudocolor reference bar provides a convenient visual guide to the composite mapping. As can be seen in Figs. 6.16(c) and (d), the interactively specified mapping functions transform the black-to-white gray scale to hues between blue and red, with yellow reserved for white. The yellow, of course, corresponds to weld cracks and porosities, which are the important features in this example. ■

EXAMPLE 6.6: Color balancing.

■ Figure 6.17 shows an application involving a full-color image, in which it is advantageous to map an image's color components independently. Commonly called *color balancing* or *color correction*, this type of mapping has been a mainstay of high-end color reproduction systems but now can be performed on most desktop computers. One important use is photo enhancement. Although color imbalances can be determined objectively by analyzing—with a color spectrometer—a known color in an image, accurate visual assessments are possible when white areas, where the RGB or CMY components should be equal, are present. As can be seen in Fig. 6.17, skin tones also are excellent samples for visual assessments because humans are highly perceptive of proper skin color.

Figure 6.17(a) shows a CMY scan of a mother and her child with an excess of magenta (keep in mind that only an RGB version of the image can be displayed by MATLAB). For simplicity and compatibility with MATLAB, function `ice` accepts only RGB (and monochrome) inputs as well—but can

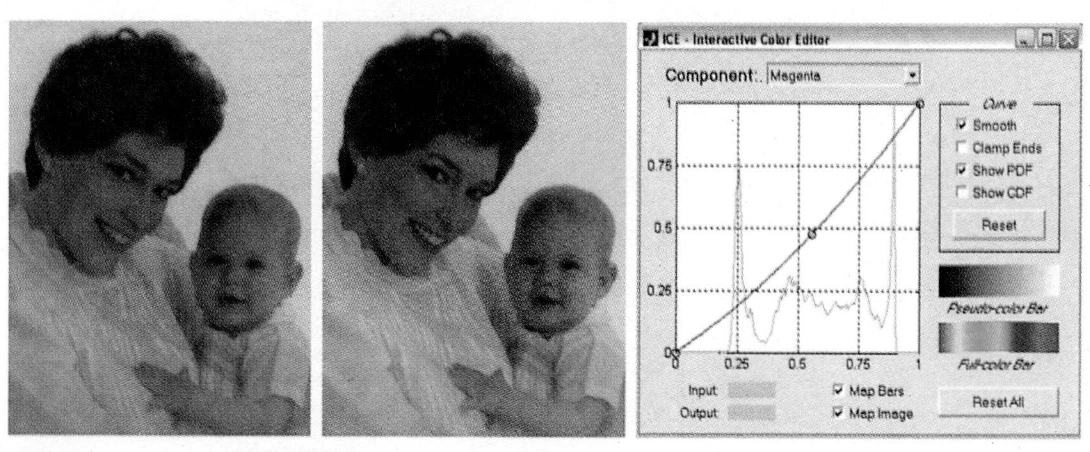

a b c

FIGURE 6.17 Using function `ice` for color balancing: (a) an image heavy in magenta; (b) the corrected image; and (c) the mapping function used to correct the imbalance.

process the input in a variety of color spaces, as detailed in Table 6.4. To interactively modify the CMY components of RGB image `f1`, for example, the appropriate `ice` call is

```
>> f2 = ice('image', f1, 'space', 'CMY');
```

As Fig. 6.17 shows, a small decrease in magenta had a significant impact on image color. ■

EXAMPLE 6.7: Histogram based mappings.

■ Histogram equalization is a gray-level mapping process that seeks to produce monochrome images with uniform intensity histograms. As discussed in Section 3.3.2, the required mapping function is the cumulative distribution function (CDF) of the gray levels in the input image. Because color images have multiple components, the gray-scale technique must be modified to handle more than one component and associated histogram. As might be expected, it is unwise to histogram equalize the components of a color image independently. The result usually is erroneous color. A more logical approach is to spread color intensities uniformly, leaving the colors themselves (i.e., the hues) unchanged.

Figure 6.18(a) shows a color image of a caster stand containing cruets and shakers. The transformed image in Fig. 6.18(b), which was produced using the HSI transformations in Figs. 6.18(c) and (d), is significantly brighter. Several of the moldings and the grain of the wood table on which the caster is resting are now visible. The intensity component was mapped using the function in Fig. 6.18(c), which closely approximates the CDF of that component (also displayed in the figure). The hue mapping function in Fig. 6.18(d) was selected to improve the overall color perception of the intensity-equalized result. Note that the histograms of the input and output image's hue, saturation, and intensity components are shown in Figs. 6.18(e) and (f), respectively. The hue components are virtually identical (which is desirable), while the intensity and saturation components were altered. Finally note that, to process an RGB image in the HSI color space, we included the input property name/value pair `'space'/'hsi'` in the call to `ice`. ■

The output images generated in the preceding examples in this section are of type RGB and class `uint8`. For monochrome results, as in Example 6.3, all three components of the RGB output are identical. A more compact representation can be obtained via the `rgb2gray` function of Table 6.3 or by using the command

```
f3 = f2(:, :, 1);
```

where `f2` is an RGB image generated by `ice` and `f3` is a standard MATLAB monochrome image.

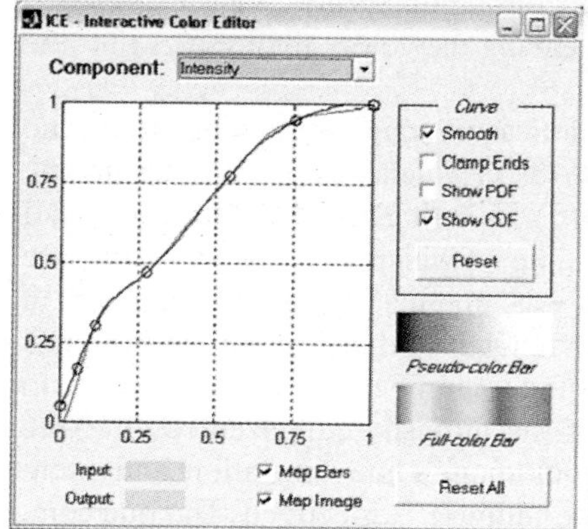

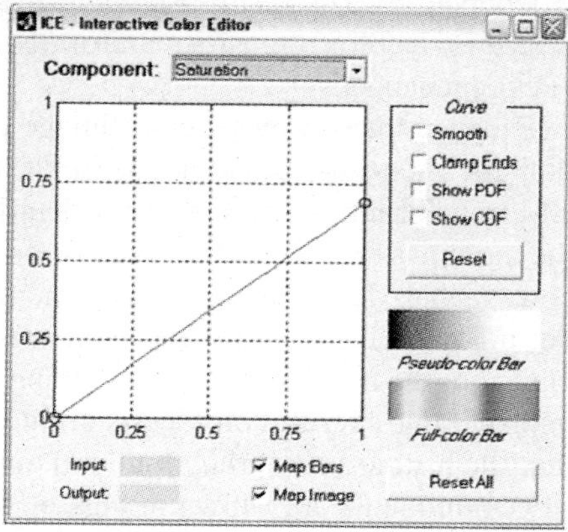

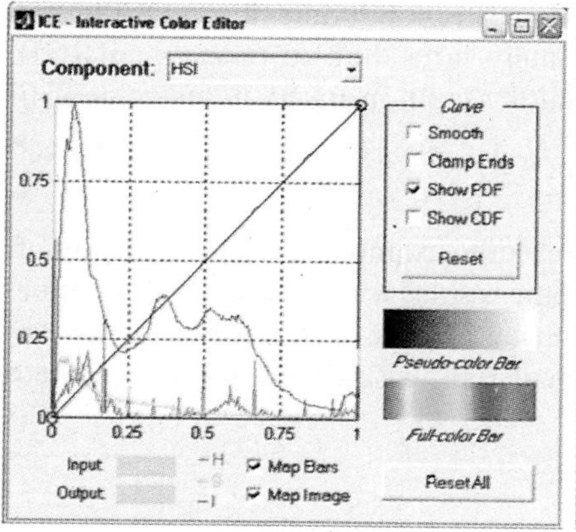

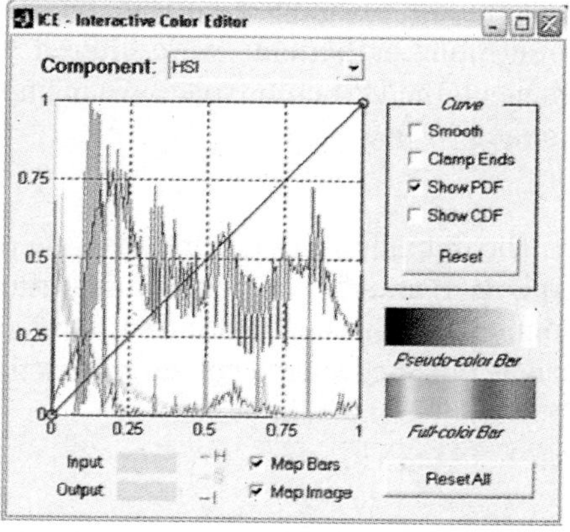

a b
c d
e f

FIGURE 6.18 Histogram equalization followed by saturation adjustment in the HSI color space: (a) input image; (b) mapped result; (c) intensity component mapping function and cumulative distribution function; (d) saturation component mapping function; (e) input image's component histograms; and (f) mapped result's component histograms.

6.5 Spatial Filtering of Color Images

The material in Section 6.4 deals with color transformations performed on single image pixels of single color component planes. The next level of complexity involves performing spatial neighborhood processing, also on single image planes. This breakdown is analogous to the discussion on intensity transformations in Section 3.2, and the discussion on spatial filtering in Sections 3.4 and 3.5. We introduce spatial filtering of color images by concentrating mostly on RGB images, but the basic concepts are applicable to other color models as well. We illustrate spatial processing of color images by two examples of linear filtering: image smoothing and image sharpening.

6.5.1 Color Image Smoothing

With reference to Fig. 6.10(a) and the discussion in Sections 3.4 and 3.5, smoothing (spatial averaging) of a monochrome image can be accomplished by multiplying all pixel values by the corresponding coefficients in the spatial mask (which are all 1s) and dividing by the total number of elements in the mask. The process of smoothing a full-color image using spatial masks is shown in Fig. 6.10(b). The process (in RGB space for example) is formulated in the same way as for gray-scale images, except that instead of single pixels we now deal with vector values in the form shown in Section 6.3.

Let S_{xy} denote the set of coordinates defining a neighborhood centered at (x, y) in a color image. The average of the RGB vectors in this neighborhood is

$$\bar{\mathbf{c}}(x, y) = \frac{1}{K} \sum_{(s,t)\in S_{xy}} \mathbf{c}(s, t)$$

where K is the number of pixels in the neighborhood. It follows from the discussion in Section 6.3 and the properties of vector addition that

$$\bar{\mathbf{c}}(x, y) = \begin{bmatrix} \dfrac{1}{K} \sum_{(s,t)\in S_{xy}} R(s, t) \\ \dfrac{1}{K} \sum_{(s,t)\in S_{xy}} G(s, t) \\ \dfrac{1}{K} \sum_{(s,t)\in S_{xy}} B(s, t) \end{bmatrix}$$

We recognize each component of this vector as the result that we would obtain by performing neighborhood averaging on each individual component image, using standard gray-scale neighborhood processing. Thus, we conclude that smoothing by neighborhood averaging can be carried out on an independent component basis. The results would be the same as if neighborhood averaging were carried out directly in color vector space.

As discussed in Section 3.5.1, IPT linear spatial filters for image smoothing are generated with function `fspecial`, with one of three options: `'average'`, `'disk'`, and `'gaussian'` (see Table 3.4). Once a filter has been generated, filtering is performed by using function `imfilter`, introduced in Section 3.4.1.

Conceptually, smoothing an RGB color image, `fc`, with a linear spatial filter consists of the following steps:

1. Extract the three component images:

```
>> fR = fc(:, :, 1); fG = fc(:, :, 2); fB = fc(:, :, 3);
```

2. Filter each component image individually. For example, letting `w` represent a smoothing filter generated using `fspecial`, we smooth the red component image as follows:

```
>> fR_filtered = imfilter(fR, w);
```

and similarly for the other two component images.

3. Reconstruct the filtered RGB image:

```
>> fc_filtered = cat(3, fR_filtered, fG_filtered, fB_filtered);
```

However, we can perform linear filtering of RGB images in MATLAB using the same syntax employed for monochrome images, allowing us to combine the preceding three steps into one:

```
>> fc_filtered = imfilter(fc, w);
```

EXAMPLE 6.8: Color image smoothing.

■ Figure 6.19(a) shows an RGB image of size 1197×1197 pixels and Figs. 6.19(b) through (d) are its RGB component images, extracted using the procedure described in the previous paragraph. Figures 6.20(a) through (c) show the three HSI component images of Fig. 6.19(a), obtained using function `rgb2hsi`.

Figure 6.21(a) shows the result of smoothing the image in Fig. 6.19(a) using function `imfilter` with the `'replicate'` option and an `'average'` filter of size 25×25 pixels. The averaging filter was large enough to produce a significant degree of blurring. A filter of this size was selected to demonstrate the difference between smoothing in RGB space and attempting to achieve a similar result using only the intensity component of the image after it had been converted to the HSI color space. Figure 6.21(b) was obtained using the commands:

```
>> h = rgb2hsi(fc);
>> H = h(:, :, 1); S = h(:, :, 2); I =  h(:, :, 3);
>> w = fspecial('average', 25);
>> I_filtered = imfilter(I, w, 'replicate');
>> h = cat(3, H, S, I_filtered);
>> f = hsi2rgb(h);
>> f = min(f, 1); % RGB images must have values in the range [0, 1].
>> imshow(f)
```

a b
c d

FIGURE 6.19 (a) RGB image; (b) through (d) are the red, green and blue component images, respectively.

a b c

FIGURE 6.20 From left to right: hue, saturation, and intensity components of Fig. 6.19(a).

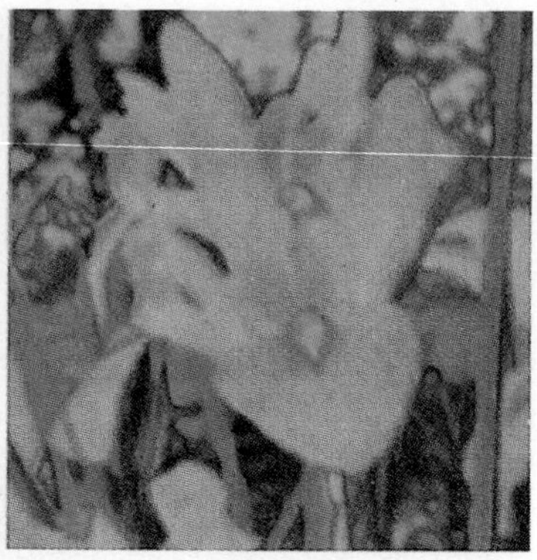

a b c

FIGURE 6.21 (a) Smoothed RGB image obtained by smoothing the R, G, and B image planes separately. (b) Result of smoothing only the intensity component of the HSI equivalent image. (c) Result of smoothing all three HSI components equally.

Clearly, the two filtered results are quite different. For example, in addition to the image being less blurred, note the green border on the top part of the flower in Fig. 6.21(b). The reason for this is simply that the hue and saturation components were not changed while the variability of values of the intensity components was reduced significantly by the smoothing process. A logical thing to try would be to smooth all three components using the same filter. However, this would change the relative relationship between values of the hue and saturation, thus producing nonsensical colors, as Fig. 6.21(c) shows.

In general, as the size of the mask decreases, the differences obtained when filtering the RGB component images and the intensity component of the HSI equivalent image also decrease. ■

6.5.2 Color Image Sharpening

Sharpening an RGB color image with a linear spatial filter follows the same procedure outlined in the previous section, but using a sharpening filter instead. In this section we consider image sharpening using the Laplacian (see Section 3.5.1). From vector analysis, we know that the Laplacian of a vector is defined as a vector whose components are equal to the Laplacian of the individual scalar components of the input vector. In the RGB color system, the Laplacian of vector **c** introduced in Section 6.3 is

$$\nabla^2[\mathbf{c}(x, y)] = \begin{bmatrix} \nabla^2 R(x, y) \\ \nabla^2 G(x, y) \\ \nabla^2 B(x, y) \end{bmatrix}$$

which, as in the previous section, tells us that we can compute the Laplacian of a full-color image by computing the Laplacian of each component image separately.

a b

FIGURE 6.22
(a) Blurred image.
(b) Image enhanced using the Laplacian, followed by contrast enhancement using function `ice`.

EXAMPLE 6.9: Color image sharpening.

■ Figure 6.22(a) shows a slightly blurred version, `fb`, of the image in Fig. 6.19(a), obtained using a 5×5 averaging filter. To sharpen this image we used the Laplacian filter mask

```
>> lapmask = [1 1 1; 1 -8 1; 1 1 1];
```

Then, as in Example 3.9, the enhanced image was computed and displayed using the commands

```
>> fen = imsubtract(fb, imfilter(fb, lapmask, 'replicate'));
>> imshow(fen)
```

where we combined the two required steps into a single command. As in the previous section, RGB images were treated exactly as monochrome images (i.e., with the same calling syntax) when using `imfilter`. Figure 6.22(b) shows the result. Note the significant increase in sharpness of features such as the water droplets, the veins in the leaves, the yellow centers of the flowers, and the green vegetation in the foreground. ■

6.6 Working Directly in RGB Vector Space

As mentioned in Section 6.3, there are cases in which processes based on individual color planes are not equivalent to working directly in RGB vector space. This is demonstrated in this section, where we illustrate vector processing by considering two important applications in color image processing: color edge detection and region segmentation.

6.6.1 Color Edge Detection Using the Gradient

The gradient of a 2-D function, $f(x, y)$, is defined as the vector

$$\nabla\mathbf{f} = \begin{bmatrix} G_x \\ G_y \end{bmatrix} = \begin{bmatrix} \dfrac{\partial f}{\partial x} \\ \dfrac{\partial f}{\partial y} \end{bmatrix}$$

The magnitude of this vector is

$$\begin{aligned} \nabla f &= \text{mag}(\nabla\mathbf{f}) = [G_x^2 + G_y^2]^{1/2} \\ &= [(\partial f/\partial x)^2 + (\partial f/\partial y)^2]^{1/2} \end{aligned}$$

Often, this quantity is approximated by absolute values:

$$\nabla f \approx |G_x| + |G_y|$$

This approximation avoids the square and square root computations, but still behaves as a derivative (i.e., it is zero in constant areas, and has a magnitude proportional to the degree of change in areas whose pixel values are variable). It is common practice to refer to the magnitude of the gradient simply as "the gradient."

A fundamental property of the gradient vector is that it points in the direction of the maximum rate of change of f at coordinates (x, y). The angle at which this maximum rate of change occurs is

$$\alpha(x, y) = \tan^{-1}\left(\frac{G_y}{G_x}\right)$$

It is customary to approximate the derivatives by differences of pixel values over small neighborhoods in an image. Figure 6.23(a) shows a neighborhood of size 3 × 3, where the z's indicate pixel values. An approximation of the partial derivatives in the x (vertical) direction at the center point of the region (i.e., z_5) is given by the difference

$$G_x = (z_7 + 2z_8 + z_9) - (z_1 + 2z_2 + z_3)$$

z_1	z_2	z_3
z_4	z_5	z_6
z_7	z_8	z_9

−1	−2	−1
0	0	0
1	2	1

−1	0	1
−2	0	2
−1	0	1

a b c

FIGURE 6.23 (a) A small neighborhood. (b) and (c) Sobel masks used to compute the gradient in the x (vertical) and y (horizontal) directions, respectively, with respect to the center point of the neighborhood.

Similarly, the derivative in the y direction is approximated by the difference

$$G_y = (z_3 + 2z_6 + z_9) - (z_1 + 2z_4 + z_7)$$

These two quantities are easily computed at all points in an image by convolving (using `imfilter`) the image separately with the two masks shown in Figs. 6.23(b) and (c), respectively. Then, an approximation of the corresponding gradient image is obtained by summing the absolute value of the two filtered images. The masks just discussed are the Sobel masks mentioned in Table 3.4, which can be generated using function `fspecial`.

The gradient computed in the manner just described is one of the most frequently-used methods for edge detection in gray-scale images, as discussed in more detail in Chapter 10. Our interest at the moment is in computing the gradient in RGB color space. However, the method just derived is applicable in 2-D space but does not extend to higher dimensions. The only way to apply it to RGB images would be to compute the gradient of each component color image and then combine the results. Unfortunately, as we show later in this section, this is not the same as computing edges in RGB vector space directly.

The problem, then, is to define the gradient (magnitude and direction) of the vector **c** defined in Section 6.3. The following is one of the various ways in which the concept of a gradient can be extended to vector functions. Recall that for a scalar function, $f(x, y)$, the gradient is a vector pointing in the direction of maximum rate of change of f at coordinates (x, y).

Let **r**, **g**, and **b** be unit vectors along the R, G, and B axis of RGB color space (see Fig. 6.2), and define the vectors

$$\mathbf{u} = \frac{\partial R}{\partial x}\mathbf{r} + \frac{\partial G}{\partial x}\mathbf{g} + \frac{\partial B}{\partial x}\mathbf{b}$$

and

$$\mathbf{v} = \frac{\partial R}{\partial y}\mathbf{r} + \frac{\partial G}{\partial y}\mathbf{g} + \frac{\partial B}{\partial y}\mathbf{b}$$

Let the quantities g_{xx}, g_{yy}, and g_{xy} be defined in terms of the dot product of these vectors, as follows:

$$g_{xx} = \mathbf{u}\cdot\mathbf{u} = \mathbf{u}^T\mathbf{u} = \left|\frac{\partial R}{\partial x}\right|^2 + \left|\frac{\partial G}{\partial x}\right|^2 + \left|\frac{\partial B}{\partial x}\right|^2$$

$$g_{yy} = \mathbf{v}\cdot\mathbf{v} = \mathbf{v}^T\mathbf{v} = \left|\frac{\partial R}{\partial y}\right|^2 + \left|\frac{\partial G}{\partial y}\right|^2 + \left|\frac{\partial B}{\partial y}\right|^2$$

and

$$g_{xy} = \mathbf{u}\cdot\mathbf{v} = \mathbf{u}^T\mathbf{v} = \frac{\partial R}{\partial x}\frac{\partial R}{\partial y} + \frac{\partial G}{\partial x}\frac{\partial G}{\partial y} + \frac{\partial B}{\partial x}\frac{\partial B}{\partial y}$$

Keep in mind that R, G, and B, and consequently the g's, are functions of x and y. Using this notation, it can be shown (Di Zenzo [1986]) that the direction of maximum rate of change of $\mathbf{c}(x, y)$ as a function (x, y) is given by the angle

$$\theta(x, y) = \frac{1}{2}\tan^{-1}\left[\frac{2g_{xy}}{(g_{xx} - g_{yy})}\right]$$

and that the value of the rate of change (i.e., the magnitude of the gradient) in the directions given by the elements of $\theta(x, y)$ is given by

$$F_\theta(x, y) = \left\{\frac{1}{2}[(g_{xx} + g_{yy}) + (g_{xx} - g_{yy})\cos 2\theta + 2g_{xy}\sin 2\theta]\right\}^{1/2}$$

Note that $\theta(x, y)$ and $F_\theta(x, y)$ are images of the same size as the input image. The elements of $\theta(x, y)$ are simply the angles at each point that the gradient is calculated, and $F_\theta(x, y)$ is the gradient image.

Because $\tan(\alpha) = \tan(\alpha \pm \pi)$, if θ_0 is a solution to the preceding $\tan^{-1}$ equation, so is $\theta_0 \pm \pi/2$. Furthermore, $F_\theta(x, y) = F_{\theta+\pi}(x, y)$, so F needs to be computed only for values of θ in the half-open interval $[0, \pi)$. The fact that the $\tan^{-1}$ equation provides two values 90° apart means that this equation associates with each point (x, y) a pair of orthogonal directions. Along one of those directions F is maximum, and it is minimum along the other, so the final result is generated by selecting the maximum at each point. The derivation of these results is rather lengthy, and we would gain little in terms of the fundamental objective of our current discussion by detailing it here. The interested reader should consult the paper by Di Zenzo [1986] for details. The partial derivatives required for implementing the preceding equations can be computed using, for example, the Sobel operators discussed earlier in this section.

The following function implements the color gradient for RGB images (see Appendix C for the code):

colorgrad

```
[VG, A, PPG] = colorgrad(f, T)
```

where `f` is an RGB image, `T` is an optional threshold in the range [0, 1] (the default is 0); `VG` is the RGB vector gradient $F_\theta(x, y)$; `A` is the angle image $\theta(x, y)$, in radians; and `PPG` is the gradient formed by summing the 2-D gradients of the individual color planes (generated for comparison purposes). These latter gradients are $\nabla R(x, y)$, $\nabla G(x, y)$, and $\nabla B(x, y)$, where the ∇ operator is as defined earlier in this section. All the derivatives required to implement the preceding equations are implemented in function `colorgrad` using Sobel operators. The outputs `VG` and `PPG` are normalized to the range [0, 1] by `colorgrad` and they are thresholded so that `VG(x, y) = 0` for values less than or equal to `T` and `VG(x, y) = VG(x, y)` otherwise. Similar comments apply to `PPG`.

EXAMPLE 6.10: RGB edge detection using function `colorgrad`.

■ Figures 6.24(a) through (c) show three simple monochrome images which, when used as RGB planes, produced the color image in Fig. 6.24(d). The objectives of this example are (1) to illustrate the use of function `colorgrad`, and (2) to show that computing the gradient of a color image by combining the gradients of its individual color planes is quite different from computing the gradient directly in RGB vector space using the method just explained.

a b c
d e f

FIGURE 6.24 (a) through (c) RGB component images (black is 0 and white is 255). (d) Corresponding color image. (e) Gradient computed directly in RGB vector space. (f) Composite gradient obtained by computing the 2-D gradient of each RGB component image separately and adding the results.

Letting `f` represent the RGB image in Fig. 6.24(d), the command

```
>> [VG, A, PPG] = colorgrad(f);
```

produced the images `VG` and `PPG` shown in Figs. 6.24(e) and (f). The most important difference between these two results is how much weaker the horizontal edge in Fig. 6.24(f) is than the corresponding edge in Fig. 6.24(e). The reason is simple: The gradients of the red and green planes [Figs. 6.24(a) and (b)] produce two vertical edges, while the gradient of the blue plane yields a single horizontal edge. Adding these three gradients to form `PPG` produces a vertical edge with twice the intensity as the horizontal edge.

On the other hand, when the gradient of the color image is computed directly in vector space [Fig. 6.24(e)], the ratio of the values of the vertical and horizontal edges is $\sqrt{2}$ instead of 2. The reason again is simple: With reference to the color cube in Fig. 6.2(a) and the image in Fig. 6.24(d) we see that the vertical edge in the color image is between a blue and white square and a black and yellow square. The distance between these colors in the color cube is $\sqrt{2}$, but the distance between black and blue and yellow and white (the horizontal edge) is only 1. Thus the ratio of the vertical to the horizontal differences is $\sqrt{2}$. If edge accuracy is an issue, and especially when a threshold is used, then the difference between these two approaches can be significant. For example, if we had used a threshold of 0.6, the horizontal line in Fig. 6.24(f) would have disappeared.

In practice, when interest is mostly on edge detection with no regard for accuracy, the two approaches just discussed generally yield comparable results. For example, Figs. 6.25(b) and (c) are analogous to Figs. 6.24(e) and (f). They were obtained by applying function `colorgrad` to the image in Fig. 6.25(a). Figure 6.25(d) is the difference of the two gradient images, scaled to the range [0, 1]. The maximum absolute difference between the two images is 0.2, which translates to 51 gray levels on the familiar 8-bit range [0, 255]. However, these two gradient images are quite close in visual appearance, with Fig. 6.25(b) being slightly brighter in some places (for reasons similar to those explained in the previous paragraph). Thus, for this type of analysis, the simpler approach of computing the gradient of each individual component generally is acceptable. In other circumstances (as in the inspection of color differences in automated machine inspection of painted products), the more accurate vector approach may be necessary. ■

a b
c d

FIGURE 6.25 (a) RGB image. (b) Gradient computed in RGB vector space. (c) Gradient computed as in Fig. 6.24(f). (d) Absolute difference between (b) and (c), scaled to the range [0, 1].

6.6.2 Image Segmentation in RGB Vector Space

Segmentation is a process that partitions an image into regions. Although segmentation is the topic of Chapter 10, we consider color region segmentation briefly here for the sake of continuity. The reader will have no difficulty following the discussion.

Color region segmentation using RGB color vectors is straightforward. Suppose that the objective is to segment objects of a specified color range in an RGB image. Given a set of sample color points representative of a color (or range of colors) of interest, we obtain an estimate of the "average" or "mean" color that we wish to segment. Let this average color be denoted by the RGB column vector **m**. The objective of segmentation is to classify each RGB pixel in an image as having a color in the specified range or not. To perform this comparison, we need a measure of similarity. One of the simplest measures is the Euclidean distance. Let **z** denote an arbitrary point in RGB space. We say that **z** is *similar* to **m** if the distance between them is less than a specified threshold, T. The Euclidean distance between **z** and **m** is given by

$$\begin{aligned} D(\mathbf{z}, \mathbf{m}) &= \|\mathbf{z} - \mathbf{m}\| \\ &= \left[(\mathbf{z} - \mathbf{m})^T(\mathbf{z} - \mathbf{m})\right]^{1/2} \\ &= \left[(z_R - m_R)^2 + (z_G - m_G)^2 + (z_B - m_B)^2\right]^{1/2} \end{aligned}$$

where $\|\cdot\|$ is the norm of the argument, and the subscripts R, G, and B, denote the RGB components of vectors **m** and **z**. The locus of points such that $D(\mathbf{z}, \mathbf{m}) \le T$ is a solid sphere of radius T, as illustrated in Fig. 6.26(a). By definition, points contained within, or on the surface of, the sphere satisfy the specified color criterion; points outside the sphere do not. Coding these two sets of points in the image with, say, black and white, produces a binary, segmented image.

Following convention, we use a superscript, T, to indicate vector or matrix transposition and a normal, inline, T to denote a threshold value. Care should be exercised not to confuse these unrelated uses of the same variable.

A useful generalization of the preceding equation is a distance measure of the form

$$D(\mathbf{z}, \mathbf{m}) = \left[(\mathbf{z} - \mathbf{m})^T\mathbf{C}^{-1}(\mathbf{z} - \mathbf{m})\right]^{1/2}$$

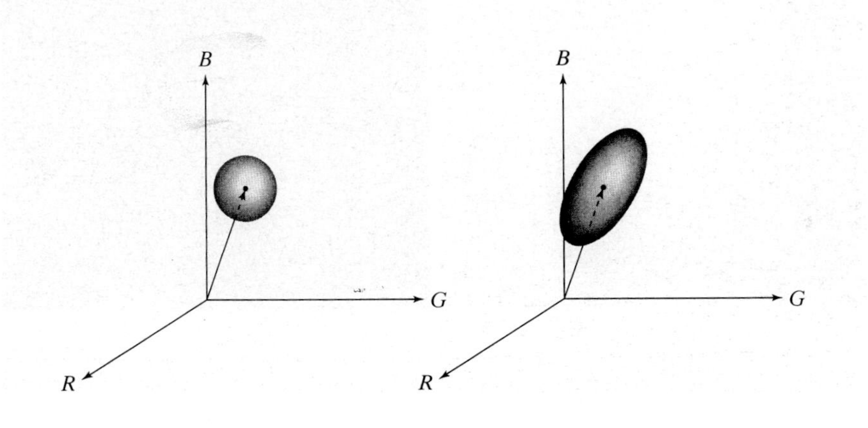

a b

FIGURE 6.26 Two approaches for enclosing data in RGB vector space for the purpose of segmentation.

See Section 12.2 for a detailed discussion on efficient implementations for computing the Euclidean and Mahalanobis distances.

where **C** is the covariance matrix† of the samples representative of the color we wish to segment. This distance is commonly referred to as the *Mahalanobis distance*. The locus of points such that $D(\mathbf{z}, \mathbf{m}) \leq T$ describes a solid 3-D elliptical body [see Fig. 6.26(b)] with the important property that its principal axes are oriented in the direction of maximum data spread. When $\mathbf{C} = \mathbf{I}$, the identity matrix, the Mahalanobis distance reduces to the Euclidean distance. Segmentation is as described in the preceding paragraph, except that the data are now enclosed by an ellipsoid instead of a sphere.

Segmentation in the manner just described is implemented by function `colorseg` (see Appendix C for the code), which has the syntax

`colorseg`

```
S = colorseg(method, f, T, parameters)
```

where `method` is either `'euclidean'` or `'mahalanobis'`, `f` is the RGB image to be segmented, and `T` is the threshold described above. The input parameters are either `m` if `'euclidean'` is chosen, or `m` and `C` if `'mahalanobis'` is chosen. Parameter `m` is the vector, **m**, described above, in either a row or column format, and `C` is the 3×3 covariance matrix, **C**. The output, `S`, is a two-level image (of the same size as the original) containing 0s in the points failing the threshold test, and 1s in the locations that passed the test. The 1s indicate the regions segmented from `f` based on color content.

EXAMPLE 6.11: RGB color image segmentation.

■ Figure 6.27(a) shows a pseudocolor image of a region on the surface of the Jupiter Moon Io. In this image, the reddish colors depict materials newly ejected from an active volcano, and the surrounding yellow materials are older sulfur deposits. This example illustrates segmentation of the reddish region using both options in function `colorseg`.

First we obtain samples representing the range of colors to be segmented. One simple way to obtain such a region of interest (ROI) is to use function `roipoly` described in Section 5.2.4, which produces a binary mask of a region selected interactively. Thus, letting `f` denote the color image in Fig. 6.27(a), the region in Fig. 6.27(b) was obtained using the commands

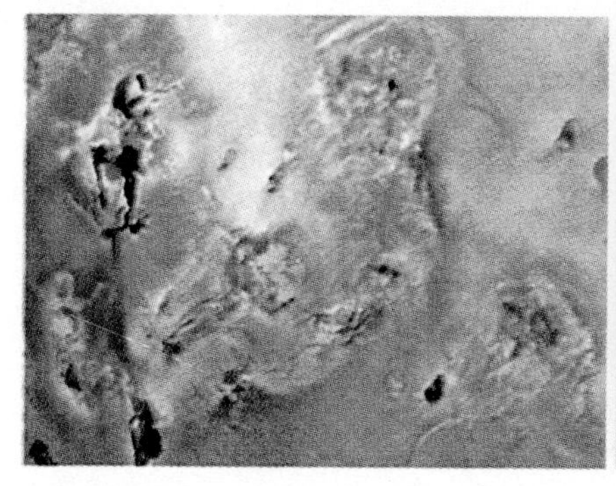
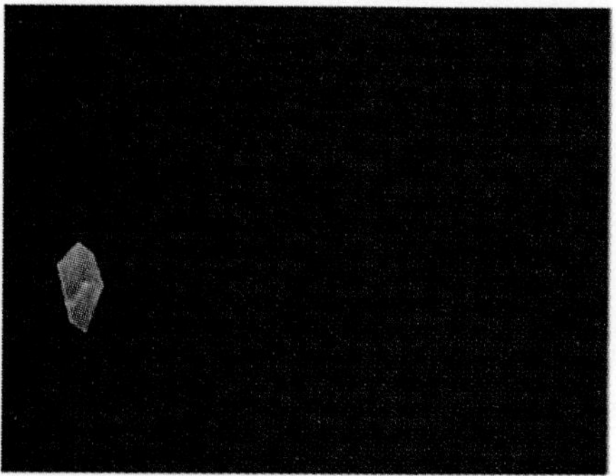

a b

FIGURE 6.27 (a) Pseudocolor of the surface of Jupiter's Moon Io. (b) Region of interest extracted interactively using function `roipoly`. (Original image courtesy of NASA.)

†Computation of the covariance matrix of a set of vector samples is discussed in Section 11.5.

```
>> mask = roipoly(f);          % Select region interactively.
>> red = immultiply(mask, f(:, :, 1));
>> green = immultiply(mask, f(:, :, 2));
>> blue = immultiply(mask, f(:, :, 3));
>> g = cat(3, red, green, blue);
>> figure, imshow(g)
```

where mask is a binary image (the same size as f) with 0s in the background and 1s in the region selected interactively.

Next, we compute the mean vector and covariance matrix of the points in the ROI, but first the coordinates of the points in the ROI must be extracted.

```
>> [M, N, K] = size(g);
>> I = reshape(g, M * N, 3); % reshape is discussed in Sec. 8.2.2.
>> idx = find(mask);
>> I = double(I(idx, 1:3));
>> [C, m] = covmatrix(I); % See Sec. 11.5 for details on covmatrix.
```

The second statement rearranges the color pixels in g as rows of I, and the third statement finds the row indices of the color pixels that are not black. These are the non-background pixels of the masked image in Fig. 6.27(b).

The final preliminary computation is to determine a value for T. A good starting point is to let T be a multiple of the standard deviation of one of the color components. The main diagonal of C contains the variances of the RGB components, so all we have to do is extract these elements and compute their square roots:

```
>> d = diag(C);
>> sd = sqrt(d)'

    22.0643    24.2442    16.1806
```

diag

d = diag(C) *returns in vector* d *the main diagonal of matrix* C.

The first element of sd is the standard deviation of the red component of the color pixels in the ROI, and similarly for the other two components.

We now proceed to segment the image using values of T equal to multiples of 25, which is an approximation to the largest standard deviation: $T = 25, 50, 75, 100$. For the 'euclidean' option with $T = 25$, we use

```
>> E25 = colorseg('euclidean', f, 25, m);
```

Figure 6.28(a) shows the result, and Figs. 6.28(b) through (d) show the segmentation results with $T = 50, 75, 100$. Similarly, Figs. 6.29(a) through (d) show the results obtained using the 'mahalanobis' option with the same sequence of threshold values.

Meaningful results [depending on what we consider as *red* in Fig. 6.27(a)] were obtained with the 'euclidean' option when $T = 25$ and 50, but $T = 75$ and 100 produced significant oversegmentation. On the other hand, the results with the 'mahalanobis' option make a more sensible transition

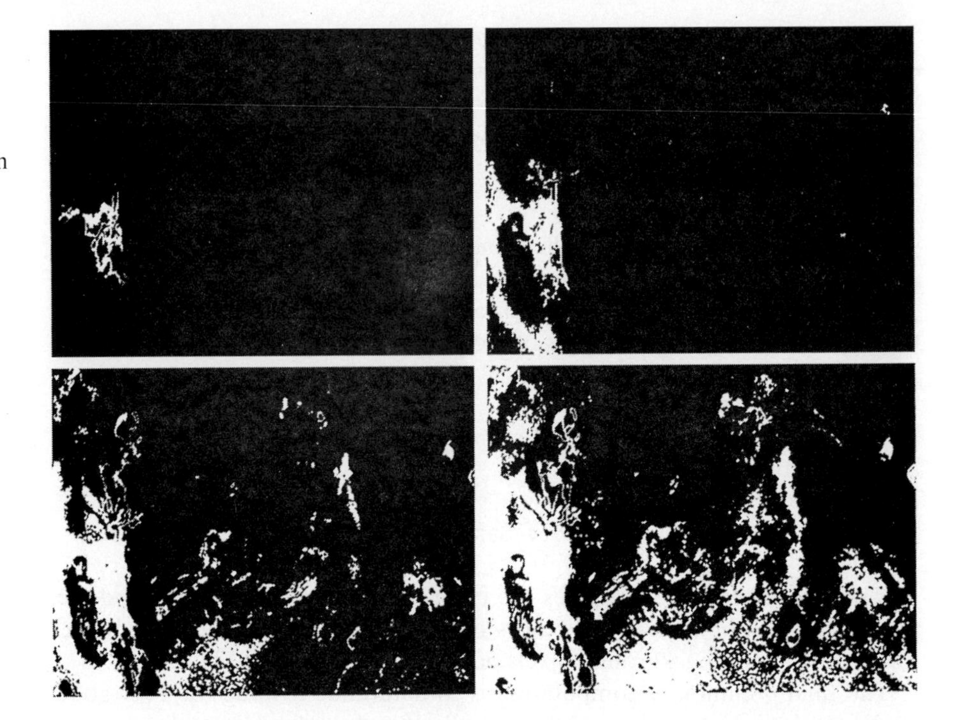

a b
c d

FIGURE 6.28 (a) through (d) Segmentation of Fig. 6.27(a) using option `'euclidean'` in function `colorseg` with $T = 25, 50, 75,$ and 100, respectively.

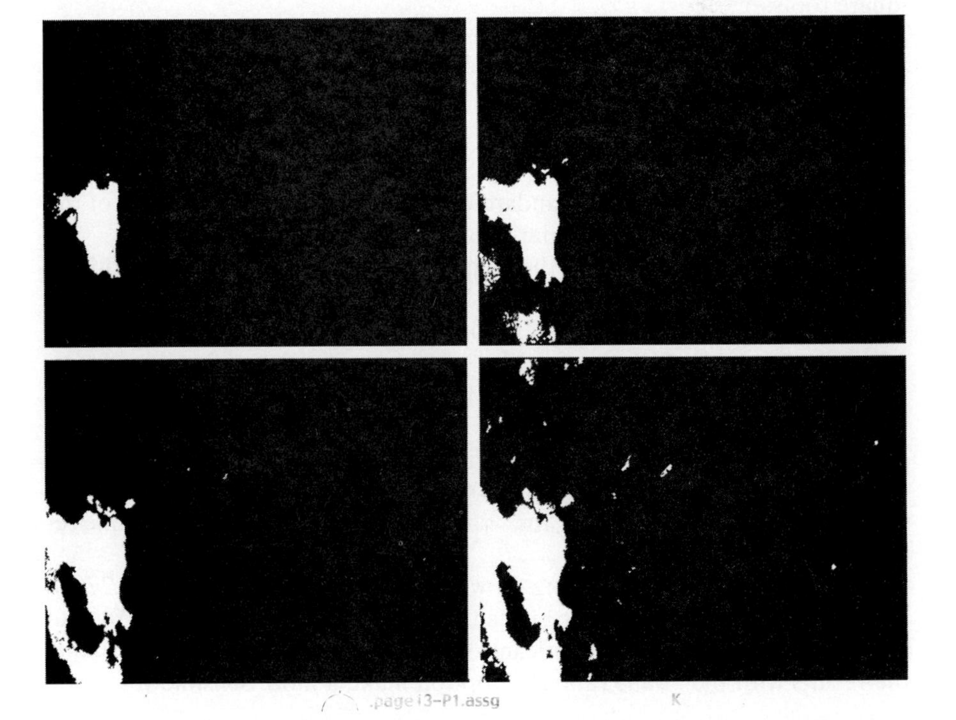

a b
c d

FIGURE 6.29 (a) through (d) Segmentation of Fig. 6.27(a) using option `'mahalanobis'` in function `colorseg` with $T = 25, 50, 75,$ and 100, respectively. Compare with Fig. 6.28.

for increasing values of T. The reason is that the 3-D color data spread in the ROI is fitted much better in this case with an ellipsoid than with a sphere. Note that in both methods increasing T allowed weaker shades of red to be included in the segmented regions, as expected. ■

Summary

The material in this chapter is an introduction to basic topics in the application and use of color in image processing, and on the implementation of these concepts using MATLAB, IPT, and the new functions developed in the preceding sections. The area of color models is broad enough so that entire books have been written on just this topic. The models discussed here were selected for their usefulness in image processing, and also because they provide a good foundation for further study in this area.

The material on pseudocolor and full-color processing on individual color planes provides a tie to the image processing techniques developed in the previous chapters for monochrome images. The material on color vector space is a departure from the methods discussed in those chapters, and highlights some important differences between gray-scale and full-color image processing. The techniques for color-vector processing discussed in the previous section are representative of vector-based processes that include median and other order filters, adaptive and morphological filters, image restoration, image compression, and many others.

7 Wavelets

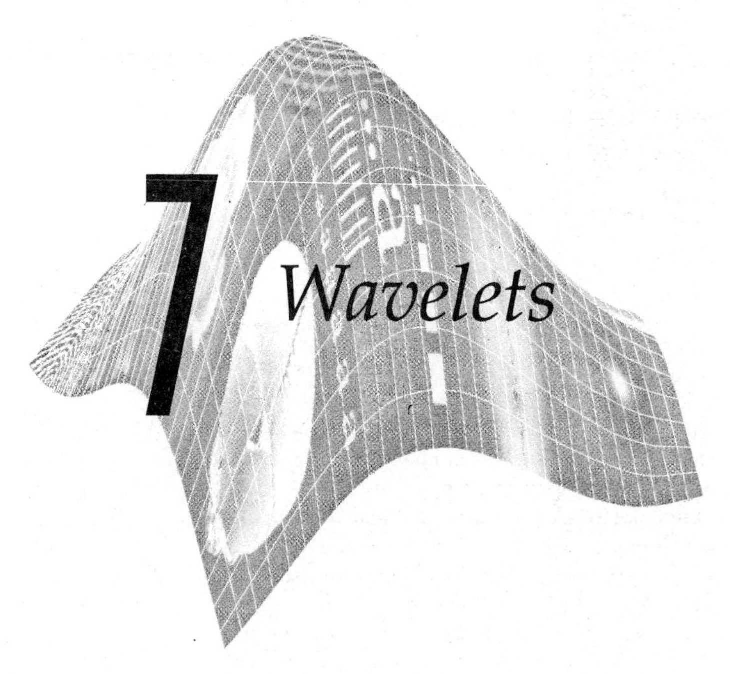

Preview

When digital images are to be viewed or processed at multiple resolutions, the *discrete wavelet transform* (DWT) is the mathematical tool of choice. In addition to being an efficient, highly intuitive framework for the representation and storage of *multiresolution* images, the DWT provides powerful insight into an image's spatial and frequency characteristics. The Fourier transform, on the other hand, reveals only an image's frequency attributes.

In this chapter, we explore both the computation and use of the discrete wavelet transform. We introduce the *Wavelet Toolbox*, a collection of MathWorks' functions designed for wavelet analysis but not included in MATLAB's *Image Processing Toolbox* (IPT), and develop a compatible set of routines that allow basic wavelet-based processing using IPT alone; that is, without the Wavelet Toolbox. These custom functions, in combination with IPT, provide the tools needed to implement all the concepts discussed in Chapter 7 of *Digital Image Processing* by Gonzalez and Woods [2002]. They are applied in much the same way—and provide a similar range of capabilities—as IPT functions `fft2` and `ifft2` in Chapter 4.

7.1 Background

Consider an image $f(x, y)$ of size $M \times N$ whose forward, discrete transform, $T(u, v, \dots)$, can be expressed in terms of the general relation

$$T(u, v, \dots) = \sum_{x, y} f(x, y) g_{u, v, \dots}(x, y)$$

where x and y are spatial variables and $u, v, \dots$ are *transform domain variables*. Given $T(u, v, \dots)$, $f(x, y)$ can be obtained using the generalized inverse discrete transform

$$f(x, y) = \sum_{u, v, \ldots} T(u, v, \ldots) h_{u, v, \ldots}(x, y)$$

The $g_{u, v, \ldots}$ and $h_{u, v, \ldots}$ in these equations are called *forward* and *inverse transformation kernels*, respectively. They determine the nature, computational complexity, and ultimate usefulness of the *transform pair. Transform coefficients* $T(u, v, \ldots)$ can be viewed as the *expansion coefficients* of a series expansion of f with respect to $\{h_{u, v, \ldots}\}$. That is, the inverse transformation kernel defines a set of *expansion functions* for the series expansion of f.

The discrete Fourier transform (DFT) of Chapter 4 fits this series expansion formulation well.[†] In this case

$$h_{u, v}(x, y) = g^*_{u, v}(x, y) = \frac{1}{\sqrt{MN}} e^{j2\pi(ux/M + vy/N)}$$

where $j = \sqrt{-1}$, $*$ is the complex conjugate operator, $u = 0, 1, \ldots, M - 1$, and $v = 0, 1, \ldots, N - 1$. Transform domain variables v and u represent horizontal and vertical frequency, respectively. The kernels are *separable* since

$$h_{u, v}(x, y) = h_u(x) h_v(y)$$

for

$$h_u(x) = \frac{1}{\sqrt{M}} e^{j2\pi ux/M} \quad \text{and} \quad h_v(y) = \frac{1}{\sqrt{N}} e^{j2\pi vy/N}$$

and *orthonormal* since

$$\langle h_r, h_s \rangle = \delta_{rs} = \begin{cases} 1 & r = s \\ 0 & \text{otherwise} \end{cases}$$

where $\langle\,\rangle$ is the inner product operator. The separability of the kernels simplifies the computation of the 2-D transform by allowing row-column or column-row passes of a 1-D transform to be used; orthonormality causes the forward and inverse kernels to be the complex conjugates of one another (they would be identical if the functions were real).

Unlike the discrete Fourier transform, which can be completely defined by two straightforward equations that revolve around a single pair of transformation kernels (given previously), the term *discrete wavelet transform* refers to a class of transformations that differ not only in the transformation kernels employed (and thus the expansion functions used), but also in the fundamental nature of those functions (e.g., whether they constitute an orthonormal or biorthogonal basis) and in the way in which they are applied (e.g., how many different resolutions are computed). Since the DWT encompasses a variety of unique but related transformations, we cannot write a single equation that

---

[†] In the DFT formulation of Chapter 4, a $1/MN$ term is placed in the inverse transform equation. Equivalently, it can be incorporated into the forward transform only, or split, as we do here, between the forward and inverse transformations as $1/\sqrt{MN}$.

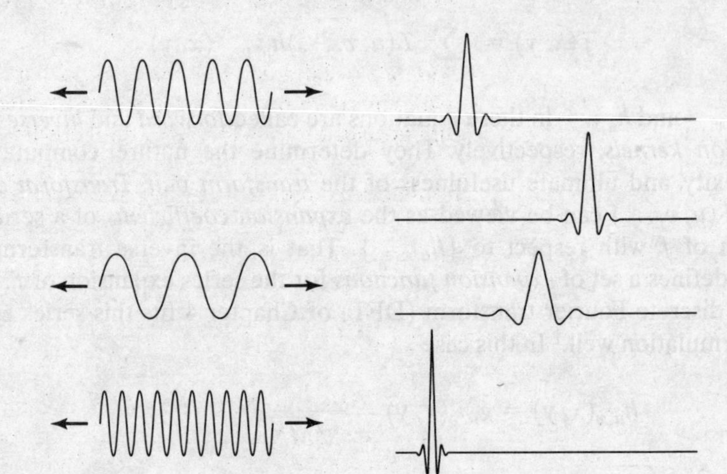

a b

FIGURE 7.1 (a) The familiar Fourier expansion functions are sinusoids of varying frequency and infinite duration. (b) DWT expansion functions are "small waves" of finite duration and varying frequency.

completely describes them all. Instead, we characterize each DWT by a transform kernel pair *or* set of parameters that defines the pair. The various transforms are related by the fact that their expansion functions are "small waves" (hence the name *wavelets*) of varying frequency and limited duration [see Fig. 7.1(b)]. In the remainder of the chapter, we introduce a number of these "small wave" kernels. Each possesses the following general properties:

Property 1: *Separability, Scalability, and Translatability*. The kernels can be represented as three separable 2-D *wavelets*

$$\psi^H(x, y) = \psi(x)\varphi(y)$$
$$\psi^V(x, y) = \varphi(x)\psi(y)$$
$$\psi^D(x, y) = \psi(x)\psi(y)$$

where $\psi^H(x, y)$, $\psi^V(x, y)$, and $\psi^D(x, y)$ are called *horizontal, vertical*, and *diagonal wavelets*, respectively, and one separable 2-D *scaling function*

$$\varphi(x, y) = \varphi(x)\varphi(y)$$

Each of these 2-D functions is the product of two 1-D real, square-integrable scaling and wavelet functions

$$\varphi_{j,k}(x) = 2^{j/2}\varphi(2^j x - k)$$
$$\psi_{j,k}(x) = 2^{j/2}\psi(2^j x - k)$$

Translation k determines the position of these 1-D functions along the x-axis, *scale j* determines their width—how broad or narrow they are along x—and $2^{j/2}$ controls their height or amplitude. Note that the associated expansion functions are binary scalings and integer translates of *mother* wavelet $\psi(x) = \psi_{0,0}(x)$ and scaling function $\varphi(x) = \varphi_{0,0}(x)$.

Property 2: *Multiresolution Compatibility*. The 1-D scaling function just introduced satisfies the following requirements of multiresolution analysis:

a. $\varphi_{j,k}$ is orthogonal to its integer translates.
b. The set of functions that can be represented as a series expansion of $\varphi_{j,k}$ at low scales or resolutions (i.e., small j) is contained within those that can be represented at higher scales.
c. The only function that can be represented at every scale is $f(x) = 0$.
d. Any function can be represented with arbitrary precision as $j \to \infty$.

When these conditions are met, there is a companion wavelet $\psi_{j,k}$ that, together with its integer translates and binary scalings, spans—that is, can represent—the difference between any two sets of $\varphi_{j,k}$-representable functions at adjacent scales.

Property 3: *Orthogonality*. The expansion functions [i.e.,$\{\varphi_{j,k}(x)\}$] form an orthonormal or biorthogonal *basis* for the set of 1-D measurable, square-integrable functions. To be called a basis, there must be a unique set of expansion coefficients for every representable function. As was noted in the introductory remarks on Fourier kernels, $g_{u,v,...} = h_{u,v,...}$ for real, orthonormal kernels. For the biorthogonal case,

$$\langle h_r, g_s \rangle = \delta_{rs} = \begin{cases} 1 & r = s \\ 0 & \text{otherwise} \end{cases}$$

and g is called the *dual* of h. For a biorthogonal wavelet transform with scaling and wavelet functions $\varphi_{j,k}(x)$ and $\psi_{j,k}(x)$, the duals are denoted $\tilde{\varphi}_{j,k}(x)$ and $\tilde{\psi}_{j,k}(x)$, respectively.

7.2 The Fast Wavelet Transform

An important consequence of the above properties is that both $\varphi(x)$ and $\psi(x)$ can be expressed as linear combinations of double-resolution copies of themselves. That is, via the series expansions

$$\varphi(x) = \sum_n h_\varphi(n)\sqrt{2}\varphi(2x - n)$$

$$\psi(x) = \sum_n h_\psi(n)\sqrt{2}\varphi(2x - n)$$

where h_φ and h_ψ—the expansion coefficients—are called *scaling* and *wavelet vectors*, respectively. They are the filter coefficients of the *fast wavelet transform* (FWT), an iterative computational approach to the DWT shown in Fig. 7.2. The $W_\varphi(j, m, n)$ and $\{W_\psi^i(j, m, n)$ for $i = H, V, D\}$ outputs in this figure are the DWT coefficients at scale j. Blocks containing time-reversed scaling and wavelet vectors—the $h_\varphi(-n)$ and $h_\psi(-m)$—are *lowpass* and *highpass decomposition filters*, respectively. Finally, blocks containing a 2 and a down arrow represent *downsampling*—extracting every other point from a sequence of points. Mathematically, the series of filtering and downsampling operations used to compute $W_\psi^H(j, m, n)$ in Fig. 7.2 is, for example,

$$W_\psi^H(j, m, n) = h_\psi(-m) * \left[h_\varphi(-n) * W_\varphi(j + 1, m, n)\big|_{n=2k,\, k\geq 0}\right]\Big|_{m=2k,\, k\geq 0}$$

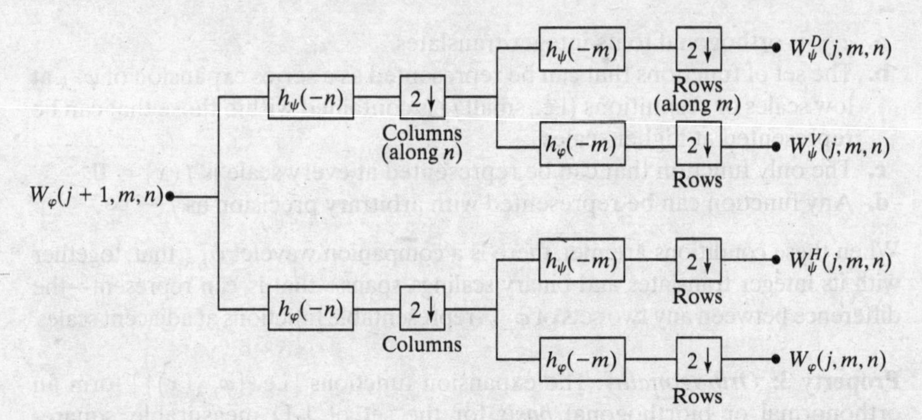

FIGURE 7.2 The 2-D fast wavelet transform (FWT) filter bank. Each pass generates one DWT scale. In the first iteration, $W_\varphi(j+1,m,n) = f(x,y)$.

where * denotes convolution. Evaluating convolutions at nonnegative, even indices is equivalent to filtering and downsampling by 2.

Each pass through the filter bank in Fig. 7.2 decomposes the input into four lower resolution (or lower scale) components. The W_φ coefficients are created via two lowpass (i.e., h_φ-based) filters and are thus called *approximation coefficients*; $\{W_\psi^i \text{ for } i = H, V, D\}$ are *horizontal, vertical*, and *diagonal detail coefficients*, respectively. Since $f(x,y)$ is the highest resolution representation of the image being transformed, it serves as the $W_\varphi(j+1,m,n)$ input for the first iteration. Note that the operations in Fig. 7.2 use *neither* wavelets nor scaling functions—only their associated wavelet and scaling vectors. In addition, three transform domain variables are involved—scale j and horizontal and vertical translation, n and m. These variables correspond to $u, v, \ldots$ in the first two equations of Section 7.1.

7.2.1 FWTs Using the Wavelet Toolbox

In this section, we use MATLAB's Wavelet Toolbox to compute the FWT of a simple 4×4 test image. In the next section, we will develop custom functions to do this without the Wavelet Toolbox (i.e., with IPT alone). The material here lays the groundwork for their development.

The Wavelet Toolbox provides decomposition filters for a wide variety of fast wavelet transforms. The filters associated with a specific transform are accessed via the function `wfilters`, which has the following general syntax:

W wfilters

```
[Lo_D, Hi_D, Lo_R, Hi_R] = wfilters(wname)
```

The W on the icon is used to denote a MATLAB Wavelet Toolbox function, as opposed to a MATLAB or Image Processing Toolbox function.

Here, input parameter `wname` determines the returned filter coefficients in accordance with Table 7.1; outputs `Lo_D`, `Hi_D`, `Lo_R`, and `Hi_R` are row vectors that return the lowpass decomposition, highpass decomposition, lowpass reconstruction, and highpass reconstruction filters, respectively. (Reconstruction filters are discussed in Section 7.4.) Frequently coupled filter pairs can alternately be retrieved using

```
[F1, F2] = wfilters(wname, type)
```

TABLE 7.1 Wavelet Toolbox FWT filters and filter family names.

Wavelet	wfamily	wname
Haar	'haar'	'haar'
Daubechies	'db'	'db2', 'db3', ..., 'db45'
Coiflets	'coif'	'coif1', 'coif2', ..., 'coif5'
Symlets	'sym'	'sym2', 'sym3', ..., 'sym45'
Discrete Meyer	'dmey'	'dmey'
Biorthogonal	'bior'	'bior1.1', 'bior1.3', 'bior1.5', 'bior2.2', 'bior2.4', 'bior2.6', 'bior2.8', 'bior3.1', 'bior3.3', 'bior3.5', 'bior3.7', 'bior3.9', 'bior4.4', 'bior5.5', 'bior6.8'
Reverse Biorthogonal	'rbio'	'rbio1.1', 'rbio1.3', 'rbio1.5', 'rbio2.2', 'rbio2.4', 'rbio2.6', 'rbio2.8', 'rbio3.1', 'rbio3.3', 'rbio3.5', 'rbio3.7', 'rbio3.9', 'rbio4.4', 'rbio5.5', 'rbio6.8'

with `type` set to `'d'`, `'r'`, `'l'`, or `'h'` to obtain a pair of decomposition, reconstruction, lowpass, or highpass filters, respectively. If this syntax is employed, a decomposition or lowpass filter is returned in `F1`, and its companion is placed in `F2`.

Table 7.1 lists the FWT filters included in the Wavelet Toolbox. Their properties—and other useful information on the associated scaling and wavelet functions—is available in the literature on digital filtering and multiresolution analysis. Some of the more important properties are provided by the Wavelet Toolbox's `waveinfo` and `wavefun` functions. To print a written description of wavelet family `wfamily` (see Table 7.1) on MATLAB's Command Window, for example, enter

```
waveinfo(wfamily)
```

at the MATLAB prompt. To obtain a digital approximation of an orthonormal transform's scaling and/or wavelet functions, type

```
[phi, psi, xval] = wavefun(wname, iter)
```

which returns approximation vectors, `phi` and `psi`, and evaluation vector `xval`. Positive integer `iter` determines the accuracy of the approximations by controlling the number of iterations used in their computation. For biorthogonal transforms, the appropriate syntax is

```
[phi1, psi1, phi2, psi2, xval] = wavefun(wname, iter)
```

where `phi1` and `psi1` are decomposition functions and `phi2` and `psi2` are reconstruction functions.

EXAMPLE 7.1: Haar filters, scaling, and wavelet functions.

■ The oldest and simplest wavelet transform is based on the Haar scaling and wavelet functions. The decomposition and reconstruction filters for a Haar-based transform are of length 2 and can be obtained as follows:

```
>> [Lo_D, Hi_D, Lo_R, Hi_R] = wfilters('haar')
Lo_D =
    0.7071    0.7071
Hi_D =
   -0.7071    0.7071
Lo_R =
    0.7071    0.7071
Hi_R =
    0.7071   -0.7071
```

Their key properties (as reported by the `waveinfo` function) and plots of the associated scaling and wavelet functions can be obtained using

```
>> waveinfo('haar');
HAARINFO Information on Haar wavelet.
     Haar Wavelet
     General characteristics: Compactly supported
     wavelet, the oldest and the simplest wavelet.
     scaling function phi = 1 on [0 1] and 0 otherwise.
     wavelet function psi = 1 on [0 0.5], = -1 on [0.5 1] and 0
     otherwise.
     Family                      Haar
     Short name                  haar
     Examples                    haar is the same as db1
     Orthogonal                  yes
     Biorthogonal                yes
     Compact support             yes
     DWT                         possible
     CWT                         possible
     Support width               1
     Filters length              2
     Regularity                  haar is not continuous
     Symmetry                    yes
     Number of vanishing
     moments for psi             1
     Reference: I. Daubechies,
     Ten lectures on wavelets,
     CBMS, SIAM, 61, 1994, 194-202.
>> [phi, psi, xval] = wavefun('haar', 10);
>> xaxis = zeros(size(xval));
>> subplot(121); plot(xval, phi, 'k', xval, xaxis, '--k');
>> axis([0 1 -1.5 1.5]); axis square;
>> title('Haar Scaling Function');
```

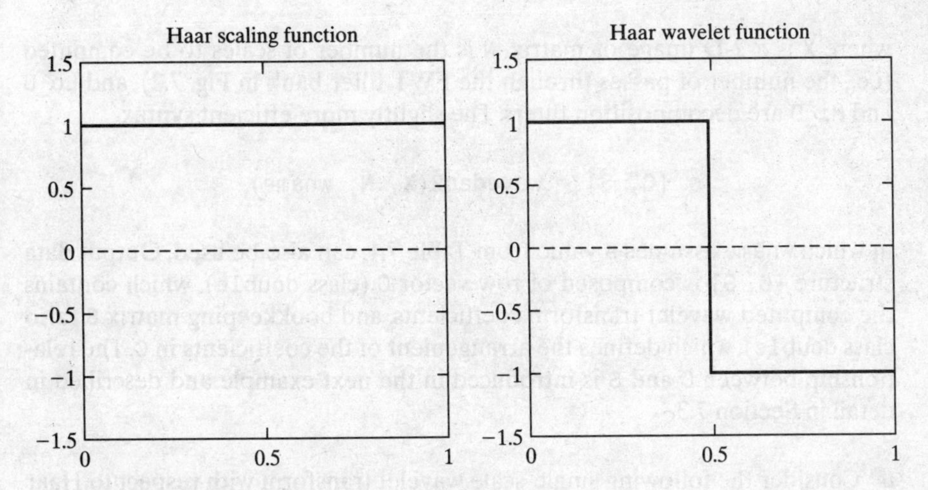

FIGURE 7.3 The Haar scaling and wavelet functions.

```
>> subplot(122); plot(xval, psi, 'k', xval, xaxis, '--k');
>> axis([0 1 −1.5 1.5]); axis square;
>> title('Haar Wavelet Function');
```

Figure 7.3 shows the display generated by the final six commands. Functions `title`, `axis`, and `plot` were described in Chapters 2 and 3; function `subplot` is used to subdivide the figure window into an array of axes or subplots. It has the following generic syntax:

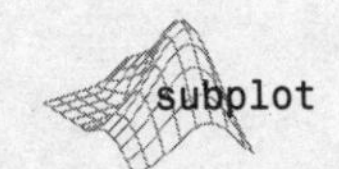


```
H = subplot(m, n, p) or H = subplot(mnp)
```

where `m` and `n` are the number of rows and columns in the subplot array, respectively. Both `m` and `n` must be greater than 1. Optional output variable `H` is the handle of the subplot (i.e., axes) selected by `p`, with incremental values of `p` (beginning at 1) selecting axes along the top row of the figure window, then the second row, and so on. With or without `H`, the `p`th axes is made the current plot. Thus, the `subplot(122)` function in the commands given previously selects the plot in row 1 and column 2 of a 1×2 subplot array as the current plot; the subsequent `axis` and `title` functions then apply only to it.

The Haar scaling and wavelet functions shown in Figure 7.3 are discontinuous and *compactly supported*, which means they are 0 outside a finite interval called the *support*. Note that the support is 1. In addition, the `waveinfo` data reveals that the Haar expansion functions are orthogonal, so that the forward and inverse transformation kernels are identical. ■

Given a set of decomposition filters, whether user provided or generated by the `wfilters` function, the simplest way of computing the associated wavelet transform is through the Wavelet Toolbox's `wavedec2` function. It is invoked using

```
[C, S] = wavedec2(X, N, Lo_D, Hi_D)
```

where X is a 2-D image or matrix, N is the number of scales to be computed (i.e., the number of passes through the FWT filter bank in Fig. 7.2), and Lo_D and Hi_D are decomposition filters. The slightly more efficient syntax

```
[C, S] = wavedec2(X, N, wname)
```

in which wname assumes a value from Table 7.1, can also be used. Output data structure [C, S] is composed of row vector C (class double), which contains the computed wavelet transform coefficients, and bookkeeping matrix S (also class double), which defines the arrangement of the coefficients in C. The relationship between C and S is introduced in the next example and described in detail in Section 7.3.

EXAMPLE 7.2: A simple FWT using Haar filters.

■ Consider the following single-scale wavelet transform with respect to Haar wavelets:

```
>> f = magic(4)
f =
    16     2     3    13
     5    11    10     8
     9     7     6    12
     4    14    15     1

>> [c1, s1] = wavedec2(f, 1, 'haar')
c1 =
  Columns 1 through 9
     17.0000     17.0000     17.0000     17.0000      1.0000
     -1.0000     -1.0000      1.0000      4.0000

  Columns 10 through 16
     -4.0000     -4.0000      4.0000     10.0000      6.0000
     -6.0000    -10.0000

s1 =
     2     2
     2     2
     4     4
```

Here, a 4×4 magic square f is transformed into a 1×16 wavelet decomposition vector c1 and 3×2 bookkeeping matrix s1. The entire transformation is performed with a single execution (with f used as the input) of the operations depicted in Fig. 7.2. Four 2×2 outputs—a downsampled approximation and three directional (horizontal, vertical, and diagonal) detail matrices—are generated. Function wavedec2 concatenates these 2×2 matrices columnwise in row vector c1 beginning with the approximation coefficients and continuing with the horizontal, vertical, and diagonal details. That is, c1(1) through c1(4) are approximation coefficients $W_\varphi(1, 0, 0)$, $W_\varphi(1, 1, 0)$, $W_\varphi(1, 0, 1)$, and $W_\varphi(1, 1, 1)$ from Fig. 7.2 with the scale of f assumed arbitrarily to be 2; c1(5) through c1(8) are $W_\psi^H(1, 0, 0)$, $W_\psi^H(1, 1, 0)$, $W_\psi^H(1, 0, 1)$, and $W_\psi^H(1, 1, 1)$;

and so on. It we were to extract the horizontal detail coefficient matrix from vector c1, for example, we would get

$$W_{\psi}^{H} = \begin{bmatrix} 1 & -1 \\ -1 & 1 \end{bmatrix}$$

Bookkeeping matrix s1 provides the sizes of the matrices that have been concatenated a column at a time into row vector c1—plus the size of the original image f [in vector s1(end, :)]. Vectors s1(1, :) and s1(2, :) contain the sizes of the computed approximation matrix and three detail coefficient matrices, respectively. The first element of each vector is the number of rows in the referenced detail or approximation matrix; the second element is the number of columns.

When the single-scale transform described above is extended to two scales, we get

```
>> [c2, s2] = wavedec2(f, 2, 'haar')
c2 =
  Columns 1 through 9
     34.0000          0          0     0.0000     1.0000
     -1.0000    -1.0000     1.0000     4.0000
  Columns 10 through 16
     -4.0000    -4.0000     4.0000    10.0000     6.0000
     -6.0000   -10.0000

s2 =
     1     1
     1     1
     2     2
     4     4
```

Note that c2(5:16) = c1(5:16). Elements c1(1:4), which were the approximation coefficients of the single-scale transform, have been fed into the filter bank of Fig. 7.2 to produce four 1×1 outputs: $W_{\varphi}(0, 0, 0)$, $W_{\psi}^{H}(0, 0, 0)$, $W_{\psi}^{V}(0, 0, 0)$, and $W_{\psi}^{D}(0, 0, 0)$. These outputs are concatenated columnwise (though they are 1×1 matrices here) in the same order that was used in the preceding single-scale transform and substituted for the approximation coefficients from which they were derived. Bookkeeping matrix s2 is then updated to reflect the fact that the single 2×2 approximation matrix in c1 has been replaced by four 1×1 detail and approximation matrices in c2. Thus, s2(end, :) is once again the size of the original image, s2(3, :) is the size of the three detail coefficient matrices at scale 1, s2(2, :) is the size of the three detail coefficient matrices at scale 0, and s2(1, :) is the size of the final approximation. ■

To conclude this section, we note that because the FWT is based on digital filtering techniques and thus convolution, border distortions can arise. To minimize these distortions, the border must be treated differently from the other

TABLE 7.2 Wavelet Toolbox image extension or padding modes.

STATUS	Description
`'sym'`	The image is extended by mirror reflecting it across its borders. This is the normal default mode.
`'zpd'`	The image is extended by padding with a value of 0.
`'spd'`, `'sp1'`	The image is extended by first-order derivative extrapolation—or padding with a linear extension of the outmost two border values.
`'sp0'`	The image is extended by extrapolating the border values—that is, by boundary value replication.
`'ppd'`	The image is extended by periodic padding.
`'per'`	The image is extended by periodic padding after it has been padded (if necessary) to an even size using `'sp0'` extension.

parts of the image. When filter elements fall outside the image during the convolution process, values must be assumed for the area, which is about the size of the filter, outside the image. Many Wavelet Toolbox functions, including the `wavedec2` function, extend or pad the image being processed based on global parameter `dwtmode`. To examine the active extension mode, enter `st = dwtmode('status')` or simply `dwtmode` at the MATLAB command prompt (e.g., `>> dwtmode`). To set the extension mode to `STATUS`, enter `dwtmode(STATUS)`; to make `STATUS` the default extension mode, use `dwtmode('save', STATUS)`. The supported extension modes and corresponding `STATUS` values are listed in Table 7.2.

7.2.2 FWTs without the Wavelet Toolbox

In this section, we develop a pair of custom functions, `wavefilter` and `wavefast`, to replace the Wavelet Toolbox functions, `wfilters` and `wavedec2`, of the previous section. Our goal is to provide additional insight into the mechanics of computing FWTs, and to begin the process of building a "standalone package" for basic wavelet-based image processing without the Wavelet Toolbox. This process is completed in Sections 7.3 and 7.4, and the resulting set of functions is used to generate the examples in Section 7.5.

The first step is to devise a function for generating wavelet decomposition and reconstruction filters. The following function, which we call `wavefilter`, uses a standard `switch` construct, together with `case` and `otherwise`, to do this in a readily extendable manner. Although `wavefilter` provides only the filters examined in Chapters 7 and 8 of *Digital Image Processing* (Gonzalez and Woods [2002]), other wavelet transforms can be accommodated by adding (as new "cases") the appropriate decomposition and reconstruction filters from the literature.

wavefilter

```
function [varargout] = wavefilter(wname, type)
%WAVEFILTER Create wavelet decomposition and reconstruction filters.
%   [VARARGOUT] = WAVEFILTER(WNAME, TYPE) returns the decomposition
%   and/or reconstruction filters used in the computation of the
%   forward and inverse FWT (fast wavelet transform).
```

```
%
%   EXAMPLES:
%     [ld, hd, lr, hr] = wavefilter('haar') Get the low and highpass
%                                           decomposition (ld, hd)
%                                           and reconstruction
%                                           (lr, hr) filters for
%                                           wavelet 'haar'.
%     [ld, hd] = wavefilter('haar','d')     Get decomposition filters
%                                           ld and hd.
%     [lr, hr] = wavefilter('haar','r')     Get reconstruction
%                                           filters lr and hr.
%
%   INPUTS:
%     WNAME                          Wavelet Name
%     ----------------------------------------------------------------
%     'haar' or 'db1'                Haar
%     'db4'                          4th order Daubechies
%     'sym4'                         4th order Symlets
%     'bior6.8'                      Cohen-Daubechies-Feauveau biorthogonal
%     'jpeg9.7'                      Antonini-Barlaud-Mathieu-Daubechies
%
%     TYPE                           Filter Type
%     ----------------------------------------------------------------
%     'd'                            Decomposition filters
%     'r'                            Reconstruction filters
%
%   See also WAVEFAST and WAVEBACK.

% Check the input and output arguments.
error(nargchk(1, 2, nargin));

if (nargin == 1 & nargout ~= 4) | (nargin == 2 & nargout ~= 2)
   error('Invalid number of output arguments.');
end

if nargin == 1 & ~ischar(wname)
   error('WNAME must be a string.');
end

if nargin == 2 & ~ischar(type)
   error('TYPE must be a string.');
end

% Create filters for the requested wavelet.
switch lower(wname)
case {'haar', 'db1'}
   ld = [1 1]/sqrt(2);     hd = [-1 1]/sqrt(2);
   lr = ld;                hr = -hd;

case 'db4'
   ld = [-1.059740178499728e-002 3.288301166698295e-002 ...
         3.084138183598697e-002 -1.870348117188811e-001 ...
         -2.798376941698385e-002 6.308807679295904e-001 ...
         7.148465705525415e-001 2.303778133088552e-001];
```

```
        t = (0:7);
        hd = ld;   hd(end:-1:1) = cos(pi * t) .* ld;
        lr = ld;   lr(end:-1:1) = ld;
        hr = cos(pi * t) .* ld;

    case 'sym4'
        ld = [-7.576571478927333e-002 -2.963552764599851e-002 ...
              4.976186676320155e-001 8.037387518059161e-001 ...
              2.978577956052774e-001 -9.921954357684722e-002 ...
              -1.260396726203783e-002 3.222310060404270e-002];
        t = (0:7);
        hd = ld;   hd(end:-1:1) = cos(pi * t) .* ld;
        lr = ld;   lr(end:-1:1) = ld;
        hr = cos(pi * t) .* ld;

    case 'bior6.8'
        ld = [0 1.908831736481291e-003 -1.914286129088767e-003 ...
              -1.699063986760234e-002 1.193456527972926e-002 ...
              4.973290349094079e-002 -7.726317316720414e-002 ...
              -9.405920349573646e-002 4.207962846098268e-001 ...
              8.259229974584023e-001 4.207962846098268e-001 ...
              -9.405920349573646e-002 -7.726317316720414e-002 ...
              4.973290349094079e-002 1.193456527972926e-002 ...
              -1.699063986760234e-002 -1.914286129088767e-003 ...
              1.908831736481291e-003];
        hd = [0 0 0 1.442628250562444e-002 -1.446750489679015e-002 ...
              -7.872200106262882e-002 4.036797903033992e-002 ...
              4.178491091502746e-001 -7.589077294536542e-001 ...
              4.178491091502746e-001 4.036797903033992e-002 ...
              -7.872200106262882e-002 -1.446750489679015e-002 ...
              1.442628250562444e-002 0 0 0 0];
        t = (0:17);
        lr = cos(pi * (t + 1)) .* hd;
        hr = cos(pi * t) .* ld;

    case 'jpeg9.7'
        ld = [0 0.02674875741080976 -0.01686411844287495 ...
              -0.07822326652898785 0.2668641184428723 ...
              0.6029490182363579 0.2668641184428723 ...
              -0.07822326652898785 -0.01686411844287495 ...
              0.02674875741080976];
        hd = [0 -0.09127176311424948 0.05754352622849957 ...
              0.5912717631142470 -1.115087052456994 ...
              0.5912717631142470 0.05754352622849957 ...
              -0.09127176311424948 0 0];
        t = (0:9);
        lr = cos(pi * (t + 1)) .* hd;
        hr = cos(pi * t) .* ld;

    otherwise
        error('Unrecognizable wavelet name (WNAME).');
    end
```

```
% Output the requested filters.
if (nargin == 1)
   varargout(1:4) = {ld, hd, lr, hr};
else
   switch lower(type(1))
   case 'd'
      varargout = {ld, hd};
   case 'r'
      varargout = {lr, hr};
   otherwise
      error('Unrecognizable filter TYPE.');
   end
end
```

Note that for each orthonormal filter in `wavefilter` (i.e., `'haar'`, `'db4'`, and `'sym4'`), the reconstruction filters are time-reversed versions of the decomposition filters and the highpass decomposition filter is a modulated version of its lowpass counterpart. Only the lowpass decomposition filter coefficients need to be explicitly enumerated in the code. The remaining filter cocfficients can be computed from them. In `wavefilter`, time reversal is carried out by reordering filter vector elements from last to first with statements like `lr(end:-1:1) = ld`. Modulation is accomplished by multiplying the components of a known filter by `cos(pi*t)`, which alternates between 1 and -1 as `t` increases from 0 in integer steps. For each biorthogonal filter in `wavefilter` (i.e., `'bior6.8'` and `'jpeg9.7'`), both the lowpass and highpass decomposition filters are specified; the reconstruction filters are computed as modulations of them. Finally, we note that the filters generated by `wavefilter` are of even length. Moreover, zero padding is used to ensure that the lengths of the decomposition and reconstruction filters of each wavelet are identical.

Given a pair of `wavefilter` generated decomposition filters, it is easy to write a general-purpose routine for the computation of the related fast wavelet transform. The goal is to devise an efficient algorithm based on the filtering and downsampling operations in Fig. 7.2. To maintain compatibility with the existing Wavelet Toolbox, we employ the same decomposition structure (i.e., `[C, S]` where `C` is a decomposition vector and `S` is a bookkeeping matrix). The following routine, which we call `wavefast`, uses symmetric image extension to reduce the border distortion associated with the computed FWT:

wavefast

```
function [c, s] = wavefast(x, n, varargin)
%WAVEFAST Perform multi-level 2-dimensional fast wavelet transform.
%   [C, L] = WAVEFAST(X, N, LP, HP) performs a 2D N-level FWT of
%   image (or matrix) X with respect to decomposition filters LP and
%   HP.
%
%   [C, L] = WAVEFAST(X, N, WNAME) performs the same operation but
%   fetches filters LP and HP for wavelet WNAME using WAVEFILTER.
%
%   Scale parameter N must be less than or equal to log2 of the
%   maximum image dimension. Filters LP and HP must be even. To
```

```
%   reduce border distortion, X is symmetrically extended. That is,
%   if X = [c1 c2 c3 ... cn] (in 1D), then its symmetric extension
%   would be [... c3 c2 c1 c1 c2 c3 ... cn cn cn-1 cn-2 ...].
%
%   OUTPUTS:
%     Matrix C is a coefficient decomposition vector:
%
%      C = [ a(n) h(n) v(n) d(n) h(n-1)  ... v(1) d(1) ]
%
%     where a, h, v, and d are columnwise vectors containing
%     approximation, horizontal, vertical, and diagonal coefficient
%     matrices, respectively. C has 3n + 1 sections where n is the
%     number of wavelet decompositions.
%
%     Matrix S is an (n+2) x 2 bookkeeping matrix:
%
%      S = [ sa(n, :); sd(n, :); sd(n-1, :); ...; sd(1, :); sx ]
%
%     where sa and sd are approximation and detail size entries.
%
%   See also WAVEBACK and WAVEFILTER.

%   Check the input arguments for reasonableness.
error(nargchk(3, 4, nargin));

if nargin == 3
   if ischar(varargin{1})
      [lp, hp] = wavefilter(varargin{1}, 'd');
   else
      error('Missing wavelet name.');
   end
else
      lp = varargin{1};     hp = varargin{2};
end

fl = length(lp);       sx = size(x);

if (ndims(x) ~= 2) | (min(sx) < 2) | ~isreal(x) | ~isnumeric(x)
   error('X must be a real, numeric matrix.');
end

if (ndims(lp) ~= 2) | ~isreal(lp) | ~isnumeric(lp) ...
      | (ndims(hp) ~= 2) | ~isreal(hp) | ~isnumeric(hp) ...
      | (fl ~= length(hp)) | rem(fl, 2) ~= 0
   error(['LP and HP must be even and equal length real, ' ...
         'numeric filter vectors.']);
end

if ~isreal(n) | ~isnumeric(n) | (n < 1) | (n > log2(max(sx)))
   error(['N must be a real scalar between 1 and ' ...
         'log2(max(size((X))).']);
end
```

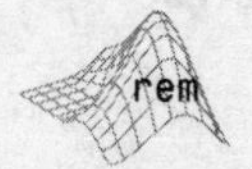

rem (X, Y) *returns the remainder of the division of* X *by* Y.

```
% Init the starting output data structures and initial approximation.
c = [];    s = sx;    app = double(x);

% For each decomposition ...
for i = 1:n
   % Extend the approximation symmetrically.
   [app, keep] = symextend(app, fl);

   % Convolve rows with HP and downsample. Then convolve columns
   % with HP and LP to get the diagonal and vertical coefficients.
   rows = symconv(app, hp, 'row', fl, keep);
   coefs = symconv(rows, hp, 'col', fl, keep);
   c = [coefs(:)' c];    s = [size(coefs); s];
   coefs = symconv(rows, lp, 'col', fl, keep);
   c = [coefs(:)' c];

   % Convolve rows with LP and downsample. Then convolve columns
   % with HP and LP to get the horizontal and next approximation
   % coefficients.
   rows = symconv(app, lp, 'row', fl, keep);
   coefs = symconv(rows, hp, 'col', fl, keep);
   c = [coefs(:)' c];
   app = symconv(rows, lp, 'col', fl, keep);
end

% Append final approximation structures.
c = [app(:)' c];    s = [size(app); s];

%-------------------------------------------------------------------%
function [y, keep] = symextend(x, fl)
% Compute the number of coefficients to keep after convolution
% and downsampling. Then extend x in both dimensions.

keep = floor((fl + size(x) - 1) / 2);
y = padarray(x, [(fl - 1) (fl - 1)], 'symmetric', 'both');

%-------------------------------------------------------------------%
function y = symconv(x, h, type, fl, keep)
% Convolve the rows or columns of x with h, downsample,
% and extract the center section since symmetrically extended.

if strcmp(type, 'row')
   y = conv2(x, h);
   y = y(:, 1:2:end);
   y = y(:, fl / 2 + 1:fl / 2 + keep(2));
else
   y = conv2(x, h');
   y = y(1:2:end, :);
   y = y(fl / 2 + 1:fl / 2 + keep(1), :);
end
```

conv2

`C = conv2(A, B)` *performs the 2-D convolution of matrices* A *and* B.

As can be seen in the main routine, only one `for` loop, which cycles through the decomposition levels (or scales) that are generated, is used to orchestrate the entire forward transform computation. For each execution of the loop, the current approximation image, `app`, which is initially set to `x`, is symmetrically extended by internal function `symextend`. This function calls `padarray`, which was introduced in Section 3.4.2, to extend `app` in two dimensions by mirror reflecting `fl - 1` of its elements (the length of the decomposition filter minus 1) across its border.

Function `symextend` returns an extended matrix of approximation coefficients and the number of pixels that should be extracted from the center of any subsequently convolved and downsampled results. The rows of the extended approximation are next convolved with highpass decomposition filter `hp` and downsampled via `symconv`. This function is described in the following paragraph. Convolved output, `rows`, is then submitted to `symconv` to convolve and downsample its columns with filters `hp` and `lp`—generating the diagonal and vertical detail coefficients of the top two branches of Fig. 7.2. These results are inserted into decomposition vector `c` (working from the last element toward the first) and the process is repeated in accordance with Fig. 7.2 to generate the horizontal detail and approximation coefficients (the bottom two branches of the figure).

Function `symconv` uses the `conv2` function to do the bulk of the transform computation work. It convolves filter `h` with the rows or columns of `x` (depending on `type`), discards the even indexed rows or columns (i.e., downsamples by 2), and extracts the center `keep` elements of each row or column. Invoking `conv2` with matrix `x` and row filter vector `h` initiates a row-by-row convolution; using column filter vector `h'` results in a columnwise convolution.

EXAMPLE 7.3: Comparing the execution times of `wavefast` and `wavedec2`.

■ The following test routine uses functions `tic` and `toc` to compare the execution times of the Wavelet Toolbox function `wavedec2` and custom function `wavefast`:

```
function [ratio, maxdiff] = fwtcompare(f, n, wname)
%FWTCOMPARE Compare wavedec2 and wavefast.
%   [RATIO, MAXDIFF] = FWTCOMPARE(F, N, WNAME) compares the operation
%   of toolbox function WAVEDEC2 and custom function WAVEFAST.
%
%   INPUTS:
%     F            Image to be transformed.
%     N            Number of scales to compute.
%     WNAME        Wavelet to use.
%
%   OUTPUTS:
%     RATIO        Execution time ratio (custom/toolbox)
%     MAXDIFF      Maximum coefficient difference.

% Get transform and computation time for wavedec2.
tic;
[c1, s1] = wavedec2(f, n, wname);
reftime = toc;
```

FIGURE 7.4 A 512 × 512 image of a vase.

```
% Get transform and computation time for wavefast.
tic;
[c2, s2] = wavefast(f, n, wname);
t2 = toc;

% Compare the results.
ratio = t2 / (reftime + eps);
maxdiff = abs(max(c1 - c2));
```

For the 512 × 512 image of Fig. 7.4 and a five-scale wavelet transform with respect to 4th order Daubechies' wavelets, `fwtcompare` yields

```
>> f = imread('Vase', 'tif');
>> [ratio, maxdifference] = fwtcompare(f, 5, 'db4')

ratio =
    0.5508
maxdifference =
    3.2969e-012
```

Note that custom function `wavefast` was almost twice as fast as its Wavelet Toolbox counterpart while producing virtually identical results. ■

7.3 Working with Wavelet Decomposition Structures

The wavelet transformation functions of the previous two sections produce *nondisplayable* data structures of the form $\{\mathbf{c}, \mathbf{S}\}$, where $\mathbf{c}$ is a transform coefficient vector and $\mathbf{S}$ is a bookkeeping matrix that defines the arrangement of coefficients in $\mathbf{c}$. To process images, we must be able to examine and/or modify $\mathbf{c}$. In this section, we formally define $\{\mathbf{c}, \mathbf{S}\}$, examine some of the Wavelet Toolbox functions for manipulating it, and develop a set of custom functions that can be used without the Wavelet Toolbox. These functions are then used to build a general purpose routine for displaying $\mathbf{c}$.

The representation scheme introduced in Example 7.2 integrates the coefficients of a multiscale two-dimensional wavelet transform into a single, one-dimensional vector

$$\mathbf{c} = [\mathbf{A}_N(:)' \quad \mathbf{H}_N(:)' \quad \cdots \quad \mathbf{H}_i(:)' \quad \mathbf{V}_i(:)' \quad \mathbf{D}_i(:)' \quad \cdots \quad \mathbf{V}_1(:)' \quad \mathbf{D}_1(:)']$$

where $\mathbf{A}_N$ is the approximation coefficient matrix of the Nth decomposition level and $\mathbf{H}_i$, $\mathbf{V}_i$, and $\mathbf{D}_i$ for $i = 1, 2, \ldots N$ are the horizontal, vertical, and diagonal transform coefficient matrices for level i. Here, $\mathbf{H}_i(:)'$, for example, is the row vector formed by concatenating the transposed columns of matrix $\mathbf{H}_i$. That is, if

$$\mathbf{H}_i = \begin{bmatrix} 3 & -2 \\ 1 & 6 \end{bmatrix}$$

then

$$\mathbf{H}_i(:) = \begin{bmatrix} 3 \\ 1 \\ -2 \\ 6 \end{bmatrix} \quad \text{and} \quad \mathbf{H}_i(:)' = [3 \quad 1 \quad -2 \quad 6]$$

Because the equation for $\mathbf{c}$ assumes N decompositions (or passes through the filter bank in Fig. 7.2), $\mathbf{c}$ contains $3N + 1$ sections—one approximation and N groups of horizontal, vertical, and diagonal details. Note that the highest scale coefficients are computed when $i = 1$; the lowest scale coefficients are associated with $i = N$. Thus, the coefficients of $\mathbf{c}$ are ordered from low to high scale.

Matrix $\mathbf{S}$ of the decomposition structure is an $(N + 2) \times 2$ bookkeeping array of the form

$$\mathbf{S} = [\mathbf{sa}_N; \quad \mathbf{sd}_N; \quad \mathbf{sd}_{N-1}; \quad \cdots \quad \mathbf{sd}_i; \quad \cdots \quad \mathbf{sd}_1; \quad \mathbf{sf}]$$

where $\mathbf{sa}_N$, $\mathbf{sd}_i$, and $\mathbf{sf}$ are 1×2 vectors containing the horizontal and vertical dimensions of Nth-level approximation $\mathbf{A}_N$, ith-level details ($\mathbf{H}_i$, $\mathbf{V}_i$, and $\mathbf{D}_i$ for $i = 1, 2, \ldots N$), and original image $\mathbf{F}$, respectively. The information in $\mathbf{S}$ can be used to locate the individual approximation and detail coefficients in $\mathbf{c}$. Note that the semicolons in the preceding equation indicate that the elements of $\mathbf{S}$ are organized as a column vector.

EXAMPLE 7.4: Wavelet Toolbox functions for manipulating transform decomposition vector $\mathbf{c}$.

■ The Wavelet Toolbox provides a variety of functions for locating, extracting, reformatting, and/or manipulating the approximation and horizontal, vertical, and diagonal coefficients of $\mathbf{c}$ as a function of decomposition level. We introduce them here to illustrate the concepts just discussed and to prepare the way for the alternative functions that will be developed in the next section. Consider, for example, the following sequence of commands:

```
>> f = magic(8);
>> [c1, s1] = wavedec2(f, 3, 'haar');
>> size(c1)
```

```
ans =
     1    64
>> s1
s1 =
     1     1
     1     1
     2     2
     4     4
     8     8
>> approx = appcoef2(c1, s1, 'haar')
approx =
   260.0000
>> horizdet2 = detcoef2('h', c1, s1, 2)
horizdet2 =
   1.0e-013 *
           0   -0.2842
           0         0
>> newc1 = wthcoef2('h', c1, s1, 2);
>> newhorizdet2 = detcoef2('h', newc1, s1, 2)
newhorizdet2 =
     0     0
     0     0
```

Here, a three-level decomposition with respect to Haar wavelets is performed on an 8×8 magic square using the wavedec2 function. The resulting coefficient vector, c1, is of size 1×64. Since s1 is 5×2, we know that the coefficients of c1 span $(N - 2) = (5 - 2) = 3$ decomposition levels. Thus, it concatenates the elements needed to populate $3N + 1 = 3(3) + 1 = 10$ approximation and detail coefficient submatrices. Based on s1, these submatrices include (a) a 1×1 approximation matrix and three 1×1 detail matrices for decomposition level 3 [see s1(1, :) and s1(2, :)], (b) three 2×2 detail matrices for level 2 [see s1(3, :)], and (c) three 4×4 detail matrices for level 1 [see s1(4, :)]. The fifth row of s1 contains the size of the original image f.

Matrix approx = 260 is extracted from c1 using toolbox function appcoef2, which has the following syntax:

```
a = appcoef2(c, s, wname)
```

Here, wname is a wavelet name from Table 7.1 and a is the returned approximation matrix. The horizontal detail coefficients at level 2 are retrieved using detcoef2, a function of similar syntax

```
d = detcoef2(o, c, s, n)
```

in which o is set to 'h', 'v', or 'd' for the horizontal, vertical, and diagonal details and n is the desired decomposition level. In this example, 2×2 matrix horizdet2 is returned. The coefficients corresponding to horizdet2 in c1 are then zeroed using wthcoef2, a wavelet thresholding function of the form

```
nc = wthcoef2(type, c, s, n, t, sorh)
```

where type is set to 'a' to threshold approximation coefficients and 'h', 'v', or 'd' to threshold horizontal, vertical, or diagonal details, respectively. Input n is a vector of decomposition levels to be thresholded based on the corresponding thresholds in vector t, while sorh is set to 's' or 'h' for soft or hard thresholding, respectively. If t is omitted, all coefficients meeting the type and n specifications are zeroed. Output nc is the modified (i.e., thresholded) decomposition vector. All three of the preceding Wavelet Toolbox functions have other syntaxes that can be examined using the MATLAB help command. ■

7.3.1 Editing Wavelet Decomposition Coefficients without the Wavelet Toolbox

Without the Wavelet Toolbox, bookkeeping matrix **S** is the key to accessing the individual approximation and detail coefficients of multiscale vector **c**. In this section, we use **S** to build a set of general-purpose routines for the manipulation of **c**. Function wavework is the foundation of the routines developed, which are based on the familiar cut-copy-paste metaphor of modern word processing applications.

wavework

```
function [varargout] = wavework(opcode, type, c, s, n, x)
%WAVEWORK is used to edit wavelet decomposition structures.
%   [VARARGOUT] = WAVEWORK(OPCODE, TYPE, C, S, N, X) gets the
%   coefficients specified by TYPE and N for access or modification
%   based on OPCODE.
%
%   INPUTS:
%     OPCODE        Operation to perform
%     --------------------------------------------------------------------
%     'copy'        [varargout] = Y = requested (via TYPE and N)
%                   coefficient matrix
%     'cut'         [varargout] = [NC, Y] = New decomposition vector
%                   (with requested coefficient matrix zeroed) AND
%                   requested coefficient matrix
%     'paste'       [varargout] = [NC] = new decomposition vector with
%                   coefficient matrix replaced by X
%
%     TYPE          Coefficient category
%     --------------------------------------------------------------------
%     'a'           Approximation coefficients
%     'h'           Horizontal details
%     'v'           Vertical details
%     'd'           Diagonal details
%
%     [C, S] is a wavelet toolbox decomposition structure.
```

```
%     N is a decomposition level (Ignored if TYPE = 'a').
%     X is a two-dimensional coefficient matrix for pasting.
%
%   See also WAVECUT, WAVECOPY, and WAVEPASTE.

error(nargchk(4, 6, nargin));

if (ndims(c) ~= 2) | (size(c, 1) ~= 1)
   error('C must be a row vector.');
end

if (ndims(s) ~= 2) | ~isreal(s) | ~isnumeric(s) | (size(s, 2) ~= 2)
   error('S must be a real, numeric two-column array.');
end

elements = prod(s, 2);          % Coefficient matrix elements.
if (length(c) < elements(end)) | ...
      ~(elements(1) + 3 * sum(elements(2:end - 1)) >= elements(end))
   error(['[C S] must form a standard wavelet decomposition ' ...
          'structure.']);
end

if strcmp(lower(opcode(1:3)), 'pas') & nargin < 6
   error('Not enough input arguments.');
end

if nargin < 5
   n = 1;          % Default level is 1.
end
nmax = size(s, 1) - 2;           % Maximum levels in [C, S].
aflag = (lower(type(1)) == 'a');
if ~aflag & (n > nmax)
   error('N exceeds the decompositions in [C, S].');
end

switch lower(type(1))            % Make pointers into C.
case 'a'
   nindex = 1;
   start = 1;    stop = elements(1);    ntst = nmax;
case {'h', 'v', 'd'}
   switch type
   case 'h', offset = 0;         % Offset to details.
   case 'v', offset = 1;
   case 'd', offset = 2;
   end
   nindex = size(s, 1) - n;      % Index to detail info.
   start = elements(1) + 3 * sum(elements(2:nmax - n + 1)) + ...
           offset * elements(nindex) + 1;
   stop = start + elements(nindex) - 1;
   ntst = n;
otherwise
   error('TYPE must begin with "a", "h", "v", or "d".');
end
```

```
switch lower(opcode)              % Do requested action.
case {'copy', 'cut'}
   y = repmat(0, s(nindex, :));
   y(:) = c(start:stop);              nc = c;
   if strcmp(lower(opcode(1:3)), 'cut')
      nc(start:stop) = 0; varargout = {nc, y};
   else
       varargout = {y};
   end
case 'paste'
   if prod(size(x)) ~= elements(end - ntst)
      error('X is not sized for the requested paste.');
   else
      nc = c;  nc(start:stop) = x(:);     varargout = {nc};
   end
otherwise
   error('Unrecognized OPCODE.');
end
```

As wavework checks its input arguments for reasonableness, the number of elements in each coefficient submatrix of c is computed via `elements = prod(s, 2)`. Recall from Section 3.4.2 that MATLAB function `Y = prod(X, DIM)` computes the products of the elements of X along dimension DIM. The first switch statement then begins the computation of a pair of pointers to the coefficients associated with input parameters type and n. For the approximation case (i.e., `case 'a'`), the computation is trivial since the coefficients are always at the start of c (so pointer start is 1); the ending index, pointer stop, is the number of elements in the approximation matrix, which is `elements(1)`. When a detail coefficient submatrix is requested, however, start is computed by summing the number of elements at all decomposition levels above n and adding `offset * elements(nindex)`; where offset is 0, 1, or 2 for the horizontal, vertical, or diagonal coefficients, respectively, and nindex is a pointer to the row of s that corresponds to input parameter n.

The second switch statement in function wavework performs the operation requested by opcode. For the `'cut'` and `'copy'` cases, the coefficients of c between start and stop are copied into y, which has been preallocated as a two-dimensional matrix whose size is determined by s. This is done using `y = repmat(0, s(nindex, :))`, in which MATLAB's "replicate matrix" function, `B = repmat(A, M, N)`, is used to create a large matrix B composed of M x N tiled copies of A. For the `'paste'` case, the elements of x are copied into nc, a copy of input c, between start and stop. For both the `'cut'` and `'paste'` operations, a new decomposition vector nc is returned.

repmat

The following three functions—wavecut, wavecopy, and wavepaste—use wavework to manipulate c using a more intuitive syntax:

wavecut

```
function [nc, y] = wavecut(type, c, s, n)
%WAVECUT Zeroes coefficients in a wavelet decomposition structure.
%   [NC, Y] = WAVECUT(TYPE, C, S, N) returns a new decomposition
%   vector whose detail or approximation coefficients (based on TYPE
```

```
%   and N) have been zeroed. The coefficients that were zeroed are
%   returned in Y.
%
%   INPUTS:
%     TYPE        Coefficient category
%     ----------------------------------------------------------------
%     'a'         Approximation coefficients
%     'h'         Horizontal details
%     'v'         Vertical details
%     'd'         Diagonal details
%
%     [C, S] is a wavelet data structure.
%     N specifies a decomposition level (ignored if TYPE = 'a').
%
%   See also WAVEWORK, WAVECOPY, and WAVEPASTE.

error(nargchk(3, 4, nargin));
if nargin == 4
   [nc, y] = wavework('cut', type, c, s, n);
else
   [nc, y] = wavework('cut', type, c, s);
end
```

wavecopy

```
function y = wavecopy(type, c, s, n)
%WAVECOPY Fetches coefficients of a wavelet decomposition structure.
%   Y = WAVECOPY(TYPE, C, S, N) returns a coefficient array based on
%   TYPE and N.
%
%   INPUTS:
%     TYPE        Coefficient category
%     ----------------------------------------------------------------
%     'a'         Approximation coefficients
%     'h'         Horizontal details
%     'v'         Vertical details
%     'd'         Diagonal details
%
%   [C, S] is a wavelet data structure.
%   N specifies a decomposition level (ignored if TYPE = 'a').
%
% See also WAVEWORK, WAVECUT, and WAVEPASTE.

error(nargchk(3, 4, nargin));
if nargin == 4
   y = wavework('copy', type, c, s, n);
else
   y = wavework('copy', type, c, s);
end
```

wavepaste

```
function nc = wavepaste(type, c, s, n, x)
%WAVEPASTE Puts coefficients in a wavelet decomposition structure.
%   NC = WAVEPASTE(TYPE, C, S, N, X) returns the new decomposition
%   structure after pasting X into it based on TYPE and N.
```

```
%
%   INPUTS:
%     TYPE      Coefficient category
%     ----------------------------------------------------------------
%     'a'       Approximation coefficients
%     'h'       Horizontal details
%     'v'       Vertical details
%     'd'       Diagonal details
%
%     [C, S] is a wavelet data structure.
%     N specifies a decomposition level (Ignored if TYPE = 'a').
%     X is a two-dimensional approximation or detail coefficient
%       matrix whose dimensions are appropriate for decomposition
%       level N.
%
%   See also WAVEWORK, WAVECUT, and WAVECOPY.

error(nargchk(5, 5, nargin))
nc = wavework('paste', type, c, s, n, x);
```

EXAMPLE 7.5: Manipulating **c** with wavecut and wavecopy.

■ Functions wavecopy and wavecut can be used to reproduce the Wavelet Toolbox based results of Example 7.4:

```
>> f = magic(8);
>> [c1, s1] = wavedec2(f, 3, 'haar');
>> approx = wavecopy('a', c1, s1)
approx =
    260.0000

>> horizdet2 = wavecopy('h', c1, s1, 2)
horizdet2 =
    1.0e-013 *

            0   -0.2842
            0         0

>> [newc1, horizdet2] = wavecut('h', c1, s1, 2);
>> newhorizdet2 = wavecopy('h', newc1, s1, 2)
newhorizdet2 =
    0    0
    0    0
```

Note that all extracted matrices are identical to those of the previous example. ■

7.3.2 Displaying Wavelet Decomposition Coefficients

As was indicated at the start of Section 7.3, the coefficients that are packed into one-dimensional wavelet decomposition vector **c** are, in reality, the coefficients of the two-dimensional output arrays from the filter bank in Fig. 7.2. For each iteration of the filter bank, four quarter-size coefficient arrays (neglecting any expansion that may result from the convolution process) are produced.

They can be arranged as a 2 × 2 array of submatrices that replace the two-dimensional input from which they are derived. Function wave2gray performs this subimage compositing—and both scales the coefficients to better reveal their differences and inserts borders to delineate the approximation and various horizontal, vertical, and diagonal detail matrices.

wave2gray

```
function w = wave2gray(c, s, scale, border)
%WAVE2GRAY Display wavelet decomposition coefficients.
%   W = WAVE2GRAY(C, S, SCALE, BORDER) displays and returns a
%   wavelet coefficient image.
%
%   EXAMPLES:
%     wave2gray(c, s);                      Display w/defaults.
%     foo = wave2gray(c, s);                Display and return.
%     foo = wave2gray(c, s, 4);             Magnify the details.
%     foo = wave2gray(c, s, -4);            Magnify absolute values.
%     foo = wave2gray(c, s, 1, 'append');   Keep border values.
%
%   INPUTS/OUTPUTS:
%     [C, S] is a wavelet decomposition vector and bookkeeping
%     matrix.
%
%     SCALE            Detail coefficient scaling
%     ----------------------------------------------------------
%     0 or 1           Maximum range (default)
%     2, 3...          Magnify default by the scale factor
%     -1, -2...        Magnify absolute values by abs(scale)
%
%     BORDER           Border between wavelet decompositions
%     ----------------------------------------------------------
%     'absorb'         Border replaces image (default)
%     'append'         Border increases width of image
%
%     Image W:   ------- ------ ------------ ---------------
%               |       |      |            |
%               | a(n)  | h(n) |            |
%               |       |      |            |
%                ------- ------     h(n-1)  |
%               |       |      |            |
%               | v(n)  | d(n) |            |      h(n-2)
%               |       |      |            |
%                ------- ------ ------------
%               |              |            |
%               |    v(n-1)    |   d(n-1)   |
%               |              |            |
%                ------------- ------------- ---------------
%               |                           |
%               |           v(n-2)          |      d(n-2)
%               |                           |
%
```

```
%     Here, n denotes the decomposition step scale and a, h, v, d are
%     approximation, horizontal, vertical, and diagonal detail
%     coefficients, respectively.
% Check input arguments for reasonableness.
error(nargchk(2, 4, nargin));

if (ndims(c) ~= 2) | (size(c, 1) ~= 1)
   error('C must be a row vector.'); end

if (ndims(s) ~= 2) | ~isreal(s) | ~isnumeric(s) | (size(s, 2) ~= 2)
   error('S must be a real, numeric two-column array.'); end

elements = prod(s, 2);
if (length(c) < elements(end)) | ...
      ~(elements(1) + 3 * sum(elements(2:end - 1)) >= elements(end))
   error(['[C S] must be a standard wavelet ' ...
          'decomposition structure.']);
end

if (nargin > 2) & (~isreal(scale) | ~isnumeric(scale))
   error('SCALE must be a real, numeric scalar.');
end

if (nargin > 3) & (~ischar(border))
   error('BORDER must be character string.');
end

if nargin == 2
   scale = 1; % Default scale.
end

if nargin < 4
   border = 'absorb';  % Default border.
end

% Scale coefficients and determine pad fill.
absflag = scale < 0;
scale = abs(scale);
if scale == 0
   scale = 1;
end

[cd, w] = wavecut('a', c, s);   w = mat2gray(w);
cdx = max(abs(cd(:))) / scale;
if absflag
   cd = mat2gray(abs(cd), [0, cdx]); fill = 0;
else
   cd = mat2gray(cd, [-cdx, cdx]); fill = 0.5;
end

% Build gray image one decomposition at a time.
for i = size(s, 1) - 2:-1:1
   ws = size(w);

   h = wavecopy('h', cd, s, i);
   pad = ws - size(h);      frontporch = round(pad / 2);
   h = padarray(h, frontporch, fill, 'pre');
   h = padarray(h, pad - frontporch, fill, 'post');
```

```
    v = wavecopy('v', cd, s, i);
    pad = ws - size(v);        frontporch = round(pad / 2);
    v = padarray(v, frontporch, fill, 'pre');
    v = padarray(v, pad - frontporch, fill, 'post');

    d = wavecopy('d', cd, s, i);
    pad = ws - size(d);        frontporch = round(pad / 2);
    d = padarray(d, frontporch, fill, 'pre');
    d = padarray(d, pad - frontporch, fill, 'post');

    % Add 1 pixel white border.
    switch lower(border)
    case 'append'
       w = padarray(w, [1 1], 1, 'post');
       h = padarray(h, [1 0], 1, 'post');
       v = padarray(v, [0 1], 1, 'post');
    case 'absorb'
       w(:, end) = 1;     w(end, :) = 1;
       h(end, :) = 1;     v(:, end) = 1;
    otherwise
       error('Unrecognized BORDER parameter.');
    end

    w = [w h; v d];                          % Concatenate coefs.
end

if nargout == 0
   imshow(w);                                % Display result.
end
```

The "help text" or header section of wave2gray details the structure of generated output image w. The subimage in the upper left corner of w, for instance, is the approximation array that results from the final decomposition step. It is surrounded—in a clockwise manner—by the horizontal, diagonal, and vertical detail coefficients that were generated during the same decomposition. The resulting 2×2 array of subimages is then surrounded (again in a clockwise manner) by the detail coefficients of the previous decomposition step; and the pattern continues until all of the scales of decomposition vector c are appended to two-dimensional matrix w.

The compositing just described takes place within the only for loop in wave2gray. After checking the inputs for consistency, wavecut is called to remove the approximation coefficients from decomposition vector c. These coefficients are then scaled for later display using mat2gray. Modified decomposition vector cd (i.e., c without the approximation coefficients) is then similarly scaled. For positive values of input scale, the detail coefficients are scaled so that a coefficient value of 0 appears as middle gray; all necessary padding is performed with a fill value of 0.5 (mid-gray). If scale is negative, the absolute values of the detail coefficients are displayed with a value of 0 corresponding to black and the pad fill value is set to 0. After the approximation and detail coefficients have been scaled for display, the first iteration of the for loop extracts the last decomposition step's detail coefficients from cd and appends them to w (after padding to make the dimensions of the four subimages match and insertion of a one-pixel white border) via the `w = [w h; v d]` statement. This process is then repeated for each scale in `c`. Note the use of `wavecopy` to extract the various detail coefficients needed to form `w`.

EXAMPLE 7.6: Transform coefficient display using `wave2gray`.

■ The following sequence of commands computes the two-scale DWT of the image in Fig. 7.4 with respect to fourth-order Daubechies' wavelets and displays the resulting coefficients:

```
>> f = imread('vase.tif');
>> [c, s] = wavefast(f, 2, 'db4');
>> wave2gray(c, s);
>> figure; wave2gray(c, s, 8);
>> figure; wave2gray(c, s, -8);
```

The images generated by the final three command lines are shown in Figs. 7.5(a) through (c), respectively. Without additional scaling, the detail coefficient differences in Fig. 7.5(a) are barely visible. In Fig. 7.5(b), the differences are accentuated by multiplying them by 8. Note the mid-gray

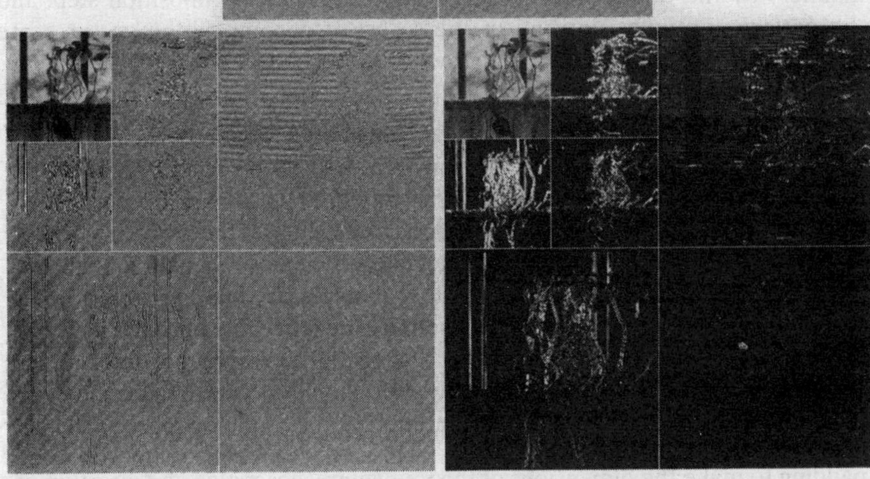

a
b c

FIGURE 7.5 Displaying a two-scale wavelet transform of the image in Fig. 7.4: (a) Automatic scaling; (b) additional scaling by 8; and (c) absolute values scaled by 8.

padding along the borders of the level 1 coefficient subimages; it was inserted to reconcile dimensional variations between transform coefficient subimages. Figure 7.5(c) shows the effect of taking the absolute values of the details. Here, all padding is done in black. ■

7.4 The Inverse Fast Wavelet Transform

Like its forward counterpart, the *inverse fast wavelet transform* can be computed iteratively using digital filters. Figure 7.6 shows the required *synthesis* or *reconstruction filter bank*, which reverses the process of the analysis or decomposition filter bank of Fig. 7.2. At each iteration, four scale j approximation and detail subimages are *upsampled* (by inserting zeroes between every element) and convolved with two one-dimension filters—one operating on the subimages' columns and the other on its rows. Addition of the results yields the scale $j + 1$ approximation, and the process is repeated until the original image is reconstructed. The filters used in the convolutions are a function of the wavelets employed in the forward transform. Recall that they can be obtained from the `wfilters` and `wavefilter` functions of Section 7.2 with input parameter `type` set to `'r'` for "reconstruction."

When using the Wavelet Toolbox, function `waverec2` is employed to compute the inverse FWT of wavelet decomposition structure `[C, S]`. It is invoked using

```
g = waverec2(C, S, wname)
```

where `g` is the resulting reconstructed two-dimensional image (of class `double`). The required reconstruction filters can be alternately supplied via syntax

```
g = waverec2(C, S, Lo_R, Hi_R)
```

W `waverec2`

The following custom routine, which we call `waveback`, can be used when the Wavelet Toolbox is unavailable. It is the final function needed to complete our wavelet-based package for processing images in conjunction with IPT (and without the Wavelet Toolbox).

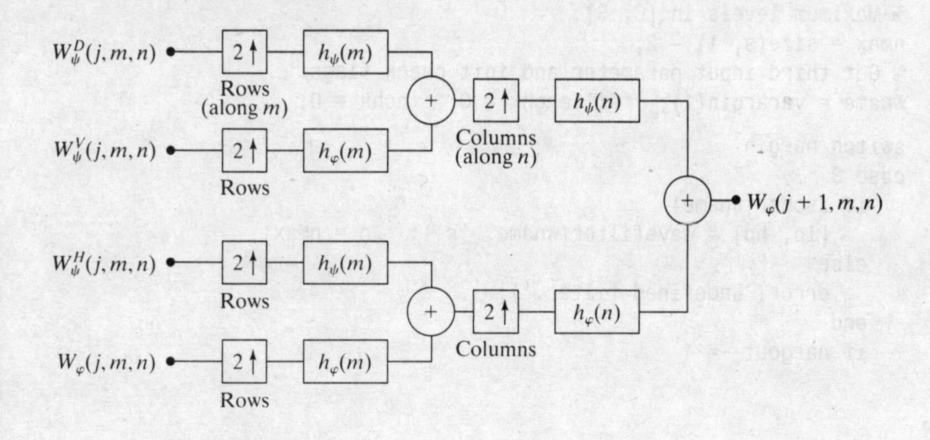

FIGURE 7.6 The 2-D FWT^{-1} filter bank. The boxes with the up arrows represent upsampling by inserting zeroes between every element.

waveback

```
function [varargout] = waveback(c, s, varargin)
%WAVEBACK Performs a multi-level two-dimensional inverse FWT.
%   [VARARGOUT] = WAVEBACK(C, S, VARARGIN) computes a 2D N-level
%   partial or complete wavelet reconstruction of decomposition
%   structure [C, S].
%
%   SYNTAX:
%     Y = WAVEBACK(C, S, 'WNAME');     Output inverse FWT matrix Y
%     Y = WAVEBACK(C, S, LR, HR);      using lowpass and highpass
%                                      reconstruction filters (LR and
%                                      HR) or filters obtained by
%                                      calling WAVEFILTER with 'WNAME'.
%
%     [NC, NS] = WAVEBACK(C, S, 'WNAME', N);  Output new wavelet
%     [NC, NS] = WAVEBACK(C, S, LR, HR, N);   decomposition structure
%                                             [NC, NS] after N step
%                                             reconstruction.
%
%   See also WAVEFAST and WAVEFILTER.

% Check the input and output arguments for reasonableness.
error(nargchk(3, 5, nargin));
error(nargchk(1, 2, nargout));

if (ndims(c) ~= 2) | (size(c, 1) ~= 1)
   error('C must be a row vector.');
end

if (ndims(s) ~= 2) | ~isreal(s) | ~isnumeric(s) | (size(s, 2) ~= 2)
   error('S must be a real, numeric two-column array.');
end

elements = prod(s, 2);
if (length(c) < elements(end)) | ...
      ~(elements(1) + 3 * sum(elements(2:end - 1)) >= elements(end))
   error(['[C S] must be a standard wavelet ' ...
          'decomposition structure.']);
end

% Maximum levels in [C, S].
nmax = size(s, 1) - 2;
% Get third input parameter and init check flags.
wname = varargin{1};   filterchk = 0;   nchk = 0;

switch nargin
case 3
   if ischar(wname)
      [lp, hp] = wavefilter(wname, 'r');   n = nmax;
   else
      error('Undefined filter.');
   end
   if nargout ~= 1
```

```
      error('Wrong number of output arguments.');
   end
case 4
   if ischar(wname)
      [lp, hp] = wavefilter(wname, 'r');
      n = varargin{2};   nchk = 1;
   else
      lp = varargin{1}; hp = varargin{2};
      filterchk = 1;     n = nmax;
      if nargout ~= 1
         error('Wrong number of output arguments.');
      end
   end
case 5
   lp = varargin{1};     hp = varargin{2};     filterchk = 1;
   n = varargin{3};      nchk = 1;
otherwise
   error('Improper number of input arguments.');
end

fl = length(lp);
if filterchk                                          % Check filters.
   if (ndims(lp) ~= 2) | ~isreal(lp) | ~isnumeric(lp) ...
         | (ndims(hp) ~= 2) | ~isreal(hp) | ~isnumeric(hp) ...
         | (fl ~= length(hp)) | rem(fl, 2) ~= 0
      error(['LP and HP must be even and equal length real, ' ...
           'numeric filter vectors.']);
   end
end

if nchk & (~isnumeric(n) | ~isreal(n))          % Check scale N.
   error('N must be a real numeric.');
end
if (n > nmax) | (n < 1)
   error('Invalid number (N) of reconstructions requested.');
end
if (n ~= nmax) & (nargout ~= 2)
   error('Not enough output arguments.');
end

nc = c;   ns = s;    nnmax = nmax;             % Init decomposition.
for i = 1:n
   % Compute a new approximation.
   a = symconvup(wavecopy('a', nc, ns), lp, lp, fl, ns(3, :)) + ...
       symconvup(wavecopy('h', nc, ns, nnmax), ...
                 hp, lp, fl, ns(3, :)) + ...
       symconvup(wavecopy('v', nc, ns, nnmax), ...
                 lp, hp, fl, ns(3, :)) + ...
       symconvup(wavecopy('d', nc, ns, nnmax), ...
                 hp, hp, fl, ns(3, :));
```

```
   % Update decomposition.
   nc = nc(4 * prod(ns(1, :)) + 1:end);     nc = [a(:)' nc];
   ns = ns(3:end, :);                       ns = [ns(1, :); ns];
   nnmax = size(ns, 1) - 2;
end

% For complete reconstructions, reformat output as 2-D.
if nargout == 1
   a = nc;     nc = repmat(0, ns(1, :));    nc(:) = a;
end

varargout{1} = nc;
if nargout == 2
   varargout{2} = ns;
end
%-------------------------------------------------------------------%
function z = symconvup(x, f1, f2, fln, keep)
% Upsample rows and convolve columns with f1; upsample columns and
% convolve rows with f2; then extract center assuming symmetrical
% extension.

y = zeros([2 1] .* size(x));            y(1:2:end, :) = x;
y = conv2(y, f1');
z = zeros([1 2] .* size(y));            z(:, 1:2:end) = y;
z = conv2(z, f2);
z = z(fln - 1:fln + keep(1) - 2, fln- 1:fln + keep(2) - 2);
```

The main routine of function waveback is a simple for loop that iterates through the requested number of decomposition levels (i.e., scales) in the desired reconstruction. As can be seen, each loop calls internal function symconvup four times and sums the returned matrices. Decomposition vector nc, which is initially set to c, is iteratively updated by replacing the four coefficient matrices passed to symconvup by the newly created approximation a. Bookkeeping matrix ns is then modified accordingly—there is now one less scale in decomposition structure [nc, ns]. This sequence of operations is slightly different than the ones outlined in Fig. 7.6, in which the top two inputs are combined to yield

$$[W_\psi^D(j, m, n)\uparrow^{2m} * h_\psi(m) + W_\psi^V(j, m, n)\uparrow^{2m} * h_\varphi(m)]\uparrow^{2n} * h_\psi(n)$$

where $\uparrow^{2m}$ and $\uparrow^{2n}$ denote upsampling along m and n, respectively. Function waveback uses the equivalent computation

$$[W_\psi^D(j, m, n)\uparrow^{2m} * h_\psi(m)]\uparrow^{2n} * h_\psi(n) + [W_\psi^V(j, m, n)\uparrow^{2m} * h_\varphi(m)]\uparrow^{2n} * h_\psi(n)$$

Function symconvup performs the convolutions and upsampling required to compute the contribution of one input of Fig. 7.6 to output $W_\varphi(j + 1, m, n)$ in accordance with the proceding equation. Input x is first upsampled in the row direction to yield y, which is convolved columnwise with filter f1. The resulting output, which replaces y, is then upsampled in the column direction and convolved row by row with f2 to produce z. Finally, the center keep elements of z (the final convolution) are returned as input x's contribution to the new approximation.

EXAMPLE 7.7: Comparing the execution times of waveback and waverec2.

■ The following test routine compares the execution times of Wavelet Toolbox function waverec2 and custom function waveback using a simple modification of the test function in Example 7.3:

```
function [ratio, maxdiff] = ifwtcompare(f, n, wname)
%IFWTCOMPARE Compare waverec2 and waveback.
%   [RATIO, MAXDIFF] = IFWTCOMPARE(F, N, WNAME) compares the
%   operation of Wavelet Toolbox function WAVEREC2 and custom function
%   WAVEBACK.
%
%   INPUTS:
%     F          Image to transform and inverse transform.
%     N          Number of scales to compute.
%     WNAME      Wavelet to use.
%
%   OUTPUTS:
%     RATIO      Execution time ratio (custom/toolbox).
%     MAXDIFF    Maximum generated image difference.

% Compute the transform and get output and computation time for
% waverec2.
[c1, s1] = wavedec2(f, n, wname);
tic;
g1 = waverec2(c1, s1, wname);
reftime = toc;

% Compute the transform and get output and computation time for
% waveback.
[c2, s2] = wavefast(f, n, wname);
tic;
g2 = waveback(c2, s2, wname);
t2 = toc;

% Compare the results.
ratio = t2 / (reftime + eps);
maxdiff = abs(max(max(g1 - g2)));
```

For a five scale transform of the 512 × 512 image in Fig. 7.4 with respect to 4th-order Daubechies' wavelets, we get

```
>> f = imread('Vase', 'tif');
>> [ratio, maxdifference] = ifwtcompare(f, 5, 'db4')

ratio =
    1.0000

maxdifference =
    3.6948e-013
```

Note that the inverse transformation times of the two functions are equivalent (i.e., the ratio is 1) and that the largest output difference is 3.6948×10^{-13}. For all practical purposes, they generate identical results in identical times. ■

7.5 Wavelets in Image Processing

As in the Fourier domain (see Section 4.3.2), the basic approach to wavelet-based image processing is to

1. Compute the two-dimensional wavelet transform of an image.
2. Alter the transform coefficients.
3. Compute the inverse transform.

Because scale in the wavelet domain is analogous to frequency in the Fourier domain, most of the Fourier-based filtering techniques of Chapter 4 have an equivalent "wavelet domain" counterpart. In this section, we use the preceding three-step procedure to give several examples of the use of wavelets in image processing. Attention is restricted to the routines developed earlier in the chapter; the Wavelet Toolbox is not needed to implement the examples given here—nor the examples in Chapter 7 of *Digital Image Processing* (Gonzalez and Woods [2002]).

EXAMPLE 7.8: Wavelet directionality and edge detection.

■ Consider the 500×500 test image in Fig. 7.7(a). This image was used in Chapter 4 to illustrate smoothing and sharpening with Fourier transforms. Here, we use it to demonstrate the directional sensitivity of the 2-D wavelet transform and its usefulness in edge detection:

```
>> f = imread('A.tif');
>> imshow(f);
>> [c, s] = wavefast(f, 1, 'sym4');
>> figure; wave2gray(c, s, -6);
>> [nc, y] = wavecut('a', c, s);
>> figure; wave2gray(nc, s, -6);
>> edges = abs(waveback(nc, s, 'sym4'));
>> figure; imshow(mat2gray(edges));
```

The horizontal, vertical, and diagonal directionality of the single-scale wavelet transform of Fig. 7.7(a) with respect to `'sym4'` wavelets is clearly visible in Fig. 7.7(b). Note, for example, that the horizontal edges of the original image are present in the horizontal detail coefficients of the upper-right quadrant of Fig. 7.7(b). The vertical edges of the image can be similarly identified in the vertical detail coefficients of the lower-left quadrant. To combine this information into a single edge image, we simply zero the approximation coefficients of the generated transform, compute its inverse, and take the absolute value. The modified transform and resulting edge image are shown in Figs. 7.7(c) and (d), respectively. A similar procedure can be used to isolate the vertical or horizontal edges alone. ■

EXAMPLE 7.9: Wavelet-based image smoothing or blurring.

■ Wavelets, like their Fourier counterparts, are effective instruments for smoothing or blurring images. Consider again the test image of Fig. 7.7(a), which is repeated in Fig. 7.8(a). Its wavelet transform with respect to fourth-

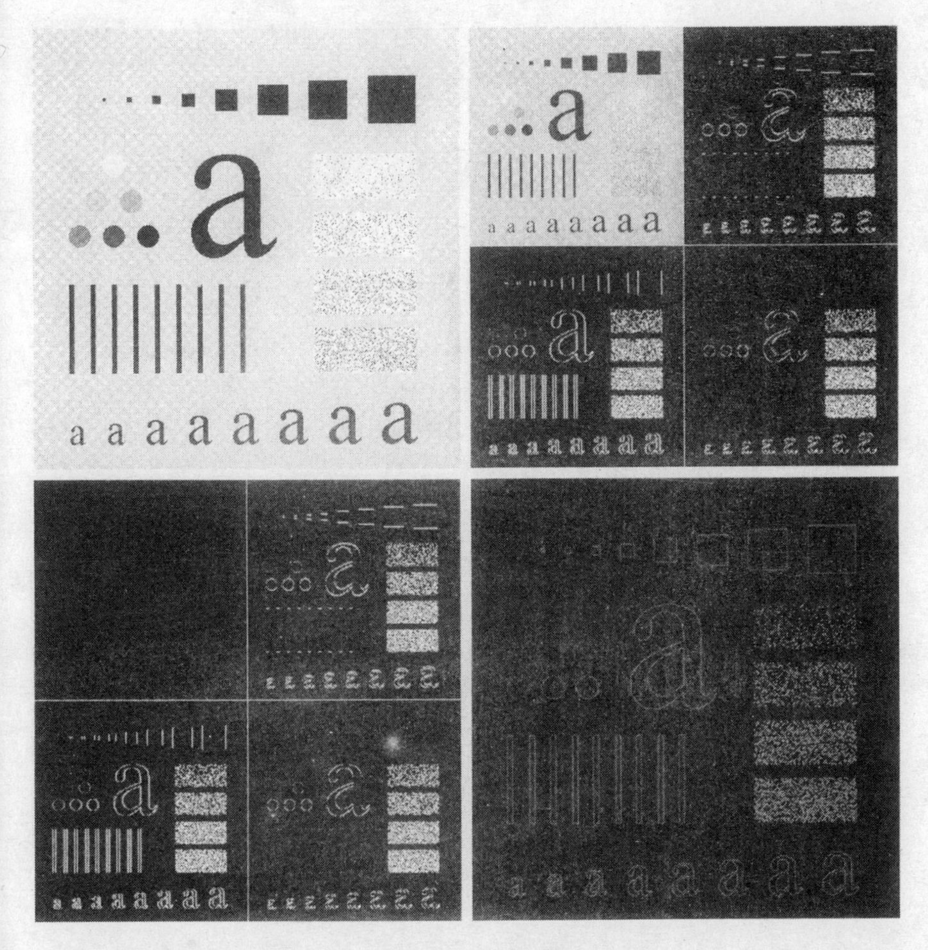

a b
c d

FIGURE 7.7 Wavelets in edge detection: (a) A simple test image; (b) its wavelet transform; (c) the transform modified by zeroing all approximation coefficients; and (d) the edge image resulting from computing the absolute value of the inverse transform.

order symlets is shown in Fig. 7.8(b), where it is clear that a four-scale decomposition has been performed. To streamline the smoothing process, we employ the following utility function:

wavezero

```
function [nc, g8] = wavezero(c, s, l, wname)
%WAVEZERO Zeroes wavelet transform detail coefficients.
%   [NC, G8] = WAVEZERO(C, S, L, WNAME) zeroes the level L detail
%   coefficients in wavelet decomposition structure [C, S] and
%   computes the resulting inverse transform with respect to WNAME
%   wavelets.

[nc, foo] = wavecut('h', c, s, l);
[nc, foo] = wavecut('v', nc, s, l);
[nc, foo] = wavecut('d', nc, s, l);
i = waveback(nc, s, wname);
g8 = im2uint8(mat2gray(i));
figure; imshow(g8);
```

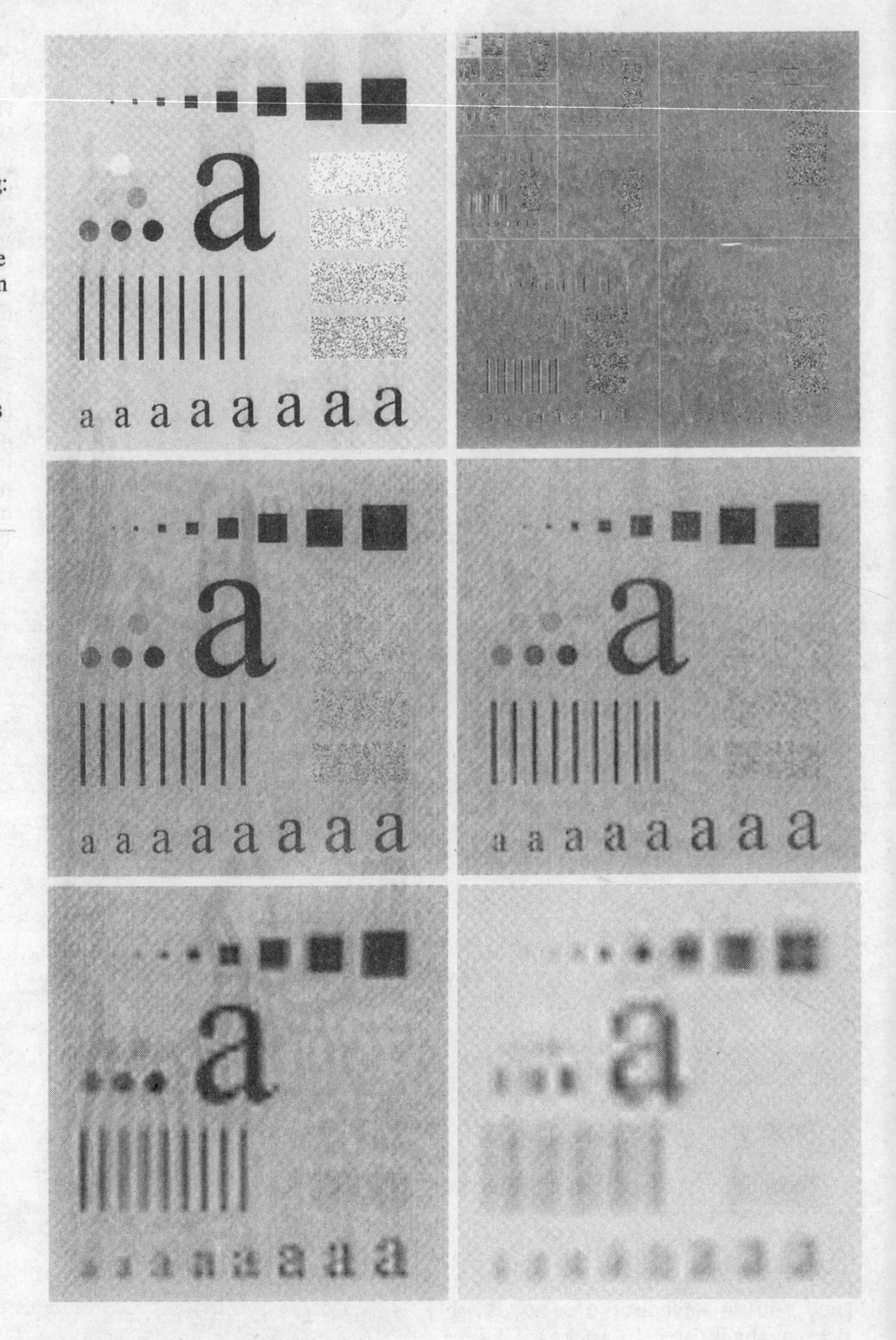

a b
c d
e f

FIGURE 7.8 Wavelet-based image smoothing: (a) A test image; (b) its wavelet transform; (c) the inverse transform after zeroing the first-level detail coefficients; and (d) through (f) similar results after zeroing the second-, third-, and fourth-level details.

Using wavezero, a series of increasingly smoothed versions of Fig. 7.8(a) can be generated with the following sequence of commands:

```
>> f = imread('A.tif');
>> [c, s] = wavefast(f, 4, 'sym4');
>> wave2gray(c, s, 20);
>> [c, g8] = wavezero(c, s, 1, 'sym4');
>> [c, g8] = wavezero(c, s, 2, 'sym4');
>> [c, g8] = wavezero(c, s, 3, 'sym4');
>> [c, g8] = wavezero(c, s, 4, 'sym4');
```

Note that the smoothed image in Fig. 7.8(c) is only slightly blurred, as it was obtained by zeroing only the first-level detail coefficients of the original image's wavelet transform (and computing the modified tranform's inverse). Additional blurring is present in the second result—Fig. 7.8(d)—which shows the effect of zeroing the second level detail coefficients as well. The coefficient zeroing process continues in Fig. 7.8(e), where the third level of details is zeroed, and concludes with Fig. 7.8(f), where all the detail coefficients have been eliminated. The gradual increase in blurring from Figs. 7.8(c) to (f) is reminiscent of similar results with Fourier transforms. It illustrates the intimate relationship between scale in the wavelet domain and frequency in the Fourier domain. ■

EXAMPLE 7.10: Progressive reconstruction.

■ Consider next the transmission and reconstruction of the four-scale wavelet transform in Fig. 7.9(a) within the context of browsing a remote image database for a specific image. Here, we deviate from the three-step procedure described at the beginning of this section and consider an application without a Fourier domain counterpart. Each image in the database is stored as a multiscale wavelet decomposition. This structure is well suited to progressive reconstruction applications, particularly when the 1-D decomposition vector used to store the transform's coefficients assumes the general format of Section 7.3. For the four-scale transform of this example, the decomposition vector is

$$[\mathbf{A}_4(:)' \quad \mathbf{H}_4(:)' \quad \cdots \quad \mathbf{H}_i(:)' \quad \mathbf{V}_i(:)' \quad \mathbf{D}_i(:)' \quad \cdots \quad \mathbf{V}_1(:)' \quad \mathbf{D}_1(:)']$$

where $\mathbf{A}_4$ is the approximation coefficient matrix of the fourth decomposition level and $\mathbf{H}_i$, $\mathbf{V}_j$, and $\mathbf{D}_i$ for $i = 1, 2, 3, 4$ are the horizontal, vertical, and diagonal transform coefficient matrices for level i. If we transmit this vector in a left-to-right manner, a remote display device can gradually build higher resolution approximations of the final high-resolution image (based on the user's needs) as the data arrives at the viewing station. For instance, when the $\mathbf{A}_4$ coefficients have been received, a low-resolution version of the image can be made available for viewing [Fig. 7.9(b)]. When $\mathbf{H}_4$, $\mathbf{V}_4$, and $\mathbf{D}_4$ have been received, a higher-resolution approximation [Fig. 7.9(c)] can be constructed, and

a
b c d e f

FIGURE 7.9 Progressive reconstruction: (a) A four-scale wavelet transform; (b) the fourth-level approximation image from the upper-left corner; (c) a refined approximation incorporating the fourth-level details; (d) through (f) further resolution improvements incorporating higher-level details.

so on. Figures 7.9(d) through (f) provide three additional reconstructions of increasing resolution. This progressive reconstruction process is easily simulated using the following MATLAB command sequence:

```
>> f = imread('Strawberries.tif');              % Generate transform
>> [c, s] = wavefast(f, 4, 'jpeg9.7');
>> wave2gray(c, s, 8);
```

```
>>
>> f = wavecopy('a', c, s);                       % Approximation 1
>> figure; imshow(mat2gray(f));
>>
>> [c, s] = waveback(c, s, 'jpeg9.7', 1);       % Approximation 2
>> f = wavecopy('a', c, s);
>> figure; imshow(mat2gray(f));
>> [c, s] = waveback(c, s, 'jpeg9.7', 1);       % Approximation 3
>> f = wavecopy('a', c, s);
>> figure; imshow(mat2gray(f));
>> [c, s] = waveback(c, s, 'jpeg9.7', 1);       % Approximation 4
>> f = wavecopy('a', c, s);
>> figure; imshow(mat2gray(f));
>> [c, s] = waveback(c, s, 'jpeg9.7', 1);       % Final image
>> f = wavecopy('a', c, s);
>> figure; imshow(mat2gray(f));
```

Note that the final four approximations use `waveback` to perform single level reconstructions. ■

Summary

The material in this chapter introduces the wavelet transform and its use in image processing. Like the Fourier transform, wavelet transforms can be used in tasks ranging from edge detection to image smoothing, both of which are considered in the material that is covered. Because they provide significant insight into both an image's spatial and frequency characteristics, wavelets can also be used in applications in which Fourier methods are not well suited, like progressive image reconstruction (see Example 7.10). Because the Image Processing Toolbox does not include routines for computing or using wavelet transforms, a significant portion of this chapter is devoted to the development of a set of functions that extend the Image Processing Toolbox to wavelet-based imaging. The functions developed were designed to be fully compatible with MATLAB's Wavelet Toolbox, which is introduced in this chapter but is not a part of the Image Processing Toolbox. In the next chapter, wavelets will be used for image compression, an area in which they have received considerable attention in the literature.

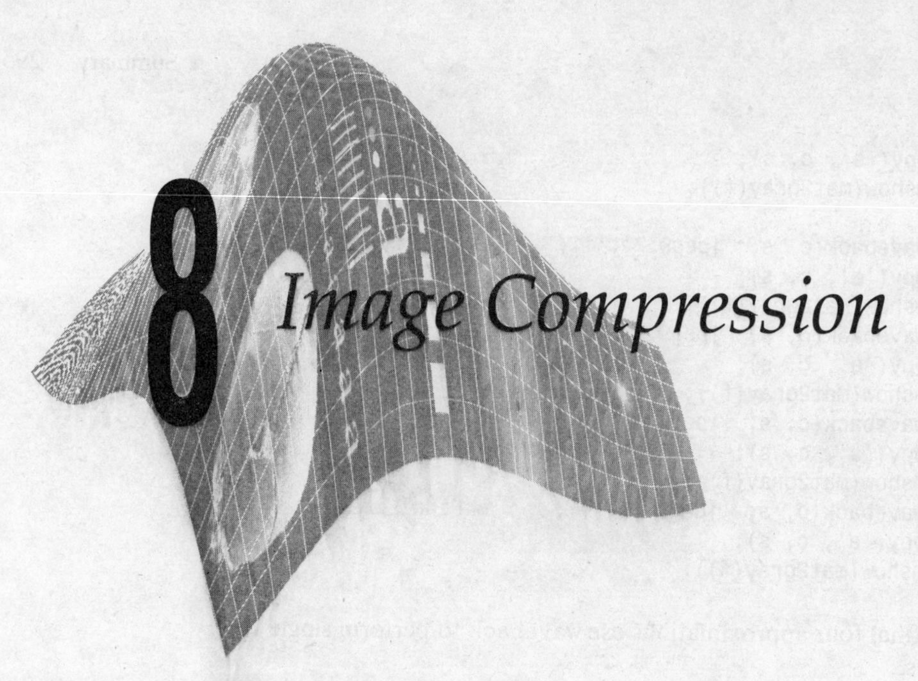

8 Image Compression

Preview

Image compression addresses the problem of reducing the amount of data required to represent a digital image. Compression is achieved by the removal of one or more of three basic data *redundancies*: (1) *coding redundancy*, which is present when less than optimal (i.e., the smallest length) code words are used; (2) *interpixel redundancy*, which results from correlations between the pixels of an image; and/or (3) *psychovisual redundancy*, which is due to data that is ignored by the human visual system (i.e., visually nonessential information). In this chapter, we examine each of these redundancies, describe a few of the many techniques that can be used to exploit them, and examine two important compression standards—JPEG and JPEG 2000. These standards unify the concepts introduced earlier in the chapter by combining techniques that collectively attack all three data redundancies.

Because the Image Processing Toolbox does not include functions for image compression, a major goal of this chapter is to provide practical ways of exploring compression techniques within the context of MATLAB. For instance, we develop a MATLAB callable C function that illustrates how to manipulate variable-length data representations at the bit level. This is important because variable-length coding is a mainstay of image compression, but MATLAB is best at processing matrices of uniform (i.e., fixed length) data. During the development of the function, we assume that the reader has a working knowledge of the C language and focus our discussion on how to make MATLAB interact with programs (both C and Fortran) external to the MATLAB environment. This is an important skill when there is a need to interface M-functions to preexisting C or Fortran programs, and when vectorized M-functions still need to be speeded up (e.g., when a `for` loop can not be adequately vectorized). In the end, the range of compression functions developed in this chapter, together with MATLAB's ability to treat C and Fortran programs as though they were conventional M-files or built-in functions, demonstrates that MATLAB can be an effective tool for prototyping image compression systems and algorithms.

8.1 Background

As can be seen in Fig. 8.1, image compression systems are composed of two distinct structural blocks: an *encoder* and a *decoder*. Image $f(x, y)$ is fed into the encoder, which creates a set of symbols from the input data and uses them to represent the image. If we let n_1 and n_2 denote the number of information carrying units (usually bits) in the original and encoded images, respectively, the compression that is achieved can be quantified numerically via the *compression ratio*

$$C_R = \frac{n_1}{n_2}$$

A compression ratio like 10 (or 10:1) indicates that the original image has 10 information carrying units (e.g., bits) for every 1 unit in the compressed data set. In MATLAB, the ratio of the number of bits used in the representation of two image files and/or variables can be computed with the following M-function:

imratio

```
function cr = imratio(f1, f2)
%IMRATIO Computes the ratio of the bytes in two images/variables.
%   CR = IMRATIO(F1, F2) returns the ratio of the number of bytes in
%   variables/files F1 and F2. If F1 and F2 are an original and
%   compressed image, respectively, CR is the compression ratio.

error(nargchk(2, 2, nargin));           % Check input arguments
cr = bytes(f1) / bytes(f2);             % Compute the ratio

%-------------------------------------------------------------------%
function b = bytes(f)
% Return the number of bytes in input f. If f is a string, assume
% that it is an image filename; if not, it is an image variable.
```

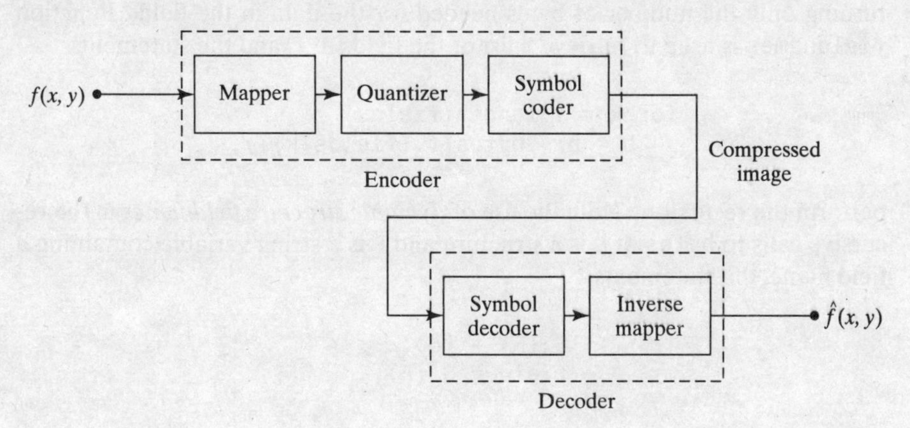

FIGURE 8.1 A general image compression system block diagram.

```
if ischar(f)
   info = dir(f);        b = info.bytes;
elseif isstruct(f)
   % MATLAB's whos function reports an extra 124 bytes of memory
   % per structure field because of the way MATLAB stores
   % structures in memory.  Don't count this extra memory; instead,
   % add up the memory associated with each field.
   b = 0;
   fields = fieldnames(f);
   for k = 1:length(fields)
      b = b + bytes(f.(fields{k}));
   end
else
   info = whos('f');     b = info.bytes;
end
```

For example, the compression of the JPEG encoded image in Fig. 2.4(c) of Chapter 2 can be computed via

```
>> r = imratio(imread('bubbles25.jpg'), 'bubbles25.jpg')
r =
    35.1612
```

Note that in function `imratio`, internal function `b = bytes(f)` is designed to return the number of bytes in (1) a file, (2) a structure variable, and/or (3) a nonstructure variable. If `f` is a nonstructure variable, function `whos`, introduced in Section 2.2, is used to get its size in bytes. If `f` is a file name, function `dir` performs a similar service. In the syntax employed, `dir` returns a structure (see Section 2.10.6 for more on structures) with fields `name`, `date`, `bytes`, and `isdir`. They contain the file's name, modification date, size in bytes, and whether or not it is a directory (`isdir` is 1 if it is and is 0 otherwise), respectively. Finally, if `f` is a structure, `bytes` calls itself recursively to sum the number of bytes allocated to each field of the structure. This eliminates the overhead associated with the structure variable itself (124 bytes per field), returning only the number of bytes needed for the data in the fields. Function `fieldnames` is used to retrieve a list of the fields in `f`, and the statements

`names = fieldnames(s)` *returns a cell array of strings containing the structure field names associated with structure* `s`.

```
for k = 1:length(fields)
   b = b + bytes(f.(fields{k}));
```

perform the recursions. Note the use of *dynamic structure fieldnames* in the recursive calls to `bytes`. If `S` is a structure and `F` is a string variable containing a field name, the statements

```
S.(F) = foo;
field = S.(F);
```

employ the dynamic structure fieldname syntax to set and/or get the contents of structure field F, respectively.

To view and/or use a compressed (i.e., encoded) image, it must be fed into a decoder (see Fig. 8.1), where a reconstructed output image, $\hat{f}(x, y)$, is generated. In general, $\hat{f}(x, y)$ may or may not be an exact representation of $f(x, y)$. If it is, the system is called *error free, information preserving*, or *lossless*; if not, some level of distortion is present in the reconstructed image. In the latter case, which is called *lossy compression*, we can define the error $e(x, y)$ between $f(x, y)$ and $\hat{f}(x, y)$, for any value of x and y as

$$e(x, y) = \hat{f}(x, y) - f(x, y)$$

so that the total error between the two images is

$$\sum_{x=0}^{M-1} \sum_{y=0}^{N-1} [\hat{f}(x, y) - f(x, y)]$$

and the *rms* (*root-mean-square*) *error* e_{rms} between $f(x, y)$ and $\hat{f}(x, y)$ is the square root of the squared error averaged over the $M \times N$ array, or

$$e_{\text{rms}} = \left[\frac{1}{MN} \sum_{x=0}^{M-1} \sum_{y=0}^{N-1} [\hat{f}(x, y) - f(x, y)]^2 \right]^{1/2}$$

The following M-function computes e_{rms} and displays (if $e_{\text{rms}} \neq 0$) both $e(x, y)$ and its histogram. Since $e(x, y)$ can contain both positive and negative values, hist rather than imhist (which handles only image data) is used to generate the histogram.

compare

```
function rmse = compare(f1, f2, scale)
%COMPARE Computes and displays the error between two matrices.
%   RMSE = COMPARE(F1, F2, SCALE) returns the root-mean-square error
%   between inputs F1 and F2, displays a histogram of the difference,
%   and displays a scaled difference image. When SCALE is omitted, a
%   scale factor of 1 is used.

% Check input arguments and set defaults.
error(nargchk(2, 3, nargin));
if nargin < 3
   scale = 1;
end

% Compute the root-mean-square error.
e = double(f1) - double(f2);
[m, n] = size(e);
rmse = sqrt(sum(e(:) .^ 2) / (m * n));

% Output error image & histogram if an error (i.e., rmse ~= 0).
if rmse
   % Form error histogram.
   emax = max(abs(e(:)));
   [h, x] = hist(e(:), emax);
```

```
    if length(h) >= 1
        figure; bar(x, h, 'k');

        % Scale the error image symmetrically and display
        emax = emax / scale;
        e = mat2gray(e, [-emax, emax]);
        figure; imshow(e);
    end
end
```

Finally, we note that the encoder of Fig. 8.1 is responsible for reducing the coding, interpixel, and/or psychovisual redundancies of the input image. In the first stage of the encoding process, the *mapper* transforms the input image into a (usually nonvisual) format designed to reduce interpixel redundancies. The second stage, or *quantizer* block, reduces the accuracy of the mapper's output in accordance with a predefined fidelity criterion—attempting to eliminate only psychovisually redundant data. This operation is irreversible and must be omitted when error-free compression is desired. In the third and final stage of the process, a *symbol coder* creates a code (that reduces coding redundancy) for the quantizer output and maps the output in accordance with the code.

The decoder in Fig. 8.1 contains only two components: a *symbol decoder* and an *inverse mapper*. These blocks perform, in reverse order, the inverse operations of the encoder's symbol coder and mapper blocks. Because quantization is irreversible, an inverse quantization block is not included.

8.2 Coding Redundancy

Let the discrete random variable r_k for $k = 1, 2, \ldots, L$ with associated probabilities $p_r(r_k)$ represent the gray levels of an L-gray-level image. As in Chapter 3, r_1 corresponds to gray level 0 (since MATLAB array indices cannot be 0) and

$$p_r(r_k) = \frac{n_k}{n} \quad k = 1, 2, \ldots, L$$

where n_k is the number of times that the kth gray level appears in the image and n is the total number of pixels in the image. If the number of bits used to represent each value of r_k is $l(r_k)$, then the average number of bits required to represent each pixel is

$$L_{\text{avg}} = \sum_{k=1}^{L} l(r_k) p_r(r_k)$$

That is, the average length of the code words assigned to the various gray-level values is found by summing the product of the number of bits used to represent each gray level and the probability that the gray level occurs. Thus the total number of bits required to code an $M \times N$ image is MNL_{avg}.

When the gray levels of an image are represented using a natural m-bit binary code, the right-hand side of the preceding equation reduces to m bits.

r_k	$p_r(r_k)$	Code 1	$l_1(r_k)$	Code 2	$l_2(r_k)$
r_1	0.1875	00	2	011	3
r_2	0.5000	01	2	1	1
r_3	0.1250	10	2	010	3
r_4	0.1875	11	2	00	2

TABLE 8.1
Illustration of coding redundancy: $L_{avg} = 2$ for Code 1; $L_{avg} \simeq 1.81$ for Code 2.

That is, $L_{avg} = m$ when m is substituted for $l(r_k)$. Then the constant m may be taken outside the summation, leaving only the sum of the $p_r(r_k)$ for $1 \leq k \leq L$, which, of course, equals 1. As is illustrated in Table 8.1, coding redundancy is almost always present when the gray levels of an image are coded using a natural binary code. In the table, both a fixed and variable-length encoding of a four-level image whose gray-level distribution is shown in column 2 is given. The 2-bit binary encoding (Code 1) in column 3 has an average length of 2 bits. The average number of bits required by Code 2 (in column 5) is

$$
\begin{aligned}
L_{avg} &= \sum_{k=1}^{4} l_2(k) p_r(r_k) \\
&= 3(0.1875) + 1(0.5) + 3(0.125) + 2(0.1875) = 1.8125
\end{aligned}
$$

and the resulting compression ratio is $C_r = 2/1.8125 \simeq 1.103$. The underlying basis for the compression achieved by Code 2 is that its code words are of varying length, allowing the shortest code words to be assigned to the gray levels that occur most frequently in the image.

The question that naturally arises is: How few bits actually are needed to represent the gray levels of an image? That is, is there a minimum amount of data that is sufficient to describe completely an image without loss of information? *Information theory* provides the mathematical framework to answer this and related questions. Its fundamental premise is that the generation of information can be modeled as a probabilistic process that can be measured in a manner that agrees with intuition. In accordance with this supposition, a random event E with probability $P(E)$ is said to contain

$$
I(E) = \log \frac{1}{P(E)} = -\log P(E)
$$

units of information. If $P(E) = 1$ (that is, the event always occurs), $I(E) = 0$ and no information is attributed to it. That is, because no uncertainty is associated with the event, no information would be transferred by communicating that the event has occurred. Given a source of random events from the discrete set of possible events $\{a_1, a_2, \dots, a_J\}$ with associated probabilities $\{P(a_1), P(a_2), \dots, P(a_J)\}$, the average information per source output, called the *entropy* of the source, is

$$
H = -\sum_{j=1}^{J} P(a_j) \log P(a_j)
$$

If an image is interpreted as a sample of a "gray-level source" that emitted it, we can model that source's symbol probabilities using the gray-level histogram of the observed image and generate an estimate, called the *first-order estimate*, $\widetilde{H}$, of the source's entropy:

$$\widetilde{H} = -\sum_{k=1}^{L} p_r(r_k) \log p_r(r_k)$$

Such an estimate is computed by the following M-function and, under the assumption that each gray level is coded independently, is a lower bound on the compression that can be achieved through the removal of coding redundancy alone.

entropy

```
function h = entropy(x, n)
%ENTROPY Computes a first-order estimate of the entropy of a matrix.
%   H = ENTROPY(X, N) returns the first-order estimate of matrix X
%   with N symbols (N = 256 if omitted) in bits/symbol. The estimate
%   assumes a statistically independent source characterized by the
%   relative frequency of occurrence of the elements in X.

error(nargchk(1, 2, nargin));        % Check input arguments
if nargin < 2
   n = 256;                          % Default for n.
end

x = double(x);                       % Make input double
xh = hist(x(:), n);                  % Compute N-bin histogram
xh = xh / sum(xh(:));                % Compute probabilities

% Make mask to eliminate 0's since log2(0) = -inf.
i = find(xh);

h = -sum(xh(i) .* log2(xh(i)));      % Compute entropy
```

Note the use of the MATLAB `find` function, which is employed to determine the indices of the nonzero elements of histogram `xh`. The statement `find(x)` is equivalent to `find(x ~= 0)`. Function `entropy` uses `find` to create a vector of indices, `i`, into histogram `xh`, which is subsequently employed to eliminate all zero-valued elements from the entropy computation in the final statement. If this were not done, the `log2` function would force output `h` to `NaN` (`0 * -inf` is *not a number*) when any symbol probability was 0.

EXAMPLE 8.1: Computing first-order entropy estimates.

■ Consider a simple 4 × 4 image whose histogram (see `p` in the following code) models the symbol probabilities in Table 8.1. The following command line sequence generates one such image and computes a first-order estimate of its entropy.

```
>> f = [119 123 168 119; 123 119 168 168];
>> f = [f; 119 119 107 119; 107 107 119 119]

f =

   119   123   168   119
```

```
    123    119    168    168
    119    119    107    119
    107    107    119    119

p = hist(f(:), 8);
p = p / sum(p)

p =
    0.1875   0.5   0.125   0   0   0   0   0.1875

h = entropy(f)

h =
    1.7806
```

Code 2 of Table 8.1, with $L_{avg} \simeq 1.81$, approaches this first-order entropy estimate and is a minimal length *binary* code for image `f`. Note that gray level 107 corresponds to r_1 and corresponding binary codeword 011_2 in Table 8.1, 119 corresponds to r_2 and code 1_2, and 123 and 168 correspond to 010_2 and 00_2, respectively. ■

8.2.1 Huffman Codes

When coding the gray levels of an image or the output of a gray-level mapping operation (pixel differences, run-lengths, and so on), *Huffman codes* contain the smallest possible number of code symbols (e.g., bits) per source symbol (e.g., gray-level value) subject to the constraint that the source symbols are coded *one at a time*.

The first step in Huffman's approach is to create a series of source reductions by ordering the probabilities of the symbols under consideration and combining the lowest probability symbols into a single symbol that replaces them in the next source reduction. Figure 8.2(a) illustrates the process for the gray-level distribution in Table 8.1. At the far left, the initial set of source symbols and their probabilities are ordered from top to bottom in terms of decreasing probability values. To form the first source reduction, the bottom two probabilities, 0.125 and 0.1875, are combined to form a "compound symbol" with probability 0.3125. This compound symbol and its associated probability are placed in the first source reduction column so that the probabilities of the reduced source are also ordered from the most to the least probable. This process is then repeated until a reduced source with two symbols (at the far right) is reached.

The second step in Huffman's procedure is to code each reduced source, starting with the smallest source and working back to the original source. The minimal length binary code for a two-symbol source, of course, consists of the symbols 0 and 1. As Fig. 8.2(b) shows, these symbols are assigned to the two symbols on the right (the assignment is arbitrary; reversing the order of the 0 and 1 would work just as well). As the reduced source symbol with probability 0.5 was generated by combining two symbols in the reduced source to its left, the 0 used to code it is now assigned to *both* of these symbols, and a 0 and 1 are arbitrarily appended to each to distinguish them from each other. This

a
b

FIGURE 8.2 Huffman (a) source reduction and (b) code assignment procedures.

Original Source		Source Reduction	
Symbol	Probability	1	2
a_2	0.5	0.5	0.5
a_4	0.1875	0.3125	0.5
a_1	0.1875	0.1875	
a_3	0.125		

Original Source			Source Reduction			
Symbol	Probability	Code	1		2	
a_2	0.5	1	0.5	1	0.5	1
a_4	0.1875	00	0.3125	01	0.5	0
a_1	0.1875	011	0.1875	00		
a_3	0.125	010				

operation is then repeated for each reduced source until the original source is reached. The final code appears at the far left (column 3) in Fig. 8.2(b).

The Huffman code in Fig. 8.2(b) (and Table 8.1) is an instantaneous uniquely decodable block code. It is a *block code* because each source symbol is mapped into a fixed sequence of code symbols. It is *instantaneous* because each code word in a string of code symbols can be decoded without referencing succeeding symbols. That is, in any given Huffman code, no code word is a prefix of any other code word. And it is *uniquely decodable* because a string of code symbols can be decoded in only one way. Thus, any string of Huffman encoded symbols can be decoded by examining the individual symbols of the string in a left-to-right manner. For the 4×4 image in Example 8.1, a top-to-bottom left-to-right encoding based on the Huffman code in Fig. 8.2(b) yields the 29-bit string 10101011010110110000011110011. Because we are using an instantaneous uniquely decodable block code, there is no need to insert delimiters between the encoded pixels. A left-to-right scan of the resulting string reveals that the first valid code word is 1, which is the code for symbol a_2 or gray level 119. The next valid code word is 010, which corresponds to gray level 123. Continuing in this manner, we eventually obtain a completely decoded image that is equivalent to `f` in the example.

The source reduction and code assignment procedures just described are implemented by the following M-function, which we call `huffman`:

huffman

```
function CODE = huffman(p)
%HUFFMAN Builds a variable-length Huffman code for a symbol source.
%   CODE = HUFFMAN(P) returns a Huffman code as binary strings in
%   cell array CODE for input symbol probability vector P. Each word
%   in CODE corresponds to a symbol whose probability is at the
```

```
%   corresponding index of P.
%
%   Based on huffman5 by Sean Danaher, University of Northumbria,
%   Newcastle UK. Available at the MATLAB Central File Exchange:
%   Category General DSP in Signal Processing and Communications.

% Check the input arguments for reasonableness.
error(nargchk(1, 1, nargin));
if (ndims(p) ~= 2) | (min(size(p)) > 1) | ~isreal(p) | ~isnumeric(p)
   error('P must be a real numeric vector.');
end

% Global variable surviving all recursions of function 'makecode'
global CODE
CODE = cell(length(p), 1);      % Init the global cell array

if length(p) > 1                % When more than one symbol ...
   p = p / sum(p);              % Normalize the input probabilities
   s = reduce(p);               % Do Huffman source symbol reductions
   makecode(s, []);             % Recursively generate the code
else
   CODE = {'1'};                % Else, trivial one symbol case!
end;
%-------------------------------------------------------------------%
function s = reduce(p);
% Create a Huffman source reduction tree in a MATLAB cell structure
% by performing source symbol reductions until there are only two
% reduced symbols remaining

s = cell(length(p), 1);

% Generate a starting tree with symbol nodes 1, 2, 3, ... to
% reference the symbol probabilities.
for i = 1:length(p)
   s{i} = i;
end

while numel(s) > 2
   [p, i] = sort(p);          % Sort the symbol probabilities
   p(2) = p(1) + p(2);        % Merge the 2 lowest probabilities
   p(1) = [];                 % and prune the lowest one

   s = s(i);                  % Reorder tree for new probabilities
   s{2} = {s{1}, s{2}};       % and merge & prune its nodes
   s(1) = [];                 % to match the probabilities
end

%-------------------------------------------------------------------%
function makecode(sc, codeword)
% Scan the nodes of a Huffman source reduction tree recursively to
% generate the indicated variable length code words.

% Global variable surviving all recursive calls
global CODE
```

```
if isa(sc, 'cell')                      % For cell array nodes,
   makecode(sc{1}, [codeword 0]);       % add a 0 if the 1st element
   makecode(sc{2}, [codeword 1]);       % or a 1 if the 2nd
else                                    % For leaf (numeric) nodes,
   CODE{sc} = char('0' + codeword);     % create a char code string
end
```

The following command line sequence uses `huffman` to generate the code in Fig. 8.2:

```
>> p = [0.1875 0.5 0.125 0.1875];
>> c = huffman(p)

c =
      '011'
      '1'
      '010'
      '00'
```

Note that the output is a variable-length character array in which each row is a string of 0s and 1s—the binary code of the correspondingly indexed symbol in `p`. For example, `'010'` (at array index 3) is the code for the gray level with probability 0.125.

In the opening lines of `huffman`, input argument `p` (the input symbol probability vector of the symbols to be encoded) is checked for reasonableness and *global variable* `CODE` is initialized as a MATLAB cell array (defined in Section 2.10.6) with `length(p)` rows and a single column. All MATLAB global variables must be declared in the functions that reference them using a statement of the form

```
global X Y Z
```

This statement makes variables `X`, `Y`, and `Z` available to the function in which they are declared. When several functions declare the same global variable, they share a single copy of that variable. In `huffman`, the main routine and internal function `makecode` share global variable `CODE`. Note that it is customary to capitalize the names of global variables. Nonglobal variables are *local variables* and are available only to the functions in which they are defined (not to other functions or the base workspace); they are typically denoted in lowercase.

In `huffman`, `CODE` is initialized using the `cell` function, whose syntax is

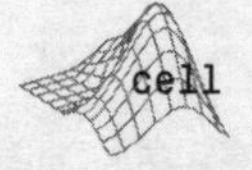


```
X = cell(m, n)
```

An equivalent expression is `X = cell([m, n])`. *For other forms, type* `>> help cell`.

It creates an $m \times n$ array of empty matrices that can be referenced by cell or by content. Parentheses, "`()`", are used for *cell indexing*; curly braces, "`{}`", are used for *content indexing*. Thus, `X(1) = []` indexes and removes element 1 from the cell array, while `X{1} = []` sets the first cell array element to the empty matrix. That is, X{1} refers to the contents of the first element (an array) of X; X(1) refers to the element itself (rather than its content). Since cell arrays can be nested within other cell arrays, the syntax X{1}{2} refers to the content of the second element of the cell array that is in the first element of cell array X.

After CODE is initialized and the input probability vector is normalized [in the p = p / sum(p) statement], the Huffman code for normalized probability vector p is created in two steps. The first step, which is initiated by the s = reduce(p) statement of the main routine, is to call internal function reduce, whose job is to perform the source reductions illustrated in Fig. 8.2(a). In reduce, the elements of an initially empty source reduction cell array s, which is sized to match CODE, are initialized to their indices. That is, s{1} = 1, s{2} = 2, and so on. The cell equivalent of a binary tree for the source reductions is then created in the while numel(s) > 2 loop. In each iteration of the loop, vector p is sorted in ascending order of probability. This is done by the sort function, whose general syntax is

```
[y, i] = sort(x)
```

where output y is the sorted elements of x and index vector i is such that y = x(i). When p has been sorted, the lowest two probabilities are merged by placing their composite probability in p(2), and p(1) is pruned. The source reduction cell array is then reordered to match p based on index vector i using s = s(i). Finally, s{2} is replaced with a two-element cell array containing the merged probability indices via s{2} = {s{1}, s{2}} (an example of content indexing), and cell indexing is employed to prune the first of the two merged elements, s(1), via s(1) = []. The process is repeated until only two elements remain in s.

Figure 8.3 shows the final output of the process for the symbol probabilities in Table 8.1 and Fig. 8.2(a). Figures 8.3(b) and (c) were generated by inserting

```
celldisp(s);
cellplot(s);
```

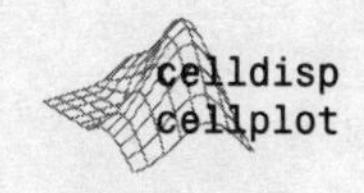

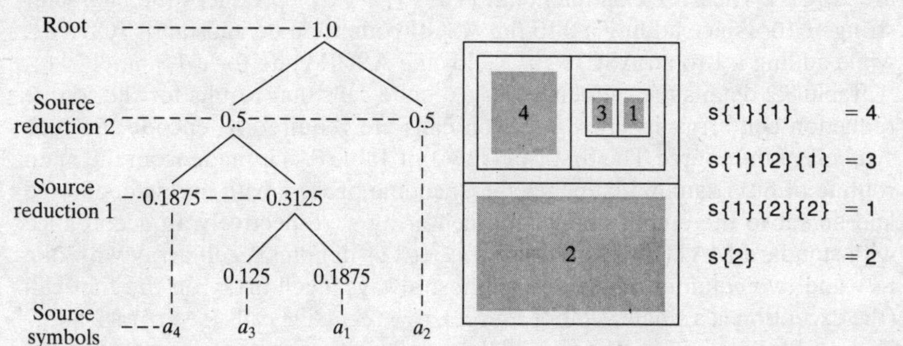

a b c

FIGURE 8.3 Source reductions of Fig 8.2(a) using function huffman: (a) binary tree equivalent; (b) display generated by cellplot(s); (c) celldisp(s) output.

between the last two statements of the huffman main routine. MATLAB function celldisp prints a cell array's contents recursively; function cellplot produces a graphical depiction of a cell array as nested boxes. Note the one-to-one correspondence between the cell array elements in Fig. 8.3(b) and the source reduction tree nodes in Fig. 8.3(a): (1) each two-way branch in the tree (which represents a source reduction) corresponds to a two-element cell array in s; and (2) each two-element cell array contains the indices of the symbols that were merged in the corresponding source reduction. For example, the merging of symbols a_3 and a_1 at the bottom of the tree produces the two-element cell array s{1}{2}, where s{1}{2}{1} = 3 and s{1}{2}{2} = 1 (the indices of symbol a_3 and a_1, respectively). The root of the tree is the top-level two-element cell array s.

The final step of the code generation process (i.e., the assignment of codes based on source reduction cell array s) is triggered by the final statement of huffman—the makecode(s, []) call. This call initiates a *recursive* code assignment process based on the procedure in Fig. 8.2(b). Although recursion generally provides no savings in storage (since a stack of values being processed must be maintained somewhere) or increase in speed, it has the advantage that the code is more compact and often easier to understand, particularly when dealing with recursively defined data structures like trees. Any MATLAB function can be used recursively; that is, it can call itself either directly or indirectly. When recursion is used, each function call generates a fresh set of local variables, independent of all previous sets.

Internal function makecode accepts two inputs: codeword, an array of 0s and 1s, and sc, a source reduction cell array element. When sc is itself a cell array, it contains the two source symbols (or composite symbols) that were joined during the source reduction process. Since they must be individually coded, a pair of recursive calls (to makecode) is issued for the elements—along with two appropriately updated code words (a 0 and 1 are appended to input codeword). When sc does not contain a cell array, it is the index of an original source symbol and is assigned a binary string created from input codeword using CODE{sc} = char('0' + codeword). As was noted in Section 2.10.5, MATLAB function char converts an array containing positive integers that represent character codes into a MATLAB character array (the first 127 codes are ASCII). Thus, for example, char('0' + [0 1 0]) produces the character string '010', since adding a 0 to the ASCII code for a 0 yields an ASCII '0', while adding a 1 to an ASCII '0' yields the ASCII code for a 1, namely '1'.

Table 8.2 details the sequence of makecode calls that results for the source reduction cell array in Fig. 8.3. Seven calls are required to encode the four symbols of the source. The first call (row 1 of Table 8.2) is made from the main routine of huffman and launches the encoding process with inputs codeword and sc set to the empty matrix and cell array s, respectively. In accordance with standard MATLAB notation, {1x2 cell} denotes a cell array with one row and two columns. Since sc is almost always a cell array on the first call (the exception is a single symbol source), two recursive calls (see rows 2 and 7 of the table) are issued. The first of these calls initiates two more calls (rows 3 and 4) and the second of these initiates two additional calls (rows 5 and 6).

Call	Origin	sc	codeword
1	main routine	{1x2 cell}	[]
		[2]	
2	makecode	[4] {1x2 cell}	0
3	makecode	4	0 0
4	makecode	[3] [1]	0 1
5	makecode	3	0 1 0
6	makecode	1	0 1 1
7	makecode	2	1

TABLE 8.2 Code assignment process for the source reduction cell array in Fig. 8.3.

Anytime that sc is not a cell array, as in rows 3, 5, 6, and 7 of the table, additional recursions are unnecessary; a code string is created from codeword and assigned to the source symbol whose index was passed as sc.

8.2.2 Huffman Encoding

Huffman code generation is not (in and of itself) compression. To realize the compression that is built into a Huffman code, the symbols for which the code was created, whether they are gray levels, run lengths, or the output of some other gray-level mapping operation, must be transformed or mapped (i.e., encoded) in accordance with the generated code.

EXAMPLE 8.2: Variable-length code mappings in MATLAB.

■ Consider the simple 16-byte 4×4 image:

```
>> f2 = uint8([2 3 4 2; 3 2 4 4; 2 2 1 2; 1 1 2 2])
f2 =
    2    3    4    2
    3    2    4    4
    2    2    1    2
    1    1    2    2
>> whos('f2')
   Name      Size      Bytes      Class
   f2        4x4       16         uint8 array
Grand total is 16 elements using 16 bytes
```

Each pixel in f2 is an 8-bit byte; 16 bytes are used to represent the entire image. Because the gray levels of f2 are not equiprobable, a variable-length code (as was indicated in the last section) will reduce the amount of memory required to represent the image. Function huffman computes one such code:

```
>> c = huffman(hist(double(f2(:)), 4))
c =
    '011'
    '1'
    '010'
    '00'
```

Since Huffman codes are based on the relative frequency of occurrence of the source symbols being coded (not the symbols themselves), `c` is identical to the code that was constructed for the image in Example 8.1. In fact, image `f2` can be obtained from `f` in Example 8.1 by mapping gray levels 107, 119, 123, and 168 to 1, 2, 3, and 4, respectively. For either image, `p = [0.1875 0.5 0.125 0.1875]`.

A simple way to encode `f2` based on code `c` is to perform a straightforward lookup operation:

```
>> h1f2 = c(f2(:))'

h1f2 =

    Columns 1 through 9

    '1'    '010'    '1'    '011'    '010'    '1'    '1'    '011'    '00'

    Columns 10 through 16

    '00'    '011'    '1'    '1'    '00'    '1'    '1'

>> whos('h1f2')
    Name      Size      Bytes      Class

    h1f2      1x16      1530       cell array

Grand total is 45 elements using 1530 bytes
```

Here, `f2` (a two-dimensional array of class UINT8) is transformed into a 1×16 cell array, `h1f2` (the transpose compacts the display). The elements of `h1f2` are strings of varying length and correspond to the pixels of `f2` in a top-to-bottom left-to-right (i.e., columnwise) scan. As can be seen, the encoded image uses 1530 bytes of storage—almost 100 times the memory required by `f2`!

The use of a cell array for `h1f2` is logical because it is one of two standard MATLAB data structures (see Section 2.10.6) for dealing with arrays of dissimilar data. In the case of `h1f2`, the dissimilarity is the length of the character strings and the price paid for transparently handling it via the cell array is the memory overhead (inherent in the cell array) that is required to track the position of the variable-length elements. We can eliminate this overhead by transforming `h1f2` into a conventional two-dimensional character array:

```
>> h2f2 = char(h1f2)'

h2f2 =
    1010011000011011
     1 11  1001   0
     0 10   1   1

>> whos('h2f2')
    Name      Size      Bytes      Class

    h2f2      3x16      96         char array

Grand total is 48 elements using 96 bytes
```

Here, cell array h1f2 is transformed into a 3×16 character array, h2f2. Each column of h2f2 corresponds to a pixel of f2 in a top-to-bottom left-to-right (i.e., columnwise) scan. Note that blanks are inserted to size the array properly and, since two bytes are required for each '0' or '1' of a code word, the total memory used by h2f2 is 96 bytes—still six times greater than the original 16 bytes needed for f2. We can eliminate the inserted blanks using

```
>> h2f2 = h2f2(:);
>> h2f2(h2f2 == ' ') = [];
>> whos('h2f2')
   Name      Size      Bytes    Class

   h2f2      29x1      58       char array

Grand total is 29 elements using 58 bytes
```

but the required memory is still greater than f2's original 16 bytes.

To compress f2, code c must be applied at the bit level, with several encoded pixels packed into a single byte:

```
>> h3f2 = mat2huff(f2)

h3f2 =
   size: [4 4]
    min: 32769
   hist: [3 8 2 3]
   code: [43867 1944]

>> whos('h3f2')
   Name      Size      Bytes    Class

   h3f2      1x1       518      struct array

Grand total is 13 elements using 518 bytes
```

Although function mat2huff returns a structure, h3f2, requiring 518 bytes of memory, most of it is associated with either (1) structure variable overhead (recall from the Section 8.1 discussion of imratio that MATLAB uses 124 bytes of overhead per structure field) or (2) mat2huff generated information to facilitate future decoding. Neglecting this overhead, which is negligible when considering practical (i.e., normal size) images, mat2huff compresses f2 by a factor of 4:1. The 16 8-bit pixels of f2 are compressed into two 16-bit words—the elements in field code of h3f2:

```
>> hcode = h3f2.code;
>> whos('hcode')
   Name      Size      Bytes    Class

   hcode     1x2       4        uint16 array

Grand total is 2 elements using 4 bytes
```

Converts a decimal integer to a binary string. For more information, type `>> help dec2bin`.

```
>> dec2bin(double(hcode))
ans =
    1010101101011011
    0000011110011000
```

Note that `dec2bin` has been employed to display the individual bits of `h3f2.code`. Neglecting the terminating modulo-16 pad bits (i.e., the final three 0s), the 32-bit encoding is equivalent to the previously generated (see Section 8.2.1) 29-bit instantaneous uniquely decodable block code, 10101011010110110000011110011. ■

As was noted in the preceding example, function `mat2huff` embeds the information needed to decode an encoded input array (e.g., its original dimensions and symbol probabilities) in a single MATLAB structure variable. The information in this structure is documented in the help text section of `mat2huff` itself:

mat2huff

```
function y = mat2huff(x)
%MAT2HUFF Huffman encodes a matrix.
%   Y = MAT2HUFF(X) Huffman encodes matrix X using symbol
%   probabilities in unit-width histogram bins between X's minimum
%   and maximum values. The encoded data is returned as a structure
%   Y:
%      Y.code   The Huffman-encoded values of X, stored in
%               a uint16 vector. The other fields of Y contain
%               additional decoding information, including:
%      Y.min    The minimum value of X plus 32768
%      Y.size   The size of X
%      Y.hist   The histogram of X
%
%   If X is logical, uint8, uint16, uint32, int8, int16, or double,
%   with integer values, it can be input directly to MAT2HUFF. The
%   minimum value of X must be representable as an int16.
%
%   If X is double with non-integer values---for example, an image
%   with values between 0 and 1---first scale X to an appropriate
%   integer range before the call. For example, use Y =
%   MAT2HUFF(255*X) for 256 gray level encoding.
%
%   NOTE: The number of Huffman code words is round(max(X(:))) -
%   round(min(X(:))) + 1. You may need to scale input X to generate
%   codes of reasonable length. The maximum row or column dimension
%   of X is 65535.
%
% See also HUFF2MAT.

if ndims(x) ~= 2 | ~isreal(x) | (~isnumeric(x) & ~islogical(x))
   error('X must be a 2-D real numeric or logical matrix.');
end
```

```
% Store the size of input x.
y.size = uint32(size(x));

% Find the range of x values and store its minimum value biased
% by +32768 as a UINT16.
x = round(double(x));
xmin = min(x(:));
xmax = max(x(:));
pmin = double(int16(xmin));
pmin = uint16(pmin + 32768);     y.min = pmin;

% Compute the input histogram between xmin and xmax with unit
% width bins, scale to UINT16, and store.
x = x(:)';
h = histc(x, xmin:xmax);
if max(h) > 65535
   h = 65535 * h / max(h);
end
h = uint16(h);   y.hist = h;

% Code the input matrix and store the result.
map = huffman(double(h));                % Make Huffman code map
hx = map(x(:) - xmin + 1);               % Map image
hx = char(hx)';                          % Convert to char array
hx = hx(:)';
hx(hx == ' ') = [];                      % Remove blanks
ysize = ceil(length(hx) / 16);           % Compute encoded size
hx16 = repmat('0', 1, ysize * 16);       % Pre-allocate modulo-16 vector
hx16(1:length(hx)) = hx;                 % Make hx modulo-16 in length
hx16 = reshape(hx16, 16, ysize);         % Reshape to 16-character words
hx16 = hx16' - '0';                      % Convert binary string to decimal
twos = pow2(15:-1:0);
y.code = uint16(sum(hx16 .* twos(ones(ysize, 1), :), 2))';
```

This function is similar to `hist`. *For more details, type* `>> help histc`.

Note that the statement `y = mat2huff(x)` Huffman encodes input matrix `x` using unit-width histogram bins between the minimum and maximum values of `x`. When the encoded data in `y.code` is later decoded, the Huffman code needed to decode it must be re-created from `y.min`, the minimum value of `x`, and `y.hist`, the histogram of `x`. Rather than preserving the Huffman code itself, `mat2huff` keeps the probability information needed to regenerate it. With this, and the original dimensions of matrix `x`, which is stored in `y.size`, function `huff2mat` of Section 8.2.3 (the next section) can decode `y.code` to reconstruct `x`.

The steps involved in the generation of `y.code` are summarized as follows:

1. Compute the histogram, `h`, of input `x` between the minimum and maximum values of `x` using unit-width bins and scale it to fit in a UINT16 vector.
2. Use `huffman` to create a Huffman code, called `map`, based on the scaled histogram, `h`.

3. Map input x using map (this creates a cell array) and convert it to a character array, hx, removing the blanks that are inserted like in h2f2 of Example 8.2.
4. Construct a version of vector hx that arranges its characters into 16-character segments. This is done by creating a modulo-16 character vector that will hold it (hx16 in the code), copying the elements of hx into it, and reshaping it into a 16 row by ysize array, where ysize = ceil(length(hx) / 16). Recall from Section 4.2 that the ceil function rounds a number toward positive infinity. The generalized MATLAB function

```
y = reshape(x, m, n)
```

returns an m by n matrix whose elements are taken column wise from x. An error is returned if x does not have m*n elements.

5. Convert the 16-character elements of hx16 to 16-bit binary numbers (i.e., unit16's). Three statements are substituted for the more compact y = uint16(bin2dec(hx16')). They are the core of bin2dec, which returns the decimal equivalent of a binary string (e.g., bin2dec('101') returns 5) but are faster because of decreased generality. MATLAB function pow2(y) is used to return an array whose elements are 2 raised to the y power. That is, twos = pow2(15: -1: 0) creates the array [32768 16384 8192 ... 8 4 2 1].

EXAMPLE 8.3: Encoding with mat2huff.

■ To illustrate further the compression performance of Huffman encoding, consider the 512 × 512 8-bit monochrome image of Fig. 8.4(a). The compression of this image using mat2huff is carried out by the following command sequence:

```
>> f = imread('Tracy.tif');
>> c = mat2huff(f);
>> cr1 = imratio(f, c)

cr1 =
    1.2191
```

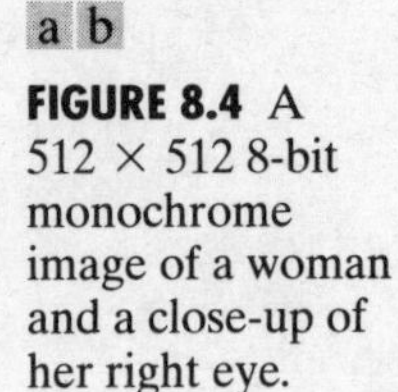

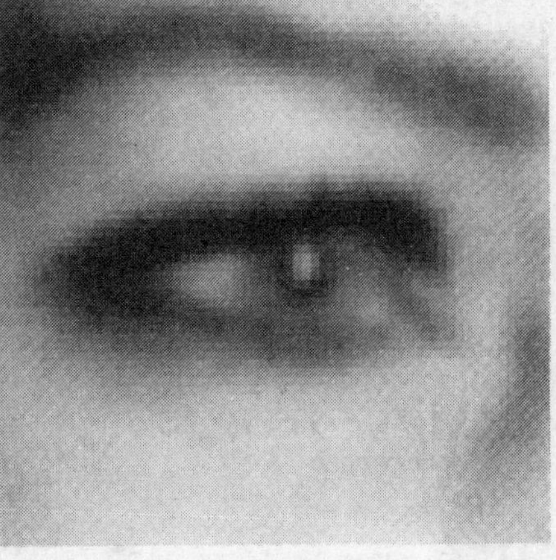

a b

FIGURE 8.4 A 512 × 512 8-bit monochrome image of a woman and a close-up of her right eye.

By removing the coding redundancy associated with its conventional 8-bit binary encoding, the image has been compressed to about 80% of its original size (even with the inclusion of the decoding overhead information).

Since the output of `mat2huff` is a structure, we write it to disk using the `save` function:

```
>> save SqueezeTracy c;
>> cr2 = imratio('Tracy.tif', 'SqueezeTracy.mat')

cr2 =
    1.2365
```

Function syntax `save file var` *stores workspace variable* `var` *to disk as a MATLAB data file called* `'file.mat'`.

The `save` function, like the **Save Workspace As** and **Save Selection As** menu commands in Section 1.7.4, appends a `.mat` extension to the file that is created. The resulting file—in this case, `SqueezeTracy.mat`, is called a *MAT-file*. It is a binary data file containing workspace variable names and values. Here, it contains the single workspace variable `c`. Finally, we note that the small difference in compression ratios `cr1` and `cr2` computed previously is due to MATLAB data file overhead. ■

8.2.3 Huffman Decoding

Huffman encoded images are of little use unless they can be decoded to recreate the original images from which they were derived. For output `y = mat2huff(x)` of the previous section, the decoder must first compute the Huffman code used to encode `x` (based on its histogram and related information in `y`) and then inverse map the encoded data (also extracted from `y`) to rebuild `x`. As can be seen in the following listing of function `x = huff2mat(y)`, this process can be broken into five basic steps:

1. Extract dimensions `m` and `n`, and minimum value `xmin` (of eventual output `x`) from input structure `y`.
2. Re-create the Huffman code that was used to encode `x` by passing its histogram to function `huffman`. The generated code is called `map` in the listing.
3. Build a data structure (transition and output table `link`) to streamline the decoding of the encoded data in `y.code` through a series of computationally efficient binary searches.
4. Pass the data structure and the encoded data [i.e., `link` and `y.code`] to C function `unravel`. This function minimizes the time required to perform the binary searches, creating decoded output vector `x` of class `double`.
5. Add `xmin` to each element of `x` and reshape it to match the dimensions of the original `x` (i.e., `m` rows and `n` columns).

A unique feature of `huff2mat` is the incorporation of MATLAB callable C function `unravel` (see Step 4), which makes the decoding of most normal resolution images nearly instantaneous.

huff2mat

```
function x = huff2mat(y)
%HUFF2MAT decodes a Huffman encoded matrix.
%   X = HUFF2MAT(Y) decodes a Huffman encoded structure Y with uint16
%   fields:
```

```
%     Y.min      Minimum value of X plus 32768
%     Y.size     Size of X
%     Y.hist     Histogram of X
%     Y.code     Huffman code
%
%   The output X is of class double.
%
%   See also MAT2HUFF.

if ~isstruct(y) | ~isfield(y, 'min') | ~isfield(y, 'size') | ...
      ~isfield(y, 'hist') | ~isfield(y, 'code')
   error('The input must be a structure as returned by MAT2HUFF.');
end
sz = double(y.size);    m = sz(1);    n = sz(2);
xmin = double(y.min) - 32768;                 % Get X minimum
map = huffman(double(y.hist));                % Get Huffman code (cell)
% Create a binary search table for the Huffman decoding process.
% 'code' contains source symbol strings corresponding to 'link'
% nodes, while 'link' contains the addresses (+) to node pairs for
% node symbol strings plus '0' and '1' or addresses (-) to decoded
% Huffman codewords in 'map'. Array 'left' is a list of nodes yet to
% be processed for 'link' entries.
code = cellstr(char('', '0', '1'));           % Set starting conditions as
link = [2; 0; 0];     left = [2 3];           % 3 nodes w/2 unprocessed
found = 0;    tofind = length(map);           % Tracking variables
while length(left) & (found < tofind)
   look = find(strcmp(map, code{left(1)}));         % Is string in map?
   if look                                    % Yes
      link(left(1)) = -look;                  % Point to Huffman map
      left = left(2:end);                     % Delete current node
      found = found + 1;                      % Increment codes found
   else                                       % No, add 2 nodes & pointers
      len = length(code);                     % Put pointers in node
      link(left(1)) = len + 1;

      link = [link; 0; 0];                    % Add unprocessed nodes
      code{end + 1} = strcat(code{left(1)}, '0');
      code{end + 1} = strcat(code{left(1)}, '1');
      left = left(2:end);                     % Remove processed node
      left = [left len + 1 len + 2];          % Add 2 unprocessed nodes
   end
end
x = unravel(y.code', link, m * n);            % Decode using C 'unravel'
x = x + xmin - 1;                             % X minimum offset adjust
x = reshape(x, m, n);                         % Make vector an array
```

As indicated earlier, huff2mat-based decoding is built on a series of binary searches or two-outcome decoding decisions. Each element of a sequentially scanned Huffman encoded string—which must of course be a `'0'` or a `'1'`—triggers a binary decoding decision based on transition and output table `link`. The construction of `link` begins with its initialization in statement `link = [2; 0; 0]`. Each element in the starting three-state `link` array corresponds to a Huffman encoded binary string in the corresponding cell array `code`; initially, `code = cellstr(char('', '0', '1'))`. The null string, `code(1)`, is the starting point (or initial decoding state) for all Huffman string decoding. The associated 2 in `link(1)` identifies the two possible decoding states that follow from appending a `'0'` and `'1'` to the null string. If the next encountered Huffman encoded bit is a `'0'`, the next decoding state is `link(2)` [since `code(2) = '0'`, the null string concatenated with `'0'`]; if it is a `'1'`, the new state is `link(3)` [at index $(2 + 1)$ or 3, with `code(3) = '1'`]. Note that the corresponding `link` array entries are 0—indicating that they have not yet been processed to reflect the proper decisions for Huffman code `map`. During the construction of `link`, if either string (i.e., the `'0'` or `'1'`) is found in `map` (i.e., it is a valid Huffman code word), the corresponding 0 in `link` is replaced by the negative of the corresponding `map` index (which is the decoded value). Otherwise, a new (positive valued) `link` index is inserted to point to the two new states (possible Huffman code words) that logically follow (either `'00'` and `'01'` or `'10'` and `'11'`). These new and as yet unprocessed `link` elements expand the size of `link` (cell array `code` must also be updated), and the construction process is continued until there are no unprocessed elements left in `link`. Rather than continually scanning `link` for unprocessed elements, however, `huff2mat` maintains a tracking array, called `left`, which is initialized to `[2, 3]` and updated to contain the indices of the link elements that have not been examined.

Table 8.3 shows the `link` table that is generated for the Huffman code in Example 8.2. If each `link` index is viewed as a decoding state, `i`, each binary coding decision (in a left-to-right scan of an encoded string) and/or Huffman decoded output is determined by `link(i)`:

1. If `link(i) < 0` (i.e., negative), a Huffman code word has been decoded. The decoded output is `|link(i)|`, where `| |` denotes the absolute value.
2. If `link(i) > 0` (i.e., positive) and the next encoded bit to be processed is a 0, the next decoding state is index `link(i)`. That is, we let `i = link(i)`.
3. If `link(i) > 0` and the next encoded bit to be processed is a 1, the next decoding state is index `link(i) + 1`. That is, `i = link(i) + 1`.

TABLE 8.3 Decoding table for the source reduction cell array in Fig. 8.3.

Index i	Value in link(i)
1	2
2	4
3	−2
4	−4
5	6
6	−3
7	−1

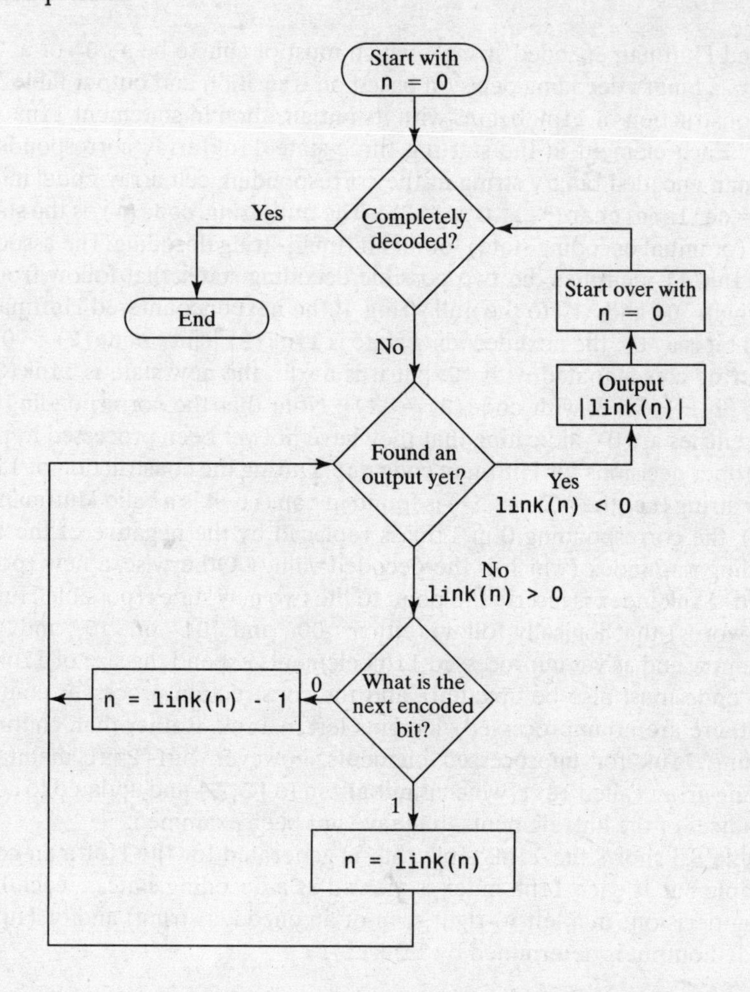

FIGURE 8.5 Flow diagram for C function `unravel`.

As noted previously, positive `link` entries correspond to binary decoding transitions, while negative entries determine decoded output values. As each Huffman code word is decoded, a new binary search is started at `link` index `i = 1`. For encoded string 101010110101... of Example 8.2, the resulting state transition sequence is `i = 1, 3, 1, 2, 5, 6, 1, ...`; the corresponding output sequence is `–, |–2|, –, –, –, |–3|, – ...`, where – is used to denote the absence of an output. Decoded output values 2 and 3 are the first two pixels of the first column of test image `f2` in Example 8.2.

C function `unravel` accepts the link structure just described and uses it to drive the binary searches required to decode input `hx`. Figure 8.5 diagrams its basic operation, which follows the decision-making process that was described in conjunction with Table 8.3. Note, however, that modifications are needed to compensate for the fact that C arrays are indexed from 0 rather than 1.

Both C and Fortran functions can be incorporated into MATLAB and serve two main purposes: (1) They allow large preexisting C and Fortran programs to be called from MATLAB without having to be rewritten as M-files, and (2) they streamline bottleneck computations that do not run fast enough as MATLAB M-files but can be coded in C or Fortran for increased efficiency. Whether C or Fortran is used, the resulting functions are referred to as *MEX-files*; they behave as though they are M-files or ordinary MATLAB functions. Unlike M-files, however, they must be compiled and linked using MATLAB's `mex` script before they can be called. To compile and link `unravel` on a Windows platform from the MATLAB command line prompt, for example, we type

A MATLAB external function produced from C or Fortran source code. It has a platform-dependent extension (e.g., `.dll` for Windows).

```
>> mex unravel.c
```

A MEX-file named `unravel.dll` with extension `.dll` will be created. Any help text, if desired, must be provided as a separate M-file of the same name (it will have a `.m` extension).

The source code for *C MEX-file* `unravel` has a `.c` extension and follows:

C MEX-file

The C source code used to build a MEX-file.

`unravel.c`

```
/*==================================================================
* unravel.c
* Decodes a variable length coded bit sequence (a vector of
* 16-bit integers) using a binary sort from the MSB to the LSB
* (across word boundaries) based on a transition table.
*==================================================================*/
#include "mex.h"

void unravel(unsigned short *hx, double *link, double *x,
    double xsz, int hxsz)
{
  int  i = 15, j = 0, k = 0, n = 0;     /* Start at root node, 1st */
                                        /* hx bit and x element */
  while (xsz - k)   {                   /* Do until x is filled */
    if (*(link + n)  > 0)   {           /* Is there a link? */
      if ((*(hx + j) >> i) & 0x0001)    /* Is bit a 1? */
          n = *(link + n);              /* Yes, get new node */
      else n = *(link + n) - 1;         /* It's 0 so get new node */
      if (i) i--; else {j++; i = 15;}   /* Set i, j to next bit */
      if (j > hxsz)                     /* Bits left to decode? */
         mexErrMsgTxt("Out of code bits ???");
      }
    else    {                           /* It must be a leaf node */
      *(x + k++) = -*(link + n);        /* Output value */
      n = 0;  }                         /* Start over at root */
  }
  if (k == xsz - 1)                     /* Is one left over? */
    *(x + k++) = -*(link + n);
}

void mexFunction(int nlhs, mxArray *plhs[],
                 int nrhs, const mxArray *prhs[])
{
  double *link, *x, xsz;
```

```
    unsigned short *hx;
    int hxsz;

    /* Check inputs for reasonableness */
    if(nrhs != 3)
       mexErrMsgTxt("Three inputs required.");
    else if (nlhs > 1)
       mexErrMsgTxt("Too many output arguments.");

    /* Is last input argument a scalar? */
    if (!mxIsDouble(prhs[2])  || mxIsComplex(prhs[2])  ||
          mxGetN(prhs[2]) * mxGetM(prhs[2]) != 1)
       mexErrMsgTxt("Input XSIZE must be a scalar.");

    /* Create input matrix pointers and get scalar */
    hx = mxGetPr(prhs[0]);                          /* UINT16 */
    link = mxGetPr(prhs[1]);                        /* DOUBLE */
    xsz = mxGetScalar(prhs[2]);                     /* DOUBLE */

    /* Get the number of elements in hx */
    hxsz = mxGetM(prhs[0]);

    /* Create 'xsz' x 1 output matrix */
    plhs[0] = mxCreateDoubleMatrix(xsz, 1, mxREAL);

    /* Get C pointer to a copy of the output matrix */
    x = mxGetPr(plhs[0]);

    /* Call the C subroutine */
    unravel(hx, link, x, xsz, hxsz);
}
```

The companion help text is provided in M-file `unravel.m`:

unravel.m

```
%UNRAVEL Decodes a variable-length bit stream.
%   X = UNRAVEL(Y, LINK, XLEN) decodes UINT16 input vector Y based on
%   transition and output table LINK. The elements of Y are
%   considered to be a contiguous stream of encoded bits--i.e., the
%   MSB of one element follows the LSB of the previous element. Input
%   XLEN is the number code words in Y, and thus the size of output
%   vector X (class DOUBLE). Input LINK is a transition and output
%   table (that drives a series of binary searches):
%
%    1. LINK(0) is the entry point for decoding, i.e., state n = 0.
%    2. If LINK(n) < 0, the decoded output is |LINK(n)|; set n = 0.
%    3. If LINK(n) > 0, get the next encoded bit and transition to
%       state [LINK(n) - 1] if the bit is 0, else LINK(n).
```

Like all C MEX-files, C MEX-file `unravel.c` consists of two distinct parts: a *computational routine* and a *gateway routine*. The computational routine, also named unravel, contains the C code that implements the link-based decoding process of Fig. 8.5. The gateway routine, which must always be named mexFunction, interfaces C computational routine unravel to M-file calling function, huff2mat. It uses MATLAB's standard MEX-file interface, which is based on the following:

1. Four standardized input/output parameters—nlhs, plhs, nrhs, and prhs. These parameters are the number of left-hand-side output arguments (an integer), an array of pointers to the left-hand-side output arguments (all MATLAB arrays), the number of right-hand-side input arguments (another integer), and an array of pointers to the right-hand-side input arguments (also MATLAB arrays), respectively.
2. A MATLAB provided set of *Application Program Interface* (API) functions. API functions that are prefixed with mx are used to create, access, manipulate, and/or destroy structures of class mxArray. For example,

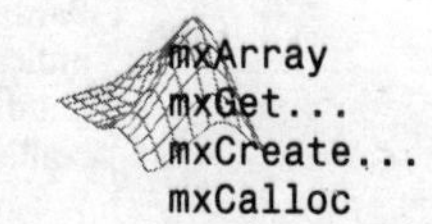

   - mxCalloc dynamically allocates memory like a standard C *calloc* function. Related functions include mxMalloc and mxRealloc that are used in place of the C *malloc* and *realloc* functions.
   - mxGetScalar extracts a scalar from input array prhs. Other mxGet... functions, like mxGetM, mxGetN, and mxGetString, extract other types of data.
   - mxCreateDoubleMatrix creates a MATLAB output array for plhs. Other mxCreate... functions, like mxCreateString and mxCreateNumericArray, facilitate the creation of other data types.

   API functions prefixed by mex perform operations in the MATLAB environment. For example, mexErrMsgTxt outputs a message to the MATLAB workspace.

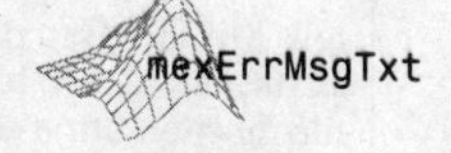

Function prototypes for the API mex and mx routines noted in item 2 of the preceding list are maintained in MATLAB header files mex.h and matrix.h, respectively. Both are located in the <matlab>/extern/include directory, where <matlab> denotes the top-level directory where MATLAB is installed on your system. Header mex.h, which must be included at the beginning of all MEX-files (note the C file inclusion statement #include "mex.h " at the start of MEX-file unravel), includes header file matrix.h. The prototypes of the mex and mx interface routines that are contained in these files define the parameters that they use and provide valuable clues about their general operation. Additional information is available in MATLAB's *External Interfaces* reference manual.

Figure 8.6 summarizes the preceding discussion, details the overall structure of C MEX-file unravel, and describes the flow of information between it and M-file huff2mat. Though constructed in the context of Huffman decoding, the concepts illustrated are easily extended to other C- and/or Fortran-based MATLAB functions.

M-file `unravel.m`

Help text for C MEX-file `unravel`:

Contains text that is displayed in response to
`>> help unravel`

MATLAB passes `y`, `link`, and `m * n` to the C MEX file:

```
prhs [0] = y
prhs [1] = link
prhs [2] = m * n
nrhs = 3
nrhs = 1
```

Parameters `nlhs` and `nrhs` are integers indicating the number of left- and right-hand arguments, and `prhs` is a vector containing *pointers* to MATLAB arrays `y`, `link`, and `m * n`.

M-file `huff2mat`

•
•
•

In M-file `huff2mat`, the statement

```
x = unravel(y, ...
    link, m * n)
```

tells MATLAB to pass `y`, `link`, and `m * n` to C MEX-file function `unravel`.

On return, `plhs(0)` is assigned to `x`.

•
•
•

MATLAB passes MEX-file output `plhs[0]` to M-file `huff2mat`.

C MEX-file `unravel.c`

In C MEX-file `unravel`, execution begins and ends in *gateway routine* `mexFunction`, which calls C *computational routine* `unravel`. To declare the entry point and interface routines, use

```
#include "mex.h"
```

C function `mexFunction`

MEX-file *gateway routine*:

```
void mexFunction(
int nlhs, mxArray *plhs[],
int nrhs, const mxArray
    *prhs[])
```

where integers `nlhs` and `nrhs` indicate the number of left- and right-hand arguments and vectors `plhs` and `prhs` contain *pointers* to input and output arguments of type `mxArray`. The `mxArray` type is MATLAB's internal array representation.

The MATLAB API provides routines to handle the data types it supports. Here, we

1. Use `mxGetM`, `mxGetN`, `mxIsDouble`, `mxIsComplex`, and `mexErrMsgTxt` to check the input and output arguments.

2. Use `mxGetPr` to get pointers to `prhs[0]` (the Huffman code) and `prhs[1]` (the decoding table) and save as C pointers `hx` and `link`, respectively.

3. Use `mxGetScalar` to get the output array size from `prhs[2]` and save as `xsz`.

4. Use `mxGetM` to get the number of elements in `prhs[0]` (the Huffman code) and save as `hxsz`.

5. Use `mxCreateDoubleMatrix` and `mxGetPr` to create an output array pointer (for the decode) and assign it to `plhs[0]`.

6. Call *computational routine* `unravel`, passing the arguments formed in Steps 2-5.

C function `unravel`

MEX-file *computational routine*:

```
void unravel(
    unsigned short *hx
    double *link, double *x,
    double xsz, int hxsz)
```

which contains the C code for decoding `hx` based on `link` and putting the result in `x`.

FIGURE 8.6 The interaction of M-file `huff2mat` and MATLAB callable C function `unravel`. Note that C MEX-file `unravel` contains two functions: gateway routine `mexFunction` and computational routine `unravel`. Help text for MEX-file `unravel` is contained in a separate M-file, also named `unravel`.

EXAMPLE 8.4: Decoding with `huff2mat`.

■ The Huffman encoded image of Example 8.3 can be decoded with the following sequence of commands:

```
>> load SqueezeTracy;
>> g = huff2mat(c);
>> f = imread('Tracy.tif');
>> rmse = compare(f, g)

rmse =
     0
```

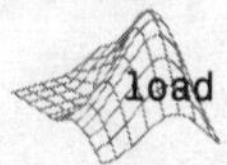

Function `load file` *reads MATLAB variables from disk file* `'file.mat'` *and loads them into the workspace. The variable names are maintained through a* `save/load` *sequence.*

Note that the overall encoding-decoding process is information preserving; the root-mean-square error between the original and decompressed images is 0. Because such a large part of the decoding job is done in C MEX-file `unravel`, `huff2mat` is slightly faster than its encoding counterpart, `mat2huff`. Note the use of the `load` function to retrieve the MAT-file encoded output from Example 8.3. ■

8.3 Interpixel Redundancy

Consider the images shown in Figs. 8.7(a) and (c). As Figs. 8.7(b) and (d) show, they have virtually identical histograms. Note also that the histograms are trimodal, indicating the presence of three dominant ranges of gray-level values.

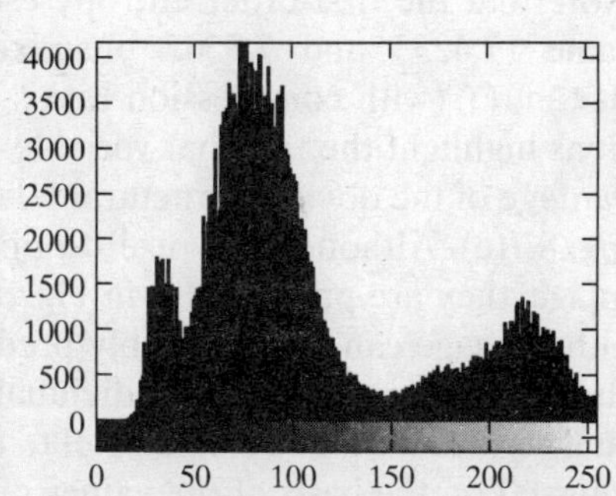

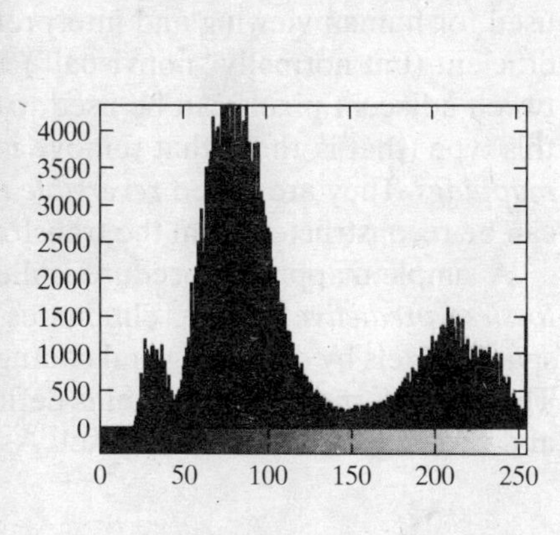

a b
c d

FIGURE 8.7 Two images and their gray-level histograms.

Because the gray levels of the images are not equally probable, variable-length coding can be used to reduce the coding redundancy that would result from a natural binary coding of their pixels:

```
>> f1 = imread('Random Matches.tif');
>> c1 = mat2huff(f1);
>> entropy(f1)

ans =
    7.4253

>> imratio(f1, c1)

ans =
    1.0704

>> f2 = imread('Aligned Matches.tif');
>> c2 = mat2huff(f2);
>> entropy(f2)

ans =
    7.3505

>> imratio(f2, c2)

ans =
    1.0821
```

Note that the first-order entropy estimates of the two images are about the same (7.4253 and 7.3505 bits/pixel); they are compressed similarly by `mat2huff` (with compression ratios of 1.0704 versus 1.0821). These observations highlight the fact that variable-length coding is not designed to take advantage of the obvious structural relationships between the aligned matches in Fig. 8.7(c). Although the pixel-to-pixel correlations are more evident in that image, they are present also in Fig. 8.7(a). Because the values of the pixels in either image can be reasonably predicted from the values of their neighbors, the information carried by individual pixels is relatively small. Much of the visual contribution of a single pixel to an image is redundant; it could have been guessed on the basis of the values of its neighbors. These correlations are the underlying basis of interpixel redundancy.

In order to reduce interpixel redundancies, the 2-D pixel array normally used for human viewing and interpretation must be transformed into a more efficient (but normally "nonvisual") format. For example, the differences between adjacent pixels can be used to represent an image. Transformations of this type (that is, those that remove interpixel redundancy) are referred to as *mappings*. They are called *reversible mappings* if the original image elements can be reconstructed from the transformed data set.

A simple mapping procedure is illustrated in Fig. 8.8. The approach, called *lossless predictive coding*, eliminates the interpixel redundancies of closely spaced pixels by extracting and coding only the new information in each pixel. The *new information* of a pixel is defined as the difference between the actual and predicted value of that pixel. As can be seen, the system consists of an

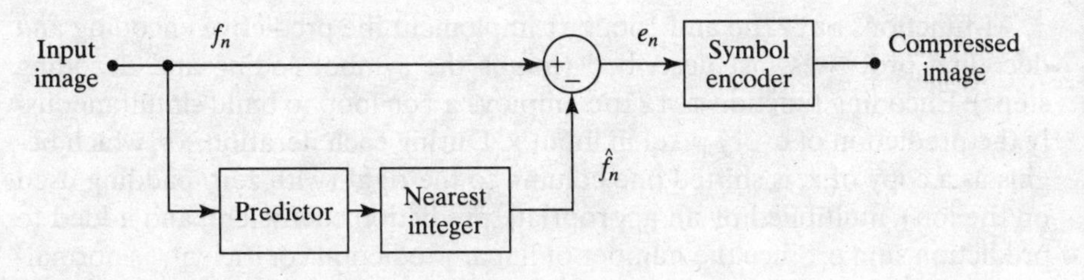

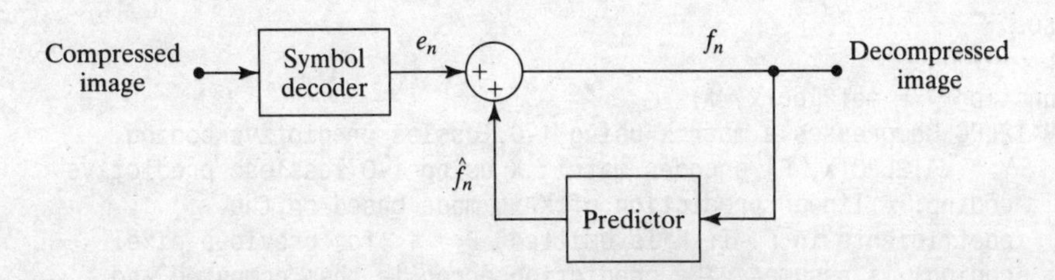

a
b

FIGURE 8.8 A lossless predictive coding model: (a) encoder and (b) decoder.

encoder and decoder, each contining an identical *predictor*. As each successive pixel of the input image, denoted f_n, is introduced to the encoder, the predictor generates the anticipated value of that pixel based on some number of past inputs. The output of the predictor is then rounded to the nearest integer, denoted $\hat{f}_n$, and used to form the difference or prediction error

$$e_n = f_n - \hat{f}_n$$

which is coded using a variable-length code (by the symbol encoder) to generate the next element of the compressed data stream. The decoder of Fig. 8.8(b) reconstructs e_n from the received variable-length code words and performs the inverse operation

$$f_n = e_n + \hat{f}_n$$

Various local, global, and adaptive methods can be used to generate $\hat{f}_n$. In most cases, however, the prediction is formed by a linear combination of m previous pixels. That is,

$$\hat{f}_n = \text{round}\left[\sum_{i=1}^{m} \alpha_i f_{n-i}\right]$$

where m is the order of the linear predictor, "round" is a function used to denote the rounding or nearest integer operation (like function `round` in MATLAB), and the α_i for $i = 1, 2, \ldots, m$ are prediction coefficients. For 1-D linear predictive coding, this equation can be rewritten

$$\hat{f}(x, y) = \text{round}\left[\sum_{i=1}^{m} \alpha_i f(x, y - i)\right]$$

where each subscripted variable is now expressed explicitly as a function of spatial coordinates x and y. Note that prediction $\hat{f}(x, y)$ is a function of the previous pixels on the current scan line alone.

M-functions mat2lpc and lpc2mat implement the predictive encoding and decoding processes just described (minus the symbol coding and decoding steps). Encoding function mat2lpc employs a for loop to build simultaneously the prediction of every pixel in input x. During each iteration, xs, which begins as a copy of x, is shifted one column to the right (with zero padding used on the left), multiplied by an appropriate prediction coefficient, and added to prediction sum p. Since the number of linear prediction coefficients is normally small, the overall process is fast. Note in the following listing that if prediction filter f is not specified, a single element filter with a coefficient of 1 is used.

mat2lpc

```
function y = mat2lpc(x, f)
%MAT2LPC Compresses a matrix using 1-D lossles predictive coding.
%   Y = MAT2LPC(X, F) encodes matrix X using 1-D lossless predictive
%   coding. A linear prediction of X is made based on the
%   coefficients in F. If F is omitted, F = 1 (for previous pixel
%   coding) is assumed. The prediction error is then computed and
%   output as encoded matrix Y.
%
%   See also LPC2MAT.
error(nargchk(1, 2, nargin));         % Check input arguments
if nargin < 2                          % Set default filter if omitted
   f = 1;
end

x = double(x);                         % Ensure double for computations
[m, n] = size(x);                      % Get dimensions of input matrix
p = zeros(m, n);                       % Init linear prediction to 0
xs = x;       zc = zeros(m, 1);        % Prepare for input shift and pad

for j = 1:length(f)                    % For each filter coefficient ...
   xs = [zc xs(:, 1:end - 1)];         % Shift and zero pad x
   p = p + f(j) * xs;                  % Form partial prediction sums
end

y = x - round(p);                      % Compute the prediction error
```

Decoding function lpc2mat performs the inverse operations of encoding counterpart mat2lpc. As can be seen in the following listing, it employs an n iteration for loop, where n is the number of columns in encoded input matrix y. Each iteration computes only one column of decoded output x, since each decoded column is required for the computation of all subsequent columns. To decrease the time spent in the for loop, x is preallocated to its maximum padded size before starting the loop. Note also that the computations employed to generate predictions are done in the same order as they were in lpc2mat to avoid floating point round-off error.

lpc2mat

```
function x = lpc2mat(y, f)
%LPC2MAT Decompresses a 1-D lossless predictive encoded matrix.
%   X = LPC2MAT(Y, F) decodes input matrix Y based on linear
```

```
%   prediction coefficients in F and the assumption of 1-D lossless
%   predictive coding. If F is omitted, filter F = 1 (for previous
%   pixel coding) is assumed.
%
%   See also MAT2LPC.

error(nargchk(1, 2, nargin));         % Check input arguments
if nargin < 2                          % Set default filter if omitted
   f = 1;
end

f = f(end:-1:1);                       % Reverse the filter coefficients
[m, n] = size(y);                      % Get dimensions of output matrix
order = length(f);                     % Get order of linear predictor
f = repmat(f, m, 1);                   % Duplicate filter for vectorizing
x = zeros(m, n + order);               % Pad for 1st 'order' column decodes
% Decode the output one column at a time. Compute a prediction based
% on the 'order' previous elements and add it to the prediction
% error. The result is appended to the output matrix being built.
for j = 1:n
   jj = j + order;
   x(:, jj) = y(:, j) + round(sum(f(:, order:-1:1) .* ...
                        x(:, (jj - 1):-1:(jj - order)), 2));
end

x = x(:, order + 1:end);               % Remove left padding
```

EXAMPLE 8.5: Lossless predictive coding.

■ Consider encoding the image of Fig. 8.7(c) using the simple first-order linear predictor

$$\hat{f}(x, y) = \text{round}[\alpha f(x, y - 1)]$$

A predictor of this form commonly is called a *previous pixel* predictor, and the corresponding predictive coding procedure is referred to as *differential coding* or *previous pixel coding*. Figure 8.9(a) shows the prediction error image that results with $\alpha = 1$. Here, gray level 128 corresponds to a prediction error of 0,

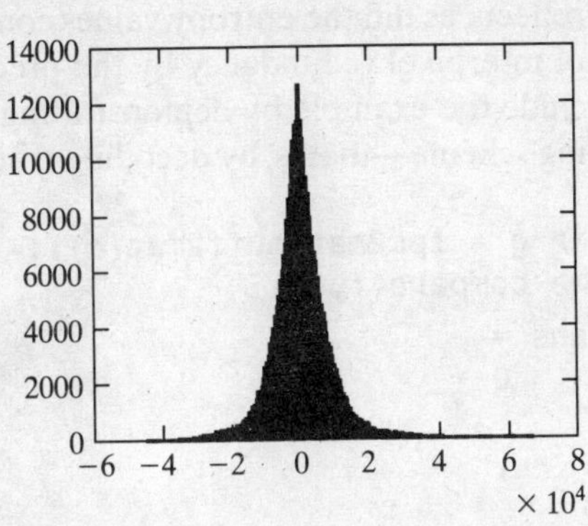

a b

FIGURE 8.9 (a) The prediction error image for Fig. 8.7(c) with `f = [1]`. (b) Histogram of the prediction error.

while nonzero positive and negative errors (under- and overestimates) are scaled by mat2gray to become lighter or darker shades of gray, respectively:

```
>> f = imread('Aligned Matches.tif');
>> e = mat2lpc(f);
>> imshow(mat2gray(e));
>> entropy(e)

ans =
    5.9727
```

Note that the entropy of the prediction error, e, is substantially lower than the entropy of the original image, f. The entropy has been reduced from the 7.3505 bits/pixel (computed at the beginning of this section) to 5.9727 bits/pixel, despite the fact that for m-bit images, $(m + 1)$-bit numbers are needed to represent accurately the resulting error sequence. This reduction in entropy means that the prediction error image can be coded more efficiently that the original image—which, of course, is the goal of the mapping. Thus, we get

```
>> c = mat2huff(e);
>> cr = imratio(f, c)

cr =
    1.3311
```

and see that the compression ratio has, as expected, increased from 1.0821 (when Huffman coding the gray levels directly) to 1.3311.

The histogram of prediction error e is shown in Fig. 8.9(b)—and computed as follows:

```
>> [h, x] = hist(e(:) * 512, 512);
>> figure; bar(x, h, 'k');
```

Note that it is highly peaked around 0 and has a relatively small variance in comparison to the input images gray-level distribution [see Fig. 8.7(d)]. This reflects, as did the entropy values computed earlier, the removal of a great deal of interpixel redundancy by the prediction and differencing process. We conclude the example by demonstrating the lossless nature of the predictive coding scheme—that is, by decoding c and comparing it to starting image f:

```
>> g = lpc2mat(huff2mat(c));
>> compare(f, g)

ans =
    0
```

■

8.4 Psychovisual Redundancy

Unlike coding and interpixel redundancy, psychovisual redundancy is associated with real or quantifiable visual information. Its elimination is desirable because the information itself is not essential for normal visual processing. Since the elimination of psychovisually redundant data results in a loss of quantitative information, it is called *quantization*. This terminology is consistent with the normal usage of the word, which generally means the mapping of a broad range of input values to a limited number of output values. As it is an irreversible operation (i.e., visual information is lost), quantization results in lossy data compression.

EXAMPLE 8.6: Compression by quantization.

■ Consider the images in Fig. 8.10. Figure 8.10(a) shows a monochrome image with 256 gray levels. Figure 8.10(b) is the same image after uniform quantization to four bits or 16 possible levels. The resulting compression ratio is 2:1. Note that false contouring is present in the previously smooth regions of the original image. This is the natural visual effect of more coarsely representing the gray levels of the image.

Figure 8.10(c) illustrates the significant improvements possible with quantization that takes advantage of the peculiarities of the human visual system. Although the compression resulting from this second quantization also is 2:1, false contouring is greatly reduced at the expense of some additional but less objectionable graininess. Note that in either case, decompression is both unnecessary and impossible (i.e., quantization is an irreversible operation). ■

a b c

FIGURE 8.10 (a) Original image. (b) Uniform quantization to 16 levels. (c) IGS quantization to 16 levels.

The method used to produce Fig. 8.10(c) is called *improved gray-scale* (IGS) *quantization*. It recognizes the eye's inherent sensitivity to edges and breaks them up by adding to each pixel a pseudorandom number, which is generated from the low-order bits of neighboring pixels, before quantizing the result. Because the low-order bits are fairly random, this amounts to adding a level of randomness (that depends on the local characteristics of the image) to the artificial edges normally associated with false contouring. Function `quantize`, listed next, performs both IGS quantization and the traditional low-order bit truncation. Note that the IGS implementation is vectorized so that input `x` is processed one column at a time. To generate a column of the 4-bit result in Fig. 8.10(c), a column sum s—initially set to all zeros—is formed as the sum of one column of `x` and the four least significant bits of the existing (previously generated) sums. If the four most significant bits of any `x` value are 1111_2, however, 0000_2 is added instead. The four most significant bits of the resulting sums are then used as the coded pixel values for the column being processed.

quantize

```
function y = quantize(x, b, type)
%QUANTIZE Quantizes the elements of a UINT8 matrix.
%   Y = QUANTIZE(X, B, TYPE) quantizes X to B bits. Truncation is
%   used unless TYPE is 'igs' for Improved Gray Scale quantization.

error(nargchk(2, 3, nargin));              % Check input arguments
if ndims(x) ~= 2 | ~isreal(x) | ...
      ~isnumeric(x) | ~isa(x, 'uint8')
   error('The input must be a UINT8 numeric matrix.');
end

% Create bit masks for the quantization
lo = uint8(2 ^ (8 - b) - 1);
hi = uint8(2 ^ 8 - double(lo) - 1);

% Perform standard quantization unless IGS is specified
if nargin < 3 | ~strcmpi (type, 'igs')
   y = bitand(x, hi);

% Else IGS quantization. Process column-wise. If the MSB's of the
% pixel are all 1's, the sum is set to the pixel value. Else, add
% the pixel value to the LSB's of the previous sum. Then take the
% MSB's of the sum as the quantized value.
else
   [m, n] = size(x);                              s = zeros(m, 1);
   hitest = double(bitand(x, hi) ~= hi);          x = double(x);
   for j = 1:n
      s = x(:, j) + hitest(:, j) .* double(bitand(uint8(s), lo));
      y(:, j) = bitand(uint8(s), hi);
   end
end
```

strcmpi

To compare string `s1` *and* `s2` *ignoring case, use* `s = strcmpi(s1, s2)`.

Improved gray-scale quantization is typical of a large group of quantization procedures that operate directly on the gray levels of the image to be compressed. They usually entail a decrease in the image's spatial and/or gray-scale resolution. If the image is first mapped to reduce interpixel redundancies, however, the quantization can lead to other types of image degradation—like blurred edges (high-frequency detail loss) when a 2-D frequency transform is used to decorrelate the data.

EXAMPLE 8.7: Combining IGS quantization with lossless predictive and Huffman coding.

■ Although the quantization used to produce Fig. 8.10(c) removes a great deal of psychovisual redundancy with little impact on perceived image quality, further compression can be achieved by employing the techniques of the previous two sections to reduce the resulting image's interpixel and coding redundancies. In fact, we can more than double the 2:1 compression of IGS quantization alone. The following sequence of commands combines IGS quantization, lossless predictive coding, and Huffman coding to compress the image of Fig. 8.10(a) to less than a quarter of its original size:

```
>> f = imread('Brushes.tif');
>> q = quantize(f, 4, 'igs');
>> qs = double(q) / 16;
>> e = mat2lpc(qs);
>> c = mat2huff(e);
>> imratio(f, c)

ans =
    4.1420
```

Encoded result c can be decompressed by the inverse sequence of operations (without 'inverse quantization'):

```
>> ne = huff2mat(c);
>> nqs = lpc2mat(ne);
>> nq = 16 * nqs;
>> compare(q, nq)
ans =
    0
>> rmse = compare(f, nq)

rmse =
    6.8382
```

Note that the root-mean-square error of the decompressed image is about 7 gray levels—and that this error results from the quantization step alone. ■

8.5 JPEG Compression

The techniques of the previous sections operate directly on the pixels of an image and thus are *spatial domain methods*. In this section, we consider a family of popular compression standards that are based on modifying the transform of an image. Our objectives are to introduce the use of 2-D transforms in image compression, to provide additional examples of how to reduce the image redundancies discussed in Section 8.2 through 8.4, and to give the reader a feel for the state of the art in image compression. The standards presented (although we consider only approximations of them) are designed to handle a wide range of image types and compression requirements.

In *transform coding*, a reversible, linear transform like the DFT of Chapter 4 or the *discrete cosine transform* (DCT)

$$T(u, v) = \sum_{x=0}^{M-1} \sum_{y=0}^{N-1} f(x, y)\alpha(u)\alpha(v) \cos\left[\frac{(2x+1)u\pi}{2M}\right] \cos\left[\frac{(2y+1)v\pi}{2N}\right]$$

where

$$\alpha(u) = \begin{cases} \sqrt{\dfrac{1}{M}} & u = 0 \\ \sqrt{\dfrac{2}{M}} & u = 1, 2, \ldots, M-1 \end{cases}$$

[and similarly for $\alpha(v)$] is used to map an image into a set of transform coefficients, which are then quantized and coded. For most natural images, a significant number of the coefficients have small magnitudes and can be coarsely quantized (or discarded entirely) with little image distortion.

8.5.1 JPEG

One of the most popular and comprehensive continuous tone, still frame compression standards is the JPEG (for Joint Photographic Experts Group) standard. In the JPEG *baseline coding system*, which is based on the discrete cosine transform and is adequate for most compression applications, the input and output images are limited to 8 bits, while the quantized DCT coefficient values are restricted to 11 bits. As can be seen in the simplified block diagram of Fig. 8.11(a), the compression itself is performed in four sequential steps: 8×8 subimage extraction, DCT computation, quantization, and variable-length code assignment.

The first step in the JPEG compression process is to subdivide the input image into nonoverlapping pixel blocks of size 8×8. They are subsequently processed left to right, top to bottom. As each 8×8 block or subimage is processed, its 64 pixels are level shifted by subtracting 2^{m-1}, where 2^m is the number of gray levels in the image, and its 2-D discrete cosine transform is

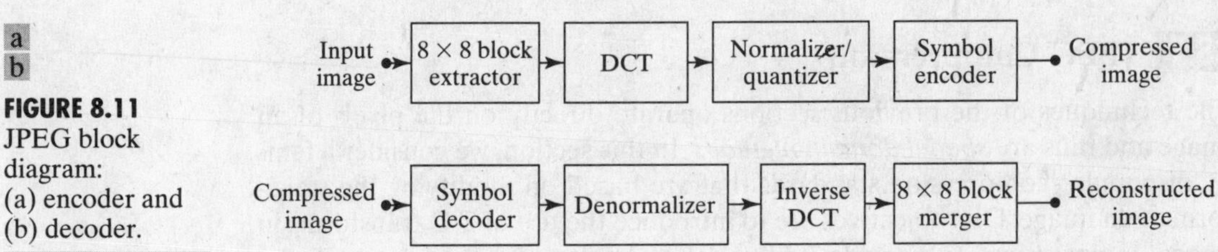

FIGURE 8.11 JPEG block diagram: (a) encoder and (b) decoder.

16	11	10	16	24	40	51	61
12	12	14	19	26	58	60	55
14	13	16	24	40	57	69	56
14	17	22	29	51	87	80	62
18	22	37	56	68	109	103	77
24	35	55	64	81	104	113	92
49	64	78	87	103	121	120	101
72	92	95	98	112	100	103	99

0	1	5	6	14	15	27	28
2	4	7	13	16	26	29	42
3	8	12	17	25	30	41	43
9	11	18	24	31	40	44	53
10	19	23	32	39	45	52	54
20	22	33	38	46	51	55	60
21	34	37	47	50	56	59	61
35	36	48	49	57	58	62	63

a b

FIGURE 8.12 (a) The default JPEG normalization array. (b) The JPEG zigzag coefficient ordering sequence.

computed. The resulting coefficients are then simultaneously normalized and quantized in accordance with

$$\hat{T}(u, v) = \text{round}\left[\frac{T(u, v)}{Z(u, v)}\right]$$

where $\hat{T}(u, v)$ for $u, v = 0, 1, \ldots, 7$ are the resulting normalized and quantized coefficients, $T(u, v)$ is the DCT of an 8×8 block of image $f(x, y)$, and $Z(u, v)$ is a *transform normalization array* like that of Fig. 8.12(a). By scaling $Z(u, v)$, a variety of compression rates and reconstructed image qualities can be achieved.

After each block's DCT coefficients are quantized, the elements of $\hat{T}(u, v)$ are reordered in accordance with the zigzag pattern of Fig. 8.12(b). Since the resulting one-dimensionally reordered array (of quantized coefficients) is qualitatively arranged according to increasing spatial frequency, the symbol encoder of Fig. 8.11(a) is designed to take advantage of the long runs of zeros that normally result from the reordering. In particular, the nonzero AC coefficients [i.e., all $\hat{T}(u, v)$ except $u = v = 0$] are coded using a variable-length code that defines the coefficient's value *and* number of preceding zeros. The DC coefficient [i.e., $\hat{T}(0, 0)$] is difference coded relative to the DC coefficient of the previous subimage. Default AC and DC Huffman coding tables are provided by the standard, but the user is free to construct custom tables, as well as normalization arrays, which may in fact be adapted to the characteristics of the image being compressed.

While a full implementation of the JPEG standard is beyond the scope of this chapter, the following M-file approximates the baseline coding process:

im2jpeg

```
function y = im2jpeg(x, quality)
%IM2JPEG Compresses an image using a JPEG approximation.
%   Y = IM2JPEG(X, QUALITY) compresses image X based on 8 x 8 DCT
%   transforms, coefficient quantization, and Huffman symbol
%   coding. Input QUALITY determines the amount of information that
%   is lost and compression achieved. Y is an encoding structure
%   containing fields:
%
```

```
%     Y.size        Size of X
%     Y.numblocks   Number of 8-by-8 encoded blocks
%     Y.quality     Quality factor (as percent)
%     Y.huffman     Huffman encoding structure, as returned by
%                   MAT2HUFF
%
%   See also JPEG2IM.
error(nargchk(1, 2, nargin));                  % Check input arguments
if ndims(x) ~= 2 | ~isreal(x) | ~isnumeric(x) | ~isa(x, 'uint8')
   error('The input must be a UINT8 image.');
end
if nargin < 2
   quality = 1;    % Default value for quality.
end
m = [16 11  10  16  24  40  51  61             % JPEG normalizing array
     12  12  14  19  26  58  60  55            % and zig-zag redordering
     14  13  16  24  40  57  69  56            % pattern.
     14  17  22  29  51  87  80  62
     18  22  37  56  68  109 103 77
     24  35  55  64  81  104 113 92
     49  64  78  87  103 121 120 101
     72  92  95  98  112 100 103 99] * quality;
order = [1 9  2  3  10 17 25 18 11 4  5  12 19 26 33  ...
        41 34 27 20 13 6  7  14 21 28 35 42 49 57 50  ...
        43 36 29 22 15 8  16 23 30 37 44 51 58 59 52  ...
        45 38 31 24 32 39 46 53 60 61 54 47 40 48 55  ...
        62 63 56 64];
[xm, xn] = size(x);                     % Get input size.
x = double(x) - 128;                    % Level shift input
t = dctmtx(8);                          % Compute 8 x 8 DCT matrix
% Compute DCTs of 8x8 blocks and quantize the coefficients.
y = blkproc(x, [8 8], 'P1 * x * P2', t, t');
y = blkproc(y, [8 8], 'round(x ./ P1)', m);
y = im2col(y, [8 8], 'distinct');   % Break 8x8 blocks into columns
xb = size(y, 2);                        % Get number of blocks
y = y(order, :);                        % Reorder column elements
eob = max(x(:)) + 1;                    % Create end-of-block symbol
r = zeros(numel(y) + size(y, 2), 1);
count = 0;
for j = 1:xb                            % Process 1 block (col) at a time
   i = max(find(y(:, j)));              % Find last non-zero element
   if isempty(i)                        % No nonzero block values
      i = 0;
   end
   p = count + 1;
   q = p + i;
   r(p:q) = [y(1:i, j); eob];           % Truncate trailing 0's, add EOB,
   count = count + i + 1;               % and add to output vector
end
```

```
r((count + 1):end) = [];            % Delete unusued portion of r

y.size       = uint16([xm xn]);
y.numblocks  = uint16(xb);
y.quality    = uint16(quality * 100);
y.huffman    = mat2huff(r);
```

In accordance with the block diagram of Fig. 8.11(a), function `im2jpeg` processes distinct 8 × 8 sections or *blocks* of input image `x` one block at a time (rather than the entire image at once). Two specialized block processing functions—`blkproc` and `im2col`—are used to simplify the computations. Function `blkproc`, whose standard syntax is

```
B = blkproc(A, [M N], FUN, P1, P2, ...),
```

streamlines or automates the entire process of dealing with images in blocks. It accepts an input image `A`, along with the size (`[M N]`) of the blocks to be processed, a function (`FUN`) to use in processing them, and some number of optional input parameters `P1, P2, ...` for block processing function `FUN`. Function `blkproc` then breaks `A` into `M x N` blocks (including any zero padding that may be necessary), calls function `FUN` with each block and parameters `P1, P2, ...`, and reassembles the results into output image `B`.

The second specialized block processing function used by `im2jpeg` is function `im2col`. When `blkproc` is not appropriate for implementing a specific block-oriented operation, `im2col` can often be used to rearrange the input so that the operation can be coded in a simpler and more efficient manner (e.g., by allowing the operation to be vectorized). The output of `im2col` is a matrix in which each column contains the elements of one distinct block of the input image. Its standardized format is

```
B = im2col(A, [M N], 'distinct')
```

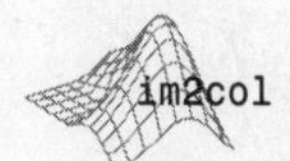

where parameters `A`, `B`, and `[M N]` are as were defined previously for function `blkproc`. String `'distinct'` tells `im2col` that the blocks to be processed are nonoverlapping; alternative string `'sliding'` signals the creation of one column in `B` for every pixel in `A` (as though a block were slid across the image).

In `im2jpeg`, function `blkproc` is used to facilitate both DCT computation and coefficient denormalization and quantization, while `im2col` is used to simplify the quantized coefficient reordering and zero run detection. Unlike the JPEG standard, `im2jpeg` detects only the final run of zeros in each reordered coefficient block, replacing the entire run with the single `eob` symbol. Finally, we note that although MATLAB provides an efficient FFT-based function for large image DCTs (refer to MATLAB's help for function `dct2`), `im2jpeg` uses an alternate matrix formulation:

To compute the DCT of `f` in 8 × 8 nonoverlapping blocks using the matrix operation `h*f*h'`, let `h = dctmtx(8)`.

$$\mathbf{T} = \mathbf{H}\mathbf{F}\mathbf{H}^T$$

where $\mathbf{F}$ is an 8 × 8 block of image $f(x, y)$, $\mathbf{H}$ is an 8 × 8 DCT transformation matrix generated by `dctmtx(8)`, and $\mathbf{T}$ is the resulting DCT of $\mathbf{F}$. Note that

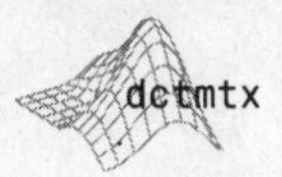

the ^{T} is used to denote the transpose operation. In the absence of quantization, the inverse DCT of **T** is

$$\mathbf{F} = \mathbf{H}^T\mathbf{T}\mathbf{H}$$

This formulation is particularly effective when transforming small square images (like JPEG's 8 × 8 DCTs). Thus, the statement

```
y = blkproc(x, [8 8], 'P1 * x * P2', h, h')
```

computes the DCTs of image x in 8 × 8 blocks, using DCT transform matrix h and transpose h' as parameters P1 and P2 of the DCT matrix multiplication, P1 * x * P2.

Similar block processing and matrix-based transformations [see Fig. 8.11(b)] are required to decompress an im2jpeg compressed image. Function jpeg2im, listed next, performs the necessary sequence of inverse operations (with the obvious exception of quantization). It uses generic function

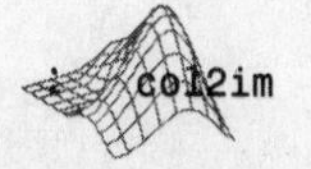


```
A = col2im(B, [M N], [MM NN], 'distinct')
```

jpeg2im

```
function x = jpeg2im(y)
%JPEG2IM Decodes an IM2JPEG compressed image.
%   X = JPEG2IM(Y) decodes compressed image Y, generating
%   reconstructed approximation X. Y is a structure generated by
%   IM2JPEG.
%
%   See also IM2JPEG.

error(nargchk(1, 1, nargin));                    % Check input arguments

m = [16 11  10  16  24  40  51  61               % JPEG normalizing array
     12 12  14  19  26  58  60  55               % and zig-zag reordering
     14 13  16  24  40  57  69  56               % pattern.
     14 17  22  29  51  87  80  62
     18 22  37  56  68  109 103 77
     24 35  55  64  81  104 113 92
     49 64  78  87  103 121 120 101
     72 92  95  98  112 100 103 99];

order = [1 9  2  3  10 17 25 18 11 4  5  12 19 26 33  ...
        41 34 27 20 13 6  7  14 21 28 35 42 49 57 50  ...
        43 36 29 22 15 8  16 23 30 37 44 51 58 59 52  ...
        45 38 31 24 32 39 46 53 60 61 54 47 40 48 55  ...
        62 63 56 64];
rev = order;                                     % Compute inverse ordering
for k = 1:length(order)
   rev(k) = find(order == k);
end

m = double(y.quality) / 100 * m;                 % Get encoding quality.
xb = double(y.numblocks);                        % Get x blocks.
sz = double(y.size);
```

```
xn = sz(2);                            % Get x columns.
xm = sz(1);                            % Get x rows.
x = huff2mat(y.huffman);               % Huffman decode.
eob = max(x(:));                       % Get end-of-block symbol

z = zeros(64, xb);   k = 1;            % Form block columns by copying
for j = 1:xb                           % successive values from x into
   for i = 1:64                        % columns of z, while changing
      if x(k) == eob                   % to the next column whenever
         k = k + 1;  break;            % an EOB symbol is found.
      else
         z(i, j) = x(k);
         k = k + 1;
      end
   end
end

z = z(rev, :);                                    % Restore order
x = col2im(z, [8 8], [xm xn], 'distinct');        % Form matrix blocks
x = blkproc(x, [8 8], 'x .* P1', m);              % Denormalize DCT
t = dctmtx(8);                                    % Get 8 x 8 DCT matrix
x = blkproc(x, [8 8], 'P1 * x * P2', t', t);      % Compute block DCT-1
x = uint8(x + 128);                               % Level shift
```

to re-create a 2-D image from the columns of matrix z, where each 64-element column is an 8×8 block of the reconstructed image. Parameters A, B, [M N], and 'distinct' are as defined for function im2col, while array [MM NN] specifies the dimensions of output image A.

EXAMPLE 8.8: JPEG compression.

■ Figures 8.13(a) and (b) show two JPEG coded and subsequently decoded approximations of the monochrome image in Fig. 8.4(a). The first result, which provides a compression ratio of about 18 to 1, was obtained by direct application of the normalization array in Fig. 8.12(a). The second, which compresses the original image by a ratio of 42 to 1, was generated by multiplying (scaling) the normalization array by 4.

The differences between the original image of Fig. 8.4(a) and the reconstructed images of Figs. 8.13(a) and (b) are shown in Figs. 8.13(c) and (d), respectively. Both images have been scaled to make the errors more visible. The corresponding rms errors are 2.5 and 4.4 gray levels. The impact of these errors on picture quality is more visible in the zoomed images of Figs. 8.13(e) and (f). These images show a magnified section of Figs. 8.13(a) and (b), respectively, and allow a better assessment of the subtle differences between the reconstructed images. [Figure 8.4(b) shows the zoomed original.] Note the *blocking artifact* that is present in both zoomed approximations.

The images in Fig. 8.13 and the numerical results just discussed were generated with the following sequence of commands:

```
>> f = imread('Tracy.tif');
>> c1 = im2jpeg(f);
>> f1 = jpeg2im(c1);
>> imratio(f, c1)
```

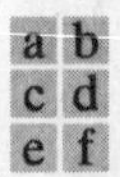

FIGURE 8.13 Left column: Approximations of Fig. 8.4 using the DCT and normalization array of Fig. 8.12(a). Right column: Similar results with the normalization array scaled by a factor of 4.

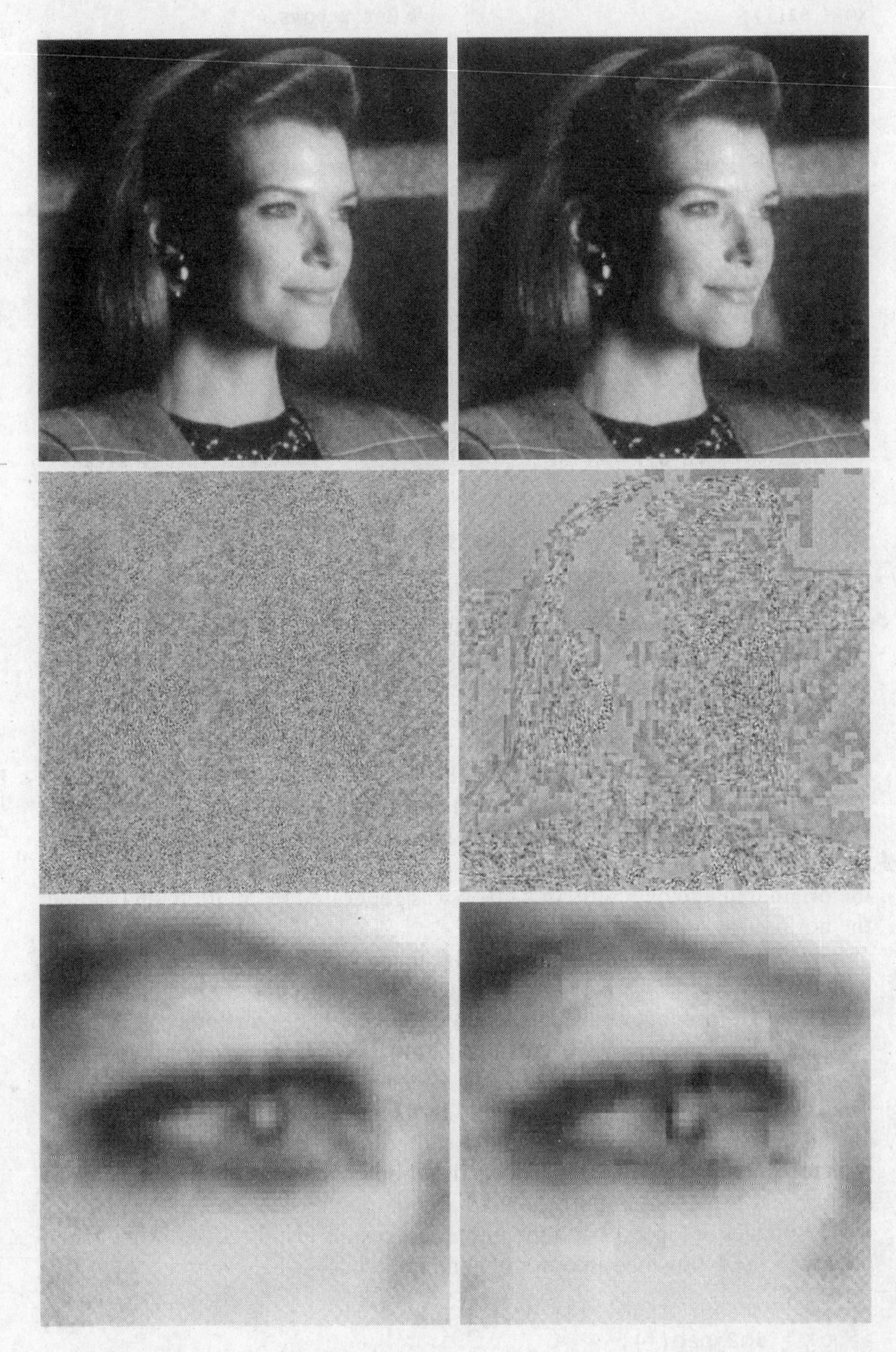

```
ans =
    18.2450
>> compare(f, f1, 3)
ans =
    2.4675
>> c4 = im2jpeg(f, 4);
>> f4 = jpeg2im(c4);
>> imratio(f, c4)
ans =
    41.7826
>> compare(f, f4, 3)
ans =
    4.4184
```

These results differ from those that would be obtained in a real JPEG baseline coding environment because `im2jpeg` approximates the JPEG standard's Huffman encoding process. Two principal differences are noteworthy: (1) In the standard, all runs of coefficient zeros are Huffman coded, while `im2jpeg` encodes only the terminating run of each block; and (2) the encoder and decoder of the standard are based on a known (default) Huffman code, while `im2jpeg` carries the information needed to reconstruct the encoding Huffman code words on an image to image basis. Using the standard, the compression ratios noted above would be approximately doubled. ■

8.5.2 JPEG 2000

Like the initial JPEG release of the previous section, JPEG 2000 is based on the idea that the coefficients of a transform that decorrelates the pixels of an image can be coded more efficiently than the original pixels themselves. If the transform's basis functions—wavelets in the JPEG 2000 case—pack most of the important visual information into a small number of coefficients, the remaining coefficients can be quantized coarsely or truncated to zero with little image distortion.

Figure 8.14 shows a simplified JPEG 2000 coding system (absent several optional operations). The first step of the encoding process, as in the original JPEG standard, is to level shift the pixels of the image by subtracting 2^{m-1}, where 2^m is the number of gray levels in the image. The one-dimensional discrete wavelet transform of the rows and the columns of the image can then be

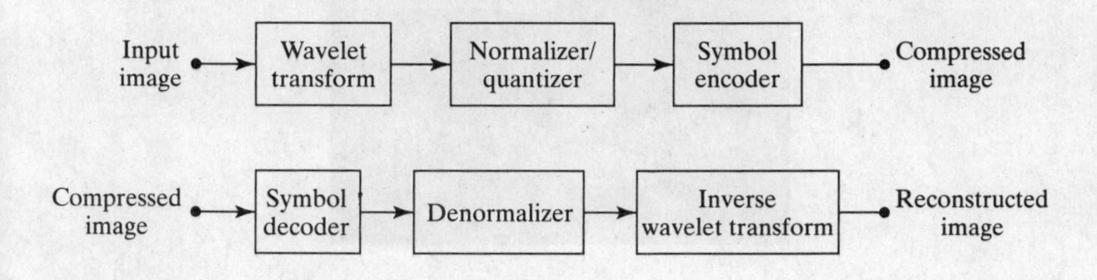

a
b

FIGURE 8.14 JPEG 2000 block diagram: (a) encoder and (b) decoder.

computed. For error-free compression, the transform used is biorthogonal, with a 5-3 coefficient scaling and wavelet vector. In lossy applications, a 9-7 coefficient scaling-wavelet vector (see the `wavefilter` function of Chapter 7) is employed. In either case, the initial decomposition results in four subbands—a low-resolution approximation of the image and the image's horizontal, vertical, and diagonal frequency characteristics.

Repeating the decomposition process N_L times, with subsequent iterations restricted to the previous decomposition's approximation coefficients, produces an N_L-scale wavelet transform. Adjacent scales are related spatially by powers of 2, and the lowest scale contains the only explicitly defined approximation of the original image. As can be surmised from Fig. 8.15, where the notation of the standard is summarized for the case of $N_L = 2$, a general N_L-scale transform contains $3N_L + 1$ subbands whose coefficients are denoted a_b for $b = N_LLL, N_LHL, \ldots, 1HL, 1LH, 1HH$. The standard does not specify the number of scales to be computed.

After the N_L-scale wavelet transform has been computed, the total number of transform coefficients is equal to the number of samples in the original image—but the important visual information is concentrated in a few coefficients. To reduce the number of bits needed to represent them, coefficient $a_b(u, v)$ of subband b is quantized to value $q_b(u, v)$ using

$$q_b(u, v) = \text{sign}[a_b(u, v)] \cdot \text{floor}\left[\frac{|a_b(u, v)|}{\Delta_b}\right]$$

For each element of `x`, `sign(x)` *returns* `1` *if the element is greater than zero,* `0` *if it equals zero, and* `-1` *if it is less than zero.*

where the "sign" and "floor" operators behave like MATLAB functions of the same name (i.e., functions `sign` and `floor`). Quantization step size Δ_b is

$$\Delta_b = 2^{R_b - \varepsilon_b}\left(1 + \frac{\mu_b}{2^{11}}\right)$$

where R_b is the nominal dynamic range of subband b, and ε_b and μ_b are the number of bits allotted to the exponent and mantissa of the subband's coefficients. The nominal dynamic range of subband b is the sum of the number of

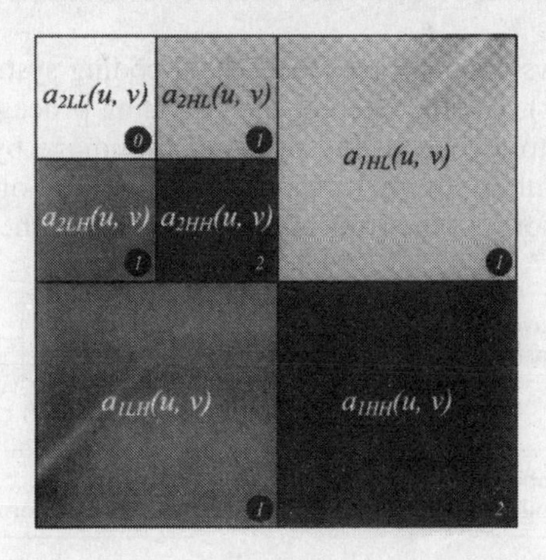

FIGURE 8.15 JPEG 2000 two-scale wavelet transform coefficient notation and analysis gain (in the circles).

bits used to represent the original image and the analysis gain bits for subband b. Subband analysis gain bits follow the simple pattern shown in Fig. 8.15. For example, there are two analysis gain bits for subband $b = 1HH$.

For error-free compression, $\mu_b = 0$ and $R_b = \varepsilon_b$ so that $\Delta_b = 1$. For irreversible compression, no particular quantization step size is specified. Instead, the number of exponent and mantissa bits must be provided to the decoder on a subband basis, called *explicit quantization*, or for the N_LLL subband only, called *implicit quantization*. In the latter case, the remaining subbands are quantized using extrapolated N_LLL subband parameters. Letting ε_0 and μ_0 be the number of bits allocated to the N_LLL subband, the extrapolated parameters for subband b are

$$\begin{aligned}\mu_b &= \mu_0 \\ \varepsilon_b &= \varepsilon_0 + nsd_b - nsd_0\end{aligned}$$

where nsd_b denotes the number of subband decomposition levels from the original image to subband b. The final step of the encoding process is to code the quantized coefficients arithmetically on a bit-plane basis. Although not discussed in the chapter, *arithmetic coding* is a variable-length coding procedure that, like Huffman coding, is designed to reduce coding redundancy.

Custom function `im2jpeg2k` approximates the JPEG 2000 coding process of Fig. 8.14(a) with the exception of the arithmetic symbol coding. As can be seen in the following listing, Huffman encoding augmented by zero run-length coding is substituted for simplicity.

im2jpeg2k

```
function y = im2jpeg2k(x, n, q)
%IM2JPEG2K Compresses an image using a JPEG 2000 approximation.
%   Y = IM2JPEG2K(X, N, Q) compresses image X using an N-scale JPEG
%   2K wavelet transform, implicit or explicit coefficient
%   quantization, and Huffman symbol coding augmented by zero
%   run-length coding. If quantization vector Q contains two
%   elements, they are assumed to be implicit quantization
%   parameters; else, it is assumed to contain explicit subband step
%   sizes. Y is an encoding structure containing Huffman-encoded
%   data and additional parameters needed by JPEG2K2IM for decoding.
%
%   See also JPEG2K2IM.

global RUNS

error(nargchk(3, 3, nargin));          % Check input arguments

if ndims(x) ~= 2 | ~isreal(x) | ~isnumeric(x) | ~isa(x, 'uint8')
   error('The input must be a UINT8 image.');
end

if length(q) ~= 2 & length(q) ~= 3 * n + 1
   error('The quantization step size vector is bad.');
end
```

```
% Level shift the input and compute its wavelet transform.
x = double(x) - 128;
[c, s] = wavefast(x, n, 'jpeg9.7');

% Quantize the wavelet coefficients.
q = stepsize(n, q);
sgn = sign(c);      sgn(find(sgn == 0)) = 1;     c = abs(c);
for k = 1:n
   qi = 3 * k - 2;
   c = wavepaste('h', c, s, k, wavecopy('h', c, s, k) / q(qi));
   c = wavepaste('v', c, s, k, wavecopy('v', c, s, k) / q(qi + 1));
   c = wavepaste('d', c, s, k, wavecopy('d', c, s, k) / q(qi + 2));
end
c = wavepaste('a', c, s, k, wavecopy('a', c, s, k) / q(qi + 3));
c = floor(c);          c = c .* sgn;

% Run-length code zero runs of more than 10. Begin by creating
% a special code for 0 runs ('zrc') and end-of-code ('eoc') and
% making a run-length table.
zrc = min(c(:)) - 1;     eoc = zrc - 1;     RUNS = [65535];

% Find the run transition points: 'plus' contains the index of the
% start of a zero run; the corresponding 'minus' is its end + 1.
z = c == 0;                  z = z - [0 z(1:end - 1)];
plus = find(z == 1);         minus = find(z == -1);

% Remove any terminating zero run from 'c'.
if length(plus) ~= length(minus)
   c(plus(end):end) = [];    c = [c eoc];
end

% Remove all other zero runs (based on 'plus' and 'minus') from 'c'.
for i = length(minus):-1:1
   run = minus(i) - plus(i);
   if run > 10
      ovrflo = floor(run / 65535);     run = run - ovrflo * 65535;
      c = [c(1:plus(i) - 1) repmat([zrc 1], 1, ovrflo) zrc ...
           runcode(run) c(minus(i):end)];
   end
end

% Huffman encode and add misc. information for decoding.
y.runs     = uint16(RUNS);
y.s        = uint16(s(:));
y.zrc      = uint16(-zrc);
y.q        = uint16(100 * q');
y.n        = uint16(n);
y.huffman  = mat2huff(c);

%-------------------------------------------------------------------%
function y = runcode(x)
% Find a zero run in the run-length table. If not found, create a
% new entry in the table. Return the index of the run.
```

```
global RUNS
y = find(RUNS == x);
if length(y) ~= 1
   RUNS = [RUNS; x];
   y = length(RUNS);
end

%-------------------------------------------------------------------%
function q = stepsize(n, p)
% Create a subband quantization array of step sizes ordered by
% decomposition (first to last) and subband (horizontal, vertical,
% diagonal, and for final decomposition the approximation subband).

if length(p) == 2                    % Implicit Quantization
   q = [];
   qn = 2 ^ (8 - p(2) + n) * (1 + p(1) / 2 ^ 11);
   for k = 1:n
      qk = 2 ^ -k * qn;
      q = [q (2 * qk) (2 * qk) (4 * qk)];
   end
   q = [q qk];
else                                 % Explicit Quantization
   q = p;
end

q = round(q * 100) / 100;            % Round to 1/100th place
if any(100 * q > 65535)
   error('The quantizing steps are not UINT16 representable.');
end
if any(q == 0)
   error('A quantizing step of 0 is not allowed.');
end
```

JPEG 2000 decoders simply invert the operations described previously. After decoding the arithmetically coded coefficients, a user-selected number of the original image's subbands are reconstructed. Although the encoder may have arithmetically encoded M_b bit-planes for a particular subband, the user—due to the embedded nature of the codestream—may choose to decode only N_b bit-planes. This amounts to quantizing the coefficients using a step size of $2^{M_b - N_b} \cdot \Delta_b$. Any non-decoded bits are set to zero and the resulting coefficients, denoted $\overline{q}_b(u, v)$, are denormalized using

$$R_{q_b}(u, v) = \begin{cases} \left(\overline{q}_b(u, v) + 2^{M_b - N_b(u, v)}\right) \cdot \Delta_b & \overline{q}_b(u, v) > 0 \\ \left(\overline{q}_b(u, v) - 2^{M_b - N_b(u, v)}\right) \cdot \Delta_b & \overline{q}_b(u, v) < 0 \\ 0 & \overline{q}_b(u, v) = 0 \end{cases}$$

where $R_{q_b}(u, v)$ denotes a denormalized transform coefficient and $N_b(u, v)$ is the number of decoded bit-planes for $\overline{q}_b(u, v)$. The denormalized coefficients are then inverse transformed and level shifted to yield an approximation of the original image. Custom function `jpeg2k2im` approximates this process, reversing the compression of `im2jpeg2k` introduced earlier.

jpeg2k2im

```
function x = jpeg2k2im(y)
%JPEG2K2IM Decodes an IM2JPEG2K compressed image.
%   X = JPEG2K2IM(Y) decodes compressed image Y, reconstructing an
%   approximation of the original image X.  Y is an encoding
%   structure returned by IM2JPEG2K.
%
% See also IM2JPEG2K.

error(nargchk(1, 1, nargin));          % Check input arguments

% Get decoding parameters: scale, quantization vector, run-length
% table size, zero run code, end-of-data code, wavelet bookkeeping
% array, and run-length table.
n = double(y.n);
q = double(y.q) / 100;
runs = double(y.runs);
rlen = length(runs);
zrc = -double(y.zrc);
eoc = zrc - 1;
s = double(y.s);
s = reshape(s, n + 2, 2);

% Compute the size of the wavelet transform.
cl = prod(s(1, :));
for i = 2:n + 1
   cl = cl + 3 * prod(s(i, :));
end

% Perform Huffman decoding followed by zero run decoding.
r = huff2mat(y.huffman);

c = [];    zi = find(r == zrc);    i = 1;
for j = 1:length(zi)
   c = [c r(i:zi(j) - 1) zeros(1, runs(r(zi(j) + 1)))];
   i = zi(j) + 2;
end

zi = find(r == eoc);                   % Undo terminating zero run
if length(zi) == 1                     % or last non-zero run.
   c = [c r(i:zi - 1)];
   c = [c zeros(1, cl - length(c))];
else
   c = [c r(i:end)];
end

% Denormalize the coefficients.
c = c + (c > 0) - (c < 0);
for k = 1:n
   qi = 3 * k - 2;
   c = wavepaste('h', c, s, k, wavecopy('h', c, s, k) * q(qi));
   c = wavepaste('v', c, s, k, wavecopy('v', c, s, k) * q(qi + 1));
   c = wavepaste('d', c, s, k, wavecopy('d', c, s, k) * q(qi + 2));
end
c = wavepaste('a', c, s, k, wavecopy('a', c, s, k) * q(qi + 3));
```

```
% Compute the inverse wavelet transform and level shift.
x = waveback(c, s, 'jpeg9.7', n);
x = uint8(x + 128);
```

The principal difference between the wavelet-based JPEG 2000 system of Fig. 8.14 and the DCT-based JPEG system of Fig. 8.11 is the omission of the latter's subimage processing stages. Because wavelet transforms are both computationally efficient and inherently local (i.e., their basis functions are limited in duration), subdivision of the image into blocks is unnecessary. As will be seen in the following example, the removal of the subdivision step eliminates the blocking artifact that characterizes DCT-based approximations at high compression ratios.

EXAMPLE 8.9: JPEG 2000 compression.

■ Figure 8.16 shows two JPEG 2000 approximations of the monochrome image in Fig. 8.4(a). Figure 8.16(a) was reconstructed from an encoding that compressed the original image by 42:1; Fig. 8.16(b) was generated from an 88:1; encoding. The two results were obtained using a five-scale transform and implicit quantization with $\mu_0 = 8$ and $\varepsilon_0 = 8.5$ and 7, respectively. Because `im2jpeg2k` only approximates the JPEG 2000's bit-plane–oriented arithmetic coding, the compression rates just noted differ from those that would be obtained by a true JPEG 2000 encoder. In fact, the actual rates would increase by a factor of 2.

Since the 42:1; compression of the results in the left column of Fig. 8.16 is identical to the compression achieved for the images in the right column of Fig. 8.13 (Example 8.8), Figs. 8.16(a), (c), and (e) can be compared—both qualitatively and quantitatively—to the transform-based JPEG results of Figs. 8.13(b), (d), and (f). A visual comparison reveals a noticeable decrease of error in the wavelet-based JPEG 2000 images. In fact, the rms error of the JPEG 2000–based result in Fig. 8.16(a) is 3.7 gray levels, as opposed to 4.4 gray levels for the corresponding transform-based JPEG result in Fig. 8.13(b). Besides decreasing reconstruction error, JPEG 2000–based coding dramatically increased (in a subjective sense) image quality. This is particularly evident in Fig. 8.16(e). Note that the blocking artifact that dominated the corresponding transform-based result in Fig. 8.13(f) is no longer present.

When the level of compression increases to 88:1, as in Fig. 8.16(b), there is a loss of texture in the woman's clothing and blurring of her eyes. Both effects are visible in Figs. 8.16(b) and (f). The rms error of these reconstructions is about 5.9 gray levels. The results of Fig. 8.16 were generated with the following sequence of commands:

```
>> f = imread('Tracy.tif');
>> c1 = im2jpeg2k(f, 5, [8 8.5]);
>> f1 = jpeg2k2im(c1);
>> rms1 = compare(f, f1)

rms1 =
    3.6931

>> cr1 = imratio(f, c1)
```

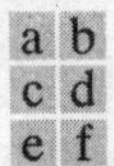

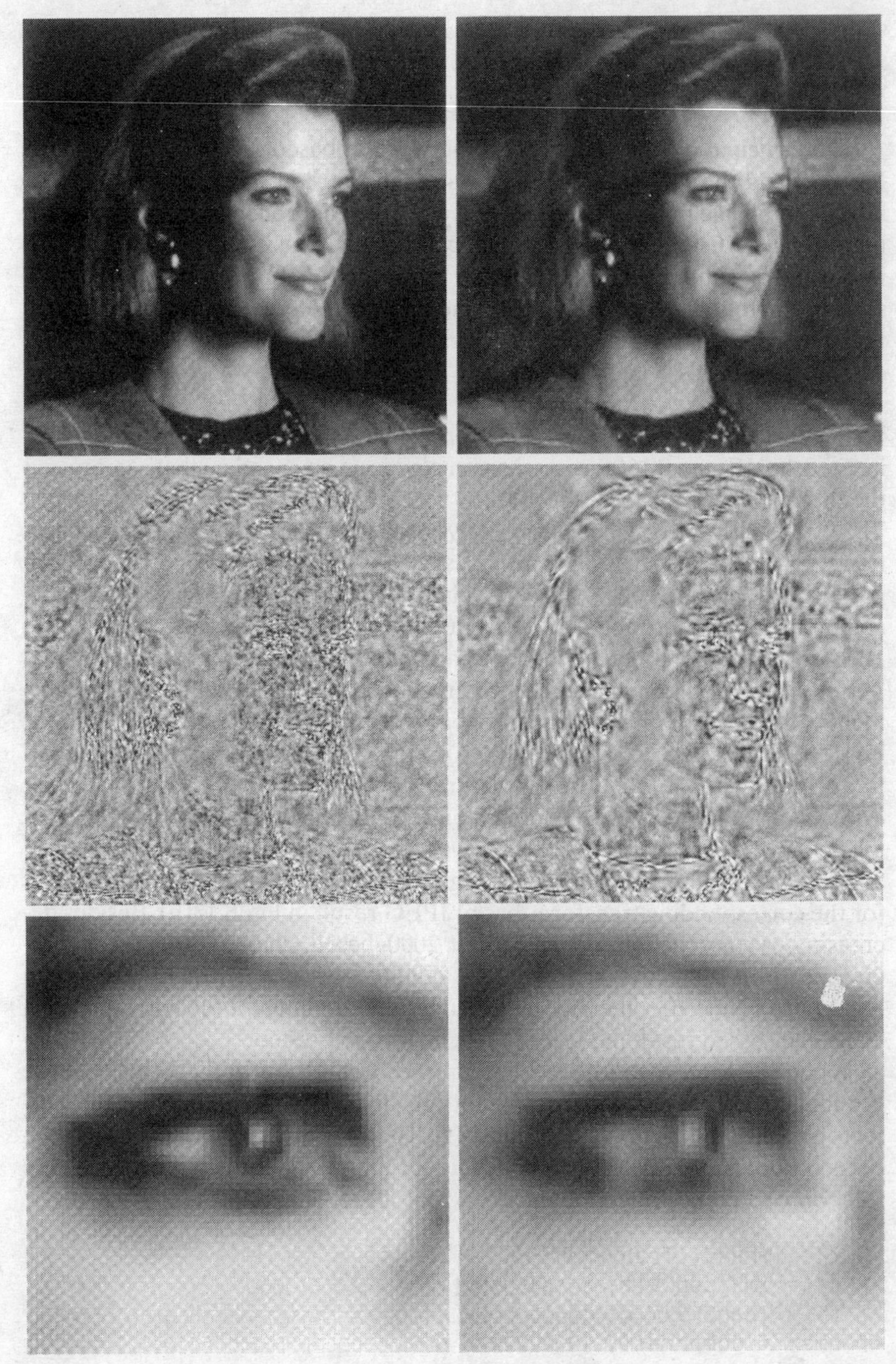

FIGURE 8.16 Left column: JPEG 2000 approximations of Fig. 8.4 using five scales and implicit quantization with $\mu_0 = 8$ and $\varepsilon_0 = 8.5$. Right column: Similar results with $\varepsilon_0 = 7$.

```
cr1 =
    42.1589

>> c2 = im2jpeg2k(f, 5, [8 7]);
>> f2 = jpeg2k2im(c2);
>> rms2 = compare(f, f2)

rms2 =
    5.9172

>> cr2 = imratio(f, c2)
cr2 =
    87.7323
```

Note that implicit quantization is used when a two-element vector is supplied as argument 3 of `im2jpeg2k`. If the length of this vector is not 2, the function assumes explicit quantization and $3N_L + 1$ step sizes (where N_L is the number of scales to be computed) must be provided. This is one for each subband of the decomposition; they must be ordered by decomposition level (first, second, third, ...) and by subband type (i.e., the horizontal, vertical, diagonal, and approximation). For example,

```
>> c3 = im2jpeg2k(f, 1, [1 1 1 1]);
```

computes a one-scale transform and employs explicit quantization—all four subbands are quantized using step size $\Delta_1 = 1$. That is, the transform coefficients are rounded to the nearest integer. This is the minimal error case for the `im2jpeg2k` implementation, and the resulting rms error and compression rate are

```
>> f3 = jpeg2k2im(c3);
>> rms3 = compare(f, f3)

rms3 =
    .1234

>> cr3 = imratio(f, c3)
cr3 =
    1.6350
```

■

Summary

The material in this chapter introduces the fundamentals of digital image compression through the removal of coding, interpixel, and psychovisual redundancy. MATLAB routines that attack each of these redundancies—and extend the Image Processing Toolbox—are developed. Finally, an overview of the popular JPEG and JPEG 2000 image compression standards is given. For additional information on the removal of image redundancies—both techniques that are not covered here and standards that address specific image subsets (like binary images)—see Chapter 8 of *Digital Image Processing* by Gonzalez and Woods [2002].

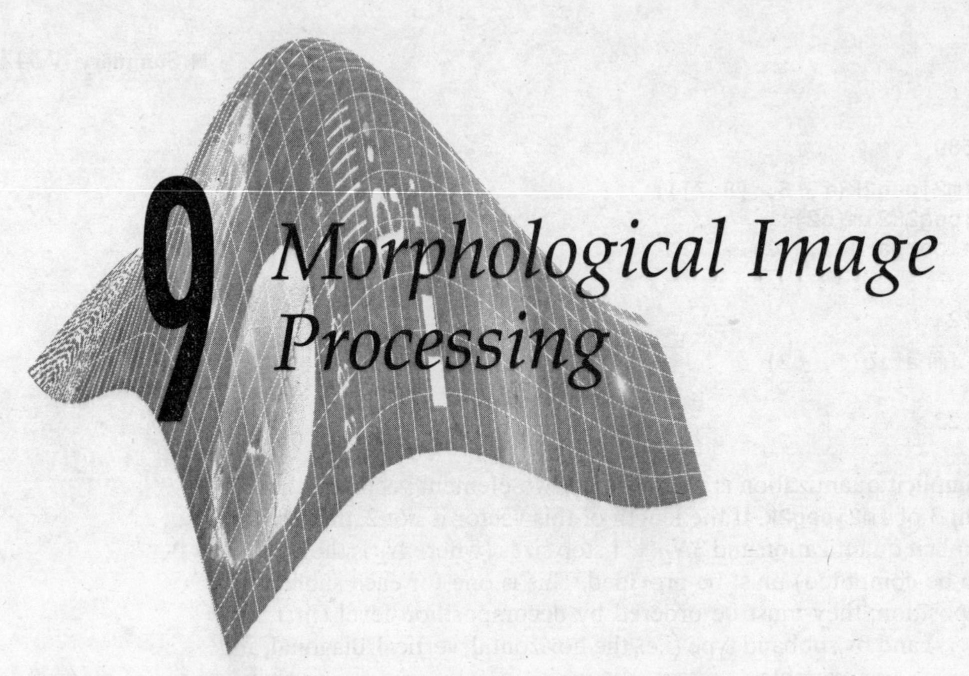

9 Morphological Image Processing

Preview

The word *morphology* commonly denotes a branch of biology that deals with the form and structure of animals and plants. We use the same word here in the context of *mathematical morphology* as a tool for extracting image components that are useful in the representation and description of region shape, such as boundaries, skeletons, and the convex hull. We are interested also in morphological techniques for pre- or postprocessing, such as morphological filtering, thinning, and pruning.

In Section 9.1 we define several set theoretic operations, introduce binary images, and discuss binary sets and logical operators. In Section 9.2 we define two fundamental morphological operations, *dilation* and *erosion*, in terms of the union (or intersection) of an image with a translated shape (*structuring element*). Section 9.3 deals with combining erosion and dilation to obtain more complex morphological operations. Section 9.4 introduces techniques for labeling connected components in an image. This is a fundamental step in extracting objects from an image for subsequent analysis.

Section 9.5 deals with *morphological reconstruction*, a morphological transformation involving two images, rather than a single image and a structuring element, as is the case in Sections 9.1 through 9.4. Section 9.6 extends morphological concepts to gray-scale images by replacing set union and intersection with maxima and minima. Most binary morphological operations have natural extensions to gray-scale processing. Some, like morphological reconstruction, have applications that are unique to gray-scale images, such as peak filtering.

The material in this chapter begins a transition from image-processing methods whose inputs and outputs are images, to image analysis methods, whose outputs in some way describe the contents of the image. Morphology is a cornerstone of the mathematical set of tools underlying the development of techniques that extract "meaning" from an image. Other approaches are developed and applied in the remaining chapters of the book.

9.1 Preliminaries

In this section we introduce some basic concepts from set theory and discuss the application of MATLAB's logical operators to binary images.

9.1.1 Some Basic Concepts from Set Theory

Let Z be the set of integers. The sampling process used to generate digital images may be viewed as partitioning the xy-plane into a grid, with the coordinates of the center of each grid being a pair of elements from the Cartesian product,† Z^2. In the terminology of set theory, a function $f(x, y)$ is said to be a *digital image* if (x, y) are integers from Z^2 and f is a mapping that assigns an intensity value (that is, a real number from the set of real numbers, R) to each distinct pair of coordinates (x, y). If the elements of R also are integers (as is usually the case in this book), a digital image then becomes a two-dimensional function whose coordinates and amplitude (i.e., intensity) values are integers.

Let A be a set in Z^2, the elements of which are pixel coordinates (x, y). If $w = (x, y)$ is an element of A, then we write

$$w \in A$$

Similarly, if w is not an element of A, we write

$$w \notin A$$

A set B of pixel coordinates that satisfy a particular condition is written as

$$B = \{w | \text{condition}\}$$

For example, the set of all pixel coordinates that do not belong to set A, denoted A^c, is given by

$$A^c = \{w | w \notin A\}$$

This set is called the *complement* of A.

The *union* of two sets, denoted by

$$C = A \cup B$$

is the set of all elements that belong to either A, B, or both. Similarly, the *intersection* of two sets A and B is the set of all elements that belong to both sets, denoted by

$$C = A \cap B$$

---

†The Cartesian product of a set of integers, Z, is the set of all ordered pairs of elements (z_i, z_j), with z_i and z_j being integers from Z. It is customary to denote this set by Z^2.

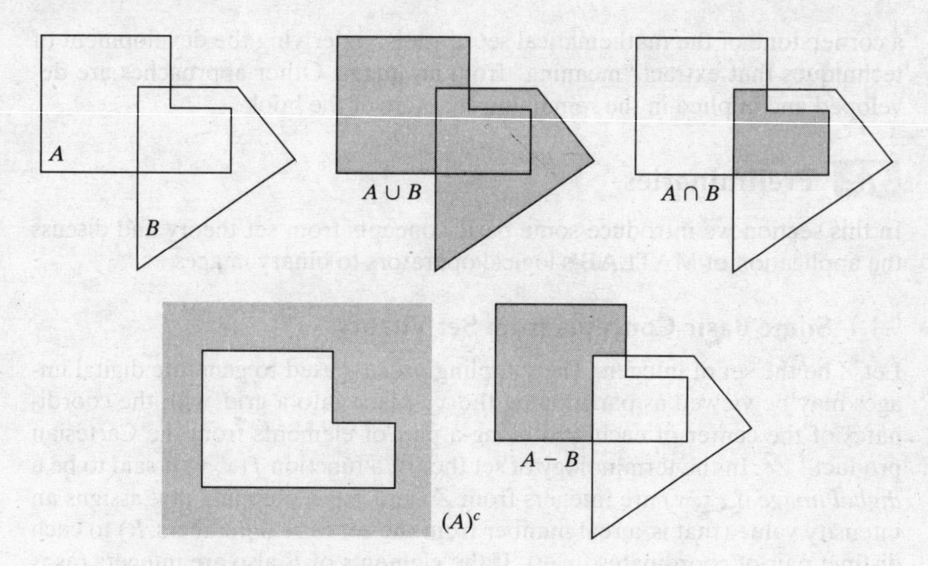

a b c
d e

FIGURE 9.1
(a) Two sets A and B. (b) The union of A and B. (c) The intersection of A and B. (d) The complement of A. (e) The difference between A and B.

The *difference* of sets A and B, denoted $A - B$, is the set of all elements that belong to A but not to B:

$$A - B = \{w | w \in A, w \notin B\}$$

Figure 9.1 illustrates these basic set operations. The result of each operation is shown in gray.

In addition to the preceding basic operations, morphological operations often require two operators that are specific to sets whose elements are pixel coordinates. The *reflection* of set B, denoted $\hat{B}$, is defined as

$$\hat{B} = \{w | w = -b, \text{ for } b \in B\}$$

The *translation* of set A by point $z = (z_1, z_2)$, denoted $(A)_z$, is defined as

$$(A)_z = \{c | c = a + z, \text{ for } a \in A\}$$

Figure 9.2 illustrates these two definitions using the sets from Fig. 9.1. The black dot identifies the *origin* assigned (arbitrarily) to the sets.

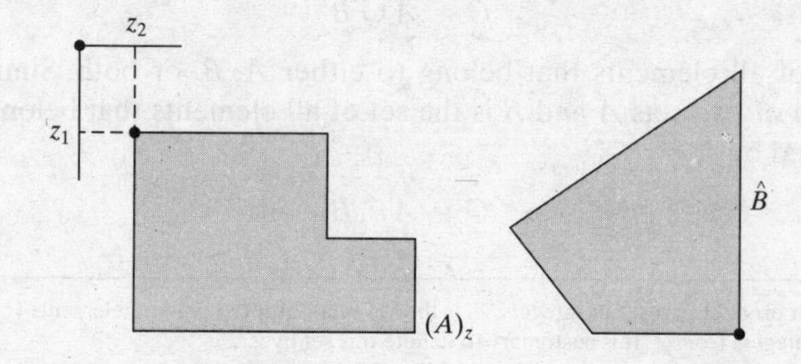

a b

FIGURE 9.2
(a) Translation of A by z.
(b) Reflection of B. The sets A and B are from Fig. 9.1.

9.1.2 Binary Images, Sets, and Logical Operators

The language and theory of mathematical morphology often present a dual view of binary images. As in the rest of the book, a binary image can be viewed as a bivalued *function* of x and y. Morphological theory views a binary image as the *set* of its foreground (1-valued) pixels, the elements of which are in Z^2. Set operations such as union and intersection can be applied directly to binary image sets. For example, if A and B are binary images, then $C = A \cup B$ is also a binary image, where a pixel in C is a foreground pixel if either or both of the corresponding pixels in A and B are foreground pixels. In the first view, that of a function, C is given by

$$C(x, y) = \begin{cases} 1 & \text{if either } A(x, y) \text{ or } B(x, y) \text{ is 1, or if both are 1} \\ 0 & \text{otherwise} \end{cases}$$

Using the set view, on the other hand, C is given by

$$C = \{(x, y) | (x, y) \in A \text{ or } (x, y) \in B \text{ or } (x, y) \in (A \text{ and } B)\}$$

The set operations defined in Fig. 9.1 can be performed on *binary* images using MATLAB's logical operators OR (|), AND (&), and NOT (~), as Table 9.1 shows.

As a simple illustration, Fig. 9.3 shows the results of applying several logical operators to two binary images containing text. (We follow the IPT convention that foreground (1-valued) pixels are displayed as white.) The image in Fig. 9.3(d) is the union of the "UTK" and "GT" images; it contains all the foreground pixels from both. By contrast, the intersection of the two images [Fig. 9.3(e)] consists of the pixels where the letters in "UTK" and "GT" overlap. Finally, the set difference image [Fig. 9.3(f)] shows the letters in "UTK" with the pixels "GT" removed.

9.2 Dilation and Erosion

The operations of *dilation* and *erosion* are fundamental to morphological image processing. Many of the algorithms presented later in this chapter are based on these operations, which are defined and illustrated in the discussion that follows.

TABLE 9.1 Using logical expressions in MATLAB to perform set operations on binary images.

Set Operation	MATLAB Expression for Binary Images	Name
$A \cap B$	`A & B`	AND
$A \cup B$	`A	B`	OR
A^c	`~A`	NOT
$A - B$	`A & ~B`	DIFFERENCE

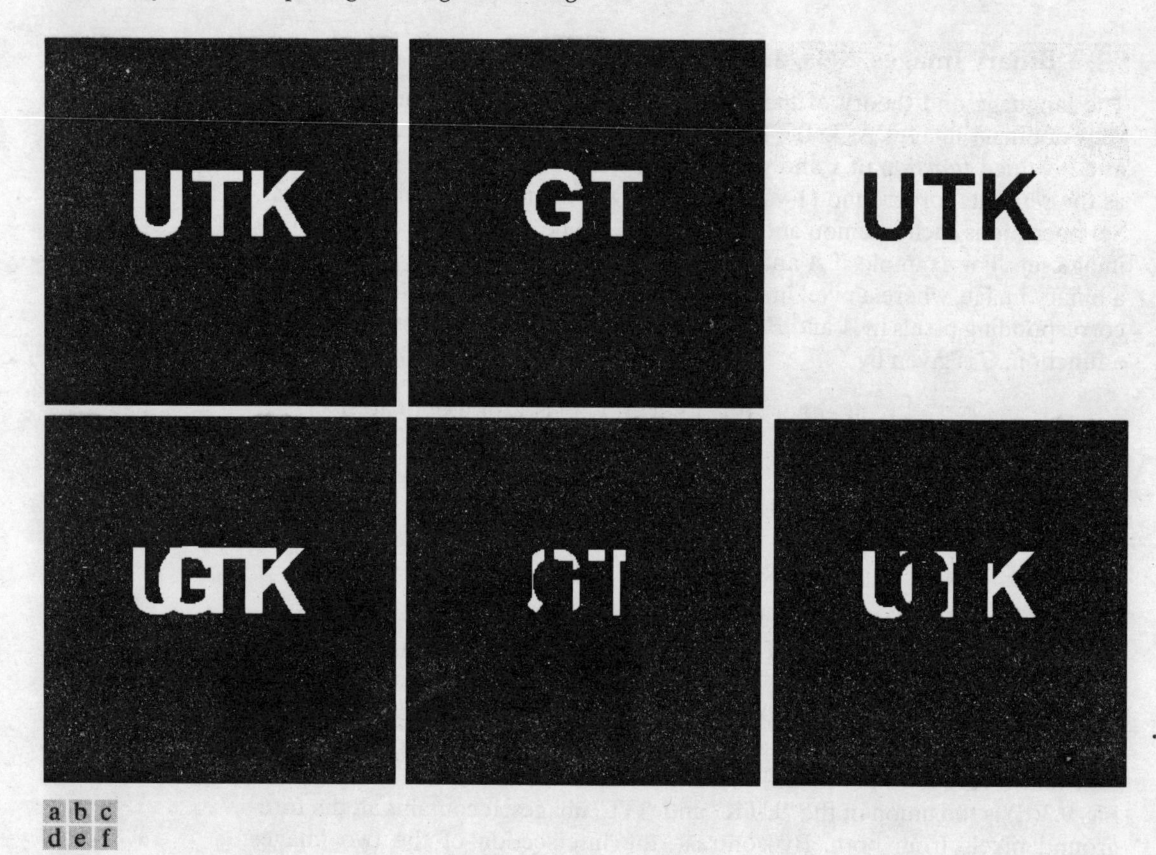

FIGURE 9.3 (a) Binary image A. (b) Binary image B. (c) Complement ~A. (d) Union A | B. (e) Intersection A & B. (f) Set difference A & ~B.

9.2.1 Dilation

Dilation is an operation that "grows" or "thickens" objects in a binary image. The specific manner and extent of this thickening is controlled by a shape referred to as a *structuring element*. Figure 9.4 illustrates how dilation works. Figure 9.4(a) shows a simple binary image containing a rectangular object. Figure 9.4(b) is a structuring element, a five-pixel-long diagonal line in this case. Computationally, structuring elements typically are represented by a matrix of 0s and 1s; sometimes it is convenient to show only the 1s, as illustrated in the figure. In addition, the origin of the structuring element must be clearly identified. Figure 9.4(b) shows the origin of the structuring element using a black outline. Figure 9.4(c) graphically depicts dilation as a process that translates the origin of the structuring element throughout the domain of the image and checks to see where it overlaps with 1-valued pixels. The output image in Fig. 9.4(d) is 1 at each location of the origin such that the structuring element overlaps at least one 1-valued pixel in the input image.

Mathematically, dilation is defined in terms of set operations. The dilation of A by B, denoted $A \oplus B$, is defined as

$$A \oplus B = \{z \mid (\hat{B})_z \cap A \neq \emptyset\}$$

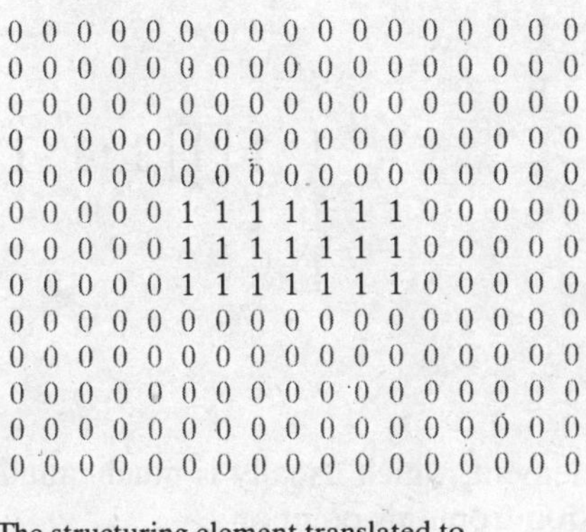

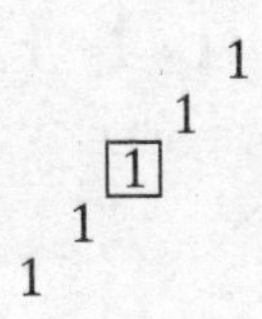

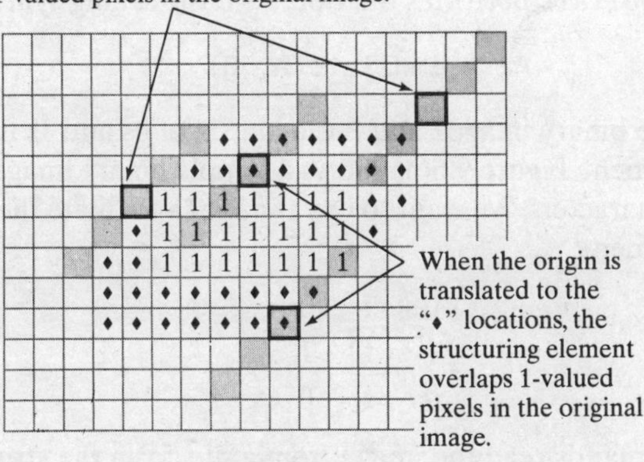
The structuring element translated to these locations does not overlap any 1-valued pixels in the original image.
When the origin is translated to the "♦" locations, the structuring element overlaps 1-valued pixels in the original image.

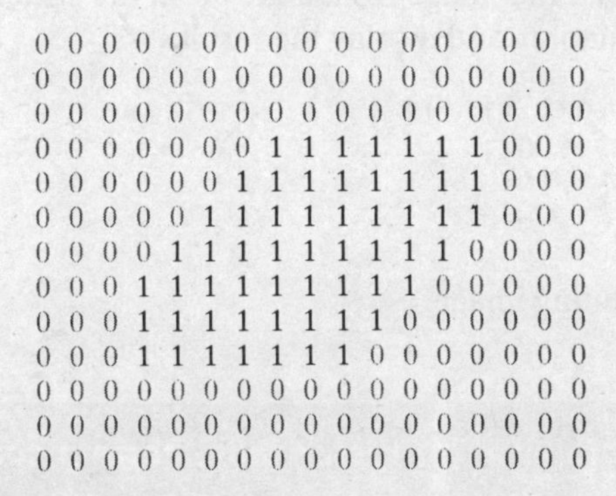

a b
c
d

FIGURE 9.4 Illustration of dilation. (a) Original image with rectangular object. (b) Structuring element with five pixels arranged in a diagonal line. The origin of the structuring element is shown with a dark border. (c) Structuring element translated to several locations on the image. (d) Output image.

where Ø is the empty set and B is the structuring element. In words, the dilation of A by B is the set consisting of all the structuring element origin locations where the reflected and translated B overlaps at least some portion of A. The translation of the structuring element in dilation is similar to the mechanics of spatial convolution discussed in Chapter 3. Figure 9.4 does not show the structuring element's reflection explicitly because the structuring element is symmetrical with respect to its origin in this case. Figure 9.5 shows a nonsymmetric structuring element and its reflection.

Dilation is *commutative*; that is, $A \oplus B = B \oplus A$. It is a convention in image processing to let the first operand of $A \oplus B$ be the image and the second

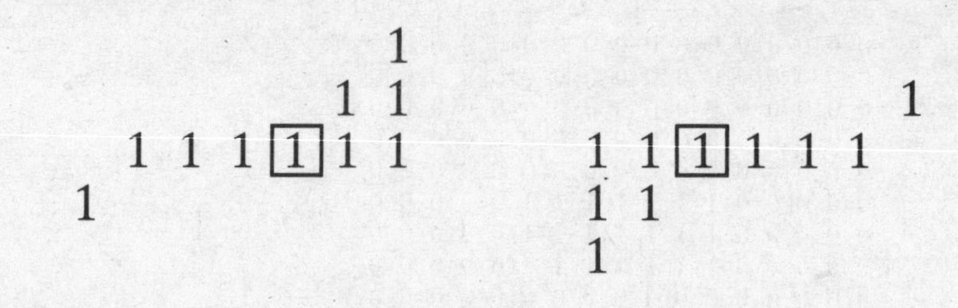

a b

FIGURE 9.5 Structuring element reflection. (a) Nonsymmetric structuring element. (b) Structuring element reflected about its origin.

operand be the structuring element, which usually is much smaller than the image. We follow this convention from this point on.

EXAMPLE 9.1: A simple application of dilation.

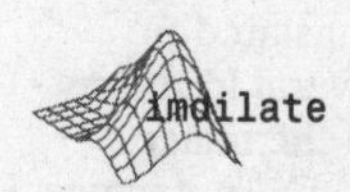

■ IPT function `imdilate` performs dilation. Its basic calling syntax is

```
A2 = imdilate(A, B)
```

where `A` and `A2` are binary images, and `B` is a matrix of 0s and 1s that specifies the structuring element. Figure 9.6(a) shows a sample binary image containing text with broken characters. We want to use `imdilate` to dilate the image with the structuring element:

$$\begin{array}{ccc} 0 & 1 & 0 \\ 1 & \boxed{1} & 1 \\ 0 & 1 & 0 \end{array}$$

The following commands read the image from a file, form the structuring element matrix, perform the dilation, and display the result.

```
>> A = imread('broken_text.tif');
>> B = [0 1 0; 1 1 1; 0 1 0];
>> A2 = imdilate(A, B);
>> imshow(A2)
```

Figure 9.6(b) shows the resulting image. ■

(a)

Historically, certain computer programs were written using only two digits rather than four to define the applicable year. Accordingly, the company's software may recognize a date using "00" as 1900 rather than the year 2000.

(b)

Historically, certain computer programs were written using only two digits rather than four to define the applicable year. Accordingly, the company's software may recognize a date using "00" as 1900 rather than the year 2000.

a b

FIGURE 9.6 A simple example of dilation. (a) Input image containing broken text. (b) Dilated image.

9.2.2 Structuring Element Decomposition

Dilation is *associative*. That is,

$$A \oplus (B \oplus C) = (A \oplus B) \oplus C$$

Suppose that a structuring element B can be represented as a dilation of two structuring elements B_1 and B_2:

$$B = B_1 \oplus B_2$$

Then $A \oplus B = A \oplus (B_1 \oplus B_2) = (A \oplus B_1) \oplus B_2$. In other words, dilating A with B is the same as first dilating A with B_1, and then dilating the result with B_2. We say that B can be *decomposed* into the structuring elements B_1 and B_2.

The associative property is important because the time required to compute dilation is proportional to the number of nonzero pixels in the structuring element. Consider, for example, dilation with a 5×5 array of 1s:

$$\begin{matrix} 1 & 1 & 1 & 1 & 1 \\ 1 & 1 & 1 & 1 & 1 \\ 1 & 1 & \boxed{1} & 1 & 1 \\ 1 & 1 & 1 & 1 & 1 \\ 1 & 1 & 1 & 1 & 1 \end{matrix}$$

This structuring element can be decomposed into a five-element row of 1s and a five-element column of 1s:

$$[1\ 1\ \boxed{1}\ 1\ 1] \oplus \begin{bmatrix} 1 \\ 1 \\ \boxed{1} \\ 1 \\ 1 \end{bmatrix}$$

The number of elements in the original structuring element is 25, but the total number of elements in the row-column decomposition is only 10. This means that dilation with the row structuring element first, followed by dilation with the column element, can be performed 2.5 times faster than dilation with the 5×5 array of 1s. In practice, the speed-up will be somewhat less because there is usually some overhead associated with each dilation operation, and at least two dilation operations are required when using the decomposed form. However, the gain in speed with the decomposed implementation is still significant.

9.2.3 The `strel` Function

IPT function `strel` constructs structuring elements with a variety of shapes and sizes. Its basic syntax is

```
se = strel(shape, parameters)
```

strel

where shape is a string specifying the desired shape, and parameters is a list of parameters that specify information about the shape, such as its size. For example, strel('diamond', 5) returns a diamond-shaped structuring element that extends ±5 pixels along the horizontal and vertical axes. Table 9.2 summarizes the various shapes that strel can create.

In addition to simplifying the generation of common structuring element shapes, function strel also has the important property of producing structuring elements in decomposed form. Function imdilate automatically uses the decomposition information to speed up the dilation process. The following example illustrates how strel returns information related to the decomposition of a structuring element.

EXAMPLE 9.2: An illustration of structuring element decomposition using strel.

■ Consider again the creation of a diamond-shaped structuring element using strel:

```
>> se = strel('diamond', 5)

se =

Flat STREL object containing 61 neighbors.

Decomposition: 4 STREL objects containing a total of 17 neighbors

Neighborhood:
     0     0     0     0     0     1     0     0     0     0     0
     0     0     0     0     1     1     1     0     0     0     0
     0     0     0     1     1     1     1     1     0     0     0
     0     0     1     1     1     1     1     1     1     0     0
     0     1     1     1     1     1     1     1     1     1     0
     1     1     1     1     1     1     1     1     1     1     1
     0     1     1     1     1     1     1     1     1     1     0
     0     0     1     1     1     1     1     1     1     0     0
     0     0     0     1     1     1     1     1     0     0     0
     0     0     0     0     1     1     1     0     0     0     0
     0     0     0     0     0     1     0     0     0     0     0
```

We see that strel does not display as a normal MATLAB matrix; it returns instead a special quantity called an *strel object*. The command-window display of an strel object includes the neighborhood (a matrix of 1s in a diamond-shaped pattern in this case); the number of 1-valued pixels in the structuring element (61); the number of structuring elements in the decomposition (4); and the total number of 1-valued pixels in the decomposed structuring elements (17). Function getsequence can be used to extract and examine separately the individual structuring elements in the decomposition.

```
>> decomp = getsequence(se);
>> whos
  Name        Size                  Bytes  Class
  decomp      4x1                    1716  strel object
  se          1x1                    3309  strel object

Grand total is 495 elements using 5025 bytes
```

TABLE 9.2
The various syntax forms of function `strel`. (The word *flat* means that the structuring element has zero height. This is meaningful only for gray-scale dilation and erosion. See Section 9.6.1.)

Syntax Forms	Description
`se = strel('diamond', R)`	Creates a flat, diamond-shaped structuring element, where `R` specifies the distance from the structuring element origin to the extreme points of the diamond.
`se = strel('disk', R)`	Creates a flat, disk-shaped structuring element with radius `R`. (Additional parameters may be specified for the disk; see the `strel` help page for details.)
`se = strel('line', LEN, DEG)`	Creates a flat, linear structuring element, where `LEN` specifies the length, and `DEG` specifies the angle (in degrees) of the line, as measured in a counterclockwise direction from the horizontal axis.
`se = strel('octagon', R)`	Creates a flat, octagonal structuring element, where `R` specifies the distance from the structuring element origin to the sides of the octagon, as measured along the horizontal and vertical axes. `R` must be a nonnegative multiple of 3.
`se = strel('pair', OFFSET)`	Creates a flat structuring element containing two members. One member is located at the origin. The second member's location is specified by the vector `OFFSET`, which must be a two-element vector of integers.
`se = strel('periodicline', P, V)`	Creates a flat structuring element containing `2*P + 1` members. `V` is a two-element vector containing integer-valued row and column offsets. One structuring element member is located at the origin. The other members are located at `1*V`, `–1*V`, `2*V`, `–2*V`, `...`, `P*V`, and `–P*V`.
`se = strel('rectangle', MN)`	Creates a flat, rectangle-shaped structuring element, where `MN` specifies the size. `MN` must be a two-element vector of nonnegative integers. The first element of `MN` is the number rows in the structuring element; the second element is the number of columns.
`se = strel('square', W)`	Creates a square structuring element whose width is `W` pixels. `W` must be a nonnegative integer scalar.
`se = strel('arbitrary', NHOOD)` `se = strel(NHOOD)`	Creates a structuring element of arbitrary shape. NHOOD is a matrix of 0s and 1s that specifies the shape. The second, simpler syntax form shown performs the same operation.

The output of whos shows that se and decomp are both strel objects, and, further, that decomp is a four-element vector of strel objects. The four structuring elements in the decomposition can be examined individually by indexing into decomp:

```
>> decomp(1)

ans =

Flat STREL object containing 5 neighbors.

Neighborhood:
     0     1     0
     1     1     1
     0     1     0

>> decomp(2)

ans =

Flat STREL object containing 4 neighbors.

Neighborhood:
     0     1     0
     1     0     1
     0     1     0

>> decomp(3)

ans =

Flat STREL object containing 4 neighbors.

Neighborhood:
     0     0     1     0     0
     0     0     0     0     0
     1     0     0     0     1
     0     0     0     0     0
     0     0     1     0     0

>> decomp(4)

ans =

Flat STREL object containing 4 neighbors.

Neighborhood:
     0     1     0
     1     0     1
     0     1     0
```

Function imdilate uses the decomposed form of a structuring element automatically, performing dilation approximately three times faster ($\approx 61/17$) than with the non-decomposed form. ■

9.2.4 Erosion

Erosion "shrinks" or "thins" objects in a binary image. As in dilation, the manner and extent of shrinking is controlled by a structuring element. Figure 9.7 illustrates the erosion process. Figure 9.7(a) is the same as Fig. 9.4(a). Figure 9.7(b) is the structuring element, a short vertical line. Figure 9.7(c) graphically depicts erosion as a process of translating the structuring element throughout the domain of the image and checking to see where it fits entirely

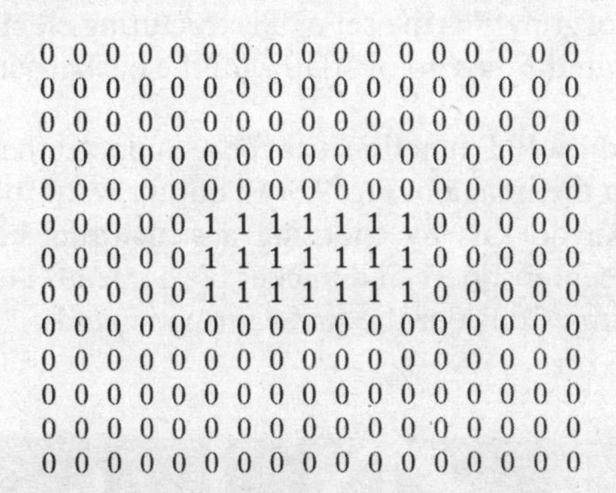

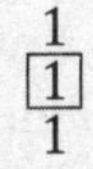

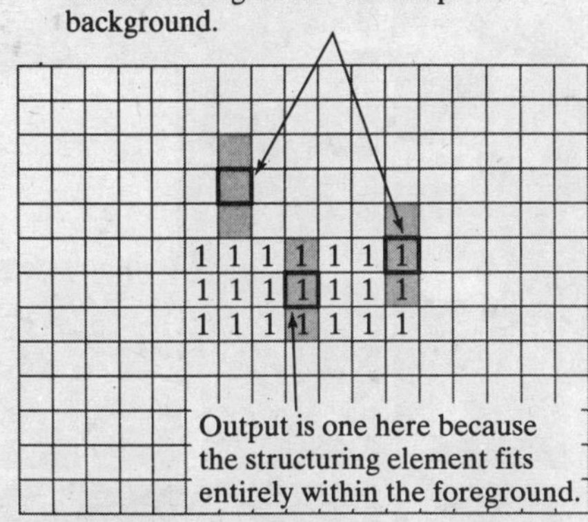

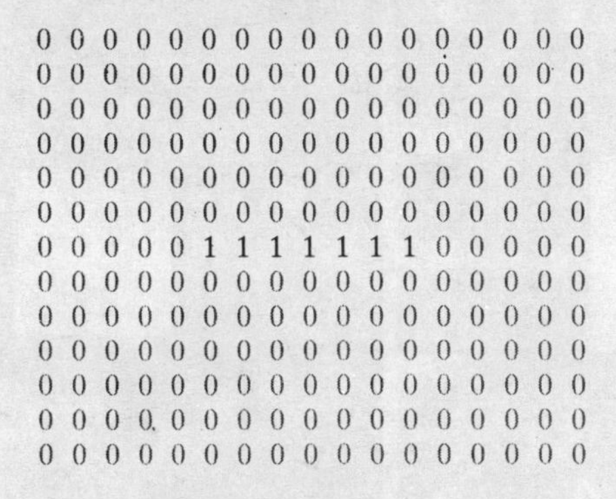

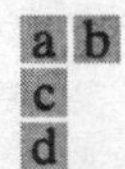

FIGURE 9.7 Illustration of erosion. (a) Original image with rectangular object. (b) Structuring element with three pixels arranged in a vertical line. The origin of the structuring element is shown with a dark border. (c) Structuring element translated to several locations on the image. (d) Output image.

within the foreground of the image. The output image in Fig. 9.7(d) has a value of 1 at each location of the origin of the structuring element, such that the element overlaps *only* 1-valued pixels of the input image (i.e., it does not overlap any of the image background).

The mathematical definition of erosion is similar to that of dilation. The erosion of A by B, denoted $A \ominus B$, is defined as

$$A \ominus B = \{z \mid (B)_z \cap A^c \neq \emptyset\}$$

In other words, erosion of A by B is the set of all structuring element origin locations where the translated B has no overlap with the background of A.

EXAMPLE 9.3: An illustration of erosion.

■ Erosion is performed by IPT function `imerode`. Suppose that we want to remove the thin wires in the image in Fig. 9.8(a), but we want to preserve the other structures. We can do this by choosing a structuring element small enough to fit within the center square and thicker border leads but too large to fit entirely within the wires. Consider the following commands:

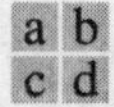

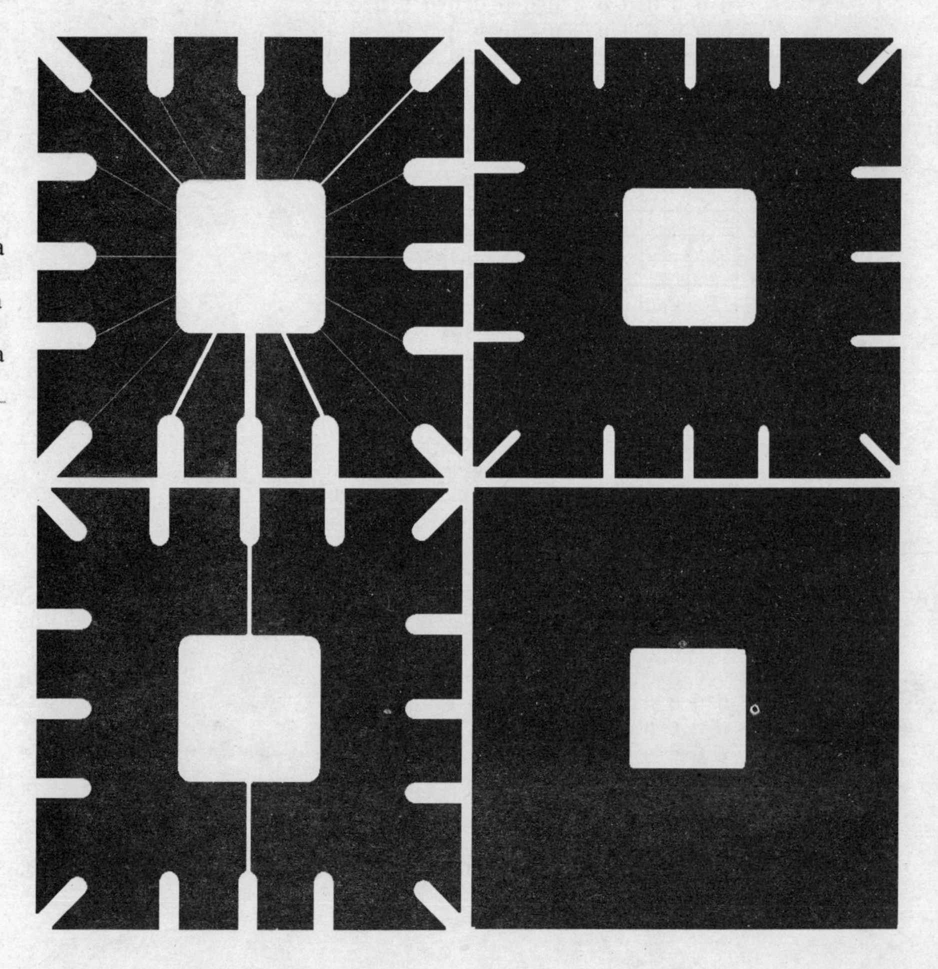

FIGURE 9.8 An illustration of erosion. (a) Original image. (b) Erosion with a disk of radius 10. (c) Erosion with a disk of radius 5. (d) Erosion with a disk of radius 20.

```
>> A = imread('wirebond_mask.tif');
>> se = strel('disk', 10);
>> A2 = imerode(A, se);
>> imshow(A2)
```

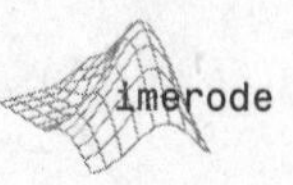

As Fig. 9.8(b) shows, these commands successfully removed the thin wires in the mask. Figure 9.8(c) shows what happens if we choose a structuring element that is too small:

```
>> se = strel('disk', 5);
>> A3 = imerode(A, se);
>> imshow(A3)
```

Some of the wire leads were not removed in this case. Figure 9.8(d) shows what happens if we choose a structuring element that is too large:

```
>> A4 = imerode(A, strel('disk', 20));
>> imshow(A4)
```

The wire leads were removed, but so were the border leads. ■

9.3 Combining Dilation and Erosion

In practical image-processing applications, dilation and erosion are used most often in various combinations. An image will undergo a series of dilations and/or erosions using the same, or sometimes different, structuring elements. In this section we consider three of the most common combinations of dilation and erosion: opening, closing, and the hit-or-miss transformation. We also introduce lookup table operations and discuss bwmorph, an IPT function that can perform a variety of practical morphological tasks.

9.3.1 Opening and Closing

The *morphological opening* of A by B, denoted $A \circ B$, is simply erosion of A by B, followed by dilation of the result by B:

$$A \circ B = (A \ominus B) \oplus B$$

An alternative mathematical formulation of opening is

$$A \circ B = \cup \{(B)_z | (B)_z \subseteq A\}$$

where $\cup\{\cdot\}$ denotes the union of all sets inside the braces, and the notation $C \subseteq D$ means that C is a subset of D. This formulation has a simple geometric interpretation: $A \circ B$ is the union of all translations of B that fit entirely within A. Figure 9.9 illustrates this interpretation. Figure 9.9(a) shows a set A and a disk-shaped structuring element B. Figure 9.9(b) shows some of the translations of B that fit *entirely* within A. The union of all such translations is the shaded region in Fig. 9.9(c); this region is the complete opening. The white regions in this figure are areas where the structuring element could not fit

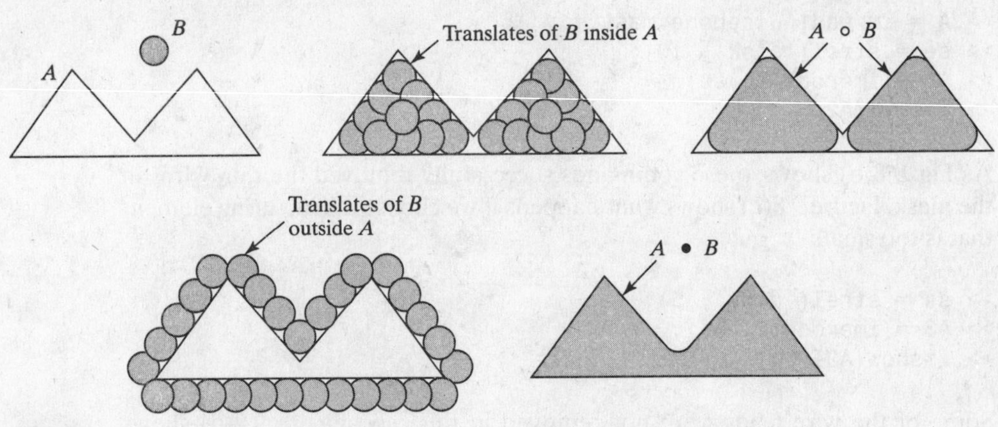

a b c
d e

FIGURE 9.9 Opening and closing as unions of translated structuring elements. (a) Set A and structuring element B. (b) Translations of B that fit entirely within set A. (c) The complete opening (shaded). (d) Translations of B outside the border of A. (e) The complete closing (shaded).

completely within A, and, therefore, are not part of the opening. Morphological opening removes completely regions of an object that cannot contain the structuring element, smoothes object contours, breaks thin connections, and removes thin protrusions.

The *morphological closing* of A by B, denoted $A \cdot B$, is a dilation followed by an erosion:

$$A \cdot B = (A \oplus B) \ominus B$$

Geometrically, $A \cdot B$ is the complement of the union of all translations of B that do not overlap A. Figure 9.9(d) illustrates several translations of B that do not overlap A. By taking the complement of the union of all such translations, we obtain the shaded region if Fig. 9.9(e), which is the complete closing. Like opening, morphological closing tends to smooth the contours of objects. Unlike opening, however, it generally joins narrow breaks, fills long thin gulfs, and fills holes smaller than the structuring element.

Opening and closing are implemented in the toolbox with functions `imopen` and `imclose`. These functions have the simple syntax forms

imopen

```
C = imopen(A, B)
```

and

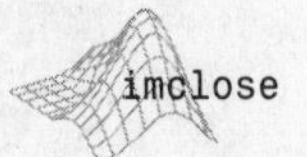


```
C = imclose(A, B)
```

where `A` is a binary image and `B` is a matrix of 0s and 1s that specifies the structuring element. A strel object, `SE`, can be used instead of `B`.

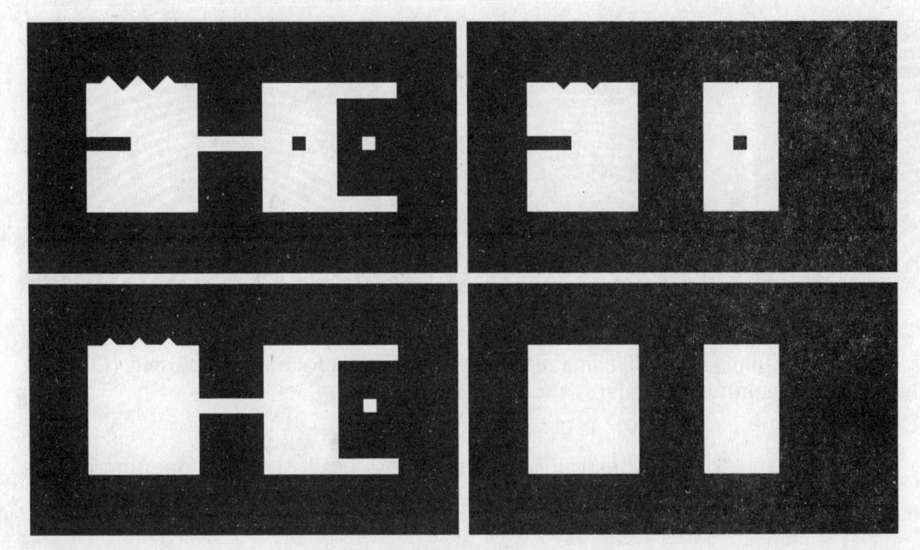

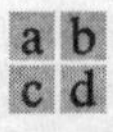

FIGURE 9.10 Illustration of opening and closing. (a) Original image. (b) Opening. (c) Closing. (d) Closing of (b).

EXAMPLE 9.4: Working with functions `imopen` and `imclose`.

■ This example illustrates the use of functions `imopen` and `imclose`. The image `shapes.tif` shown in Fig. 9.10(a) has several features designed to illustrate the characteristic effects of opening and closing, such as thin protrusions, joins, gulfs, an isolated hole, a small isolated object, and a jagged boundary. The following commands open the image with a 20×20 structuring element:

```
>> f = imread('shapes.tif');
>> se = strel('square', 20);
>> fo = imopen(f, se);
>> imshow(fo)
```

Figure 9.10(b) shows the result. Note that the thin protrusions and outward-pointing boundary irregularities were removed. The thin join and the small isolated object were removed also. The commands

```
>> fc = imclose(f, se);
>> imshow(fc)
```

produced the result in Fig. 9.10(c). Here, the thin gulf, the inward-pointing boundary irregularities, and the small hole were removed. As the next paragraph shows, combining a closing and an opening can be quite effective in removing noise. In terms of Fig. 9.10, performing a closing on the result of the earlier opening has the net effect of smoothing the object quite significantly. We close the opened image as follows:

```
>> foc = imclose(fo, se);
>> imshow(foc)
```

Figure 9.10(d) shows the resulting smoothed objects.

a b c

FIGURE 9.11 (a) Noisy fingerprint image. (b) Opening of image. (c) Opening followed by closing. (Original image courtesy of the National Institute of Standards and Technology.)

Figure 9.11 further illustrates the usefulness of closing and opening by applying these operations to a noisy fingerprint [Fig. 9.11(a)]. The commands

```
>> f = imread('fingerprint.tif');
>> se = strel('square', 3);
>> fo = imopen(f, se);
>> imshow(fo)
```

produced the image in Fig. 9.11(b). Note that noisy spots were removed by opening the image, but this process introduced numerous gaps in the ridges of the fingerprint. Many of the gaps can be filled in by following the opening with a closing:

```
>> foc = imclose(fo,se);
>> imshow(foc)
```

Figure 9.11(c) shows the final result. ■

9.3.2 The Hit-or-Miss Transformation

Often, it is useful to be able to identify specified configurations of pixels, such as isolated foreground pixels, or pixels that are end points of line segments. The *hit-or-miss transformation* is useful for applications such as these. The hit-or-miss transformation of A by B is denoted $A \otimes B$. Here, B is a structuring element pair, $B = (B_1, B_2)$, rather than a single element, as before. The hit-or-miss transformation is defined in terms of these two structuring elements as

$$A \otimes B = (A \ominus B_1) \cap (A^c \ominus B_2)$$

Figure 9.12 shows how the hit-or-miss transformation can be used to identify the locations of the following cross-shaped pixel configuration:

$$\begin{matrix} 0 & 1 & 0 \\ 1 & 1 & 1 \\ 0 & 1 & 0 \end{matrix}$$

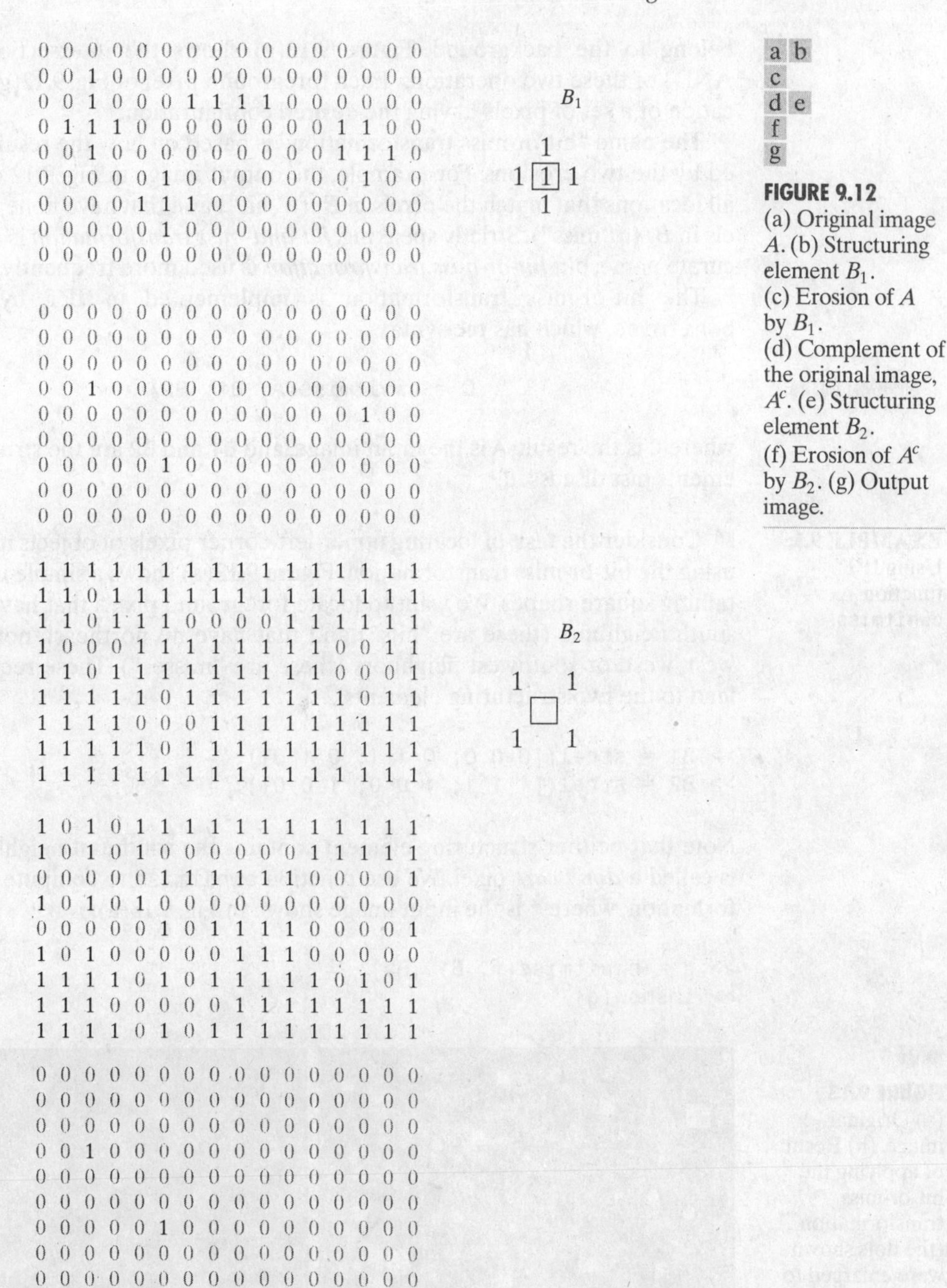

a b
c
d e
f
g

FIGURE 9.12 (a) Original image A. (b) Structuring element B_1. (c) Erosion of A by B_1. (d) Complement of the original image, A^c. (e) Structuring element B_2. (f) Erosion of A^c by B_2. (g) Output image.

Figure 9.12(a) contains this configuration of pixels in two different locations. Erosion with structuring element B_1 determines the locations of foreground pixels that have north, east, south, and west foreground neighbors. Erosion of the complement with structuring element B_2 determines the locations of all the pixels whose northeast, southeast, southwest, and northwest neighbors belong to the background. Figure 9.12(g) shows the intersection (logical AND) of these two operations. Each foreground pixel of Fig. 9.12(g) is the location of a set of pixels having the desired configuration.

The name "hit-or-miss transformation" is based on how the result is affected by the two erosions. For example, the output image in Fig. 9.12 consists of all locations that match the pixels in B_1 (a "hit") and that have none of the pixels in B_2 (a "miss"). Strictly speaking, *hit-and-miss transformation* is a more accurate name, but *hit-or-miss transformation* is used more frequently.

The hit-or-miss transformation is implemented in IPT by function `bwhitmiss`, which has the syntax

```
C = bwhitmiss(A, B1, B2)
```

where `C` is the result, `A` is the input image, and `B1` and `B2` are the structuring elements just discussed.

EXAMPLE 9.5: Using IPT function `bwhitmiss`.

■ Consider the task of locating upper-left-corner pixels of objects in an image using the hit-or-miss transformation. Figure 9.13(a) shows a simple image containing square shapes. We want to locate foreground pixels that have east and south neighbors (these are "hits") and that have no northeast, north, northwest, west, or southwest neighbors (these are "misses"). These requirements lead to the two structuring elements:

```
>> B1 = strel([0 0 0; 0 1 1; 0 1 0]);
>> B2 = strel([1 1 1; 1 0 0; 1 0 0]);
```

Note that neither structuring element contains the southeast neighbor, which is called a *don't care* pixel. We use function `bwhitmiss` to compute the transformation, where `f` is the input image shown in Fig. 9.13(a):

```
>> g = bwhitmiss(f, B1 ,B2);
>> imshow(g)
```

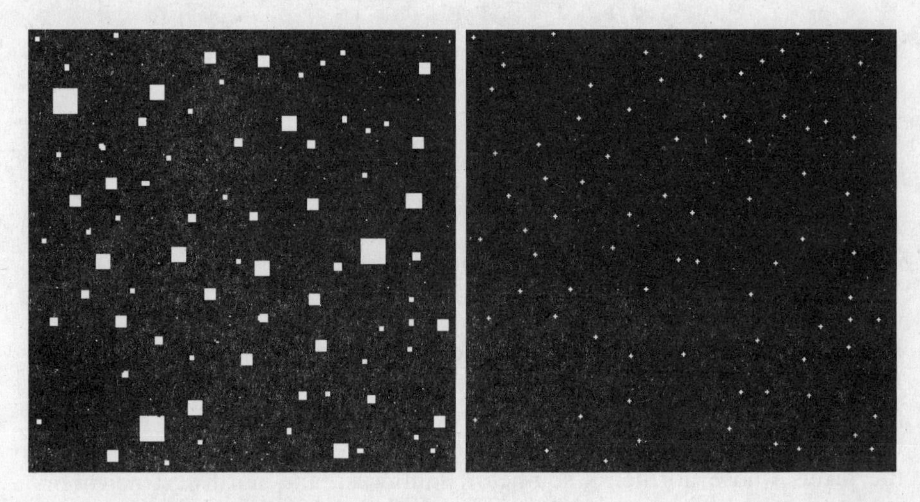

a b

FIGURE 9.13 (a) Original image. (b) Result of applying the hit-or-miss transformation (the dots shown were enlarged to facilitate viewing).

Each single-pixel dot in Fig. 9.13(b) is an upper-left-corner pixel of the objects in Fig. 9.13(a). The pixels in Fig. 9.13(b) were enlarged for clarity. ■

9.3.3 Using Lookup Tables

When the hit-or-miss structuring elements are small, a faster way to compute the hit-or-miss transformation is to use a lookup table (LUT). The technique is to precompute the output pixel value for every possible neighborhood configuration and then store the answers in a table for later use. For instance, there are $2^9 = 512$ different 3×3 configurations of pixel values in a binary image.

To make the use of lookup tables practical, we must assign a unique index to each possible configuration. A simple way to do this for, say, the 3×3 case, is to multiply each 3×3 configuration element-wise by the matrix

$$\begin{matrix} 1 & 8 & 64 \\ 2 & 16 & 128 \\ 4 & 32 & 256 \end{matrix}$$

and then sum all the products. This procedure assigns a unique value in the range $[0, 511]$ to each different 3×3 neighborhood configuration. For example, the value assigned to the neighborhood

$$\begin{matrix} 1 & 1 & 0 \\ 1 & 0 & 1 \\ 1 & 0 & 1 \end{matrix}$$

is $1(1) + 2(1) + 4(1) + 8(1) + 16(0) + 32(0) + 64(0) + 128(1) + 256(1) = 399$, where the first number in these products is a coefficient from the preceding matrix and the numbers in parentheses are the pixel values, taken columnwise.

The toolbox provides two functions, `makelut` and `applylut` (illustrated later in this section), that can be used to implement this technique. Function `makelut` constructs a lookup table based on a user-supplied function, and `applylut` processes binary images using this lookup table. Continuing with the 3×3 case, using `makelut` requires writing a function that accepts a 3×3 binary matrix and returns a single value, typically either a 0 or 1. Function `makelut` calls the user-supplied function 512 times, passing it each possible 3×3 neighborhood. It records and returns all the results in the form of a 512-element vector.

As an illustration, we write a function, `endpoints.m`, that uses `makelut` and `applylut` to detect end points in a binary image. We define an *end point* as a foreground pixel that has exactly one foreground neighbor. Function `endpoints` computes and then applies a lookup table for detecting end points in an input image. The line of code

```
persistent lut
```

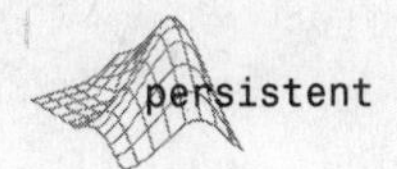

used in function `endpoints` establishes a variable called `lut` and declares it to be *persistent*. MATLAB "remembers" the value of persistent variables in between function calls. The first time function `endpoints` is called, variable `lut` is automatically initialized to the empty matrix (`[]`). When `lut` is empty, the function calls `makelut`, passing it a handle to subfunction `endpoint_fcn`. Function `applylut` then finds the end points using the lookup table. The lookup table is saved in persistent variable `lut` so that, the next time `endpoints` is called, the lookup table does not need to be recomputed.

`endpoints`

See Section 3.4.2 for a discussion of function handle, @.

```
function g = endpoints(f)
%ENDPOINTS Computes end points of a binary image.
%   G = ENDPOINTS(F) computes the end points of the binary image F
%   and returns them in the binary image G.

persistent lut

if isempty(lut)
    lut = makelut(@endpoint_fcn, 3);
end

g = applylut(f, lut);

%------------------------------------------------------------------%
function is_end_point = endpoint_fcn(nhood)
%   Determines if a pixel is an end point.
%   IS_END_POINT = ENDPOINT_FCN(NHOOD) accepts a 3-by-3 binary
%   neighborhood, NHOOD, and returns a 1 if the center element is an
%   end point; otherwise it returns a 0.

is_end_point = nhood(2, 2) & (sum(nhood(:)) == 2);
```

Figure 9.14 illustrates a typical use of function `endpoints`. Figure 9.14(a) is a binary image containing a morphological skeleton (see Section 9.3.4), and Fig. 9.14(b) shows the output of function `endpoints`.

EXAMPLE 9.6: Playing Conway's Game of Life using binary images and lookup-table-based computation.

■ An interesting application of lookup tables is Conway's "Game of Life," which involves "organisms" arranged on a rectangular grid. We include it here as another illustration of the power and simplicity of lookup tables. There are simple rules for how the organisms in Conway's game are born, survive, and die from one "generation" to the next. A binary image is a convenient representation for the game, where each foreground pixel represents a living organism in that location.

Conway's genetic rules describe how to compute the next generation (or next binary image) from the current one:

1. Every foreground pixel with two or three neighboring foreground pixels survives to the next generation.
2. Every foreground pixel with zero, one, or at least four foreground neighbors "dies" (becomes a background pixel) because of "isolation" or "overpopulation."
3. Every background pixel adjacent to exactly three foreground neighbors is a "birth" pixel and becomes a foreground pixel.

All births and deaths occur simultaneously in the process of computing the next binary image depicting the next generation.

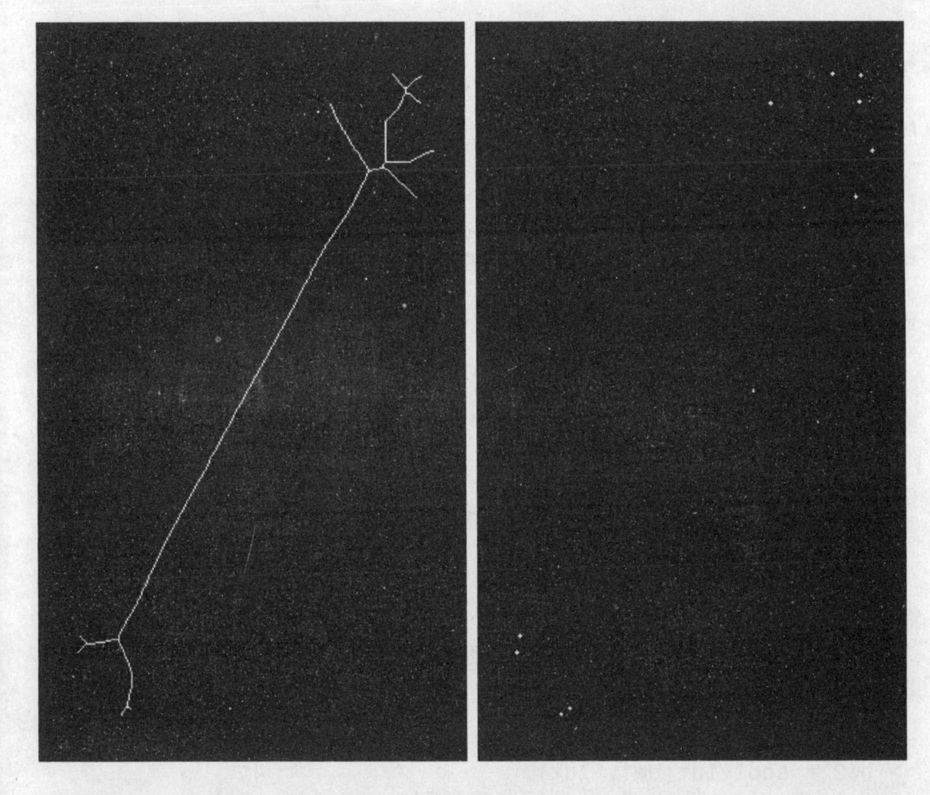

a b

FIGURE 9.14
(a) Image of a morphological skeleton.
(b) Output of function `endpoints`. The pixels in (b) were enlarged for clarity.

To implement the game of life using `makelut` and `applylut`, we first write a function that applies Conway's genetic laws to a single pixel and its 3×3 neighborhood:

conwaylaws

```
function out = conwaylaws(nhood)
%CONWAYLAWS Applies Conway's genetic laws to a single pixel.
%   OUT = CONWAYLAWS(NHOOD) applies Conway's genetic laws to a single
%   pixel and its 3-by-3 neighborhood, NHOOD.
num_neighbors = sum(nhood(:)) - nhood(2, 2);
if nhood(2, 2) == 1
   if num_neighbors <= 1
      out = 0; % Pixel dies from isolation.
   elseif num_neighbors > = 4
      out = 0; % Pixel dies from overpopulation.
   else
      out = 1; % Pixel survives.
   end
else
   if num_neighbors == 3
      out = 1; % Birth pixel.
   else
      out = 0; % Pixel remains empty.
   end
end
```

The lookup table is constructed next by calling `makelut` with a function handle to `conwaylaws`:

```
>> lut = makelut(@conwaylaws, 3);
```

Various starting images have been devised to demonstrate the effect of Conway's laws on successive generations (see Gardner, [1970, 1971]). Consider, for example, an initial image called the "Cheshire cat configuration,"

```
>> bw1 = [0 0 0 0 0 0 0 0 0 0
          0 0 0 0 0 0 0 0 0 0
          0 0 0 1 0 0 1 0 0 0
          0 0 0 1 1 1 1 0 0 0
          0 0 1 0 0 0 0 1 0 0
          0 0 1 0 1 1 0 1 0 0
          0 0 1 0 0 0 0 1 0 0
          0 0 0 1 1 1 1 0 0 0
          0 0 0 0 0 0 0 0 0 0
          0 0 0 0 0 0 0 0 0 0];
```

The following commands perform the computation and display up to the third generation:

```
>> imshow(bw1, 'n'), title('Generation 1')
>> bw2 = applylut(bw1, lut);
>> figure, imshow(bw2, 'n'); title('Generation 2')
>> bw3 = applylut(bw2, lut);
>> figure, imshow(bw3, 'n'); title('Generation 3')
```

We leave it as an exercise to show that after a few generations the cat fades to a "grin" before finally leaving a "paw print." ■

9.3.4 Function bwmorph

IPT function `bwmorph` implements a variety of useful operations based on combinations of dilations, erosions, and lookup table operations. Its calling syntax is

```
g = bwmorph(f, operation, n)
```

where `f` is an input binary image, `operation` is a string specifying the desired operation, and `n` is a positive integer specifying the number of times the operation is to be repeated. Input argument `n` is optional and can be omitted, in which case the operation is performed once. Table 9.3 describes the set of valid operations for `bwmorph`. In the rest of this section we concentrate on two of these: *thinning* and *skeletonization*.

Thinning means reducing binary objects or shapes in an image to strokes that are a single pixel wide. For example, the fingerprint ridges shown in

TABLE 9.3 Operations supported by function `bwmorph`.

Operation	Description
`bothat`	"Bottom-hat" operation using a 3 × 3 structuring element; use `imbothat` (see Section 9.6.2) for other structuring elements.
`bridge`	Connect pixels separated by single-pixel gaps.
`clean`	Remove isolated foreground pixels.
`close`	Closing using a 3 × 3 structuring element; use `imclose` for other structuring elements.
`diag`	Fill in around diagonally connected foreground pixels.
`dilate`	Dilation using a 3 × 3 structuring element; use `imdilate` for other structuring elements.
`erode`	Erosion using a 3 × 3 structuring element; use `imerode` for other structuring elements.
`fill`	Fill in single-pixel "holes" (background pixels surrounded by foreground pixels); use `imfill` (see Section 11.1.2) to fill in larger holes.
`hbreak`	Remove H-connected foreground pixels.
`majority`	Make pixel p a foreground pixel if at least five pixels in $N_8(p)$ (see Section 9.4) are foreground pixels; otherwise make p a background pixel.
`open`	Opening using a 3 × 3 structuring element; use function `imopen` for other structuring elements.
`remove`	Remove "interior" pixels (foreground pixels that have no background neighbors).
`shrink`	Shrink objects with no holes to points; shrink objects with holes to rings.
`skel`	Skeletonize an image.
`spur`	Remove spur pixels.
`thicken`	Thicken objects without joining disconnected 1s.
`thin`	Thin objects without holes to minimally connected strokes; thin objects with holes to rings.
`tophat`	"Top-hat" operation using a 3 × 3 structuring element; use `imtophat` (see Section 9.6.2) for other structuring elements.

Fig. 9.11(c) are fairly thick. It may be desirable for subsequent shape analysis to thin the ridges so that each is one pixel thick. Each application of `bwmorph`'s thinning operation removes one or two pixels from the thickness of binary image objects. The following commands, for example, display the results of applying the thinning operation one and two times.

```
>> f = imread('fingerprint_cleaned.tif');
>> g1 = bwmorph(f, 'thin', 1);
>> g2 = bwmorph(f, 'thin', 2);
>> imshow(g1), figure, imshow(g2)
```

a b c

FIGURE 9.15 (a) Fingerprint image from Fig. 9.11(c) thinned once. (b) Image thinned twice. (c) Image thinned until stability.

Figures 9.15(a) and 9.15(b), respectively, show the results. A key question is how many times to apply the thinning operation. For several operations, including thinning, `bwmorph` allows `n` to be set to infinity (`Inf`). Calling `bwmorph` with `n = Inf` instructs `bwmorph` to repeat the operation until the image stops changing. Sometimes this is called repeating an operation *until stability*. For example,

```
>> ginf = bwmorph(f, 'thin', Inf);
>> imshow(ginf)
```

As Fig. 9.15(c) shows, this is a significant improvement over Fig. 9.11(c).

Skeletonization is another way to reduce binary image objects to a set of thin strokes that retain important information about the shapes of the original objects. (Skeletonization is described in more detail in Gonzalez and Woods [2002].) Function `bwmorph` performs skeletonization when `operation` is set to `'skel'`. Let `f` denote the image of the bonelike object in Fig. 9.16(a). To compute its skeleton, we call `bwmorph`, with `n = Inf`:

```
>> fs = bwmorph(f, 'skel', Inf);
>> imshow(f), figure, imshow(fs)
```

Figure 9.16(b) shows the resulting skeleton, which is a reasonable likeness of the basic shape of the object.

Skeletonization and thinning often produce short extraneous spurs, sometimes called *parasitic components*. The process of cleaning up (or removing) these spurs is called *pruning*. Function `endpoints` (Section 9.3.3) can be used for this purpose. The method is to iteratively identify and remove endpoints. The following simple commands, for example, postprocesses the skeleton image `fs` through five iterations of endpoint removals:

```
>> for k = 1:5
      fs = fs & ~endpoints(fs);
end
```

Figure 9.16(c) shows the result.

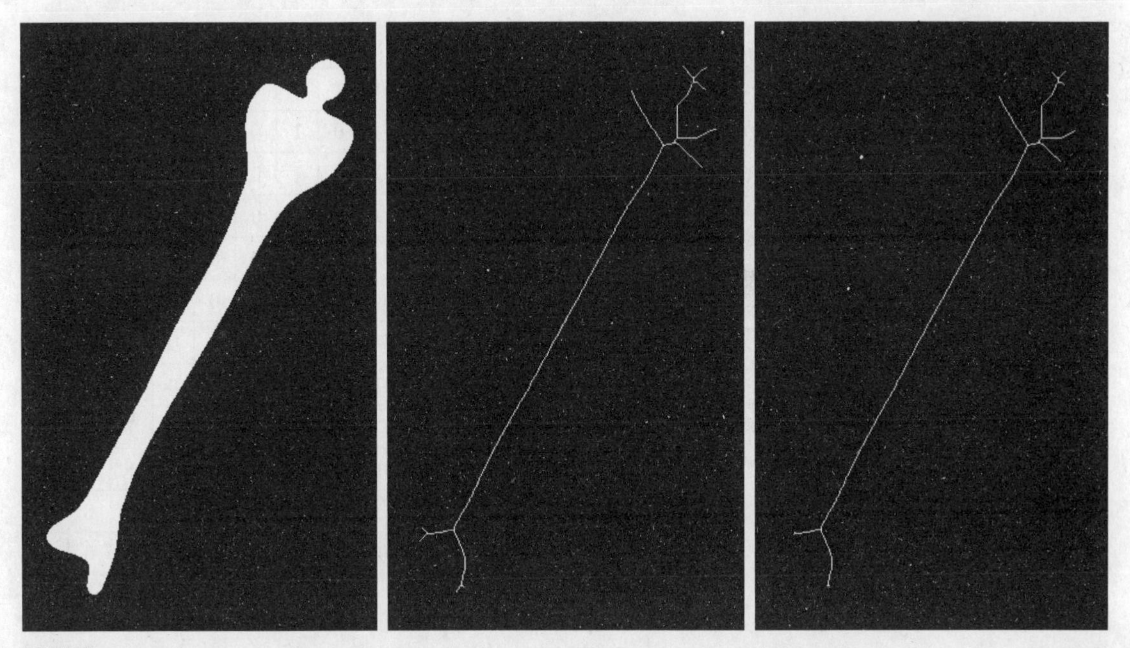

a b c

FIGURE 9.16 (a) Bone image. (b) Skeleton obtained using function `bwmorph`. (c) Resulting skeleton after pruning with function `endpoints`.

9.4 Labeling Connected Components

The concepts discussed thus far are applicable mostly to all foreground (or all background) individual pixels and their immediate neighbors. In this section we consider the important "middle ground" between individual foreground pixels and the set of all foreground pixels. This leads to the notion of *connected components*, also referred to as *objects* in the following discussion.

When asked to count the objects in Fig. 9.17(a), most people would identify ten: six characters and four simple geometric shapes. Figure 9.17(b) shows a small rectangular section of pixels in the image. How are the sixteen foreground pixels in Fig. 9.17(b) related to the ten objects in the image? Although they appear to be in two separate groups, all sixteen pixels actually belong to the letter "E" in Fig. 9.17(a). To develop computer programs that locate and operate on objects, we need a more precise set of definitions for key terms.

A pixel p at coordinates (x, y) has two horizontal and two vertical neighbors whose coordinates are $(x + 1, y)$, $(x - 1, y)$, $(x, y + 1)$ and $(x, y - 1)$. This set of *4-neighbors* of p, denoted $N_4(p)$, is shaded in Fig. 9.18(a). The four diagonal neighbors of p have coordinates $(x + 1, y + 1)$, $(x + 1, y - 1)$, $(x - 1, y + 1)$ and $(x - 1, y - 1)$. Figure 9.18(b) shows these neighbors, which are denoted $N_D(p)$. The union of $N_4(p)$ and $N_D(p)$ in Fig. 9.18(c) are the *8-neighbors* of p, denoted $N_8(p)$.

Two pixels p and q are said to be *4-adjacent* if $q \in N_4(p)$. Similarly, p and q are said to be *8-adjacent* if $q \in N_8(p)$. Figures 9.18(d) and (e) illustrate

a b

FIGURE 9.17 (a) Image containing ten objects. (b) A subset of pixels from the image.

0	1	1	1	0	0	0	0	0	0
0	0	1	1	0	0	0	0	0	0
0	0	0	1	0	0	0	0	0	0
0	0	0	1	0	0	0	0	0	0
0	0	0	1	0	0	0	0	0	0
0	0	0	0	0	0	0	0	0	0
0	0	0	0	0	0	0	0	0	0
0	0	0	0	0	0	0	0	0	0
0	0	0	0	0	0	0	0	1	0
0	0	0	0	0	0	0	1	0	0
0	0	0	0	0	0	1	1	0	0
0	0	0	0	0	0	1	1	0	0
0	0	0	0	0	1	1	0	0	0

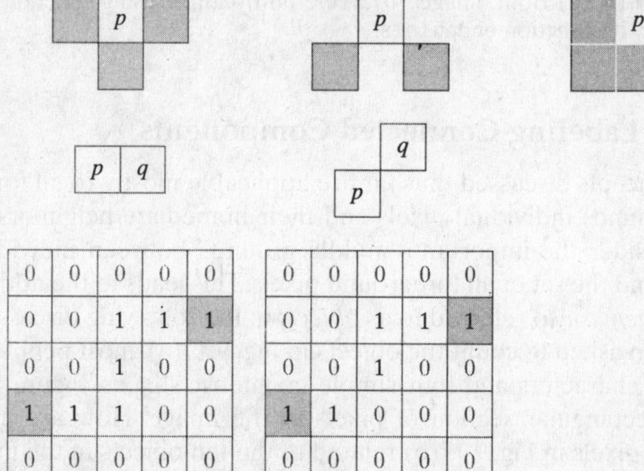

a b c
d e
f g

FIGURE 9.18 (a) Pixel p and its 4-neighbors, $N_4(p)$. (b) Pixel p and its diagonal neighbors, $N_D(p)$. (c) Pixel p and its 8-neighbors, $N_8(p)$. (d) Pixels p and q are 4-adjacent and 8-adjacent. (e) Pixels p and q are 8-adjacent but not 4-adjacent. (f) The shaded pixels are both 4-connected and 8-connected. (g) The shaded foreground pixels are 8-connected but not 4-connected.

these concepts. A *path* between pixels p_1 and p_n is a sequence of pixels $p_1, p_2, \ldots, p_{n-1}, p_n$ such that p_k is adjacent to p_{k+1}, for $1 \leq k < n$. A path can be 4-*connected* or 8-*connected*, depending on the definition of adjacency used.

Two foreground pixels p and q are said to be 4-connected if there exists a 4-connected path between them, consisting entirely of foreground pixels [Fig. 9.18(f)]. They are 8-connected if there exists an 8-connected path between them [Fig. 9.18(g)]. For any foreground pixel, p, the set of all foreground pixels connected to it is called the *connected component* containing p.

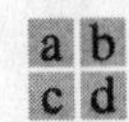

1	1	1	0	0	0	0	0
1	1	1	0	1	1	0	0
1	1	1	0	1	1	0	0
1	1	1	0	0	0	1	0
1	1	1	0	0	0	1	0
1	1	1	0	0	0	1	0
1	1	1	0	0	1	0	0
1	1	1	0	0	0	0	0

1	1	1	0	0	0	0	0
1	1	1	0	1	1	0	0
1	1	1	0	1	1	0	0
1	1	1	0	0	0	1	0
1	1	1	0	0	0	1	0
1	1	1	0	0	0	1	0
1	1	1	0	0	1	0	0
1	1	1	0	0	0	0	0

1	1	1	0	0	0	0	0
1	1	1	0	2	2	0	0
1	1	1	0	2	2	0	0
1	1	1	0	0	0	4	0
1	1	1	0	0	0	4	0
1	1	1	0	0	0	4	0
1	1	1	0	0	3	0	0
1	1	1	0	0	0	0	0

1	1	1	0	0	0	0	0
1	1	1	0	2	2	0	0
1	1	1	0	2	2	0	0
1	1	1	0	0	0	2	0
1	1	1	0	0	0	2	0
1	1	1	0	0	0	2	0
1	1	1	0	0	2	0	0
1	1	1	0	0	0	0	0

FIGURE 9.19 Connected components (a) Four 4-connected components. (b) Two 8-connected components. (c) Label matrix obtained using 4-connectivity (d) Label matrix obtained using 8-connectivity.

The term *connected component* was just defined in terms of a path, and the definition of a path in turn depends on adjacency. This implies that the nature of a connected component depends on which form of adjacency we choose, with 4- and 8-adjacency being the most common. Figure 9.19 illustrates the effect that adjacency can have on determining the number of connected components in an image. Figure 9.19(a) shows a small binary image with four 4-connected components. Figure 9.19(b) shows that choosing 8-adjacency reduces the number of connected components to two.

IPT function `bwlabel` computes all the connected components in a binary image. The calling syntax is

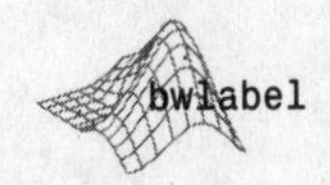


```
[L, num] = bwlabel(f, conn)
```

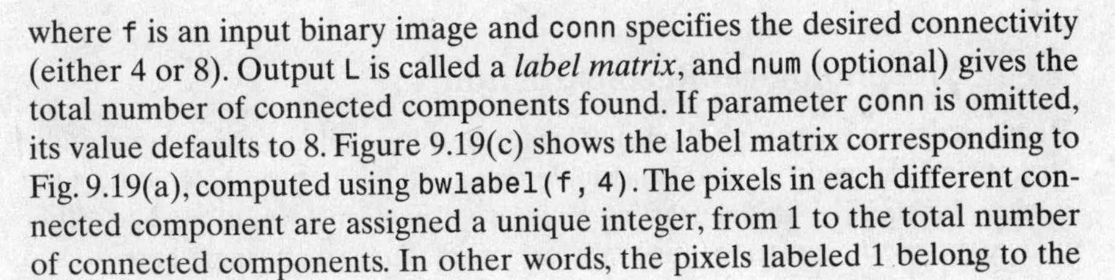

where `f` is an input binary image and `conn` specifies the desired connectivity (either 4 or 8). Output `L` is called a *label matrix*, and `num` (optional) gives the total number of connected components found. If parameter conn is omitted, its value defaults to 8. Figure 9.19(c) shows the label matrix corresponding to Fig. 9.19(a), computed using `bwlabel(f, 4)`. The pixels in each different connected component are assigned a unique integer, from 1 to the total number of connected components. In other words, the pixels labeled 1 belong to the first connected component; the pixels labeled 2 belong to the second connected component; and so on. Background pixels are labeled 0. Figure 9.19(d) shows the label matrix corresponding to Fig. 9.19(a), computed using `bwlabel(f, 8)`.

EXAMPLE 9.7: Computing and displaying the center of mass of connected components.

■ This example shows how to compute and display the center of mass of each connected component in Fig. 9.17(a). First, we use `bwlabel` to compute the 8-connected components:

```
>> f = imread('objects.tif');
>> [L, n] = bwlabel(f);
```

Function `find` (Section 5.2.2) is useful when working with label matrices. For example, the following call to `find` returns the row and column indices for all the pixels belonging to the third object:

```
>> [r, c] = find(L == 3);
```

Function `mean` with `r` and `c` as inputs then computes the center of mass of this object.

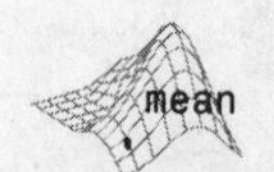

If `A` *is a vector,* `mean(A)` *computes the average value of its elements. If* `A` *is a matrix,* `mean(A)` *treats the columns of* `A` *as vectors, returning a row vector of mean values. The syntax* `mean(A, dim)` *returns the mean values of the elements along the dimension of* `A` *specified by scalar* `dim`.

```
>> rbar = mean(r);
>> cbar = mean(c);
```

A loop can be used to compute and display the centers of mass of all the objects in the image. To make the centers of mass visible when superimposed on the image, we display them using a white " * " marker on top of a black-filled circle marker, as follows:

```
>> imshow(f)
>> hold on   % So later plotting commands plot on top of the image.
>> for k = 1:n
   [r, c] = find(L == k);
   rbar = mean(r);
   cbar = mean(c);
   plot(cbar, rbar, 'Marker', 'o', 'MarkerEdgeColor', 'k',...
        'MarkerFaceColor', 'k', 'MarkerSize', 10)
   plot(cbar, rbar, 'Marker', '*', 'MarkerEdgeColor', 'w')
end
```

Figure 9.20 shows the result. ■

9.5 Morphological Reconstruction

Reconstruction is a morphological transformation involving two images and a structuring element (instead of a single image and structuring element). One image, the *marker*, is the starting point for the transformation. The other image, the *mask*, constrains the transformation. The structuring element used

FIGURE 9.20 Centers of mass (white asterisks) shown superimposed on their corresponding connected components.

defines connectivity. In this section we use 8-connectivity (the default), which implies that B in the following discussion is a 3×3 matrix of 1s, with the center defined at coordinates $(2, 2)$.

If g is the mask and f is the marker, the reconstruction of g from f, denoted $R_g(f)$, is defined by the following iterative procedure:

1. Initialize h_1 to be the marker image f.
2. Create the structuring element: `B = ones(3)`.
3. Repeat:

$$h_{k+1} = (h_k \oplus B) \cap g$$

until $h_{k+1} = h_k$.

Marker f must be a subset of g; that is,

$$f \subseteq g$$

Figure 9.21 illustrates the preceding iterative procedure. Note that, although this iterative formulation is useful conceptually, much faster computational algorithms exist. IPT function `imreconstruct` uses the "fast hybrid reconstruction" algorithm described in Vincent [1993]. The calling syntax for `imreconstruct` is

```
out = imreconstruct(marker, mask)
```

where `marker` and `mask` are as defined at the beginning of this section.

9.5.1 Opening by Reconstruction

See Sections 10.4.2 and 10.4.3 for additional applications of morphological reconstruction.

In morphological opening, erosion typically removes small objects, and the subsequent dilation tends to restore the shape of the objects that remain. However, the accuracy of this restoration depends on the similarity between

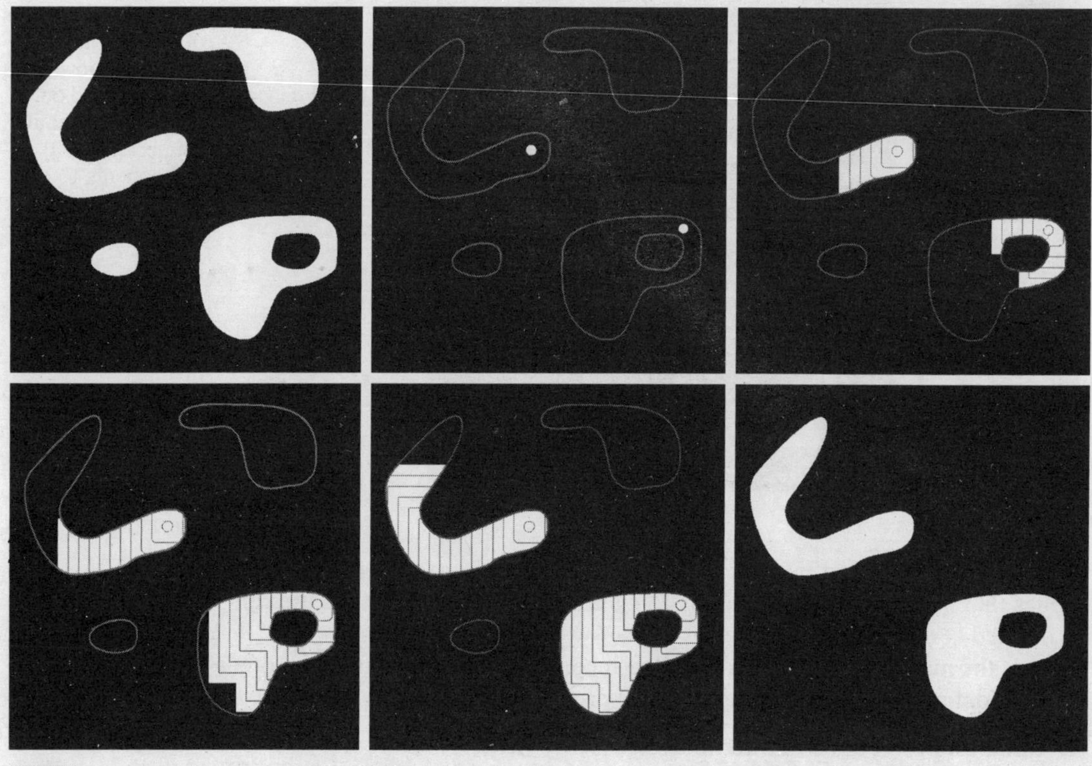

a b c
d e f

FIGURE 9.21 Morphological reconstruction. (a) Original image (the mask). (b) Marker image. (c)–(e) Intermediate result after 100, 200, and 300 iterations, respectively. (f) Final result. [The outlines of the objects in the mask image are superimposed on (b)–(e) as visual references.]

the shapes and the structuring element. The method discussed in this section, *opening by reconstruction*, restores exactly the shapes of the objects that remain after erosion. The opening by reconstruction of f, using structuring element B, is defined as $R_f(f \ominus B)$.

EXAMPLE 9.8: Opening by reconstruction.

■ A comparison between opening and opening by reconstruction for an image containing text is shown in Fig. 9.22. In this example, we are interested in extracting from Fig. 9.22(a) the characters that contain long vertical strokes. Since opening by reconstruction requires an eroded image, we perform that step first, using a thin, vertical structuring element of length proportional to the height of the characters:

```
>> f = imread('book_text_bw.tif');
>> fe = imerode(f, ones(51, 1));
```

Figure 9.22(b) shows the result. The opening, shown in Fig. 9.22(c), is computed using `imopen`:

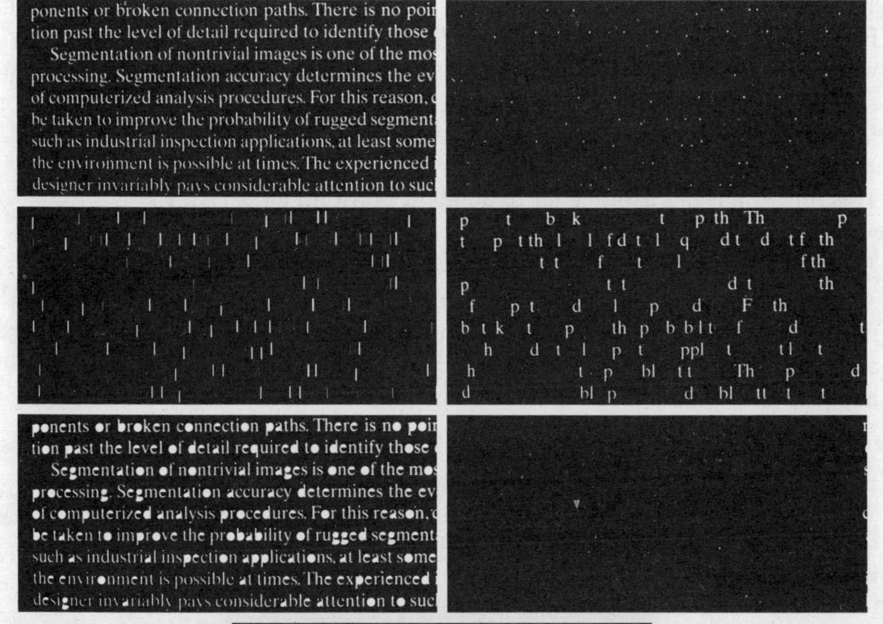

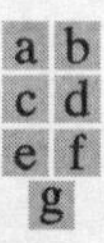

FIGURE 9.22 Morphological reconstruction: (a) Original image. (b) Eroded with vertical line. (c) Opened with a vertical line. (d) Opened by reconstruction with a vertical line. (e) Holes filled. (f) Characters touching the border (see right border). (g) Border characters removed.

```
>> fo = imopen(f, ones(51, 1));
```

Note that the vertical strokes were restored, but not the rest of the characters containing the strokes. Finally, we obtain the reconstruction:

```
>> fobr = imreconstruct(fe, f);
```

The result in Fig. 9.22(d) shows that characters containing long vertical strokes were restored exactly; all other characters were removed. The remaining parts of Fig. 9.22 are explained in the following two sections. ■

9.5.2 Filling Holes

Morphological reconstruction has a broad spectrum of practical applications, each determined by the selection of the marker and mask images. For example, suppose that we choose the marker image, f_m, to be 0 everywhere except on the image border, where it is set to $1 - f$:

$$f_m(x, y) = \begin{cases} 1 - f(x, y) & \text{if } (x, y) \text{ is on the border of } f \\ 0 & \text{otherwise} \end{cases}$$

Then $g = [R_{f^c}(f_m)]^c$ has the effect of filling the holes in f, as illustrated in Fig. 9.22(e). IPT function `imfill` performs this computation automatically when the optional argument `'holes'` is used:

```
g = imfill(f, 'holes')
```

This function is discussed in more detail in Section 11.1.2.

9.5.3 Clearing Border Objects

Another useful application of reconstruction is removing objects that touch the border of an image. Again, the key task is to select the appropriate marker and mask images to achieve the desired effect. In this case, we use the original image as the mask, and the marker image, f_m, is defined as

$$f_m(x, y) = \begin{cases} f(x, y) & \text{if } (x, y) \text{ is on the border of } f \\ 0 & \text{otherwise} \end{cases}$$

Figure 9.22(f) shows that the reconstruction, $R_f(f_m)$, contains only the objects touching the border. The set difference $f - R_f(f_m)$, shown in Fig. 9.22(g), contains only the objects from the original image that do not touch the border. IPT function `imclearborder` performs this entire procedure automatically. Its syntax is

```
g = imclearborder(f, conn)
```

where `f` is the input image and `g` is the result. The value of `conn` can be either 4 or 8 (the default). This function suppresses structures that are lighter than their surroundings and that are connected to the image border. Input `f` can be a gray-scale or binary image. The output image is a gray-scale or binary image, respectively.

9.6 Gray-Scale Morphology

All the binary morphological operations discussed in this chapter, with the exception of the hit-or-miss transform, have natural extensions to gray-scale images. In this section, as in the binary case, we start with dilation and erosion, which for gray-scale images are defined in terms of minima and maxima of pixel neighborhoods.

9.6.1 Dilation and Erosion

The *gray-scale dilation* of f by structuring element b, denoted $f \oplus b$, is defined as

$$(f \oplus b)(x, y) = \max\{f(x - x', y - y') + b(x', y') \mid (x', y') \in D_b\}$$

where D_b is the domain of b, and $f(x, y)$ is assumed to equal $-\infty$ outside the domain of f. This equation implements a process similar to the concept of spatial convolution, explained in Section 3.4.1. Conceptually, we can think of rotating the structuring element about its origin and translating it to all locations in the image, just as the convolution kernel is rotated and then translated about the image. At each translated location, the rotated structuring element values are added to the image pixel values and the maximum is computed.

One important difference between convolution and gray-scale dilation is that, in the latter, D_b, a binary matrix, defines which locations in the neighborhood are included in the max operation. In other words, for an arbitrary pair of coordinates (x_0, y_0) in the domain of D_b, the sum $f(x - x_0, y - y_0) + b(x_0, y_0)$ is included in the max computation only if D_b is 1 at those coordinates. If D_b is 0 at (x_0, y_0), the sum is not considered in the max operation. This is repeated for all coordinates $(x', y') \in D_b$ each time that coordinates (x, y) change. Plotting $b(x', y')$ as a function of coordinates x' and y' would look like a digital "surface" with the height at any pair of coordinates being given by the value of b at those coordinates.

In practice, gray-scale dilation usually is performed using *flat* structuring elements (see Table 9.2) in which the value (height) of b is 0 at all coordinates over which D_b is defined. That is,

$$b(x', y') = 0 \quad \text{for } (x', y') \in D_b$$

In this case, the max operation is specified completely by the pattern of 0s and 1s in binary matrix D_b, and the gray-scale dilation equation simplifies to

$$(f \oplus b)(x, y) = \max\{f(x - x', y - y') \mid (x', y') \in D_b\}$$

Thus, flat gray-scale dilation is a local-maximum operator, where the maximum is taken over a set of pixel neighbors determined by the shape of D_b.

Nonflat structuring elements are created with `strel` by passing it two matrices: (1) a matrix of 0s and 1s specifying the structuring element domain, D_b, and (2) a second matrix specifying height values, $b(x', y')$. For example,

```
>> b = strel([1 1 1], [1 2 1])
   b =
   Nonflat STREL object containing 3 neighbors.
   Neighborhood:
     1     1     1
   Height:
     1     2     1
```

creates a 1×3 structuring element whose height values are $b(0, -1) = 1$, $b(0, 0) = 2$, and $b(0, 1) = 1$.

Flat structuring elements for gray-scale images are created using `strel` in the same way as for binary images. For example, the following commands show how to dilate the image `f` in Fig. 9.23(a) using a flat 3×3 structuring element:

```
>> se = strel('square', 3);
>> gd = imdilate(f, se);
```

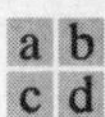

FIGURE 9.23 Dilation and erosion. (a) Original image. (b) Dilated image. (c) Eroded image. (d) Morphological gradient. (Original image courtesy of NASA.)

Figure 9.23(b) shows the result. As expected, the image is slightly blurred. The rest of this figure is explained in the following discussion.

The *gray-scale erosion* of f by structuring element b, denoted $f \ominus b$, is defined as

$$(f \ominus b)(x, y) = \min\{f(x + x', y + y') - b(x', y') \mid (x', y') \in D_b\}$$

where D_b is the domain of b and $f(x, y)$ is assumed to be $+\infty$ outside the domain of f. Conceptually, we again can think of translating the structuring element to all locations in the image. At each translated location, the structuring element values are subtracted from the image pixel values and the minimum is taken.

As with dilation, gray-scale erosion is most often performed using flat structuring elements. The equation for flat gray-scale erosion can then be simplified to

$$(f \ominus b)(x, y) = \min\{f(x + x', y + y') \mid (x', y') \in D_b\}$$

Thus, flat gray-scale erosion is a local-minimum operator, in which the minimum is taken over a set of pixel neighbors determined by the shape of D_b.

Figure 9.23(c) shows the result of using imerode with the same structuring element used for Fig. 9.23(b):

```
>> ge = imerode(f, se);
```

Dilation and erosion can be combined to achieve a variety of effects. For instance, subtracting an eroded image from its dilated version produces a "morphological gradient," which is a measure of local gray-level variation in the image. For example, letting

Computing the morphological gradient requires a different procedure for nonsymmetric structuring elements. Specifically, a reflected structuring element must be used in the dilation step.

```
>> morph_grad = imsubtract(gd, ge);
```

produced the image in Fig. 9.23(d), which is the morphological gradient of the image in Fig. 9.23(a). This image has edge-enhancement characteristics similar to those that would be obtained using the gradient operations discussed in Sections 6.6.1 and later in Section 10.1.3.

9.6.2 Opening and Closing

The expressions for opening and closing gray-scale images have the same form as their binary counterparts. The opening of image f by structuring element b, denoted $f \circ b$, is defined as

$$f \circ b = (f \ominus b) \oplus b$$

As before, this is simply the erosion of f by b, followed by the dilation of the result by b. Similarly, the closing of f by b, denoted $f \bullet b$, is dilation followed by erosion:

$$f \bullet b = (f \oplus b) \ominus b$$

Both operations have simple geometric interpretations. Suppose that an image function $f(x, y)$ is viewed as a 3-D surface; that is, its intensity values are interpreted as height values over the xy-plane. Then the opening of f by b can be interpreted geometrically as pushing structuring element b up against the underside of the surface and translating it across the entire domain of f. The opening is constructed by finding the highest points reached by any part of the structuring element as it slides against the undersurface of f.

Figure 9.24 illustrates the concept in one dimension. Consider the curve in Fig. 9.24(a) to be the values along a single row of an image. Figure 9.24(b) shows a flat structuring element in several positions, pushed up against the bottom of the curve. The complete opening is shown as the curve along the top of the shaded region in Fig. 9.24(c). Since the structuring element is too large to fit inside the upward peak on the middle of the curve, that peak is removed by the opening. In general, openings are used to remove small bright details while leaving the overall gray levels and larger bright features relatively undisturbed.

Figure 9.24(d) provides a graphical illustration of closing. Note that the structuring element is pushed down on top of the curve while being translated

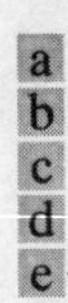

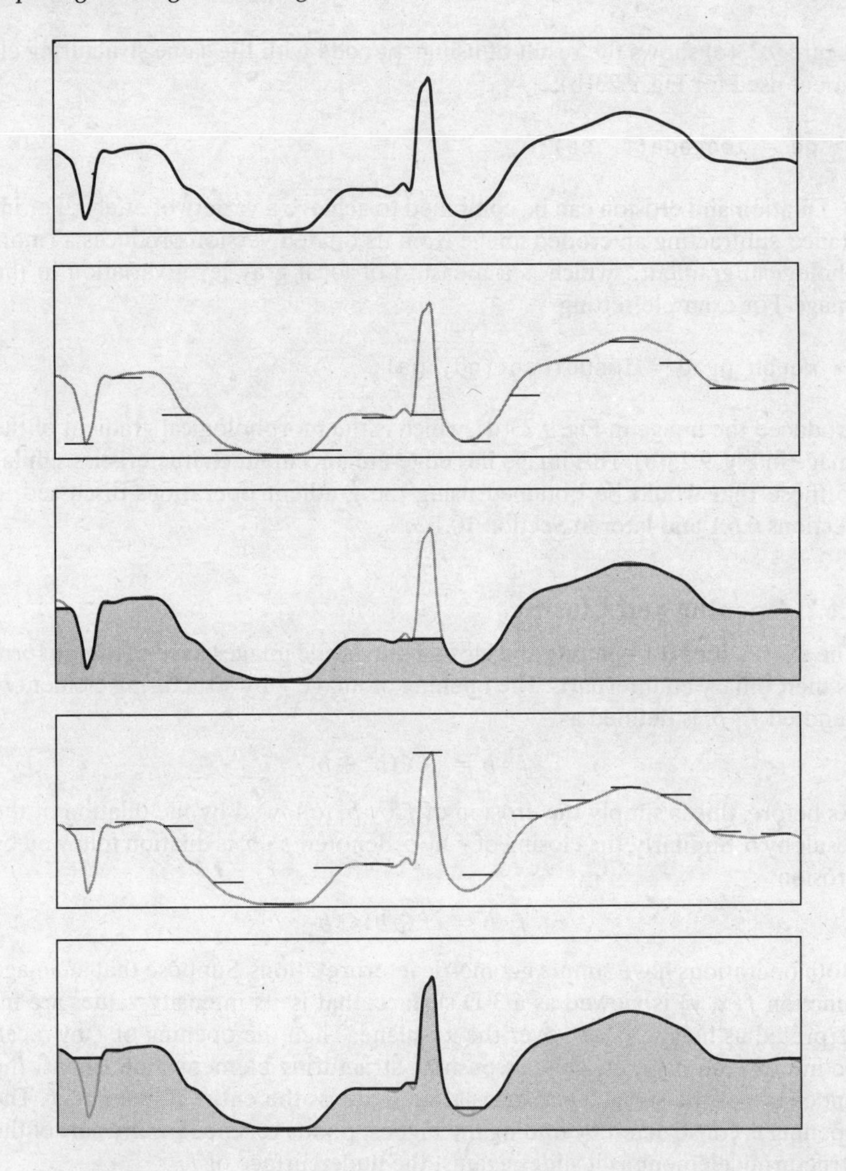

FIGURE 9.24 Opening and closing in one dimension. (a) Original 1-D signal. (b) Flat structuring element pushed up underneath the signal. (c) Opening. (d) Flat structuring element pushed down along the top of the signal. (e) Closing.

to all locations. The closing, shown in Fig. 9.24(e), is constructed by finding the lowest points reached by any part of the structuring element as it slides against the upper side of the curve. Here, we see that closing suppresses dark details smaller than the structuring element.

EXAMPLE 9.9: Morphological smoothing using openings and closings.

■ Because opening suppresses bright details smaller than the structuring element, and closing suppresses dark details smaller than the structuring element, they are used often in combination for image smoothing and noise removal. In this example we use `imopen` and `imclose` to smooth the image of wood dowel plugs shown in Fig. 9.25(a):

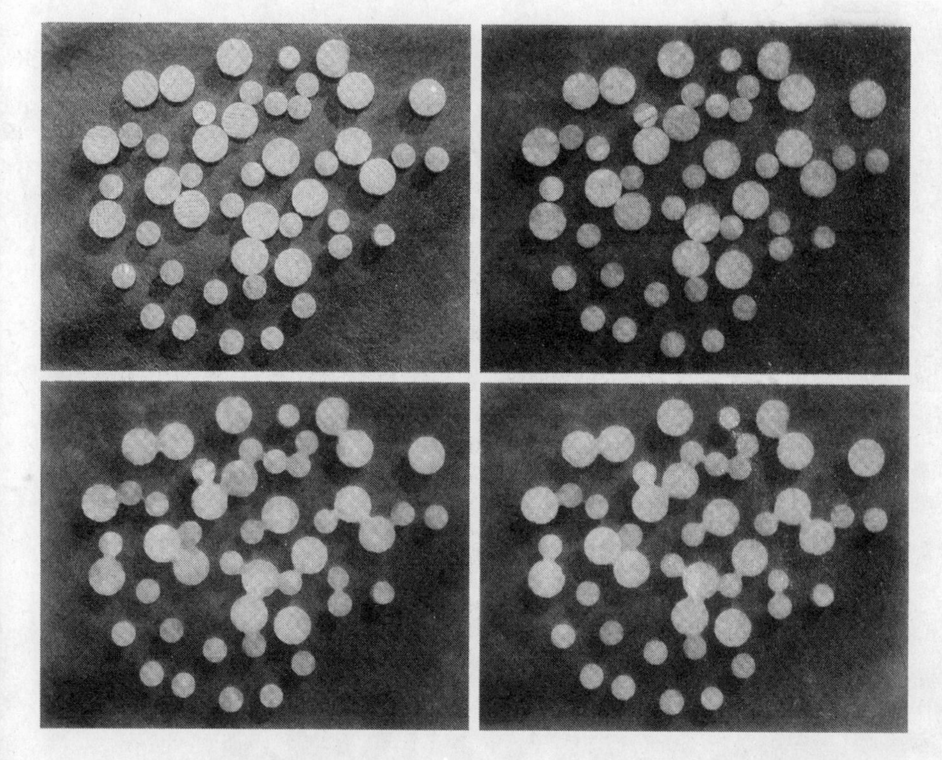

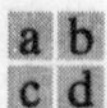

FIGURE 9.25 Smoothing using openings and closings. (a) Original image of wood dowel plugs. (b) Image opened using a disk of radius 5. (c) Closing of the opening. (d) Alternating sequential filter result.

```
>> f = imread('plugs.jpg');
>> se = strel('disk', 5);
>> fo = imopen(f, se);
>> foc = imclose(fo, se);
```

Figure 9.25(b) shows the opened image, `fo`, and Fig. 9.25(c) shows the closing of the opening, `foc`. Note the smoothing of the background and of the details in the objects. This procedure is often called *open-close filtering. Close-open filtering* produces similar results.

Another way to use openings and closings in combination is in *alternating sequential filtering*. One form of alternating sequential filtering is to perform open-close filtering with a series of structuring elements of increasing size. The following commands illustrate this process, which begins with a small structuring element and increases its size until it is the same as the structuring element used to obtain Figs. 9.25(b) and (c):

```
>> fasf = f;
>> for k = 2:5
      se = strel('disk', k);
      fasf = imclose(imopen(fasf, se), se);
   end
```

The result, shown in Fig. 9.25(d), yielded slightly smoother results than using a single open-close filter, at the expense of additional processing. ■

a b c
d e

FIGURE 9.26 Top-hat transformation. (a) Original image. (b) Thresholded image. (c) Opened image. (d) Top-hat transformation. (e) Thresholded top-hat image. (Original image courtesy of The MathWorks, Inc.)

EXAMPLE 9.10: Using the tophat transformation.

■ Openings can be used to compensate for nonuniform background illumination. Figure 9.26(a) shows an image, `f`, of rice grains in which the background is darker towards the bottom than in the upper portion of the image. The uneven illumination makes image thresholding (Section 10.3) difficult. Figure 9.26(b), for example, is a thresholded version in which grains at the top of the image are well separated from the background, but grains at the bottom are improperly extracted from the background. Opening the image can produce a reasonable estimate of the background across the image, as long as the structuring element is large enough so that it does not fit entirely within the rice grains. For example, the commands

```
>> se = strel('disk', 10);
>> fo = imopen(f, se);
```

resulted in the opened image in Fig. 9.26(c). By subtracting this image from the original image, we can produce an image of the grains with a reasonably even background:

```
>> f2 = imsubtract(f, fo);
```

Figure 9.26(d) shows the result, and Fig. 9.26(e) shows the new thresholded image. The improvement is apparent.

Subtracting an opened image from the original is called a *top-hat* transformation. IPT function `imtophat` performs this operation in a single step:

```
>> f2 = imtophat(f, se);
```

imtophat

Function `imtophat` can also be called as `g = imtophat(f, NHOOD)`, where `NHOOD` is an array of 0s and 1s that specifies the size and shape of the structuring element. This syntax is the same as using the call `imtophat(f, strel(NHOOD))`.

A related function, `imbothat`, performs a *bottom-hat* transformation, defined as the closing of the image minus the image. Its syntax is the same as for function `imtophat`. These two functions can be used together for contrast enhancement using commands such as

imbothat

```
>> se = strel('disk', 3);
>> g = imsubtract(imadd(f, imtophat(f, se)), imbothat(f , se)); ■
```

EXAMPLE 9.11: Granulometry.

■ Techniques for determining the size distribution of particles in an image are an important part of the field of *granulometry*. Morphological techniques can be used to measure particle size distribution indirectly; that is, without identifying explicitly and measuring every particle. For particles with regular shapes that are lighter than the background, the basic approach is to apply morphological openings of increasing size. For each opening, the sum of all the pixel values in the opening is computed; this sum sometimes is called the *surface area* of the image. The following commands apply disk-shaped openings with radii 0 to 35 to the image in Fig. 9.25(a):

```
>> f = imread('plugs.jpg');
>> sumpixels = zeros(1, 36);
>> for k = 0:35
      se = strel('disk', k);
      fo = imopen(f, se);
      sumpixels(k + 1) = sum(fo(:));
   end
>> plot(0:35, sumpixels), xlabel('k'), ylabel('Surface area')
```

If `v` *is a vector, then* `diff(v)` *returns a vector, one element shorter than* `v`, *of differences between adjacent elements. If* `X` *is a matrix, then* `diff(X)` *returns a matrix of row differences:* `X(2:end, :) - X(1:end-1, :)`.

Figure 9.27(a) shows the resulting plot of `sumpixels` versus `k`. More interesting is the reduction in surface area between successive openings:

```
>> plot(-diff(sumpixels))
>> xlabel('k')
>> ylabel('Surface area reduction')
```

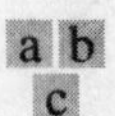

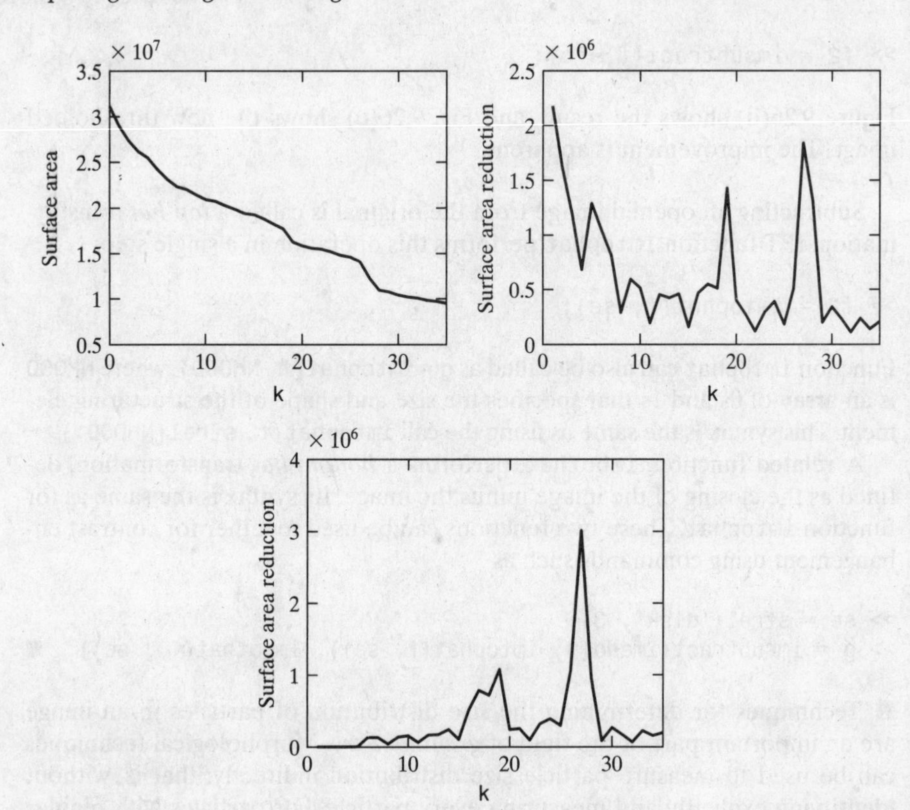

FIGURE 9.27 Granulometry. (a) Surface area versus structuring element radius. (b) Reduction in surface area versus radius. (c) Reduction in surface area versus radius for a smoothed image.

Peaks in the plot in Fig. 9.27(b) indicate the presence of a large number of objects having that radius. Since the plot is quite noisy, we repeat this procedure with the smoothed version of the plugs image in Fig. 9.25(d). The result, shown in Fig. 9.27(c), more clearly indicates the two different sizes of objects in the original image. ■

9.6.3 Reconstruction

Gray-scale morphological reconstruction is defined by the same iterative procedure given in Section 9.5. Figure 9.28 shows how reconstruction works in one dimension. The top curve of Fig. 9.28(a) is the mask while the bottom, gray curve is the marker. In this case the marker is formed by subtracting a constant from the mask, but in general any signal can be used for the marker as long as none of its values exceed the corresponding value in the mask. Each iteration of the reconstruction procedure spreads the peaks in the marker curve until they are forced downward by the mask curve [Fig. 9.28(b)].

The final reconstruction is the black curve in Fig. 9.28(c). Notice that the two smaller peaks were eliminated in the reconstruction, but the two taller peaks, although they are now shorter, remain. When a marker image is formed by subtracting a constant h from the mask image, the reconstruction is called the *h-minima transform*. The h-minima transform, computed by IPT function `imhmin`, is used to suppress small peaks.

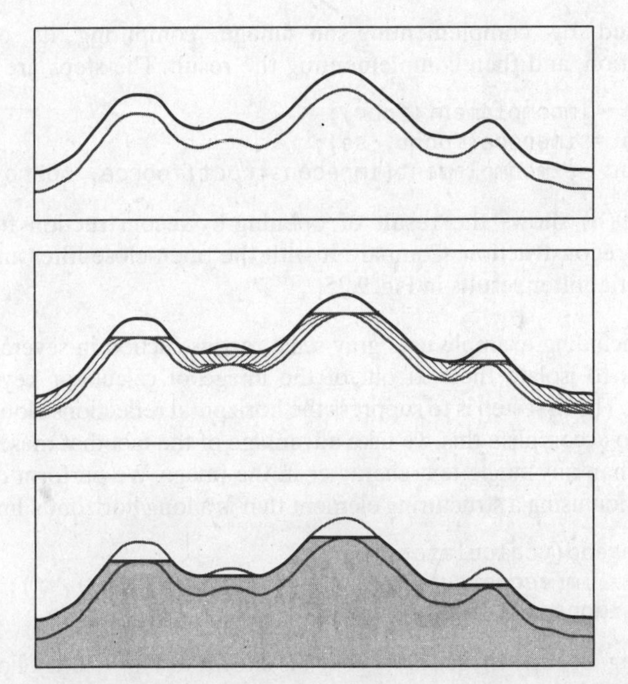

FIGURE 9.28 Gray-scale morphological reconstruction in one dimension. (a) Mask (top) and marker curves. (b) Iterative computation of the reconstruction. (c) Reconstruction result (black curve).

Another useful gray-scale reconstruction technique is *opening-by-reconstruction*, in which an image is first eroded, just as in standard morphological opening. However, instead of following the opening by a closing, the eroded image is used as the marker image in a reconstruction. The original image is used as the mask. Figure 9.29(a) shows an example of opening-by-reconstruction, obtained using the commands

```
>> f = imread('plugs.jpg');
>> se = strel('disk', 5);
>> fe = imerode(f, se);
>> fobr = imreconstruct(fe, f);
```

Reconstruction can be used to clean up image `fobr` further by applying to it a technique called *closing-by-reconstruction*. Closing-by-reconstruction is

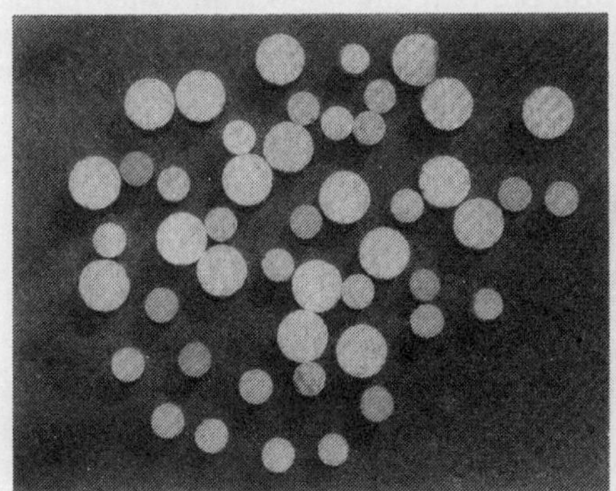

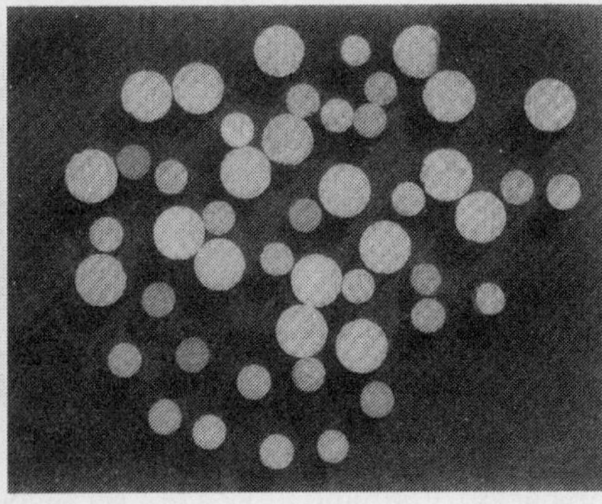

FIGURE 9.29 (a) Opening-by-reconstruction. (b) Opening-by-reconstruction followed by closing-by-reconstruction.

implemented by complementing an image, computing its opening-by-reconstruction, and then complementing the result. The steps are as follows:

```
>> fobrc = imcomplement(fobr);
>> fobrce = imerode(fobrc, se);
>> fobrcbr = imcomplement(imreconstruct(fobrce, fobrc));
```

Figure 9.29(b) shows the result of opening-by-reconstruction followed by closing-by-reconstruction. Compare it with the open-close filter and alternating sequential filter results in Fig. 9.25.

EXAMPLE 9.12: Using reconstruction to remove a complex image background.

■ Our concluding example uses gray-scale reconstruction in several steps. The objective is to isolate the text out of the image of calculator keys shown in Fig. 9.30(a). The first step is to suppress the horizontal reflections along the top of each key. To accomplish this, we take advantage of the fact that these reflections are wider than any single text character in the image. We perform opening-by-reconstruction using a structuring element that is a long horizontal line:

```
>> f = imread('calculator.jpg');
>> f_obr = imreconstruct(imerode(f, ones(1, 71)), f);
>> f_o = imopen(f, ones(1, 71));  % For comparison.
```

The opening-by-reconstruction (`f_obr`) is shown in Fig. 9.30(b). For comparison, Fig. 9.30(c) shows the standard opening (`f_o`). Opening-by-reconstruction did a better job of extracting the background between horizontally adjacent keys. Subtracting the opening-by-reconstruction from the original image is called *tophat-by-reconstruction*, and is shown in Fig. 9.30(d):

```
>> f_thr = imsubtract(f, f_obr);
>> f_th = imsubtract(f, f_o);  % Or imtophat(f, ones(1, 71))
```

Figure 9.30(e) shows the standard top-hat computation (i.e., `f_th`).

Next, we suppress the vertical reflections on the right edges of the keys in Fig. 9.30(d). This is done by performing opening-by-reconstruction with a small horizontal line:

```
>> g_obr = imreconstruct(imerode(f_thr, ones(1, 11)), f_thr);
```

In the result [Fig. 9.30(f)], the vertical reflections are gone, but so are thin-vertical-stroke characters, such as the slash on the percent symbol and the "I" in ASIN. We take advantage of the fact that the characters that have been suppressed in error are very close to other characters still present by first performing a dilation [Fig. 9.30(g)],

```
>> g_obrd = imdilate(g_obr, ones(1, 21));
```

followed by a final reconstruction with `f_thr` as the mask and `min(g_obrd, f_thr)` as the marker:

```
>> f2 = imreconstruct(min(g_obrd, f_thr), f_thr);
```

Figure 9.30(h) shows the final result. Note that the shading and reflections on the background and keys were removed successfully. ■

FIGURE 9.30 An application of gray-scale reconstruction. (a) Original image. (b) Opening-by-reconstruction. (c) Opening. (d) Tophat-by-reconstruction. (e) Tophat. (f) Opening-by-reconstruction of (d) using a horizontal line. (g) Dilation of (f) using a horizontal line. (h) Final reconstruction result.

Summary

The morphological concepts and techniques introduced in this chapter constitute a powerful set of tools for extracting features from an image. The basic operators of erosion, dilation, and reconstruction—defined for both binary and gray-scale image processing—can be used in combination to perform a wide variety of tasks. As shown in the following chapter, morphological techniques can be used for image segmentation. Moreover, they play a major role in algorithms for image description, as discussed in Chapter 11.

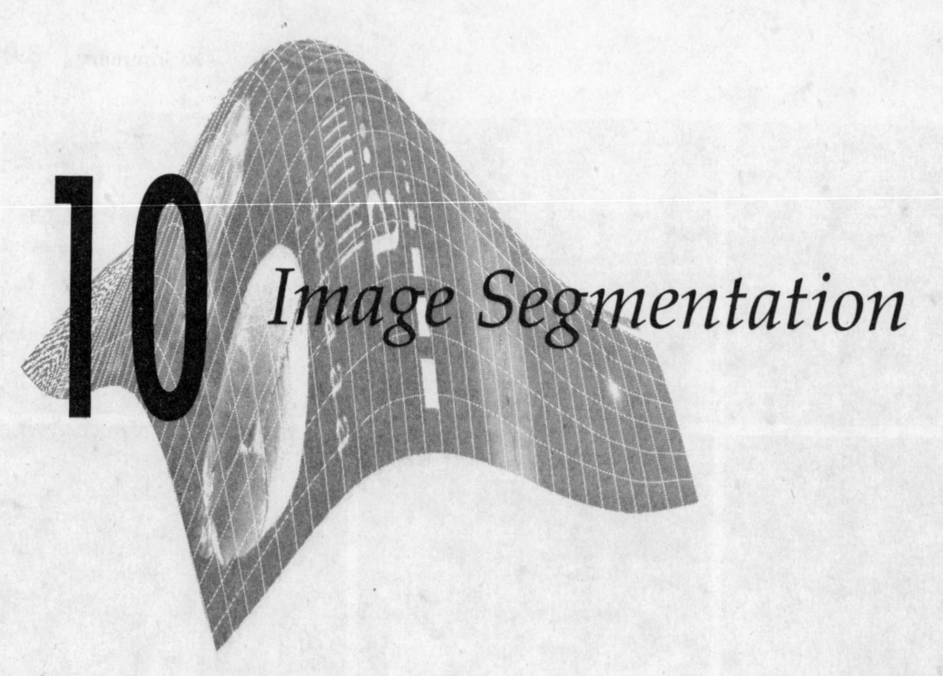

10 Image Segmentation

Preview

The material in the previous chapter began a transition from image processing methods whose inputs and outputs are images to methods in which the inputs are images, but the outputs are attributes extracted from those images. Segmentation is another major step in that direction.

Segmentation subdivides an image into its constituent regions or objects. The level to which the subdivision is carried depends on the problem being solved. That is, segmentation should stop when the objects of interest in an application have been isolated. For example, in the automated inspection of electronic assemblies, interest lies in analyzing images of the products with the objective of determining the presence or absence of specific anomalies, such as missing components or broken connection paths. There is no point in carrying segmentation past the level of detail required to identify those elements.

Segmentation of nontrivial images is one of the most difficult tasks in image processing. Segmentation accuracy determines the eventual success or failure of computerized analysis procedures. For this reason, considerable care should be taken to improve the probability of rugged segmentation. In some situations, such as industrial inspection applications, at least some measure of control over the environment is possible at times. In others, as in remote sensing, user control over image acquisition is limited principally to the choice of imaging sensors.

Segmentation algorithms for monochrome images generally are based on one of two basic properties of image intensity values: discontinuity and similarity. In the first category, the approach is to partition an image based on abrupt changes in intensity, such as edges in an image. The principal approaches in the second category are based on partitioning an image into regions that are similar according to a set of predefined criteria.

In this chapter we discuss a number of approaches in the two categories just mentioned as they apply to monochrome images (edge detection and segmentation of color images are discussed in Section 6.6). We begin the development with methods suitable for detecting intensity discontinuities such as points, lines, and edges. Edge detection in particular has been a staple of segmentation algorithms for many years. In addition to edge detection per se, we also discuss detecting linear edge segments using methods based on the *Hough transform.* The discussion of edge detection is followed by the introduction to thresholding techniques. Thresholding also is a fundamental approach to segmentation that enjoys a significant degree of popularity, especially in applications where speed is an important factor. The discussion on thresholding is followed by the development of region-oriented segmentation approaches. We conclude the chapter with a discussion of a morphological approach to segmentation called *watershed segmentation.* This approach is particularly attractive because it produces closed, well-defined regions, behaves in a global fashion, and provides a framework in which a priori knowledge about the images in a particular application can be utilized to improve segmentation results.

10.1 Point, Line, and Edge Detection

In this section we discuss techniques for detecting the three basic types of intensity discontinuities in a digital image: points, lines, and edges. The most common way to look for discontinuities is to run a mask through the image in the manner described in Sections 3.4 and 3.5. For a 3 × 3 mask this procedure involves computing the sum of products of the coefficients with the intensity levels contained in the region encompassed by the mask. That is, the response, R, of the mask at any point in the image is given by

$$R = w_1 z_1 + w_2 z_2 + \cdots + w_9 z_9 = \sum_{i=1}^{9} w_i z_i$$

where z_i is the intensity of the pixel associated with mask coefficient w_i. As before, the response of the mask is defined with respect to its center.

10.1.1 Point Detection

The detection of isolated points embedded in areas of constant or nearly constant intensity in an image is straightforward in principle. Using the mask shown in Fig. 10.1, we say that an isolated point has been detected at the location on which the mask is centered if

$$|R| \geq T$$

−1	−1	−1
−1	8	−1
−1	−1	−1

FIGURE 10.1 A mask for point detection.

where T is a nonnegative threshold. Point detection is implemented in MATLAB using function `imfilter`, with the mask in Fig. 10.1, or other similar mask. The important requirements are that the strongest response of a mask must be when the mask is centered on an isolated point, and that the response be 0 in areas of constant intensity.

If `T` is given, the following command implements the point-detection approach just discussed:

```
>> g = abs(imfilter(double(f), w)) >= T;
```

where `f` is the input image, `w` is an appropriate point-detection mask [e.g., the mask in Fig. 10.1], and `g` is the resulting image. Recall from the discussion in Section 3.4.1 that `imfilter` converts its output to the class of the input, so we use `double(f)` in the filtering operation to prevent premature truncation of values if the input is of class `uint8`, and because the `abs` operation does not accept integer data. The output image `g` is of class `logical`; its values are 0 and 1. If `T` is not given, its value often is chosen based on the filtered result, in which case the previous command string is broken down into three basic steps: (1) Compute the filtered image, `abs(imfilter(double(f), w))`, (2) find the value for `T` using the data from the filtered image, and (3) compare the filtered image against `T`. This approach is illustrated in the following example.

EXAMPLE 10.1: Point detection.

■ Figure 10.2(a) shows an image with a nearly invisible black point in the dark gray area of the northeast quadrant. Letting `f` denote this image, we find the location of the point as follows:

```
>> w = [-1 -1 -1; -1 8 -1; -1 -1 -1];
>> g = abs(imfilter(double(f), w));
>> T = max(g(:));
>> g = g >= T;
>> imshow(g)
```

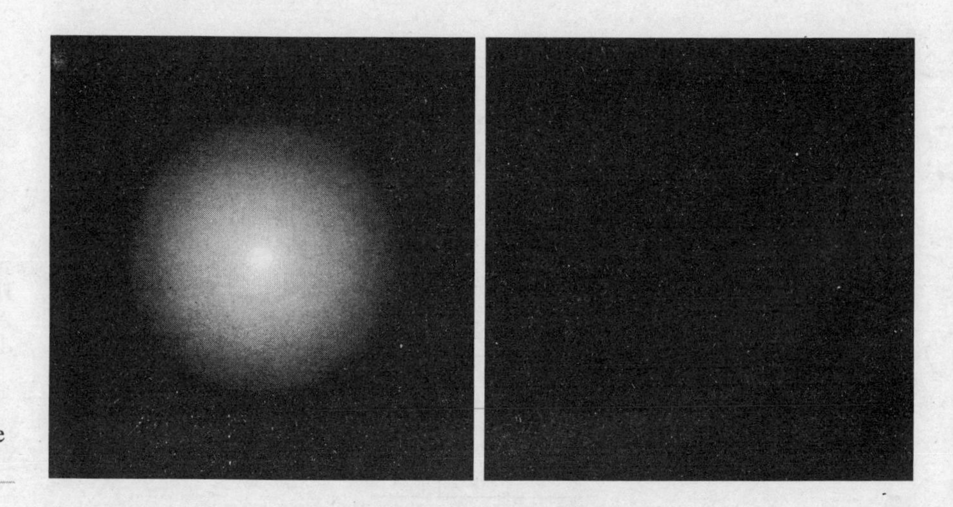

a b

FIGURE 10.2 (a) Gray-scale image with a nearly invisible isolated black point in the dark gray area of the northeast quadrant. (b) Image showing the detected point. (The point was enlarged to make it easier to see.)

By selecting T to be the maximum value in the filtered image, g, and then finding all points in g such that g >= T, we identify the points that give the largest response. The assumption is that all these points are isolated points embedded in a constant or nearly constant background. Note that the test against T was conducted using the >= operator for consistency in notation. Since T was selected in this case to be the maximum value in g, clearly there can be no points in g with values greater than T. As Fig. 10.2(b) shows, there was a single isolated point that satisfied the condition g >= T with T set to max(g(:)). ■

Another approach to point detection is to find the points in all neighborhoods of size $m \times n$ for which the difference of the maximum and minimum pixels values exceeds a specified value of T. This approach can be implemented using function ordfilt2 introduced in Section 3.5.2:

```
>> g = imsubtract(ordfilt2(f, m*n, ones(m, n)), ...
                                  ordfilt2(f, 1, ones(m, n)));
>> g = g >= T;
```

It is easily verified that choosing T = max(g(:)) yields the same result as in Fig. 10.2(b). The preceding formulation is more flexible than using the mask in Fig. 10.1. For example, if we wanted to compute the difference between the highest and the next highest pixel value in a neighborhood, we would replace the 1 on the far right of the preceding expression by m*n – 1. Other variations of this basic theme are formulated in a similar manner.

10.1.2 Line Detection

The next level of complexity is line detection. Consider the masks in Fig. 10.3. If the first mask were moved around an image, it would respond more strongly to lines (one pixel thick) oriented horizontally. With a constant background, the maximum response would result when the line passed through the middle row of the mask. Similarly, the second mask in Fig. 10.3 responds best to lines oriented at +45°; the third mask to vertical lines; and the fourth mask to lines in the −45° direction. Note that the preferred direction of each mask is weighted with a larger coefficient (i.e., 2) than other possible directions. The coefficients of each mask sum to zero, indicating a zero response from the mask in areas of constant intensity.

Horizontal:

−1	−1	−1
2	2	2
−1	−1	−1

+45°:

−1	−1	2
−1	2	−1
2	−1	−1

Vertical:

−1	2	−1
−1	2	−1
−1	2	−1

−45°:

2	−1	−1
−1	2	−1
−1	−1	2

FIGURE 10.3 Line detector masks.

Let R_1, R_2, R_3, and R_4 denote the responses of the masks in Fig. 10.3, from left to right, where the R's are given by the equation in the previous section. Suppose that the four masks are run individually through an image. If, at a certain point in the image, $|R_i| > |R_j|$, for all $j \neq i$, that point is said to be more likely associated with a line in the direction of mask i. For example, if at a point in the image, $|R_1| > |R_j|$ for $j = 2, 3, 4$, that particular point is said to be more likely associated with a horizontal line. Alternatively, we may be interested in detecting lines in a specified direction. In this case, we would use the mask associated with that direction and threshold its output, as in the equation in the previous section. In other words, if we are interested in detecting all the lines in an image in the direction defined by a given mask, we simply run the mask through the image and threshold the absolute value of the result. The points that are left are the strongest responses, which, for lines one pixel thick, correspond closest to the direction defined by the mask. The following example illustrates this procedure.

EXAMPLE 10.2: Detection of lines in a specified direction.

■ Figure 10.4(a) shows a digitized (binary) portion of a wire-bond mask for an electronic circuit. The image size is 486 × 486 pixels. Suppose that we are interested in finding all the lines that are one pixel thick, oriented at −45°. For this purpose, we use the last mask in Fig. 10.3. Figures 10.4(b) through (f) were generated using the following commands, where `f` is the image in Fig. 10.4(a):

```
>> w = [2 -1 -1 ; -1 2 -1; -1 -1 2];
>> g = imfilter(double(f), w);
>> imshow(g, [ ])  % Fig. 10.4(b)
>> gtop = g(1:120, 1:120);
>> gtop = pixeldup(gtop, 4);
>> figure, imshow(gtop, [ ]) % Fig. 10.4(c)
>> gbot = g(end-119:end, end-119:end);
>> gbot = pixeldup(gbot, 4);
>> figure, imshow(gbot, [ ]) % Fig. 10.4(d)
>> g = abs(g);
>> figure, imshow(g, [ ]) % Fig. 10.4(e)
>> T = max(g(:));
>> g = g >= T;
>> figure, imshow(g)  % Fig. 10.4(f)
```

The shades darker than the gray background in Fig. 10.4(b) correspond to negative values. There are two main segments oriented in the −45° direction, one at the top, left and one at the bottom, right [Figs. 10.4(c) and (d) show zoomed sections of these two areas]. Note how much brighter the straight line segment in Fig. 10.4(d) is than the segment in Fig. 10.4(c). The reason is that the component in the bottom, right of Fig. 10.4(a) is one pixel thick, while the one at the top, left is not. The mask response is stronger for the one-pixel-thick component.

Figure 10.4(e) shows the absolute value of Fig. 10.4(b). Since we are interested in the strongest response, we let `T` equal the maximum value in this image. Figure 10.4(f) shows in white the points whose values satisfied the

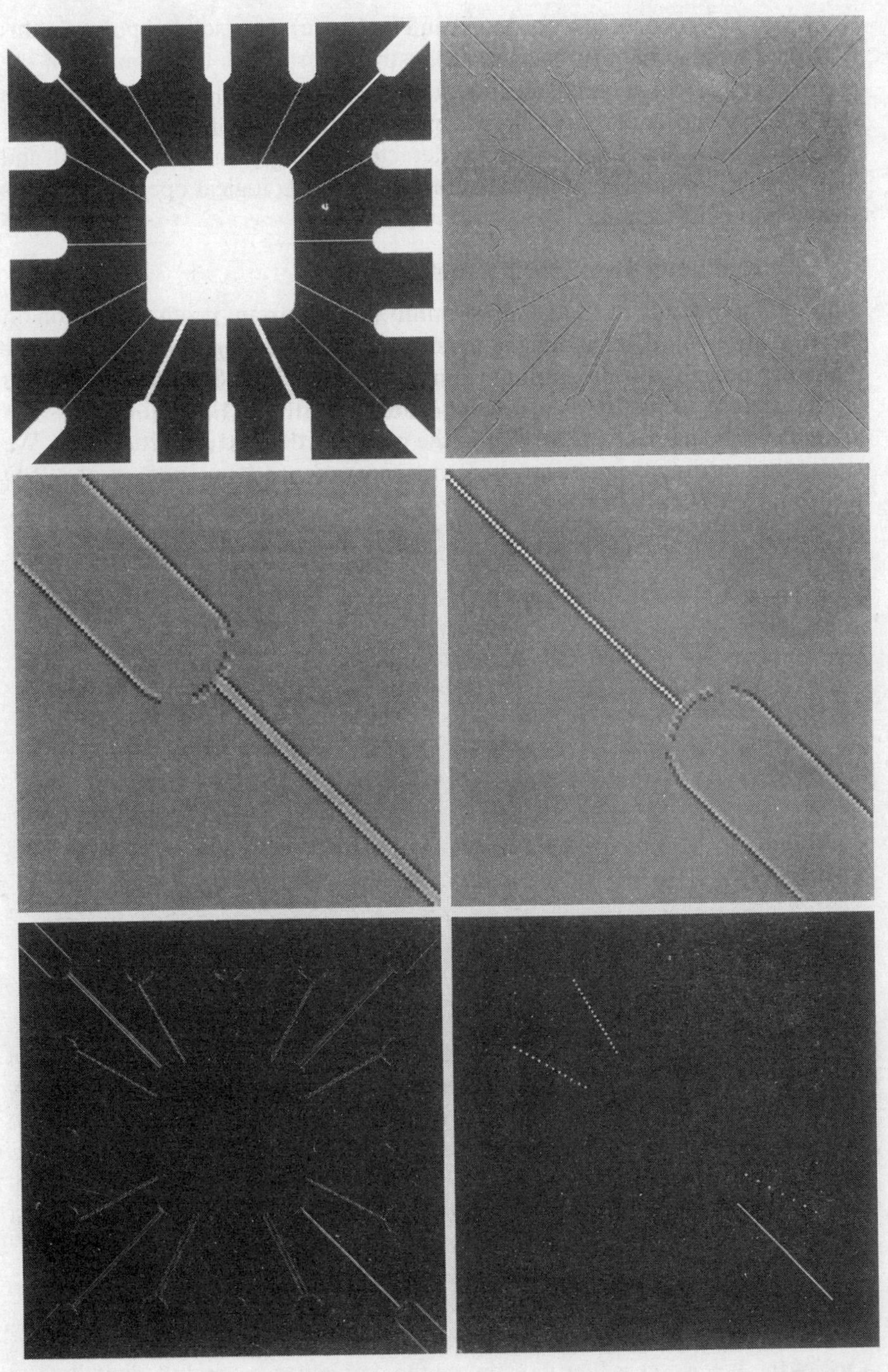

a b
c d
e f

FIGURE 10.4 (a) Image of a wire-bond mask. (b) Result of processing with the $-45°$ detector in Fig. 10.3. (c) Zoomed view of the top, left region of (b). (d) Zoomed view of the bottom, right section of (b). (e) Absolute value of (b). (f) All points (in white) whose values satisfied the condition `g >= T`, where `g` is the image in (e). (The points in (f) were enlarged slightly to make them easier to see.)

condition `g >= T`, where `g` is the image in Fig. 10.4(e). The isolated points in this figure are points that also had similarly strong responses to the mask. In the original image, these points and their immediate neighbors are oriented in such a way that the mask produced a maximum response at those isolated locations. These isolated points can be detected using the mask in Fig. 10.1 and then deleted, or they could be deleted using morphological operators, as discussed in the last chapter. ■

10.1.3 Edge Detection Using Function edge

Although point and line detection certainly are important in any discussion on image segmentation, edge detection is by far the most common approach for detecting meaningful discontinuities in intensity values. Such discontinuities are detected by using first- and second-order derivatives. The first-order derivative of choice in image processing is the gradient, defined in Section 6.6.1. We repeat the pertinent equations here for convenience. The *gradient* of a 2-D function, $f(x, y)$, is defined as the *vector*

$$\nabla \mathbf{f} = \begin{bmatrix} G_x \\ G_y \end{bmatrix} = \begin{bmatrix} \dfrac{\partial f}{\partial x} \\ \dfrac{\partial f}{\partial y} \end{bmatrix}$$

The magnitude of this vector is

$$\begin{aligned} \nabla f &= \text{mag}(\nabla \mathbf{f}) = [G_x^2 + G_y^2]^{1/2} \\ &= [(\partial f/\partial x)^2 + (\partial f/\partial y)^2]^{1/2} \end{aligned}$$

To simplify computation, this quantity is approximated sometimes by omitting the square-root operation,

$$\nabla f \approx G_x^2 + G_y^2$$

or by using absolute values,

$$\nabla f \approx |G_x| + |G_y|$$

These approximations still behave as derivatives; that is, they are zero in areas of constant intensity and their values are proportional to the degree of intensity change in areas whose pixel values are variable. It is common practice to refer to the magnitude of the gradient or its approximations simply as "the gradient."

A fundamental property of the gradient vector is that it points in the direction of the maximum rate of change of f at coordinates (x, y). The angle at which this maximum rate of change occurs is

$$\alpha(x, y) = \tan^{-1}\left(\frac{G_y}{G_x}\right)$$

One of the key issues is how to estimate the derivatives G_x and G_y digitally. The various approaches used by function `edge` are discussed later in this section.

Second-order derivatives in image processing are generally computed using the Laplacian introduced in Section 3.5.1. That is, the Laplacian of a 2-D function $f(x, y)$ is formed from second-order derivatives, as follows:

$$\nabla^2 f(x, y) = \frac{\partial^2 f(x, y)}{\partial x^2} + \frac{\partial^2 f(x, y)}{\partial y^2}$$

The Laplacian is seldom used by itself for edge detection because, as a second-order derivative, it is unacceptably sensitive to noise, its magnitude produces double edges, and it is unable to detect edge direction. However, as discussed later in this section, the Laplacian can be a powerful complement when used in combination with other edge-detection techniques. For example, although its double edges make it unsuitably for edge detection directly, this property can be used for edge *location*.

With the preceding discussion as background, the basic idea behind edge detection is to find places in an image where the intensity changes rapidly, using one of two general criteria:

1. Find places where the first derivative of the intensity is greater in magnitude than a specified threshold.
2. Find places where the second derivative of the intensity has a zero crossing.

IPT's function `edge` provides several derivative estimators based on the criteria just discussed. For some of these estimators, it is possible to specify whether the edge detector is sensitive to horizontal or vertical edges or to both. The general syntax for this function is

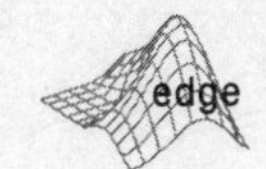


```
[g, t] = edge(f, 'method', parameters)
```

where `f` is the input image, `method` is one of the approaches listed in Table 10.1, and `parameters` are additional parameters explained in the following discussion. In the output, `g` is a logical array with 1s at the locations where edge points were detected in `f` and 0s elsewhere. Parameter `t` is optional; it gives the threshold used by `edge` to determine which gradient values are strong enough to be called edge points.

Sobel Edge Detector

The *Sobel* edge detector uses the masks in Fig. 10.5(b) to approximate digitally the first derivatives G_x and G_y. In other words, the gradient at the center point in a neighborhood is computed as follows by the Sobel detector:

$$\begin{aligned} g &= [G_x^2 + G_y^2]^{1/2} \\ &= \{[(z_7 + 2z_8 + z_9) - (z_1 + 2z_2 + z_3)]^2 \\ &\quad + [(z_3 + 2z_6 + z_9) - (z_1 + 2z_4 + z_7)]^2\}^{1/2} \end{aligned}$$

TABLE 10.1
Edge detectors available in function edge.

Edge Detector	Basic Properties
Sobel	Finds edges using the Sobel approximation to the derivatives shown in Fig. 10.5(b).
Prewitt	Finds edges using the Prewitt approximation to the derivatives shown in Fig. 10.5(c).
Roberts	Finds edges using the Roberts approximation to the derivatives shown in Fig. 10.5(d).
Laplacian of a Gaussian (LoG)	Finds edges by looking for zero crossings after filtering $f(x, y)$ with a Gaussian filter.
Zero crossings	Finds edges by looking for zero crossings after filtering $f(x, y)$ with a user-specified filter.
Canny	Finds edges by looking for local maxima of the gradient of $f(x, y)$. The gradient is calculated using the derivative of a Gaussian filter. The method uses two thresholds to detect strong and weak edges, and includes the weak edges in the output only if they are connected to strong edges. Therefore, this method is more likely to detect true weak edges.

Then, we say that a pixel at location (x, y) is an edge pixel if $g \geq T$ at that location, where T is a specified threshold.

From the discussion in Section 3.5.1, we know that Sobel edge detection can be implemented by filtering an image, `f`, (using `imfilter`) with the left mask in Fig. 10.5(b), filtering `f` again with the other mask, squaring the pixels values of each filtered image, adding the two results, and computing their square root. Similar comments apply to the second and third entries in Table 10.1. Function `edge` simply packages the preceding operations into one function call and adds other features, such as accepting a threshold value or determining a threshold automatically. In addition, `edge` contains edge detection techniques that are not implementable directly with `imfilter`.

The general calling syntax for the Sobel detector is

```
[g, t] = edge(f, 'sobel', T, dir)
```

where `f` is the input image, `T` is a specified threshold, and `dir` specifies the preferred direction of the edges detected: `'horizontal'`, `'vertical'`, or `'both'` (the default). As noted earlier, `g` is a `logical` image containing 1s at locations where edges were detected and 0s elsewhere. Parameter `t` in the output is optional. It is the threshold value used by `edge`. If `T` is specified, then `t = T`. Otherwise, if `T` is not specified (or is empty, `[ ]`), `edge` sets `t` equal to a threshold it determines automatically and then uses for edge detection. One of the principal reason for including `t` in the output argument is to get an initial value for the threshold. Function `edge` uses the Sobel detector as a default if the syntax `g = edge(f)`, or `[g, t] = edge(f)`, is used.

z_1	z_2	z_3
z_4	z_5	z_6
z_7	z_8	z_9

Image neighborhood

−1	−2	−1		−1	0	1
0	0	0		−2	0	2
1	2	1		−1	0	1

Sobel

$G_x = (z_7 + 2z_8 + z_9) - (z_1 + 2z_2 + z_3)$ $\quad G_y = (z_3 + 2z_6 + z_9) - (z_1 + 2z_4 + z_7)$

−1	−1	−1		−1	0	1
0	0	0		−1	0	1
1	1	1		−1	0	1

Prewitt

$G_x = (z_7 + z_8 + z_9) - (z_1 + z_2 + z_3)$ $\quad G_y = (z_3 + z_6 + z_9) - (z_1 + z_4 + z_7)$

−1	0		0	−1
0	1		1	0

Roberts

$G_x = z_9 - z_5$ $\quad G_y = z_8 - z_6$

a b c d

FIGURE 10.5 Some edge detector masks and the first-order derivatives they implement.

Prewitt Edge Detector

The Prewitt edge detector uses the masks in Fig. 10.5(c) to approximate digitally the first derivatives G_x and G_y. Its general calling syntax is

```
[g, t] = edge(f, 'prewitt', T, dir)
```

The parameters of this function are identical to the Sobel parameters. The Prewitt detector is slightly simpler to implement computationally than the Sobel detector, but it tends to produce somewhat noisier results. (It can be shown that the coefficient with value 2 in the Sobel detector provides smoothing.)

Roberts Edge Detector

The Roberts edge detector uses the masks in Fig. 10.5(d) to approximate digitally the first derivatives G_x and G_y. Its general calling syntax is

```
[g, t] = edge(f, 'roberts', T, dir)
```

The parameters of this function are identical to the Sobel parameters. The Roberts detector is one of the oldest edge detectors in digital image processing, and as Fig. 10.5(d) shows, it is also the simplest. This detector is used considerably less than the others in Fig. 10.5 due in part to its limited functionality (e.g., it is not symmetric and cannot be generalized to detect edges that are multiples of 45°). However, it still is used frequently in hardware implementations where simplicity and speed are dominant factors.

Laplacian of a Gaussian (LoG) Detector

Consider the Gaussian function

$$h(r) = -e^{-\frac{r^2}{2\sigma^2}}$$

where $r^2 = x^2 + y^2$ and σ is the standard deviation. This is a smoothing function which, if convolved with an image, will blur it. The degree of blurring is determined by the value of σ. The Laplacian of this function (the second derivative with respect to r) is

$$\nabla^2 h(r) = -\left[\frac{r^2 - \sigma^2}{\sigma^4}\right] e^{-\frac{r^2}{2\sigma^2}}$$

For obvious reasons, this function is called the Laplacian of a Gaussian (LoG). Because the second derivative is a linear operation, convolving (filtering) an image with $\nabla^2 h(r)$ is the same as convolving the image with the smoothing function first and then computing the Laplacian of the result. This is the key concept underlying the LoG detector. We convolve the image with $\nabla^2 h(r)$, knowing that it has two effects: It smoothes the image (thus reducing noise), and it computes the Laplacian, which yields a double-edge image. Locating edges then consists of finding the zero crossings between the double edges.

The general calling syntax for the LoG detector is

```
[g, t] = edge(f, 'log', T, sigma)
```

where `sigma` is the standard deviation and the other parameters are as explained previously. The default value for `sigma` is 2. As before, `edge` ignores any edges that are not stronger than `T`. If `T` is not provided, or it is empty, `[ ]`, `edge` chooses the value automatically. Setting `T` to 0 produces edges that are closed contours, a familiar characteristic of the LoG method.

Zero-Crossings Detector

This detector is based on the same concept as the LoG method, but the convolution is carried out using a specified filter function, `H`. The calling syntax is

```
[g, t] = edge(f, 'zerocross', T, H)
```

The other parameters are as explained for the LoG detector.

Canny Edge Detector

The Canny detector (Canny [1986]) is the most powerful edge detector provided by function `edge`. The method can be summarized as follows:

1. The image is smoothed using a Gaussian filter with a specified standard deviation, σ, to reduce noise.
2. The local gradient, $g(x, y) = [G_x^2 + G_y^2]^{1/2}$, and edge direction, $\alpha(x, y) = \tan^{-1}(G_y/G_x)$, are computed at each point. Any of the first three techniques in Table 10.1 can be used to compute G_x and G_y. An edge point is defined to be a point whose strength is locally maximum in the direction of the gradient.
3. The edge points determined in (2) give rise to ridges in the gradient magnitude image. The algorithm then tracks along the top of these ridges and sets to zero all pixels that are not actually on the ridge top so as to give a thin line in the output, a process known as *nonmaximal suppression*. The ridge pixels are then thresholded using two thresholds, $T1$ and $T2$, with $T1 < T2$. Ridge pixels with values greater than $T2$ are said to be "strong" edge pixels. Ridge pixels with values between $T1$ and $T2$ are said to be "weak" edge pixels.
4. Finally, the algorithm performs edge linking by incorporating the weak pixels that are 8-connected to the strong pixels.

The syntax for the Canny edge detector is

```
[g, t] = edge(f, 'canny', T, sigma)
```

where `T` is a vector, `T = [T1, T2]`, containing the two thresholds explained in step 3 of the preceding procedure, and `sigma` is the standard deviation of the smoothing filter. If `t` is included in the output argument, it is a two-element vector containing the two threshold values used by the algorithm. The rest of the syntax is as explained for the other methods, including the automatic computation of thresholds if `T` is not supplied. The default value for `sigma` is 1.

EXAMPLE 10.3: Edge extraction with the Sobel detector.

■ We can extract and display the vertical edges in the image, `f`, of Fig. 10.6(a) using the commands

```
>> [gv, t] = edge(f, 'sobel', 'vertical');
>> imshow(gv)
>> t

t =

   0.0516
```

As Fig. 10.6(b) shows, the predominant edges in the result are vertical (the inclined edges have vertical and horizontal components, so they are detected as well). We can clean up the weaker edges somewhat by specifying a higher threshold value. For example, Fig. 10.6(c) was generated using the command

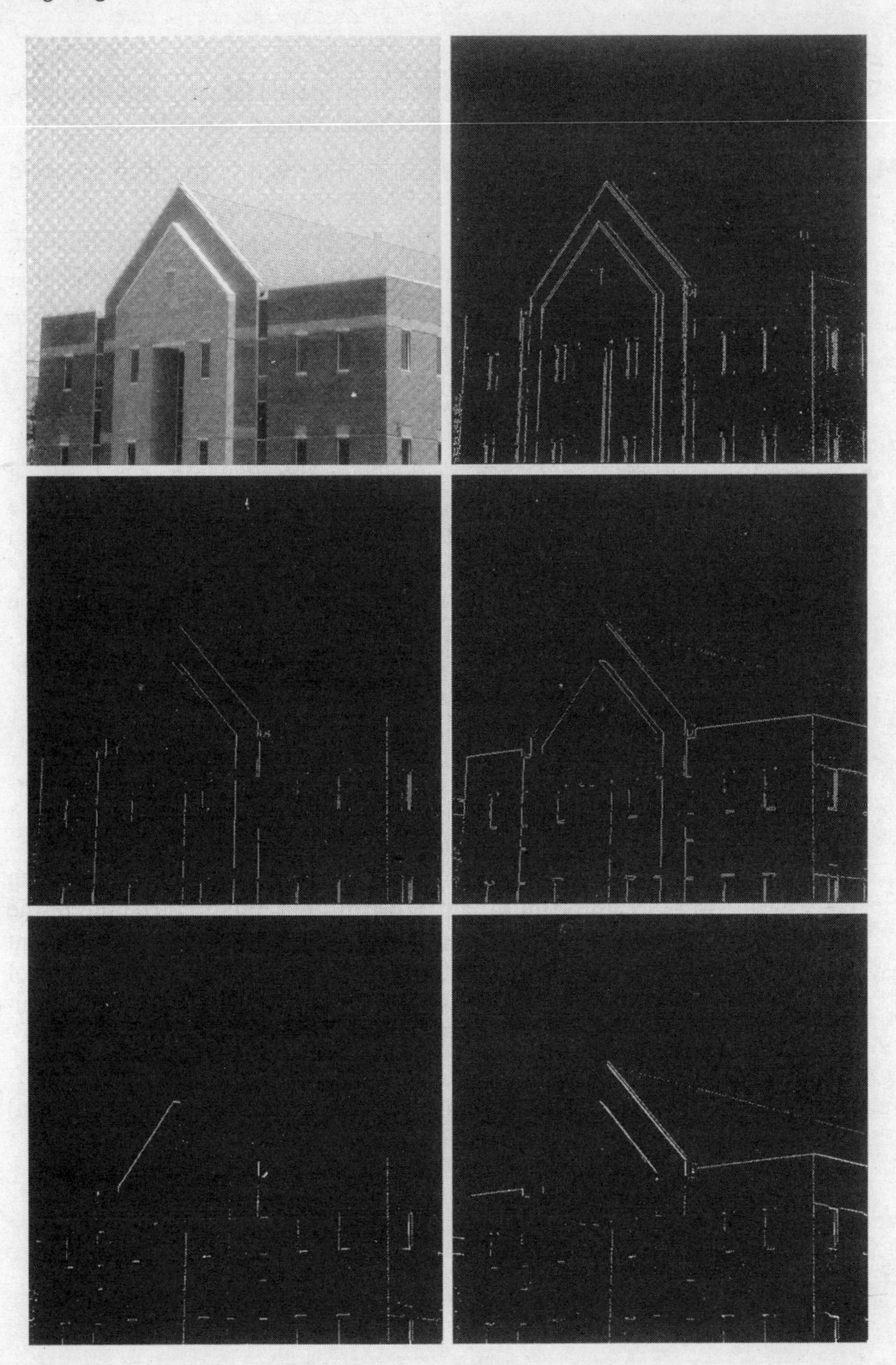

a b
c d
e f

FIGURE 10.6
(a) Original image. (b) Result of function `edge` using a vertical Sobel mask with the threshold determined automatically. (c) Result using a specified threshold. (d) Result of determining both vertical and horizontal edges with a specified threshold. (e) Result of computing edges at 45° with `imfilter` using a specified mask and a specified threshold. (f) Result of computing edges at −45° with `imfilter` using a specified mask and a specified threshold.

```
>> gv = edge(f, 'sobel', 0.15, 'vertical');
```

Using the same value of T in the command

```
>> gboth = edge(f, 'sobel', 0.15);
```

resulted in Fig. 10.6(d), which shows predominantly vertical and horizontal edges.

Function `edge` does not compute Sobel edges at ±45°. To compute such edges we need to specify the mask and use `imfilter`. For example, Fig. 10.6(e) was generated using the commands

```
>> w45 = [-2 -1 0; -1 0 1; 0 1 2]

w45 =

    -2    -1     0
    -1     0     1
     0     1     2

>> g45 = imfilter(double(f), w45, 'replicate');
>> T = 0.3*max(abs(g45(:)));
>> g45 = g45 >= T;
>> figure, imshow(g45);
```

The value of T *was chosen experimentally to show results comparable with Figs. 10(c) and 10(d).*

The strongest edge in Fig. 10.6(e) is the edge oriented at 45°. Similarly, using the mask `wm45 = [0 1 2; -1 0 1; -2 -1 0]` with the same sequence of commands resulted in the strong edges oriented at −45° shown in Fig. 10.6(f).

Using the `'prewitt'` and `'roberts'` options in function `edge` follows the same general procedure just illustrated for the Sobel edge detector. ■

EXAMPLE 10.4: Comparison of the Sobel, LoG, and Canny edge detectors.

■ In this example we compare the relative performance of the Sobel, LoG, and Canny edge detectors. The objective is to produce a clean edge "map" by extracting the principal edge features of the building image, `f`, in Fig. 10.6(a), while reducing "irrelevant" detail, such as the fine texture in the brick walls and tile roof. The principal edges of interest in this discussion are the building corners, the windows, the light-brick structure framing the entrance, the entrance itself, the roofline, and the concrete band surrounding the building about two-thirds of the distance above ground level.

The left column in Fig. 10.7 shows the edge images obtained using the default syntax for the `'sobel'`, `'log'`, and `'canny'` options:

```
>> [g_sobel_default, ts] = edge(f, 'sobel'); % Fig. 10.7(a)
>> [g_log_default, tlog] = edge(f, 'log');   % Fig. 10.7(c)
>> [g_canny_default, tc] = edge(f, 'canny'); % Fig. 10.7(e)
```

The values of the thresholds in the output argument resulting from the preceding computations were `ts = 0.074`, `tlog = 0.0025`, and `tc = [0.019, 0.047]`. The defaults values of `sigma` for the `'log'` and `'canny'` options were 2.0 and 1.0, respectively. With the exception of the Sobel image, the default results were far from the objective of producing clean edge maps.

a b
c d
e f

FIGURE 10.7 Left column: Default results for the Sobel, LoG, and Canny edge detectors. Right column: Results obtained interactively to bring out the principal features in the original image of Fig. 10.6(a) while reducing irrelevant, fine detail. The Canny edge detector produced the best results by far.

Starting with the default values, the parameters in each option were varied interactively with the objective of bringing out the principal features mentioned earlier, while reducing irrelevant detail as much as possible. The results in the right column of Fig. 10.7 were obtained with the following commands:

```
>> g_sobel_best = edge(f, 'sobel', 0.05);                 % Fig. 10.7(b)
>> g_log_best   = edge(f, 'log', 0.003, 2.25);            % Fig. 10.7(d)
>> g_canny_best = edge(f, 'canny', [0.04 0.10], 1.5); % Fig. 10.7(f)
```

As Fig. 10.7(b) shows, the Sobel result actually deviated even further from the objective when we tried to bring out the concrete band and left edge of the entrance way. The LoG result in Fig. 10.7(d) is somewhat better than the Sobel result and much better than the LoG default, but it still could not bring out the left edge of the main entrance nor the concrete band around the building. The Canny result [Fig. 10.7(f)] is superior by far to the other two results. Note in particular how the left edge of the entrance was clearly detected, as were the concrete band and other details such as the complete roof ventilation grill above the main entrance. In addition to detecting the desired features, the Canny detector also produced the cleanest edge map. ■

10.2 Line Detection Using the Hough Transform

Ideally, the methods discussed in the previous section should yield pixels lying only on edges. In practice, the resulting pixels seldom characterize an edge completely because of noise, breaks in the edge from nonuniform illumination, and other effects that introduce spurious intensity discontinuities. Thus edge-detection algorithms typically are followed by linking procedures to assemble edge pixels into meaningful edges. One approach that can be used to find and link line segments in an image is the *Hough transform* (Hough [1962]).

Given a set of points in an image (typically a binary image), suppose that we want to find subsets of these points that lie on straight lines. One possible solution is to first find all lines determined by every pair of points and then find all subsets of points that are close to particular lines. The problem with this procedure is that it involves finding $n(n-1)/2 \sim n^2$ lines and then performing $n(n(n-1))/2 \sim n^3$ comparisons of every point to all lines. This approach is computationally prohibitive in all but the most trivial applications.

With the Hough transform, on the other hand, we consider a point (x_i, y_i) and all the lines that pass through it. Infinitely many lines pass through (x_i, y_i), all of which satisfy the slope-intercept equation $y_i = ax_i + b$ for some values of a and b. Writing this equation as $b = -x_i a + y_i$ and considering the *ab*-plane (also called *parameter space*) yields the equation of a *single* line for a fixed pair (x_i, y_i). Furthermore, a second point (x_j, y_j) also has a line in parameter space associated with it, and this line intersects the line associated with (x_i, y_i) at (a', b'), where a' is the slope and b' the intercept of the line containing both (x_i, y_i) and (x_j, y_j) in the *xy*-plane. In fact, all points contained on this line have lines in parameter space that intersect at (a', b'). Figure 10.8 illustrates these concepts.

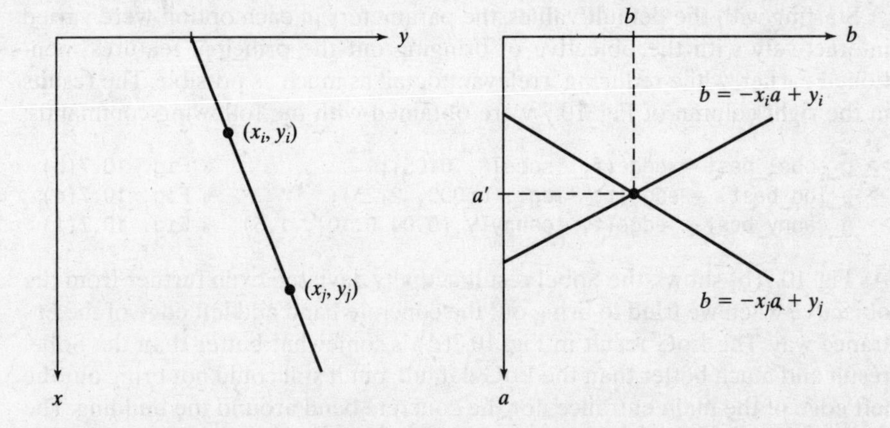

a b

FIGURE 10.8 (a) xy-plane. (b) Parameter space.

In principle, the parameter-space lines corresponding to all image points (x_i, y_i) could be plotted, and then image lines could be identified by where large numbers of parameter-space lines intersect. A practical difficulty with this approach, however, is that a (the slope of the line) approaches infinity as the line approaches the vertical direction. One way around this difficulty is to use the normal representation of a line:

$$x \cos \theta + y \sin \theta = \rho$$

Figure 10.9(a) illustrates the geometric interpretation of the parameters ρ and θ. A horizontal line has $\theta = 0°$, with ρ being equal to the positive x-intercept. Similarly, a vertical line has $\theta = 90°$, with ρ being equal to the positive y-intercept, or $\theta = -90°$, with ρ being equal to the negative y intercept. Each sinusoidal curve in Figure 10.9(b) represents the family of lines that pass through a particular point (x_i, y_i). The intersection point (ρ', θ') corresponds to the line that passes through both (x_i, y_i) and (x_j, y_j).

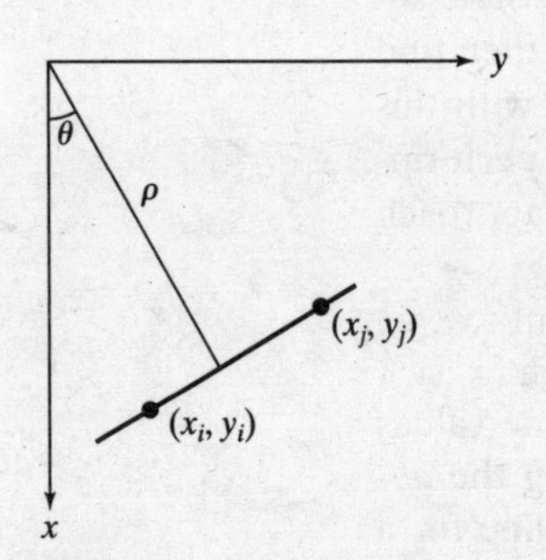

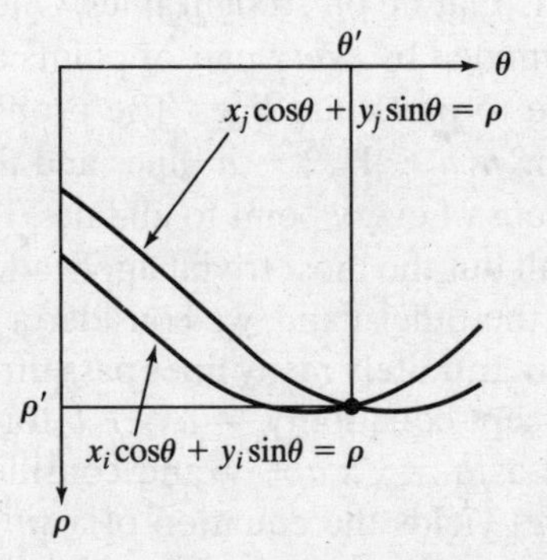

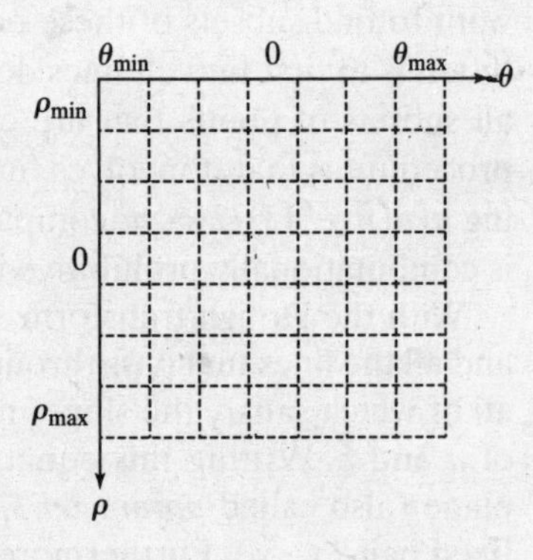

a b c

FIGURE 10.9 (a) (ρ, θ) parameterization of lines in the xy-plane. (b) Sinusoidal curves in the $\rho\theta$-plane; the point of intersection, (ρ', θ'), corresponds to the parameters of the line joining (x_i, y_i) and (x_j, y_j). (c) Division of the $\rho\theta$-plane into accumulator cells.

The computational attractiveness of the Hough transform arises from subdividing the $\rho\theta$ parameter space into so-called *accumulator cells*, as illustrated in Figure 10.9(c), where $(\rho_{\min}, \rho_{\max})$ and $(\theta_{\min}, \theta_{\max})$ are the expected ranges of the parameter values. Usually, the maximum range of values is $-90° \leq \theta \leq 90°$ and $-D \leq \rho \leq D$, where D is the distance between corners in the image. The cell at coordinates (i, j), with accumulator value $A(i, j)$, corresponds to the square associated with parameter space coordinates (ρ_i, θ_j). Initially, these cells are set to zero. Then, for every nonbackground point (x_k, y_k) in the image plane, we let θ equal each of the allowed subdivision values on the θ axis and solve for the corresponding ρ using the equation $\rho = x_k \cos\theta + y_k \sin\theta$. The resulting ρ-values are then rounded off to the nearest allowed cell value along the ρ-axis. The corresponding accumulator cell is then incremented. At the end of this procedure, a value of Q in $A(i, j)$, means that Q points in the xy-plane lie on the line $x\cos\theta_j + y\sin\theta_j = \rho_i$. The number of subdivisions in the $\rho\theta$-plane determines the accuracy of the colinearity of these points.

A function for computing the Hough transform is given next. This function makes use of *sparse* matrices, which are matrices that contain a small number of nonzero elements. This characteristic provides advantages in both matrix storage space and computation time. Given a matrix A, we convert it to sparse matrix format by using function sparse, which has the basic syntax

```
S = sparse(A)
```

For example,

```
>> A = [ 0  0  0  5
         0  2  0  0
         1  3  0  0
         0  0  4  0 ];
>> S = sparse(A)

S =

   (3,1)    1
   (2,2)    2
   (3,2)    3
   (4,3)    4
   (1,4)    5
```

This output lists the nonzero elements of S, together with their row and column indices. The elements are sorted by columns.

A syntax used more frequently with function sparse consists of five arguments:

```
S = sparse(r, c, s, m, n)
```

Here, r and c are vectors of row and column indices, respectively, of the nonzero elements of the matrix we wish to convert to sparse format. Parameter s is a vector containing the values that correspond to the index pairs (r, c), and m and n are the row and column dimensions for the resulting matrix. For instance, the matrix S in the previous example can be generated directly using the command

```
>> S = sparse([3 2 3 4 1], [1 2 2 3 4], [1 2 3 4 5], 4, 4);
```

There are a number of other syntax forms for function sparse, as detailed in the help page for this function.

Given a sparse matrix S generated by any of its applicable syntax forms, we can obtain the full matrix back by using function full, whose syntax is

```
A = full(S)
```

To explore Hough transform-based line detection in MATLAB, we first write a function, hough.m, that computes the Hough transform:

hough

```
function [h, theta, rho] = hough(f, dtheta, drho)
%HOUGH Hough transform.
%   [H, THETA, RHO] = HOUGH(F, DTHETA, DRHO) computes the Hough
%   transform of the image F. DTHETA specifies the spacing (in
%   degrees) of the Hough transform bins along the theta axis. DRHO
%   specifies the spacing of the Hough transform bins along the rho
%   axis. H is the Hough transform matrix. It is NRHO-by-NTHETA,
%   where NRHO = 2*ceil(norm(size(F))/DRHO) - 1, and NTHETA =
%   2*ceil(90/DTHETA). Note that if 90/DTHETA is not an integer, the
%   actual angle spacing will be 90 / ceil(90/DTHETA).
%
%   THETA is an NTHETA-element vector containing the angle (in
%   degrees) corresponding to each column of H. RHO is an
%   NRHO-element vector containing the value of rho corresponding to
%   each row of H.
%
%   [H, THETA, RHO] = HOUGH(F) computes the Hough transform using
%   DTHETA = 1 and DRHO = 1.

if nargin < 3
   drho = 1;
end
if nargin < 2
   dtheta = 1;
end

f = double(f);
[M,N] = size(f);
theta = linspace(-90, 0, ceil(90/dtheta) + 1);
theta = [theta -fliplr(theta(2:end - 1))];
ntheta = length(theta);
```

```
D = sqrt((M - 1)^2 + (N - 1)^2);
q = ceil(D/drho);
nrho = 2*q - 1;
rho = linspace(-q*drho, q*drho, nrho);

[x, y, val] = find(f);
x = x - 1;  y = y - 1;

% Initialize output.
h = zeros(nrho, length(theta));

% To avoid excessive memory usage, process 1000 nonzero pixel
% values at a time.
for k = 1:ceil(length(val)/1000)
   first = (k - 1)*1000 + 1;
   last = min(first+999, length(x));

   x_matrix     = repmat(x(first:last), 1, ntheta);
   y_matrix     = repmat(y(first:last), 1, ntheta);
   val_matrix   = repmat(val(first:last), 1, ntheta);
   theta_matrix = repmat(theta, size(x_matrix, 1), 1)*pi/180;

   rho_matrix = x_matrix.*cos(theta_matrix) + ...
       y_matrix.*sin(theta_matrix);
slope = (nrho - 1)/(rho(end) - rho(1));
rho_bin_index = round(slope*(rho_matrix - rho(1)) + 1);

theta_bin_index = repmat(1:ntheta, size(x_matrix, 1), 1);

% Take advantage of the fact that the SPARSE function, which
% constructs a sparse matrix, accumulates values when input
% indices are repeated. That's the behavior we want for the
% Hough transform. We want the output to be a full (nonsparse)
% matrix, however, so we call function FULL on the output of
% SPARSE.
h = h + full(sparse(rho_bin_index(:), theta_bin_index(:), ...
                    val_matrix(:), nrho, ntheta));
end
```

EXAMPLE 10.5: Illustration of the Hough transform.

■ In this example we illustrate the use of function `hough` on a simple binary image. First we construct an image containing isolated foreground pixels in several locations.

```
>> f = zeros(101, 101);
>> f(1, 1)     = 1;  f(101, 1) = 1; f(1, 101) = 1;
>> f(101, 101) = 1;  f(51, 51) = 1;
```

Figure 10.10(a) shows our test image. Next we compute and display the Hough transform.

```
>> H = hough(f);
>> imshow(H, [ ])
```

Figure 10.10(b) shows the result, displayed with `imshow` in the familiar way. However, it often is more useful to visualize Hough transforms in a larger plot,

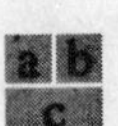

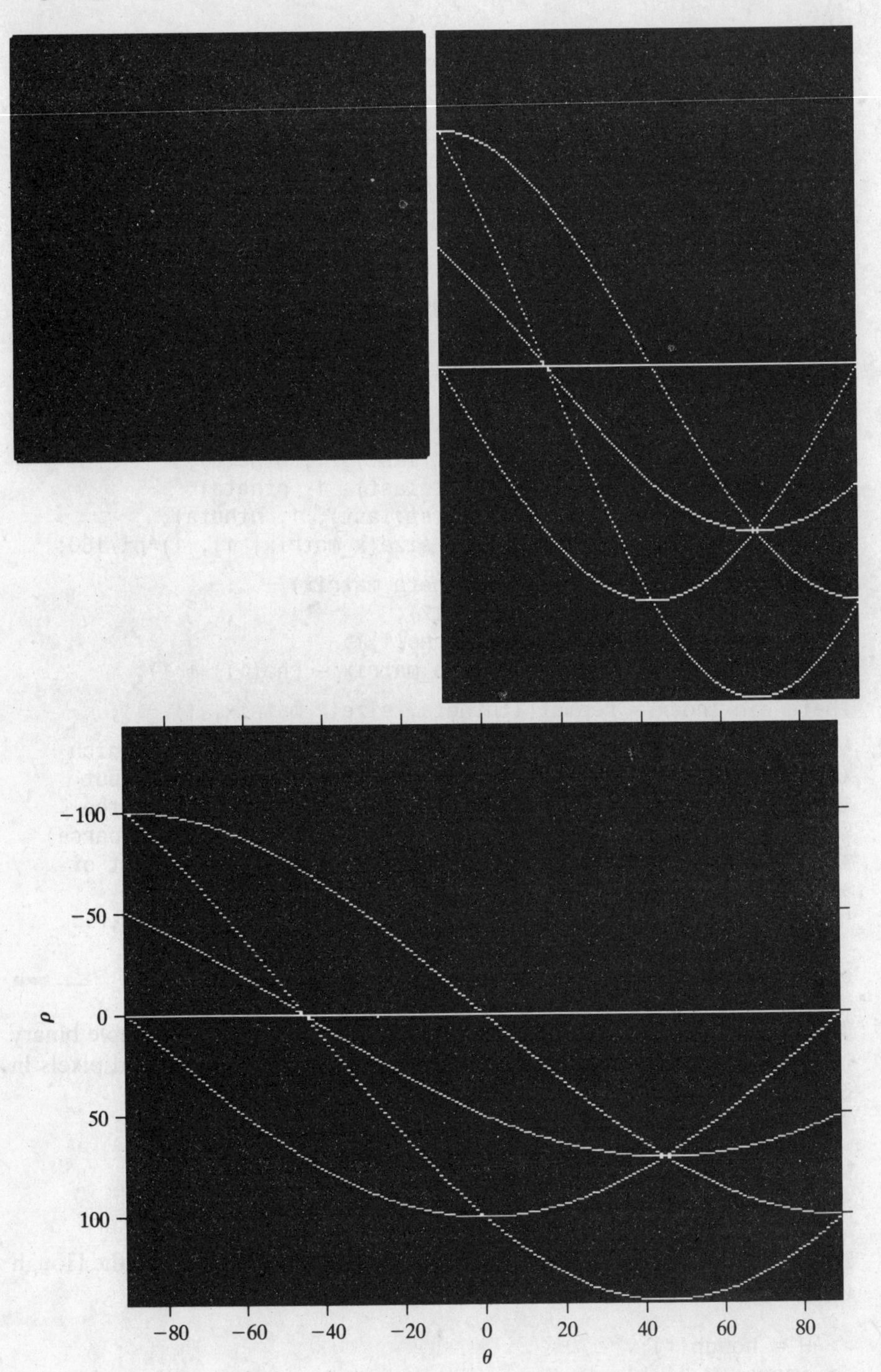

FIGURE 10.10 (a) Binary image with five dots (four of the dots are in the corners). (b) Hough transform displayed using `imshow`. (c) Alternative Hough transform display with axis labeling. (The dots in (a) were enlarged to make them easier to see.)

with labeled axes. In the next code fragment we call hough with three output arguments; the second two output arguments contain the θ and ρ values corresponding to each column and row, respectively, of the Hough transform matrix. These vectors, theta and rho, can then be passed as additional input arguments to imshow to control the horizontal and vertical axis labeling. We also pass the 'notruesize' option to imshow. The axis function is used to turn on axis labeling and to make the display fill the rectangular shape of the figure. Finally the xlabel and ylabel functions (see Section 3.3.1) are used to label the axes using a LaTeX-style notation for Greek letters.

```
>> [H, theta, rho] = hough(f);
>> imshow(theta, rho, H, [ ], 'notruesize')
>> axis on, axis normal
>> xlabel('\theta'), ylabel('\rho')
```

Figure 10.10(c) shows the labeled result. The intersections of three sinusoidal curves at $\pm 45°$ indicate that there are two sets of three collinear points in f. The intersections of two sinusoidal curves at $(\theta, \rho) = (-90, 0)$, $(-90, -100)$, $(0, 0)$, and $(0, 100)$ indicate that there are four sets of collinear points that lie along vertical and horizontal lines. ■

10.2.1 Hough Transform Peak Detection

The first step in using the Hough transform for line detection and linking is peak detection. Finding a meaningful set of distinct peaks in a Hough transform can be challenging. Because of the quantization in space of the digital image, the quantization in parameter space of the Hough transform, as well as the fact that edges in typical images are not perfectly straight, Hough transform peaks tend to lie in more than one Hough transform cell. One strategy to overcome this problem is the following:

1. Find the Hough transform cell containing the highest value and record its location.
2. Suppress (set to zero) Hough transform cells in the immediate neighborhood of the maximum found in step 1.
3. Repeat until the desired number of peaks has been found, or until a specified threshold has been reached.

Function houghpeaks implements this strategy.

houghpeaks

```
function [r, c, hnew] = houghpeaks(h, numpeaks, threshold, nhood)
%HOUGHPEAKS Detect peaks in Hough transform.
%   [R, C, HNEW] = HOUGHPEAKS(H, NUMPEAKS, THRESHOLD, NHOOD) detects
%   peaks in the Hough transform matrix H. NUMPEAKS specifies the
%   maximum number of peak locations to look for. Values of H below
%   THRESHOLD will not be considered to be peaks. NHOOD is a
%   two-element vector specifying the size of the suppression
%   neighborhood. This is the neighborhood around each peak that is
```

```
%   set to zero after the peak is identified. The elements of NHOOD
%   must be positive, odd integers. R and C are the row and column
%   coordinates of the identified peaks. HNEW is the Hough transform
%   with peak neighborhood suppressed.
%
%   If NHOOD is omitted, it defaults to the smallest odd values >=
%   size(H)/50. If THRESHOLD is omitted, it defaults to
%   0.5*max(H(:)). If NUMPEAKS is omitted, it defaults to 1.
if nargin < 4
   nhood = size(h)/50;
   % Make sure the neighborhood size is odd.
   nhood = max(2*ceil(nhood/2) + 1, 1);
end
if nargin < 3
   threshold = 0.5 * max(h(:));
end
if nargin < 2
   numpeaks = 1;
end
done = false;
hnew = h; r = []; c = [];
while ~done
   [p, q] = find(hnew == max(hnew(:)));
   p = p(1); q = q(1);
   if hnew(p, q) >= threshold
      r(end + 1) = p; c(end + 1) = q;

      % Suppress this maximum and its close neighbors.
      p1 = p - (nhood(1) - 1)/2; p2 = p + (nhood(1) - 1)/2;
      q1 = q - (nhood(2) - 1)/2; q2 = q + (nhood(2) - 1)/2;
      [pp, qq] = ndgrid(p1:p2, q1:q2);
      pp = pp(:); qq = qq(:);

      % Throw away neighbor coordinates that are out of bounds in
      % the rho direction.
      badrho = find((pp < 1) | (pp > size(h, 1)));
      pp(badrho) = []; qq(badrho) = [];

      % For coordinates that are out of bounds in the theta
      % direction, we want to consider that H is antisymmetric
      % along the rho axis for theta = +/- 90 degrees.
      theta_too_low = find(qq < 1);
      qq(theta_too_low) = size(h, 2) + qq(theta_too_low);
      pp(theta_too_low) = size(h, 1) - pp(theta_too_low) + 1;
      theta_too_high = find(qq > size(h, 2));
      qq(theta_too_high) = qq(theta_too_high) - size(h, 2);
      pp(theta_too_high) = size(h, 1) - pp(theta_too_high) + 1;

      % Convert to linear indices to zero out all the values.
      hnew(sub2ind(size(hnew), pp, qq)) = 0;
```

```
        done = length(r) == numpeaks;
    else
        done = true;
    end
end
```

Function houghpeaks is illustrated in Example 10.6.

10.2.2 Hough Transform Line Detection and Linking

Once a set of candidate peaks has been identified in the Hough transform, it remains to be determined if there are line segments associated with those peaks, as well as where they start and end. For each peak, the first step is to find the location of all nonzero pixels in the image that contributed to that peak. For this purpose, we write function houghpixels.

houghpixels

```
function [r, c] = houghpixels(f, theta, rho, rbin, cbin)
%HOUGHPIXELS Compute image pixels belonging to Hough transform bin.
%   [R, C] = HOUGHPIXELS(F, THETA, RHO, RBIN, CBIN) computes the
%   row-column indices (R, C) for nonzero pixels in image F that map
%   to a particular Hough transform bin, (RBIN, CBIN). RBIN and CBIN
%   are scalars indicating the row-column bin location in the Hough
%   transform matrix returned by function HOUGH. THETA and RHO are
%   the second and third output arguments from the HOUGH function.

[x, y, val] = find(f);
x = x - 1; y = y - 1;

theta_c = theta(cbin) * pi / 180;
rho_xy = x*cos(theta_c) + y*sin(theta_c);
nrho = length(rho);
slope = (nrho - 1)/(rho(end) - rho(1));
rho_bin_index = round(slope*(rho_xy - rho(1)) + 1);

idx = find(rho_bin_index == rbin);

r = x(idx) + 1; c = y(idx) + 1;
```

The pixels associated with the locations found using houghpixels must be grouped into line segments. Function houghlines uses the following strategy:

1. Rotate the pixel locations by $90° - \theta$ so that they lie approximately along a vertical line.
2. Sort the pixel locations by their rotated x-values.
3. Use function diff to locate gaps. Ignore small gaps; this has the effect of merging adjacent line segments that are separated by a small space.
4. Return information about line segments that are longer than some minimum length threshold.

houghlines

```
function lines = houghlines(f,theta,rho,rr,cc,fillgap,minlength)
%HOUGHLINES Extract line segments based on the Hough transform.
%   LINES = HOUGHLINES(F, THETA, RHO, RR, CC, FILLGAP, MINLENGTH)
```

```
%   extracts line segments in the image F associated with particular
%   bins in a Hough transform. THETA and RHO are vectors returned by
%   function HOUGH. Vectors RR and CC specify the rows and columns
%   of the Hough transform bins to use in searching for line
%   segments. If HOUGHLINES finds two line segments associated with
%   the same Hough transform bin that are separated by less than
%   FILLGAP pixels, HOUGHLINES merges them into a single line
%   segment. FILLGAP defaults to 20 if omitted. Merged line
%   segments less than MINLENGTH pixels long are discarded.
%   MINLENGTH defaults to 40 if omitted.
%
%   LINES is a structure array whose length equals the number of
%   merged line segments found. Each element of the structure array
%   has these fields:
%
%      point1   End-point of the line segment; two-element vector
%      point2   End-point of the line segment; two-element vector
%      length   Distance between point1 and point2
%      theta    Angle (in degrees) of the Hough transform bin
%      rho      Rho-axis position of the Hough transform bin
if nargin < 6
   fillgap = 20;
end
if nargin < 7
   minlength = 40;
end
numlines = 0; lines = struct;
for k = 1:length(rr)
   rbin = rr(k); cbin = cc(k);

   % Get all pixels associated with Hough transform cell.
   [r, c] = houghpixels(f, theta, rho, rbin, cbin);
   if isempty(r)
      continue
   end

   % Rotate the pixel locations about (1,1) so that they lie
   % approximately along a vertical line.
   omega = (90 - theta(cbin)) * pi / 180;
   T = [cos(omega) sin(omega); -sin(omega) cos(omega)];
   xy = [r - 1 c - 1] * T;
   x = sort(xy(:,1));

   % Find the gaps larger than the threshold.
   diff_x = [diff(x); Inf];
   idx = [0; find(diff_x > fillgap)];
   for p = 1:length(idx) - 1
      x1 = x(idx(p) + 1); x2 = x(idx(p + 1));
      linelength = x2 - x1;
      if linelength >= minlength
         point1 = [x1 rho(rbin)]; point2 = [x2 rho(rbin)];
```

```
        % Rotate the end-point locations back to the original
        % angle.
        Tinv = inv(T);
        point1 = point1 * Tinv; point2 = point2 * Tinv;

        numlines = numlines + 1;
        lines(numlines).point1 = point1 + 1;
        lines(numlines).point2 = point2 + 1;
        lines(numlines).length = linelength;
        lines(numlines).theta = theta(cbin);
        lines(numlines).rho = rho(rbin);
      end
    end
end
```

B = inv(A) *computes the inverse of square matrix* A.

EXAMPLE 10.6: Using the Hough transform for line detection and linking.

■ In this example we use functions hough, houghpeaks, and houghlines to find a set of line segments in the binary image, f, in Fig. 10.7(f). First, we compute and display the Hough transform, using a finer angular spacing than the default ($\Delta\theta = 0.5$ instead of 1.0).

```
>> [H, theta, rho] = hough(f, 0.5);
>> imshow(theta, rho, H, [ ], 'notruesize'), axis on, axis normal
>> xlabel('\theta'), ylabel('\rho')
```

Next we use function houghpeaks to find five Hough transform peaks that are likely to be significant.

```
>> [r, c] = houghpeaks(H, 5);
>> hold on
>> plot(theta(c), rho(r), 'linestyle', 'none', ...
       'marker', 's', 'color', 'w')
```

Figure 10.11(a) shows the Hough transform with the peak locations superimposed. Finally, we use function houghlines to find and link line segments, and

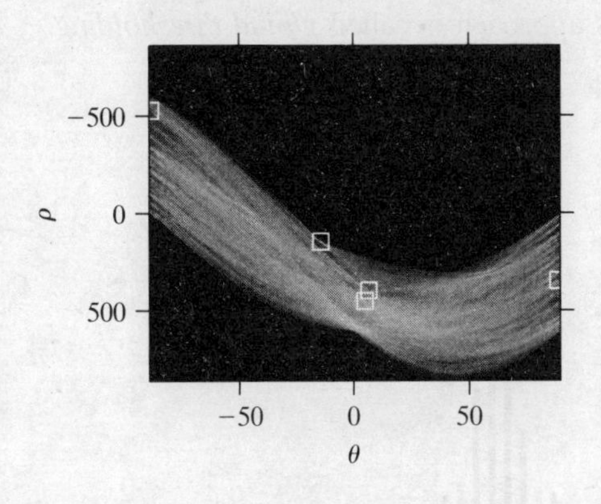

a b

FIGURE 10.11 (a) Hough transform with five peak locations selected. (b) Line segments corresponding to the Hough transform peaks.

we superimpose the line segments on the original binary image using `imshow`, `hold on`, and `plot`:

```
>> lines = houghlines(f, theta, rho, r, c)
>> figure, imshow(f), hold on
>> for k = 1:length(lines)
xy = [lines(k).point1 ; lines(k).point2];
plot(xy(:,2), xy(:,1), 'LineWidth', 4, 'Color', [.6 .6 .6]);
end
```

Figure 10.11(b) shows the resulting image with the detected segments superimposed as thick, gray lines. ■

10.3 Thresholding

Because of its intuitive properties and simplicity of implementation, image thresholding enjoys a central position in applications of image segmentation. Simple thresholding was first introduced in Section 2.7.2, and we have used it in various discussions in the preceding chapters. In this section, we discuss ways of choosing the threshold value automatically, and we consider a method for varying the threshold according to the properties of local image neighborhoods.

Suppose that the intensity histogram shown in Fig. 10.12 corresponds to an image, $f(x, y)$, composed of light objects on a dark background, in such a way that object and background pixels have intensity levels grouped into two dominant modes. One obvious way to extract the objects from the background is to select a threshold T that separates these modes. Then any point (x, y) for which $f(x, y) \geq T$ is called an *object point*; otherwise, the point is called a *background point*. In other words, the thresholded image $g(x, y)$ is defined as

$$g(x, y) = \begin{cases} 1 & \text{if } f(x, y) \geq T \\ 0 & \text{if } f(x, y) < T \end{cases}$$

Pixels labeled 1 correspond to objects, whereas pixels labeled 0 correspond to the background. When T is a constant, this approach is called *global thresholding*.

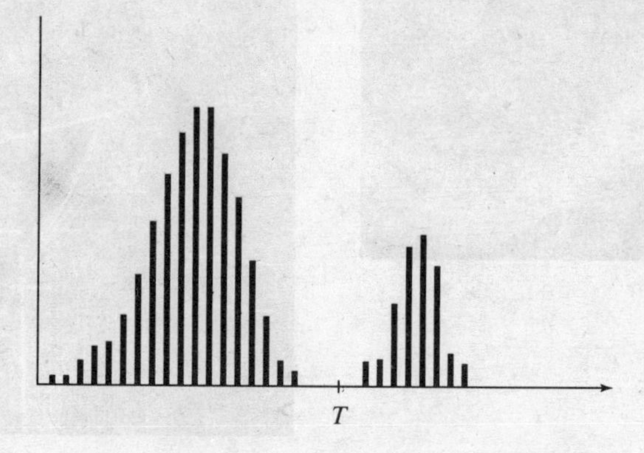

FIGURE 10.12 Selecting a threshold by visually analyzing a bimodal histogram.

Methods for choosing a global threshold are discussed in Section 10.3.1. In Section 10.3.2 we discuss allowing the threshold to vary, which is called *local thresholding.*

10.3.1 Global Thresholding

One way to choose a threshold is by visual inspection of the image histogram. The histogram in Figure 10.12 clearly has two distinct modes; as a result, it is easy to choose a threshold T that separates them. Another method of choosing T is by trial and error, picking different thresholds until one is found that produces a good result as judged by the observer. This is particularly effective in an interactive environment, such as one that allows the user to change the threshold using a *widget* (graphical control) such as a slider and see the result immediately.

For choosing a threshold automatically, Gonzalez and Woods [2002] describe the following iterative procedure:

1. Select an initial estimate for T. (A suggested initial estimate is the midpoint between the minimum and maximum intensity values in the image.)
2. Segment the image using T. This will produce two groups of pixels: G_1, consisting of all pixels with intensity values $\geq T$, and G_2, consisting of pixels with values $< T$.
3. Compute the average intensity values μ_1 and μ_2 for the pixels in regions G_1 and G_2.
4. Compute a new threshold value:

$$T = \frac{1}{2}(\mu_1 + \mu_2)$$

5. Repeat steps 2 through 4 until the difference in T in successive iterations is smaller than a predefined parameter T_o.

We show how to implement this procedure in MATLAB in Example 10.7.

The toolbox provides a function called `graythresh` that computes a threshold using Otsu's method (Otsu [1979]). To examine the formulation of this histogram-based method, we start by treating the normalized histogram as a discrete probability density function, as in

$$p_r(r_q) = \frac{n_q}{n} \qquad q = 0, 1, 2, \ldots, L - 1$$

where n is the total number of pixels in the image, n_q is the number of pixels that have intensity level r_q, and L is the total number of possible intensity levels in the image. Now suppose that a threshold k is chosen such that C_0 is the set of pixels with levels $[0, 1, \ldots, k - 1]$ and C_1 is the set of pixels with levels $[k, k + 1, \ldots, L - 1]$. Otsu's method chooses the threshold value k that maximizes the *between-class variance* σ_B^2, which is defined as

$$\sigma_B^2 = \omega_0(\mu_0 - \mu_T)^2 + \omega_1(\mu_1 - \mu_T)^2$$

where

$$\omega_0 = \sum_{q=0}^{k-1} p_q(r_q)$$

$$\omega_1 = \sum_{q=k}^{L-1} p_q(r_q)$$

$$\mu_0 = \sum_{q=0}^{k-1} qp_q(r_q)/\omega_0$$

$$\mu_1 = \sum_{q=k}^{L-1} qp_q(r_q)/\omega_1$$

$$\mu_T = \sum_{q=0}^{L-1} qp_q(r_q)$$

Function `graythresh` takes an image, computes its histogram, and then finds the threshold value that maximizes σ_B^2. The threshold is returned as a normalized value between 0.0 and 1.0. The calling syntax for `graythresh` is

```
T = graythresh(f)
```

where `f` is the input image and `T` is the resulting threshold. To segment the image we use `T` in function `im2bw` introduced in Section 2.7.2. Because the threshold is normalized to the range [0, 1], it must be scaled to the proper range before it is used. For example, if `f` is of class `uint8`, we multiply `T` by 255 before using it.

EXAMPLE 10.7: Computing global thresholds.

■ In this example we illustrate the iterative procedure described previously as well as Otsu's method on the gray-scale image, `f`, of scanned text, shown in Fig. 10.13(a). The iterative method can be implemented as follows:

```
>> T = 0.5*(double(min(f(:))) + double(max(f(:))));
>> done = false;
>> while ~done
   g = f >= T;
   Tnext = 0.5*(mean(f(g)) + mean(f(~g)));
   done = abs(T - Tnext) < 0.5;
   T = Tnext;
end
```

For this particular image, the while loop executes four times and terminates

a b

ponents or broken connection paths. There is no poi
tion past the level of detail required to identify those
Segmentation of nontrivial images is one of the mo
processing. Segmentation accuracy determines the ev
of computerized analysis procedures. For this reason,
be taken to improve the probability of rugged segment
such as industrial inspection applications, at least some
the environment is possible at times. The experienced
designer invariably pays considerable attention to suc

ponents or broken connection paths. There is no poi
tion past the level of detail required to identify those
Segmentation of nontrivial images is one of the mo
processing. Segmentation accuracy determines the ev
of computerized analysis procedures. For this reason,
be taken to improve the probability of rugged segment
such as industrial inspection applications, at least some
the environment is possible at times. The experienced
designer invariably pays considerable attention to suc

FIGURE 10.13 (a) Scanned text. (b) Thresholded text obtained using function `graythresh`.

with T equal to 101.47.

Next we compute a threshold using function graythresh:

```
>> T2 = graythresh(f)
T2 =
     0.3961
>> T2 * 255
ans =
     101
```

Thresholding using these two values produces images that are almost indistinguishable from each other. Figure 10.13(b) shows the image thresholded using T2. ■

10.3.2 Local Thresholding

Global thresholding methods can fail when the background illumination is uneven, as was illustrated in Figs. 9.26(a) and (b). A common practice in such situations is to preprocess the image to compensate for the illumination problems and then apply a global threshold to the preprocessed image. The improved thresholding result shown in Fig. 9.26(e) was computed by applying a morphological top-hat operator and then using graythresh on the result. We can show that this process is equivalent to thresholding $f(x, y)$ with a locally varying threshold function $T(x, y)$:

$$g(x, y) = \begin{cases} 1 & \text{if } f(x, y) \geq T(x, y) \\ 0 & \text{if } f(x, y) < T(x, y) \end{cases}$$

where

$$T(x, y) = f_o(x, y) + T_o$$

The image $f_o(x, y)$ is the morphological opening of f, and the constant T_o is the result of function graythresh applied to f_o.

10.4 Region-Based Segmentation

The objective of segmentation is to partition an image into regions. In Sections 10.1 and 10.2 we approached this problem by finding boundaries between regions based on discontinuities in intensity levels, whereas in Section 10.3 segmentation was accomplished via thresholds based on the distribution of pixel properties, such as intensity values. In this section we discuss segmentation techniques that are based on finding the regions directly.

10.4.1 Basic Formulation

Let R represent the entire image region. We may view segmentation as a process that partitions R into n subregions, $R_1, R_2, \ldots, R_n$, such that

(a) $\bigcup_{i=1}^{n} R_i = R$.

(b) R_i is a connected region, $i = 1, 2, \ldots, n$.

(c) $R_i \cap R_j = \emptyset$ for all i and j, $i \neq j$.
(d) $P(R_i) = \text{TRUE}$ for $i = 1, 2, \ldots, n$.
(e) $P(R_i \cup R_j) = \text{FALSE}$ for any adjacent regions R_i and R_j.

In the context of the discussion in Section 9.4, two disjoint regions, R_i and R_j, are said to be adjacent if their union forms a connected component.

Here, $P(R_i)$ is a logical predicate defined over the points in set R_i and $\emptyset$ is the null set.

Condition (a) indicates that the segmentation must be complete; that is, every pixel must be in a region. The second condition requires that points in a region be connected in some predefined sense (e.g., 4- or 8-connected). Condition (c) indicates that the regions must be disjoint. Condition (d) deals with the properties that must be satisfied by the pixels in a segmented region—for example $P(R_i) = \text{TRUE}$ if all pixels in R_i have the same gray level. Finally, condition (e) indicates that adjacent regions R_i and R_j are different in the sense of predicate P.

10.4.2 Region Growing

As its name implies, *region growing* is a procedure that groups pixels or subregions into larger regions based on predefined criteria for growth. The basic approach is to start with a set of "seed" points and from these grow regions by appending to each seed those neighboring pixels that have predefined properties similar to the seed (such as specific ranges of gray level or color).

Selecting a set of one or more starting points often can be based on the nature of the problem, as shown later in Example 10.8. When a priori information is not available, one procedure is to compute at every pixel the same set of properties that ultimately will be used to assign pixels to regions during the growing process. If the result of these computations shows clusters of values, the pixels whose properties place them near the centroid of these clusters can be used as seeds.

The selection of similarity criteria depends not only on the problem under consideration, but also on the type of image data available. For example, the analysis of land-use satellite imagery depends heavily on the use of color. This problem would be significantly more difficult, or even impossible, to handle without the inherent information available in color images. When the images are monochrome, region analysis must be carried out with a set of descriptors based on intensity levels (such as moments or texture) and spatial properties. We discuss descriptors useful for region characterization in Chapter 11.

Descriptors alone can yield misleading results if connectivity (adjacency) information is not used in the region-growing process. For example, visualize a random arrangement of pixels with only three distinct intensity values. Grouping pixels with the same intensity level to form a "region" without paying attention to connectivity would yield a segmentation result that is meaningless in the context of this discussion.

Another problem in region growing is the formulation of a stopping rule. Basically, growing a region should stop when no more pixels satisfy the criteria for inclusion in that region. Criteria such as intensity values, texture, and color, are local in nature and do not take into account the "history" of region growth. Additional criteria that increase the power of a region-growing algorithm utilize the concept of size, likeness between a candidate pixel and the pixels grown so far (such as a comparison of the intensity of a candidate and the average intensity of the grown region), and the shape of the region being grown. The use of these types of descriptors is based on the assumption that a model of expected results is at least partially available.

To illustrate the principles of how region segmentation can be handled in MATLAB, we develop next an M-function, called regiongrow, to do basic region growing. The syntax for this function is

```
[g, NR, SI, TI] = regiongrow(f, S, T)
```

where f is an image to be segmented and parameter S can be an array (the same size as f) or a scalar. If S is an array, it must contain 1s at all the coordinates where seed points are located and 0s elsewhere. Such an array can be determined by inspection, or by an external seed-finding function. If S is a scalar, it defines an intensity value such that all the points in f with that value become seed points. Similarly, T can be an array (the same size as f) or a scalar. If T is an array, it contains a threshold value for each location in f. If T is scalar, it defines a global threshold. The threshold value(s) is (are) used to test if a pixel in the image is sufficiently similar to the seed or seeds to which it is 8-connected.

For example, if S = a and T = b, and we are comparing intensities, then a pixel is said to be similar to a (in the sense of passing the threshold test) if the absolute value of the difference between its intensity and a is less than or equal to b. If, in addition, the pixel in question is 8-connected to one or more seed values, then the pixel is considered a member of one or more regions. Similar comments hold if S and T are arrays, the basic difference being that comparisons are done with the appropriate locations defined in S and corresponding values of T.

In the output, g is the segmented image, with the members of each region being labeled with an integer value. Parameter NR is the number of different regions. Parameter SI is an image containing the seed points, and parameter TI is an image containing the pixels that passed the threshold test before they were processed for connectivity. Both SI and TI are of the same size as f.

The code for function regiongrow is as follows. Note the use of Chapter 9 function bwmorph to reduce to 1 the number of connected seed points in each region in S (when S is an array) and function imreconstruct to find pixels connected to each seed.

regiongrow

```
function [g, NR, SI, TI] = regiongrow(f, S, T)
%REGIONGROW Perform segmentation by region growing.
%   [G, NR, SI, TI] = REGIONGROW(F, SR, T). S can be an array (the
%   same size as F) with a 1 at the coordinates of every seed point
%   and 0s elsewhere. S can also be a single seed value. Similarly,
%   T can be an array (the same size as F) containing a threshold
%   value for each pixel in F. T can also be a scalar, in which
%   case it becomes a global threshold.
%
```

```
%   On the output, G is the result of region growing, with each
%   region labeled by a different integer, NR is the number of
%   regions, SI is the final seed image used by the algorithm, and TI
%   is the image consisting of the pixels in F that satisfied the
%   threshold test.

f = double(f);
% If S is a scalar, obtain the seed image.
if numel(S) == 1
   SI = f == S;
   S1 = S;
else
   % S is an array. Eliminate duplicate, connected seed locations
   % to reduce the number of loop executions in the following
   % sections of code.
   SI = bwmorph(S, 'shrink', Inf);
   J = find(SI);
   S1 = f(J); % Array of seed values.
end

TI = false(size(f));
for K = 1:length(S1)
   seedvalue = S1(K);
   S = abs(f - seedvalue) <= T;
   TI = TI | S;
end

% Use function imreconstruct with SI as the marker image to
% obtain the regions corresponding to each seed in S. Function
% bwlabel assigns a different integer to each connected region.
[g, NR] = bwlabel(imreconstruct(SI, TI));
```

true / false

`true` *is equivalent to* `logical(1)`, *and* `false` *is equivalent to* `logical(0)`.

EXAMPLE 10.8: Application of region growing to weld porosity detection.

■ Figure 10.14(a) shows an X-ray image of a weld (the horizontal dark region) containing several cracks and porosities (the bright, white streaks running horizontally through the middle of the image). We wish to use function `regiongrow` to segment the regions corresponding to weld failures. These segmented regions could be used for inspection, for inclusion in a database of historical studies, for controlling an automated welding system, and for other numerous applications.

The first order of business is to determine the initial seed points. In this application, it is known that some pixels in areas of defective welds tend to have the maximum allowable digital value (255 in this case). Based in this information, we let `S = 255`. The next step is to choose a threshold or threshold array. In this particular example we used `T = 65`. This number was based on analysis of the histogram in Fig. 10.15 and represents the difference between 255 and the location of the first major valley to the left, which is representative of the highest intensity value in the dark weld region. As noted earlier, a pixel has to

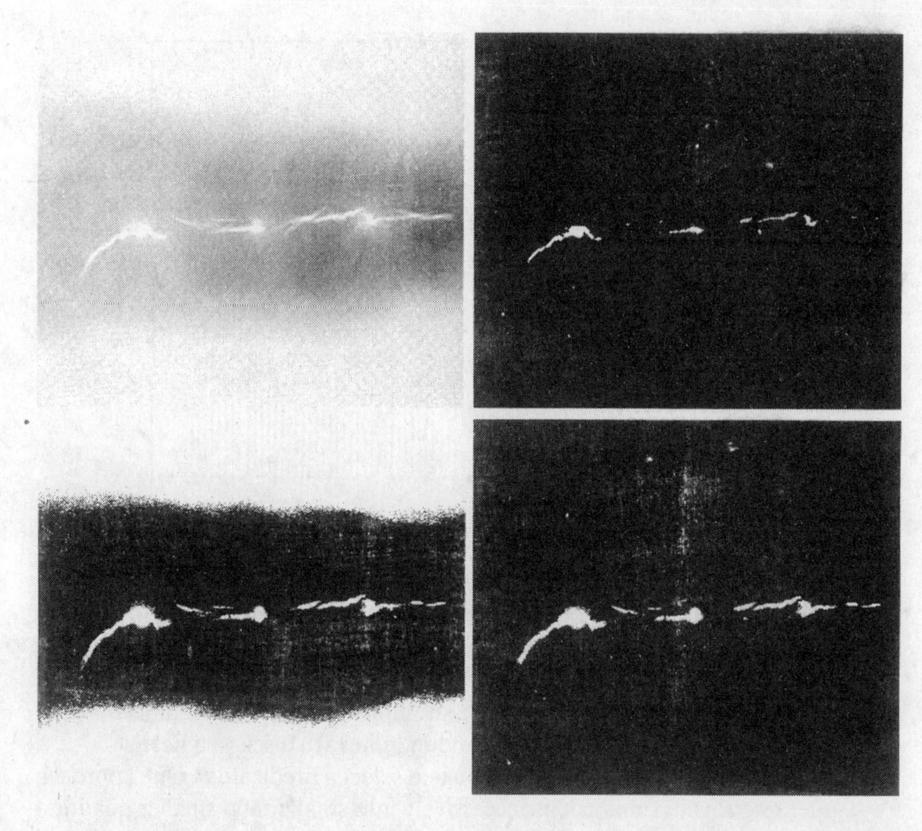

a b
c d

FIGURE 10.14 (a) Image showing defective welds. (b) Seed points. (c) Binary image showing all the pixels (in white) that passed the threshold test. (d) Result after all the pixels in (c) were analyzed for 8-connectivity to the seed points. (Original image courtesy of X-TEK Systems, Ltd.)

be 8-connected to at least one pixel in a region to be included in that region. If a pixel is found to be connected to more than one region, the regions are automatically merged by `regiongrow`.

Figure 10.14(b) shows the seed points (image `SI`). They are numerous in this case because the seeds were specified simply as all points in the image with a value of 255. Figure 10.14(c) is image `TI`. It shows all the points that passed the threshold test; that is, the points with intensity z_i, such that $|z_i - S| \leq T$. Figure 10.14(d) shows the result of extracting all the pixels in Figure 10.14(c) that were connected to the seed points. This is the segmented image, `g`. It is evident by comparing this image with the original that the region growing procedure did indeed segment the defective welds with a reasonable degree of accuracy.

Finally, we note by looking at the histogram in Fig. 10.15 that it would not have been possible to obtain the same or equivalent solution by any of the thresholding methods discussed in Section 10.3. The use of connectivity was a fundamental requirement in this case. ■

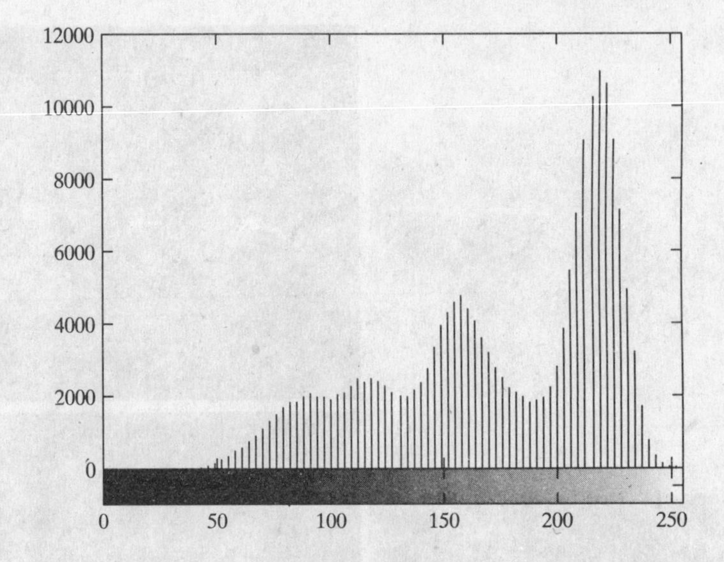

FIGURE 10.15 Histogram of Fig. 10.14(a).

10.4.3 Region Splitting and Merging

The procedure just discussed grows regions from a set of seed points. An alternative is to subdivide an image initially into a set of arbitrary, disjointed regions and then merge and/or split the regions in an attempt to satisfy the conditions stated in Section 10.4.1. The basics of splitting and merging are discussed next.

Let R represent the entire image region and select a predicate P. One approach for segmenting R is to subdivide it successively into smaller and smaller quadrant regions so that, for any region R_i, $P(R_i) = \text{TRUE}$. We start with the entire region. If $P(R) = \text{FALSE}$, we divide the image into quadrants. If P is FALSE for any quadrant, we subdivide that quadrant into subquadrants, and so on. This particular splitting technique has a convenient representation in the form of a so-called *quadtree*; that is, a tree in which each node has exactly four descendants, as illustrated in Fig. 10.16 (the subimages corresponding to the nodes of a quadtree sometimes are called *quadregions* or *quadimages*). Note that the root of the tree corresponds to the entire image and that each node corresponds to the subdivision of a node into four descendant nodes. In this case, only R_4 was subdivided further.

If only splitting is used, the final partition normally contains adjacent regions with identical properties. This drawback can be remedied by allowing merging, as

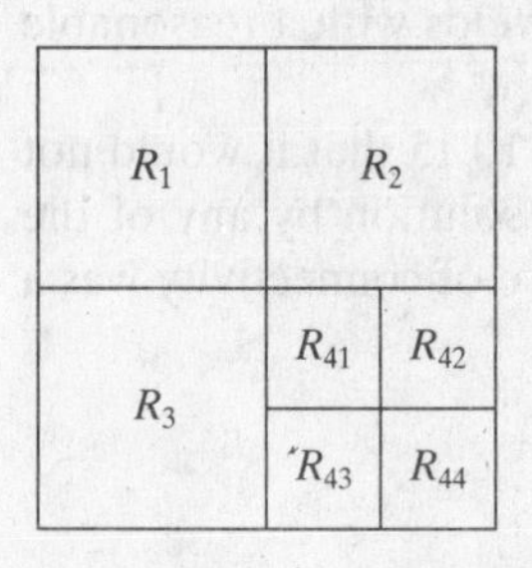

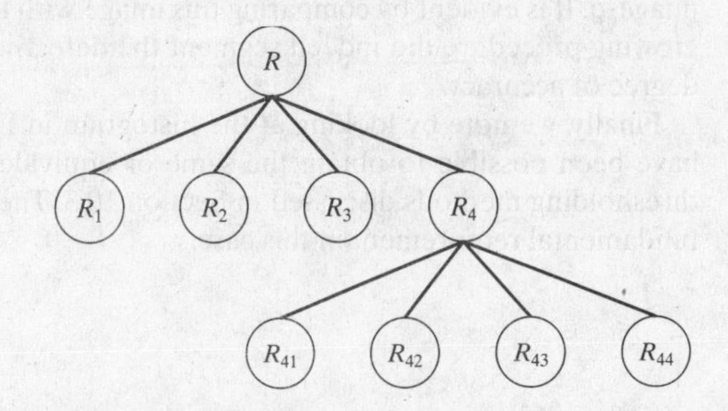

a b

FIGURE 10.16 (a) Partitioned image. (b) Corresponding quadtree.

well as splitting. Satisfying the constraints of Section 10.4.1 requires merging only adjacent regions whose combined pixels satisfy the predicate P. That is, two adjacent regions R_j and R_k are merged only if $P(R_j \cup R_k) = \text{TRUE}$.

The preceding discussion may be summarized by the following procedure in which, at any step,

1. Split into four disjoint quadrants any region R_i for which $P(R_i) = \text{FALSE}$.
2. When no further splitting is possible, merge any adjacent regions R_j and R_k for which $P(R_j \cup R_k) = \text{TRUE}$.
3. Stop when no further merging is possible.

Numerous variations of the preceding basic theme are possible. For example, a significant simplification results if we allow merging of any two adjacent regions R_i and R_j if each one satisfies the predicate individually. This results in a much simpler (and faster) algorithm because testing of the predicate is limited to individual quadregions. As Example 10.9 shows, this simplification is still capable of yielding good segmentation results in practice. Using this approach in step 2 of the procedure, all quadregions that satisfy the predicate are filled with 1s and their connectivity can be easily examined using, for example, function `imreconstruct`. This function, in effect, accomplishes the desired merging of adjacent quadregions. The quadregions that do not satisfy the predicate are filled with 0s to create a segmented image.

The function in IPT for implementing quadtree decomposition is `qtdecomp`. The syntax of interest in this section is

```
S = qtdecomp(f, @split_test, parameters)
```

where `f` is the input image and `S` is a sparse matrix containing the quadtree structure. If `S(k, m)` is nonzero, then `(k, m)` is the upper-left corner of a block in the decomposition and the size of the block is `S(k, m)`. Function `split_test` (see function `splitmerge` below for an example) is used to determine whether a region is to be split or not, and `parameters` are any additional parameters (separated by commas) required by `split_test`. The mechanics of this are similar to those discussed in Section 3.4.2 for function `coltfilt`.

Other forms of `qtdecomp` *are discussed in Section 11.2.2.*

To get the actual quadregion pixel values in a quadtree decomposition we use function `qtgetblk`, with syntax

```
[vals, r, c] = qtgetblk(f, S, m)
```

where `vals` is an array containing the values of the blocks of size `m` × `m` in the quadtree decomposition of `f`, and `S` is the sparse matrix returned by `qtdecomp`. Parameters `r` and `c` are vectors containing the row and column coordinates of the upper-left corners of the blocks.

We illustrate the use of function `qtdecomp` by writing a basic split-and-merge M-function that uses the simplification discussed earlier, in which two regions are merged if each satisfies the predicate individually. The function, which we call `splitmerge`, has the following calling syntax:

```
g = splitmerge(f, mindim, @predicate)
```

where `f` is the input image and `g` is the output image in which each connected region is labeled with a different integer. Parameter `mindim` defines the size of the smallest block allowed in the decomposition; this parameter has to be a positive integer power of 2.

Function `predicate` is a user-defined function that must be included in the MATLAB path. Its syntax is

```
flag = predicate(region)
```

This function must be written so that it returns `true` (a logical 1) if the pixels in `region` satisfy the predicate defined by the code in the function; otherwise, the value of `flag` must be `false` (a logical 0). Example 10.9 illustrates the use of this function.

Function `splitmerge` has a simple structure. First, the image is partitioned using function `qtdecomp`. Function `split_test` uses `predicate` to determine if a region should be split or not. Because when a region is split into four it is not known which (if any) of the resulting four regions will pass the predicate test individually, it is necessary to examine the regions after the fact to see which regions in the partitioned image pass the test. Function `predicate` is used for this purpose also. Any quadregion that passes the test is filled with 1s. Any that does not is filled with 0s. A marker array is created by selecting one element of each region that is filled with 1s. This array is used in conjunction with the partitioned image to determine region connectivity (adjacency); function `imreconstruct` is used for this purpose.

The code for function `splitmerge` follows. The simple predicate function shown in the comments section of the code is used in Example 10.9. Note that the size of the input image is brought up to a square whose dimensions are the minimum integer power of 2 that encompasses the image. This is a requirement of function `qtdecomp` to guarantee that splits down to size 1 are possible.

splitmerge

```
function g = splitmerge(f, mindim, fun)
%SPLITMERGE Segment an image using a split-and-merge algorithm.
%   G = SPLITMERGE(F, MINDIM, @PREDICATE) segments image F by using a
%   split-and-merge approach based on quadtree decomposition. MINDIM
%   (a positive integer power of 2) specifies the minimum dimension
%   of the quadtree regions (subimages) allowed. If necessary, the
%   program pads the input image with zeros to the nearest square
%   size that is an integer power of 2. This guarantees that the
%   algorithm used in the quadtree decomposition will be able to
%   split the image down to blocks of size 1-by-1. The result is
%   cropped back to the original size of the input image. In the
%   output, G, each connected region is labeled with a different
%   integer.
%
%   Note that in the function call we use @PREDICATE for the value of
%   fun. PREDICATE is a function in the MATLAB path, provided by the
%   user. Its syntax is
%
```

```
%      FLAG = PREDICATE(REGION) which must return TRUE if the pixels
%      in REGION satisfy the predicate defined by the code in the
%      function; otherwise, the value of FLAG must be FALSE.
%
%   The following simple example of function PREDICATE is used in
%   Example 10.9 of the book. It sets FLAG to TRUE if the
%   intensities of the pixels in REGION have a standard deviation
%   that exceeds 10, and their mean intensity is between 0 and 125.
%   Otherwise FLAG is set to false.
%
%      function flag = predicate(region)
%      sd = std2(region);
%      m = mean2(region);
%      flag = (sd > 10) & (m > 0) & (m < 125);

% Pad image with zeros to guarantee that function qtdecomp will
% split regions down to size 1-by-1.
Q = 2^nextpow2(max(size(f)));
[M, N] = size(f);
f = padarray(f, [Q - M, Q - N], 'post');

% Perform splitting first.
S = qtdecomp(f, @split_test, mindim, fun);

% Now merge by looking at each quadregion and setting all its
% elements to 1 if the block satisfies the predicate.

% Get the size of the largest block. Use full because S is sparse.
Lmax = full(max(S(:)));
% Set the output image initially to all zeros. The MARKER array is
% used later to establish connectivity.
g = zeros(size(f));
MARKER = zeros(size(f));
% Begin the merging stage.
for K = 1:Lmax
   [vals, r, c] = qtgetblk(f, S, K);
   if ~isempty(vals)
      % Check the predicate for each of the regions
      % of size K-by-K with coordinates given by vectors
      % r and c.
      for I = 1:length(r)
         xlow = r(I); ylow = c(I);
         xhigh = xlow + K - 1; yhigh = ylow + K - 1;
         region = f(xlow:xhigh, ylow:yhigh);
         flag = feval(fun, region);
         if flag
            g(xlow:xhigh, ylow:yhigh) = 1;
            MARKER(xlow, ylow) = 1;
         end
      end
   end
end
```

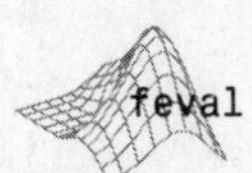

`feval(fun, param)` *evaluates function* `fun` *with parameter* `param`. *See the help page for* `feval` *for other syntax forms applicable to this function.*

```
% Finally, obtain each connected region and label it with a
% different integer value using function bwlabel.
g = bwlabel(imreconstruct(MARKER, g));

% Crop and exit
g = g(1:M, 1:N);

%-------------------------------------------------------------------%
function v = split_test(B, mindim, fun)
% THIS FUNCTION IS PART OF FUNCTION SPLIT-MERGE. IT DETERMINES
% WHETHER QUADREGIONS ARE SPLIT. The function returns in v
% logical 1s (TRUE) for the blocks that should be split and
% logical 0s (FALSE) for those that should not.

% Quadregion B, passed by qtdecomp, is the current decomposition of
% the image into k blocks of size m-by-m.

% k is the number of regions in B at this point in the procedure.
k = size(B, 3);

% Perform the split test on each block. If the predicate function
% (fun) returns TRUE, the region is split, so we set the appropriate
% element of v to TRUE. Else, the appropriate element of v is set to
% FALSE.
v(1:k) = false;
for I = 1:k
   quadregion = B(:, :, I);
   if size(quadregion, 1) <= mindim
      v(I) = false;
      continue
   end
   flag = feval(fun, quadregion);
   if flag
      v(I) = true;
   end
end
```

EXAMPLE 10.9: Image segmentation using region splitting and merging.

■ Figure 10.17(a) shows an X-ray band image of the Cygnus Loop. The image is of size 256×256 pixels. The objective of this example is to segment out of the image the "ring" of less dense matter surrounding the dense center. The region of interest has some obvious characteristics that should help in its segmentation. First, we note that the data has a random nature to it, indicating that its standard deviation should be greater than the standard deviation of the background (which is 0) and of the large central region. Similarly, the mean value (average intensity) of a region containing data from the outer ring should be greater than the mean of the background (which is 0) and less than the mean of the large, lighter central region. Thus, we should be able to segment the region of interest by using these two parameters. In fact, the predicate function shown as an example in the documentation of function `splitmerge` contains this knowledge about the problem. The parameters were determined by computing the mean and standard deviation of various regions in Fig. 10.17(a).

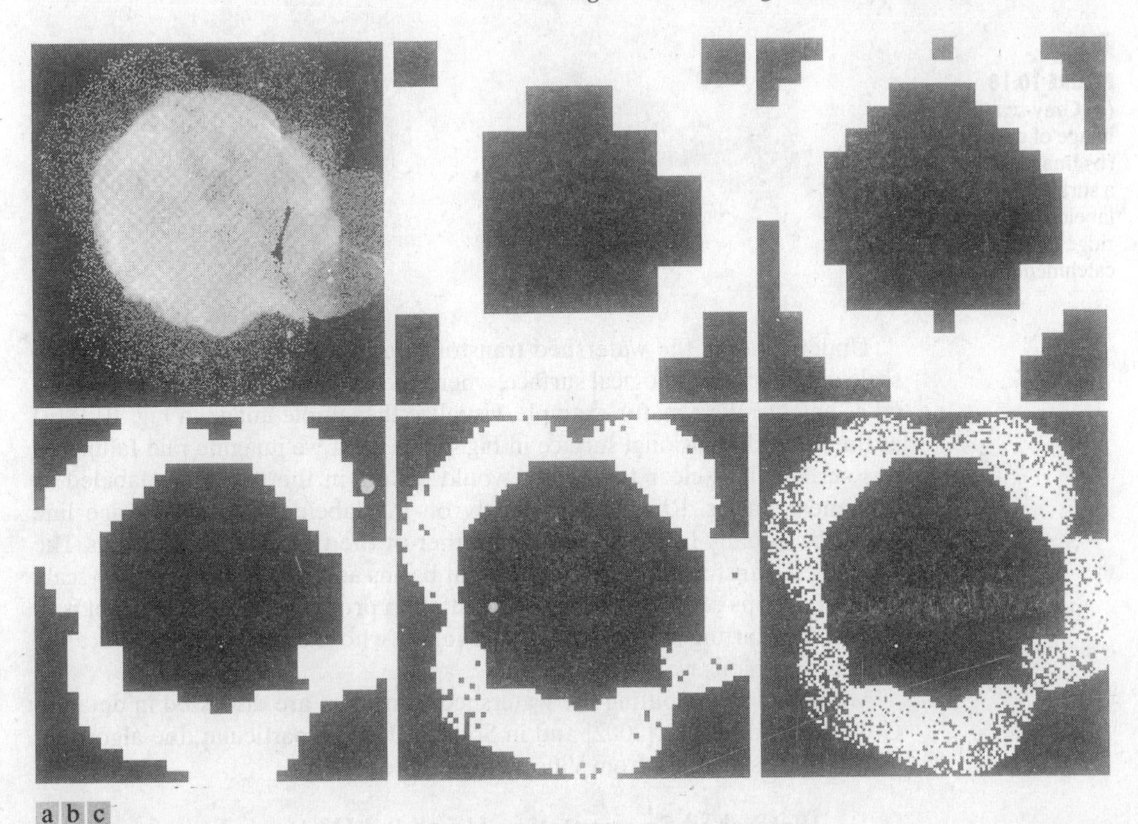

a b c
d e f

FIGURE 10.17 Image segmentation by a split-and-merge procedure. (a) Original image. (b) through (f) results of segmentation using function `splitmerge` with values of `mindim` equal to 32, 16, 8, 4, and 2, respectively. (Original image courtesy of NASA.)

Figures 10.17(b) through (f) show the results of segmenting Fig. 10.17(a) using function `splitmerge` with `mindim` values of 32, 16, 8, 4, and 2, respectively. All images show segmentation results with levels of detail that are inversely proportional to the value of `mindim`.

All results in Fig. 10.17 are reasonable segmentations. If one of these images were to be used as a mask to extract the region of interest out of the original image, then the result in Fig. 10.17(d) would be the best choice because it is the solid region with the most detail. An important aspect of the method just illustrated is its ability to "capture" in function `predicate` information about a problem domain that can help in segmentation. ■

10.5 Segmentation Using the Watershed Transform

In geography, a *watershed* is the ridge that divides areas drained by different river systems. A *catchment basin* is the geographical area draining into a river or reservoir. The *watershed transform* applies these ideas to gray-scale image processing in a way that can be used to solve a variety of image segmentation problems.

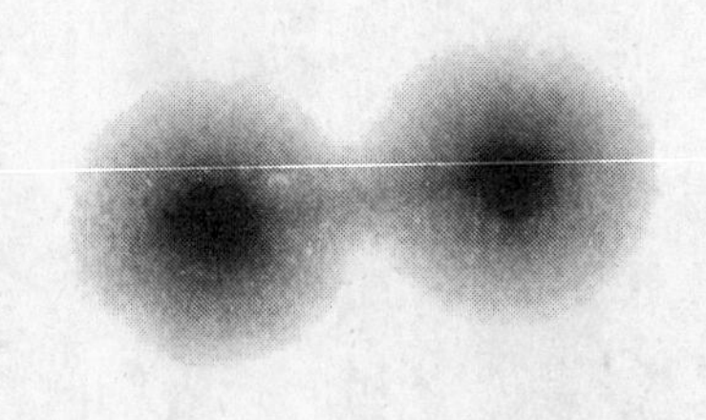

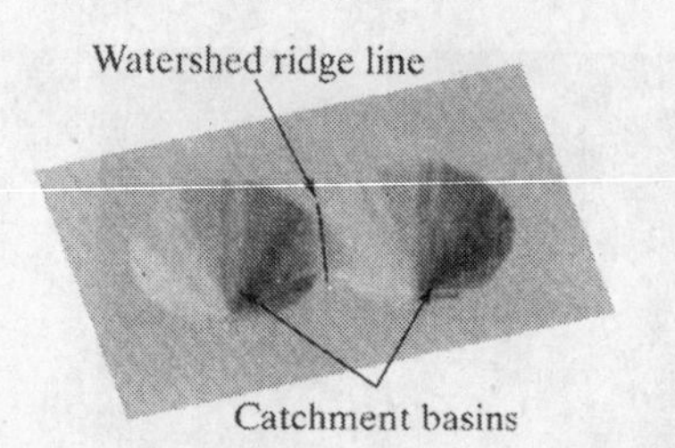

a b

FIGURE 10.18 (a) Gray-scale image of dark blobs. (b) Image viewed as a surface, with labeled watershed ridge line and catchment basins.

Understanding the watershed transform requires that we think of a gray-scale image as a topological surface, where the values of $f(x, y)$ are interpreted as heights. We can, for example, visualize the simple image in Fig. 10.18(a) as the three-dimensional surface in Fig. 10.18(b). If we imagine rain falling on this surface, it is clear that water would collect in the two areas labeled as catchment basins. Rain falling exactly on the labeled watershed ridge line would be equally likely to collect in either of the two catchment basins. The watershed transform finds the catchment basins and ridge lines in a gray-scale image. In terms of solving image segmentation problems, the key concept is to change the starting image into another image whose catchment basins are the objects or regions we want to identify.

Methods for computing the watershed transform are discussed in detail in Gonzalez and Woods [2002] and in Soille [2003]. In particular, the algorithm used in IPT is adapted from Vincent and Soille [1991].

10.5.1 Watershed Segmentation Using the Distance Transform

A tool commonly used in conjunction with the watershed transform for segmentation is the *distance transform*. The distance transform of a binary image is a relatively simple concept: It is the distance from every pixel to the nearest nonzero-valued pixel. Figure 10.19 illustrates the distance transform. Figure 10.19(a) shows a small binary image matrix. Figure 10.19(b) shows the corresponding distance transform. Note that 1-valued pixels have a distance transform value of 0. The distance transform can be computed using IPT function `bwdist`, whose calling syntax is

```
D = bwdist(f)
```

a b

FIGURE 10.19 (a) Small binary image. (b) Distance transform.

1	1	0	0	0
1	1	0	0	0
0	0	0	0	0
0	0	0	0	0
0	1	1	1	0

0.00	0.00	1.00	2.00	3.00
0.00	0.00	1.00	2.00	3.00
1.00	1.00	1.41	2.00	2.24
1.41	1.00	1.00	1.00	1.41
1.00	0.00	0.00	0.00	1.00

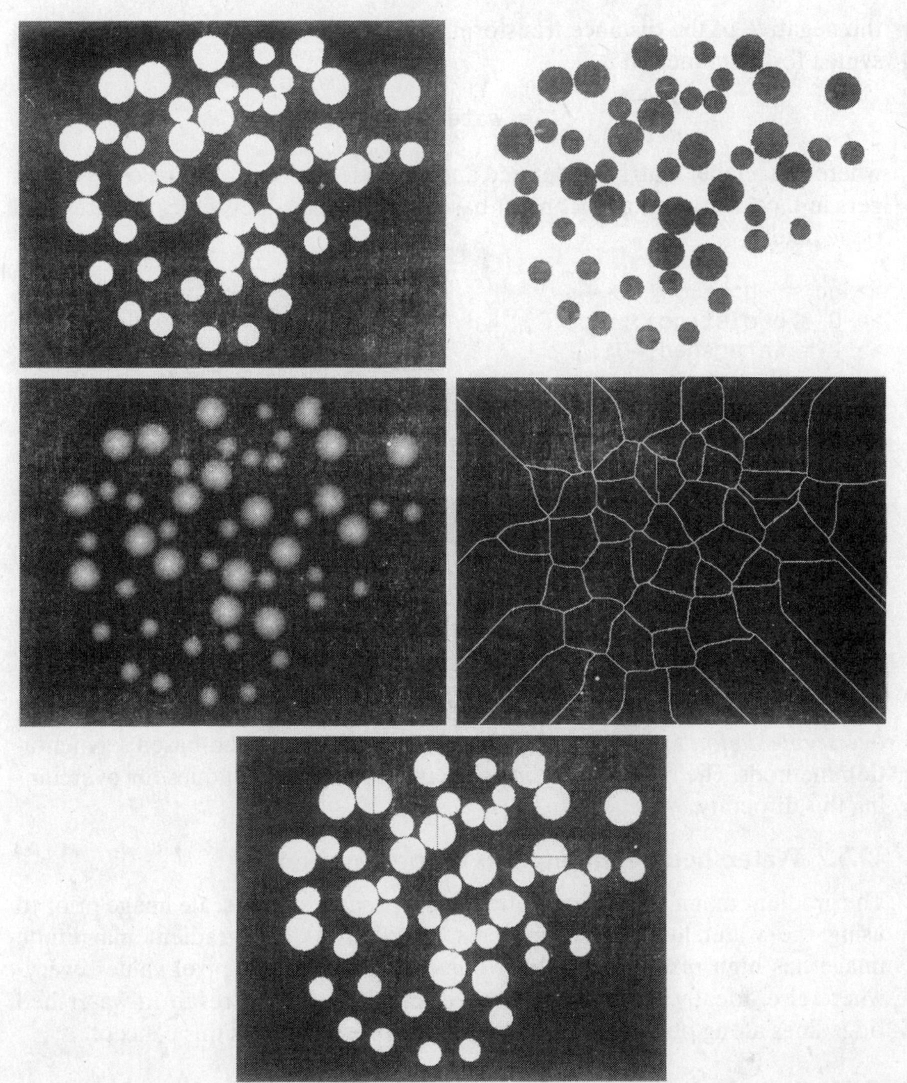

a b
c d
e

FIGURE 10.20
(a) Binary image.
(b) Complement of image in (a).
(c) Distance transform.
(d) Watershed ridge lines of the negative of the distance transform.
(e) Watershed ridge lines superimposed in black over original binary image. Some oversegmentation is evident.

EXAMPLE 10.10: Segmenting a binary image using the distance and watershed transforms.

■ In this example we show how the distance transform can be used with IPT's watershed transform to segment circular blobs, some of which are touching each other. Specifically, we want to segment the preprocessed dowel image, `f`, shown in Figure 9.29(b). First, we convert the image to binary using `im2bw` and `graythresh`, as described in Section 10.3.1.

```
>> g = im2bw(f, graythresh(f));
```

Figure 10.20(a) shows the result. The next steps are to complement the image, compute its distance transform, and then compute the watershed transform of the negative of the distance transform, using function watershed. The calling syntax for this function is

```
L = watershed(f)
```

where L is a label matrix, as defined and discussed in Section 9.4. Positive integers in L correspond to catchment basins, and zero values indicate watershed ridge pixels.

```
>> gc = ~g;
>> D = bwdist(gc);
>> L = watershed(-D);
>> w = L == 0;
```

Figures 10.20(b) and (c) show the complemented image and its distance transform. Since 0-valued pixels of L are watershed ridge pixels, the last line of the preceding code computes a binary image, w, that shows only these pixels. This watershed ridge image is shown in Fig. 10.20(d). Finally, a logical AND of the original binary image and the complement of w serves to complete the segmentation, as shown in Fig. 10.20(e).

```
>> g2 = g & ~w;
```

Note that some objects in Fig. 10.20(e) were split improperly. This is called *oversegmentation* and is a common problem with watershed-based segmentation methods. The next two sections discuss different techniques for overcoming this difficulty. ■

10.5.2 Watershed Segmentation Using Gradients

The gradient magnitude is used often to preprocess a gray-scale image prior to using the watershed transform for segmentation. The gradient magnitude image has high pixel values along object edges, and low pixel values everywhere else. Ideally, then, the watershed transform would result in watershed ridge lines along object edges. The next example illustrates this concept.

EXAMPLE 10.11: Segmenting a gray-scale image using gradients and the watershed transform.

■ Figure 10.21(a) shows an image, f, containing several dark blobs. We start by computing its gradient magnitude, using either the linear filtering methods described in Section 10.1, or using a morphological gradient as described in Section 9.6.1.

```
>> h = fspecial('sobel');
>> fd = double(f);
>> g = sqrt(imfilter(fd, h, 'replicate') .^ 2 + ...
            imfilter(fd, h','replicate') .^ 2);
```

Figure 10.21(b) shows the gradient magnitude image, g. Next we compute the watershed transform of the gradient and find the watershed ridge lines.

```
>> L = watershed(g);
>> wr = L == 0;
```

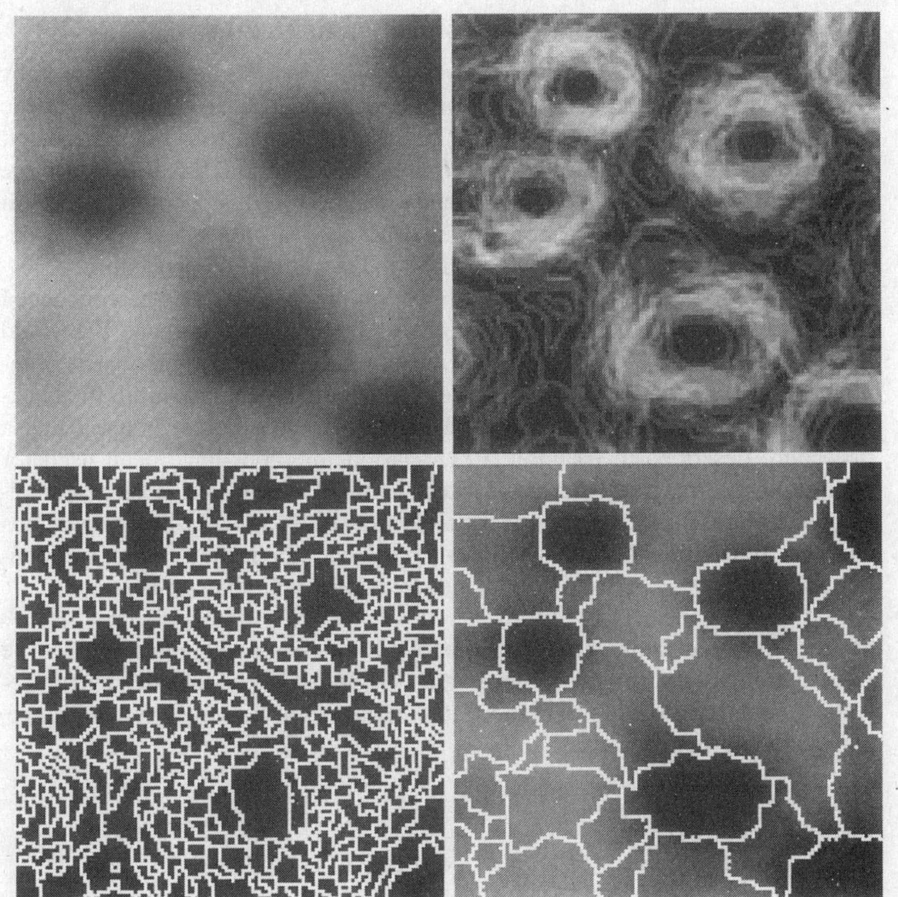

a b
c d

FIGURE 10.21 (a) Gray-scale image of small blobs. (b) Gradient magnitude image. (c) Watershed transform of (b), showing severe oversegmentation. (d) Watershed transform of the smoothed gradient image; some oversegmentation is still evident. (Original image courtesy of Dr. S. Beucher, CMM/Ecole de Mines de Paris.)

As Fig. 10.21(c) shows, this is not a good segmentation result; there are too many watershed ridge lines that do not correspond to the objects in which we are interested. This is another example of oversegmentation. One approach to this problem is to smooth the gradient image before computing its watershed transform. Here we use a close-opening, as described in Chapter 9.

```
>> g2 = imclose(imopen(g, ones(3,3)), ones(3,3));
>> L2 = watershed(g2);
>> wr2 = L2 == 0;
>> f2 = f;
>> f2(wr2) = 255;
```

The last two lines in the preceding code superimpose the watershed ridgelines in wr as white lines in the original image. Figure 10.21(d) shows the superimposed result. Although improvement over Fig. 10.21(c) was achieved, there are still some extraneous ridge lines, and it can be difficult to determine which catchment basins are actually associated with the objects of interest. The next section describes further refinements of watershed-based segmentation that deal with these difficulties. ■

10.5.3 Marker-Controlled Watershed Segmentation

Direct application of the watershed transform to a gradient image usually leads to oversegmentation due to noise and other local irregularities of the gradient. The resulting problems can be serious enough to render the result virtually useless. In the context of the present discussion, this means a large number of segmented regions. A practical solution to this problem is to limit the number of allowable regions by incorporating a preprocessing stage designed to bring additional knowledge into the segmentation procedure.

An approach used to control oversegmentation is based on the concept of markers. A *marker* is a connected component belonging to an image. We would like to have a set of *internal* markers, which are inside each of the objects of interest, as well as a set of *external* markers, which are contained within the background. These markers are then used to modify the gradient image using a procedure described in Example 10.12. Various methods have been used for computing internal and external markers, many of which involve the linear filtering, nonlinear filtering, and morphological processing described in previous chapters. Which method we choose for a particular application is highly dependent on the specific nature of the images associated with that application.

EXAMPLE 10.12: Illustration of marker-controlled watershed segmentation.

■ This example applies marker-controlled watershed segmentation to the electrophoresis gel image in Figure 10.22(a). We start by considering the results obtained from computing the watershed transform of the gradient image, without any other processing.

```
>> h = fspecial('sobel');
>> fd = double(f);
>> g = sqrt(imfilter(fd, h, 'replicate') .^ 2 + ...
           imfilter(fd, h', 'replicate') .^ 2);
>> L = watershed(g);
>> wr = L == 0;
```

We can see in Fig. 10.22(b) that the result is severely oversegmented, due in part to the large number of regional minima. IPT function `imregionalmin` computes the location of all regional minima in an image. Its calling syntax is

```
rm = imregionalmin(f)
```

where `f` is a gray-scale image and `rm` is a binary image whose foreground pixels mark the locations of regional minima. We can use `imregionalmin` on the gradient image to see why the `watershed` function produces so many small catchment basins:

```
>> rm = imregionalmin(g);
```

Most of the regional minima locations shown in Fig. 10.22(c) are very shallow and represent detail that is irrelevant to our segmentation problem. To eliminate these extraneous minima we use IPT function `imextendedmin`,

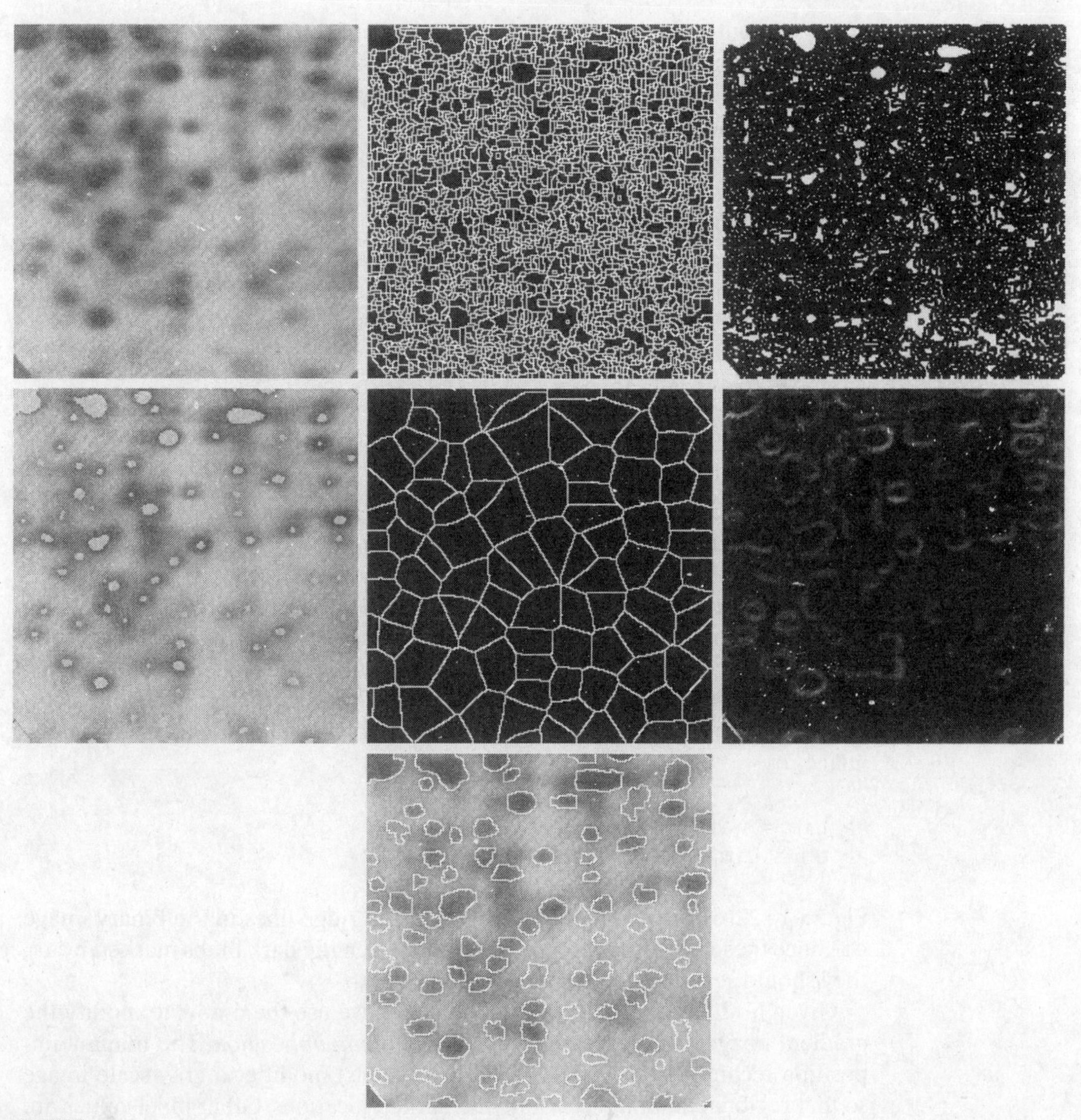

a b c
d e f
g

FIGURE 10.22 (a) Gel image. (b) Oversegmentation resulting from applying the watershed transform to the gradient magnitude image. (c) Regional minima of gradient magnitude. (d) Internal markers. (e) External markers. (f) Modified gradient magnitude. (g) Segmentation result. (Original image courtesy of Dr. S. Beucher, CMM/Ecole des Mines de Paris.)

which computes the set of "low spots" in the image that are deeper (by a certain height threshold) than their immediate surroundings. (See Soille [2003] for a detailed explanation of the *extended minima transform* and related operations.) The calling syntax for this function is

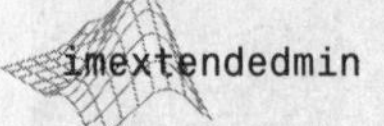


```
im = imextendedmin(f, h)
```

where `f` is a gray-scale image, `h` is the height threshold, and `im` is a binary image whose foreground pixels mark the locations of the deep regional minima. Here we use function `imextendedmin` to obtain our set of internal markers:

```
>> im = imextendedmin(f, 2);
>> fim = f;
>> fim(im) = 175;
```

The last two lines superimpose the extended minima locations as gray blobs on the original image, as shown in Fig. 10.22(d). We see that the resulting blobs do a reasonably good job of "marking" the objects we want to segment.

Next we must find external markers, or pixels that we are confident belong to the background. The approach we follow here is to mark the background by finding pixels that are exactly midway between the internal markers. Surprisingly, we do this by solving another watershed problem; specifically, we compute the watershed transform of the distance transform of the internal marker image, `im`:

```
>> Lim = watershed(bwdist(im));
>> em = Lim == 0;
```

Figure 10.22(e) shows the resulting watershed ridge lines in the binary image `em`. Since these ridgelines are midway in between the dark blobs marked by `im`, they should serve well as our external markers.

Given both internal and external markers, we use them now to modify the gradient image using a procedure called *minima imposition*. The minima imposition technique (see Soille [2003] for details) modifies a gray-scale image so that regional minima occur only in marked locations. Other pixel values are "pushed up" as necessary to remove all other regional minima. IPT function `imimposemin` implements this technique. Its calling syntax is

```
mp = imimposemin(f, mask)
```

where `f` is a gray-scale image and `mask` is a binary image whose foreground pixels mark the desired locations of regional minima in the output image, `mp`. We modify the gradient image by imposing regional minima at the locations of both the internal and the external markers:

```
>> g2 = imimposemin(g, im | em);
```

Figure 10.22(f) shows the result. We are finally ready to compute the watershed transform of the marker-modified gradient image and look at the resulting watershed ridgelines:

```
>> L2 = watershed(g2);
>> f2 = f;
>> f2(L2 == 0) = 255;
```

The last two lines superimpose the watershed ridge lines on the original image. The result, a much-improved segmentation, is shown in Fig. 10.22(g). ■

Marker selection can range from the simple procedures just described to considerably more complex methods involving size, shape, location, relative distances, texture content, and so on (see Chapter 11 regarding descriptors). The point is that using markers brings a priori knowledge to bear on the segmentation problem. Humans often aid segmentation and higher-level tasks in everyday vision by using a priori knowledge, one of the most familiar being the use of context. Thus, the fact that segmentation by watersheds offers a framework that can make effective use of this type of knowledge is a significant advantage of this method.

Summary

Image segmentation is an essential preliminary step in most automatic pictorial pattern-recognition and scene analysis problems. As indicated by the range of examples presented in this chapter, the choice of one segmentation technique over another is dictated mostly by the particular characteristics of the problem being considered. The methods discussed in this chapter, although far from exhaustive, are representative of techniques used commonly in practice.

11 Representation and Description

Preview

After an image has been segmented into regions by methods such as those discussed in Chapter 10, the next step usually is to represent and describe the aggregate of segmented, "raw" pixels in a form suitable for further computer processing. Representing a region involves two basic choices: (1) We can represent the region in terms of its external characteristics (its boundary), or (2) we can represent it in terms of its internal characteristics (the pixels comprising the region). Choosing a representation scheme, however, is only part of the task of making the data useful to a computer. The next task is to *describe* the region based on the chosen representation. For example, a region may be *represented* by its boundary, and the boundary may be *described* by features such as its length and the number of concavities it contains.

An external representation is selected when interest is on shape characteristics. An internal representation is selected when the principal focus is on regional properties, such as color and texture. Both types of representations sometimes are used in the same application to solve a problem. In either case, the features selected as descriptors should be as insensitive as possible to variations in region size, translation, and rotation. For the most part, the descriptors discussed in this chapter satisfy one or more of these properties.

11.1 Background

A *region* is a connected component, and the *boundary* (also called the *border* or *contour*) of a region is the set of pixels in the region that have one or more neighbors that are not in the region. Points not on a boundary or region are called *background* points. Initially we are interested only in binary images, so region or boundary points are represented by 1s and background points by 0s. Later in this chapter we allow pixels to have gray-scale or multispectral values.

From the definition given in the previous paragraph, it follows that a boundary is a connected set of points. The points on a boundary are said to be *ordered* if they form a clockwise or counterclockwise sequence. A boundary is said to be *minimally connected* if each of its points has exactly two 1-valued neighbors that are not 4-adjacent. An *interior point* is defined as a point anywhere in a region, except on its boundary.

The material in this chapter differs significantly from the discussions thus far in the sense that we have to be able to handle a mixture of different types of data such as boundaries, regions, topological data, and so forth. Thus, before proceeding, we pause briefly to introduce some basic MATLAB and IPT concepts and functions for use later in the chapter.

11.1.1 Cell Arrays and Structures

We begin with a discussion of MATLAB's cell arrays and structures, which were introduced briefly in Section 2.10.6.

Cell Arrays

Cell arrays provide a way to combine a mixed set of objects (e.g., numbers, characters, matrices, other cell arrays) under one variable name. For example, suppose that we are working with (1) an `uint8` image, `f`, of size 512 × 512; (2) a sequence of 2-D coordinates in the form of rows of a 188 × 2 array, `b`; and (3) a cell array containing two character names, `char_array = {'area', 'centroid'}`. These three dissimilar entities can be organized into a single variable, `C`, using cell arrays:

```
C = {f, b, char_array}
```

where the curly braces designate the contents of the cell array. Typing `C` at the prompt would output the following results:

```
>> C
C =
    [512x512 uint8]    [188x2 double]    {1x2 cell}
```

In other words, the outputs are not the values of the various variables, but a description of some of their properties instead. To address the complete contents of an element of the cell, we enclose the numerical location of that element in curly braces. For instance, to see the contents of `char_array` we type

```
>> C{3}
ans =
    'area'  'centroid'
```

or we can use function `celldisp`:

```
>> celldisp(C{3})
ans{1} =
    area
ans{2} =
    centroid
```

Using parentheses instead of curly braces on an element of C gives a description of the variable, as above:

```
>> C(3)
ans =
    {1x2 cell}
```

We can work with specified contents of a cell array by transferring them to a numeric or other pertinent from of array. For instance, to extract f from C we use

```
>> f = C{1};
```

Function size gives the size of a cell array:

```
>> size(C)
ans =
    1   3
```

Function cellfun, with syntax

```
D = cellfun('fname', C)
```

See the cellfun *help page for a list of valid entries for* fname.

applies the function fname to the elements of cell array C and returns the results in the double array D. Each element of D contains the value returned by fname for the corresponding element in C. The output array D is the same size as the cell array C. For example,

```
>> D = cellfun('length', C)
D =
    512    188    2
```

In other words, length(f) = 512, length(b) = 188 and length(char_array) = 2. Recall from Section 2.10.3 that length(A) gives the size of the longest dimension of a multidimensional array A.

Finally, keep in mind the comment made in Section 2.10.6 that cell arrays contain copies of the arguments, not pointers to those arguments. Thus, if any of the arguments of C in the preceding example were to change after C was created, that change would not be reflected in C.

EXAMPLE 11.1: A simple illustration of cell arrays.

■ Suppose that we want to write a function that outputs the average intensity of an image, its dimensions, the average intensity of its rows, and the average intensity of its columns. We can do it in the "standard" way by writing a function of the form

```
function [AI, dim, AIrows, AIcols] = image_stats(f)
dim = size(f);
AI = mean2(f);
AIrows = mean(f, 2);
AIcols = mean(f, 1);
```

where `f` is the input image and the output variables correspond to the quantities just mentioned. Using cells arrays, we would write

```
function G = image_stats(f)
G{1} = size(f);
G{2} = mean2(f);
G{3} = mean(f, 2);
G{4} = mean(f, 1);
```

Writing `G(1) = {size(f)}`, and similarly for the other terms, also is acceptable. Cell arrays can be multidimensional. For instance, the previous function could be written also as

```
function H = image_stats2(f)
H(1, 1) = {size(f)};
H(1, 2) = {mean2(f)};
H(2, 1) = {mean(f, 2)};
H(2, 2) = {mean(f, 1)};
```

Or, we could have used `H{1,1} = size(f)`, and so on for the other variables. Additional dimensions are handled in a similar manner.

Suppose that `f` is of size 512 × 512. Typing `G` and `H` at the prompt would give

```
>> G = image_stats(f)
>> H = image_stats2(f);
>> G
G =
    [1x2 double]    [1]    [512x1 double]    [1x512 double]
>> H
H =
    [  1x2 double]    [          1]
    [512x1 double]    [1x512 double]
```

If we want to work with any of the variables contained in `G`, we extract it by addressing a specific element of the cell array, as before. For instance, if we want to work with the size of `f`, we write

```
>> v = G{1}
```

or

```
>> v = H{1,1}
```

where v is a 1×2 vector. Note that we did not use the familiar command [M, N] = G{1} to obtain the size of the image. This would cause an error because only functions can produce multiple outputs. To obtain M and N we would use M = v(1) and N = v(2). ■

The economy of notation evident in the preceding example becomes even more obvious when the number of outputs is large. One drawback is the loss of clarity in the use of numerical addressing, as opposed to assigning names to the outputs. Using structures helps in this regard.

Structures

Structures are similar to cell arrays in the sense that they allow grouping of a collection of dissimilar data into a single variable. However, unlike cell arrays, where cells are addressed by numbers, the elements of structures are addressed by names called *fields*.

EXAMPLE 11.2: A simple illustration of structures.

■ Continuing with the theme of Example 11.1 will clarify these concepts. Using structures, we write

```
function s = image_stats(f)
s.dim = size(f);
s.AI = mean2(f);
s.AIrows = mean(f, 2);
s.AIcols = mean(f, 1);
```

where s is a structure. The fields of the structure in this case are AI (a scalar), dim (a 1×2 vector), AIrows (an $M \times 1$ vector), and AIcols (a $1 \times N$ vector), where M and N are the number of rows and columns of the image. Note the use of a dot to separate the structure from its various fields. The field names are arbitrary, but they must begin with a nonnumeric character.

Using the same image as in Example 11.1 and typing s and size(s) at the prompt gives the following output:

```
>> s =

s =

        dim:  [512 512]
         AI:  1
     AIrows:  [512x1 double]
     AIcols:  [1x512 double]
```

```
>> size(s)
ans =

     1     1
```

Note that s itself is a scalar, with four fields associated with it in this case.

We see in this example that the logic of the code is the same as before, but the organization of the output data is much clearer. As in the case of cell arrays, the advantage of using structures would become even more evident if we were dealing with a larger number of outputs. ■

The preceding illustration used a single structure. If, instead of one image, we had Q images organized in the form of an $M \times N \times Q$ array, the function would become

```
function s = image_stats(f)
K = size(f);
for k = 1:K(3)
   s(k).dim = size(f(:, :, k));
   s(k).AI = mean2(f(:, :, k));
   s(k).AIrows = mean(f(:, :, k), 2);
   s(k).AIcols = mean(f(:, :, k), 1);
end
```

In other words, structures themselves can be indexed. Although, like cell arrays, structures can have any number of dimensions, their most common form is a vector, as in the preceding function.

Extracting data from a field requires that the dimensions of both s and the field be kept in mind. For example, the following statement extracts all the values of AIrows and stores them in v:

```
for k = 1:length(s)
   v(:, k) = s(k).AIrows;
end
```

Note that the colon is in the first dimension of v and that k is in the second because s is of dimension $1 \times Q$ and AIrows is of dimension $M \times Q$. Thus, because k goes from 1 to Q, v is of dimension $M \times Q$. Had we been interested in extracting the values of AIcols instead, we would have used v(k, :) in the loop.

Square brackets can be used to extract the information into a vector or matrix if the field of a structure contains scalars. For example, suppose that D.Area contains the area of each of 20 regions in an image. Writing

```
>> w = [D.Area];
```

creates a 1×20 vector w in which each elements is the area of one of the regions.

As with cell arrays, when a value is assigned to a structure field, MATLAB makes a copy of that value in the structure. If the original value is changed at a later time, the change is not reflected in the structure.

11.1.2 Some Additional MATLAB and IPT Functions Used in This Chapter

Function `imfill` was mentioned briefly in Table 9.3 and in Section 9.5.2. This function performs differently for binary and intensity image inputs, so, to help clarify the notation in this section, we let `fB` and `fI` represent binary and intensity images, respectively. If the output is a binary image, we denote it by `gB`; otherwise we denote simply as `g`. The syntax

```
gB = imfill(fB, locations, conn)
```

performs a flood-fill operation on background pixels (i.e., it changes background pixels to 1) of the input binary image `fB`, starting from the points specified in `locations`. This parameter can be an $n \times 1$ vector (n is the number of locations), in which case it contains the *linear indices* (see Section 2.8.2) of the starting coordinate locations. Parameter `locations` can also be an $n \times 2$ matrix, in which each row contains the 2-D coordinates of one of the starting locations in `fB`. Parameter `conn` specifies the connectivity to be used on the background pixels: 4 (the default), or 8. If both `location` and `conn` are omitted from the input argument, the command `gB = imfill(fB)` displays the binary image, `fB`, on the screen and lets the user select the starting locations using the mouse. Click the left mouse button to add points. Press **BackSpace** or **Delete** to remove the previously selected point. A shift-click, right-click, or double-click selects a final point and then starts the fill operation. Pressing **Return** finishes the selection without adding a point.

Using the syntax

```
gB = imfill(fB, conn, 'holes')
```

fills holes in the input binary image. A *hole* is a set of background pixels that cannot be reached by filling the background from the edge of the image. As before, `conn` specifies connectivity: 4 (the default) or 8.

The syntax

```
g = imfill(fI, conn, 'holes')
```

fills holes in an input intensity image, `fI`. In this case, a hole is an area of dark pixels surrounded by lighter pixels. Parameter `conn` is as before.

See Section 5.2.2 for a discussion of function `find` *and Section 9.4 for a discussion of* `bwlabel`.

Function `find` can be used in conjunction with `bwlabel` to return vectors of coordinates for the pixels that make up a specific object. For example, if `[gB, num] = bwlabel(fB)` yields more than one connected region (i.e., `num > 1`), we obtain the coordinates of, say, the second region using

```
[r, c] = find(g == 2)
```

The 2-D *coordinates* of regions or boundaries are organized in this chapter in the form of $np \times 2$ arrays, where each row is an (x, y) coordinate pair, and np is the number of points in the region or boundary. In some cases it is necessary to sort these arrays. Function `sortrows` can be used for this purpose:

```
z = sortrows(S)
```

This function sorts the rows of `S` in ascending order. Argument `S` must be either a matrix or a column vector. In this chapter, `sortrows` is used only with $np \times 2$ arrays. If several rows have identical first coordinates, they are sorted in ascending order of the second coordinate. If we want to sort the rows of `S` and also eliminate duplicate rows, we use function `unique`, which has the syntax

```
[z, m, n] = unique(S, 'rows')
```

where `z` is the sorted array with no duplicate rows, and `m` and `n` are such that `z = S(m, :)` and `S = z(n, :)`. For example, if `S = [1 2; 6 5; 1 2; 4 3]`, then `z = [1 2;4 3;6 5]`, `m = [3; 4; 2]`, and `n = [1; 3; 1; 2]`. Note that `z` is arranged in ascending order and that `m` indicates which rows of the original array were kept.

Frequently, it is necessary to shift the rows of an array up, down, or sideways a specified number of positions. For this we use function `circshift`:

```
z = circshift(S, [ud lr])
```

where `ud` is the number of elements by which `S` is shifted up or down. If `ud` is positive, the shift is down; otherwise it is up. Similarly, if `lr` is positive, the array is shifted to the right `lr` elements; otherwise it is shifted to the left. If only up and down shifting is needed, we can use a simpler syntax

```
z = circshift(S, ud)
```

If `S` is an image, `circshift` is really nothing more than the familiar *scrolling* (up and down) or *panning* (right and left), with the image wrapping around.

11.1.3 Some Basic Utility M-Functions

Tasks such as converting between regions and boundaries, ordering boundary points in a contiguous chain of coordinates, and subsampling a boundary to simplify its representation and description are typical of the processes that are employed routinely in this chapter. The following utility M-functions are used for these purposes. To avoid a loss of focus on the main topic of this chapter, we discuss only the syntax of these functions. The documented code for each non-MATLAB function is included in Appendix C. As noted earlier, boundaries are represented as $np \times 2$ arrays in which each row represents a 2-D pair of coordinates. Many of these functions automatically convert $2 \times np$ coordinate arrays to arrays of size $np \times 2$.

Function

boundaries

```
B = boundaries(f, conn, dir)
```

traces the *exterior* boundaries of the objects in f, which is assumed to be a binary image with 0s as the background. Parameter conn specifies the desired connectivity of the output boundaries; its values can be 4 or 8 (the default). Parameter dir specifies the direction in which the boundaries are traced; its values can be 'cw' (the default) or 'ccw', indicating a clockwise or counterclockwise direction. Thus, if 8-connectivity and a 'cw' direction are acceptable, we can use the simpler syntax

```
B = boundaries(f)
```

Output B in both syntaxes is a cell array whose elements are the coordinates of the boundaries found. The first and last points in the boundaries returned by function boundaries are the same. This produces a closed boundary.

As an example to fix ideas, suppose that we want to find the boundary of the object with the longest boundary in image f (for simplicity we assume that the longest boundary is unique). We do this with the following sequence of commands:

```
>> B = boundaries(f);
>> d = cellfun('length', B);
>> [max_d, k] = max(d);
>> v = B{k(1)};
```

See Section 2.10.2 for an explanation of this use of function max.

Vector v contains the coordinates of the longest boundary in the input image, and k is the corresponding region number; array v is of size $np \times 2$. The last statement simply selects the first boundary of maximum length if there is more than one such boundary. As noted in the previous paragraph, the first and last points of every boundary computed using function boundaries are the same, so row v(1, :) is the same as row v(end, :).

Function bound2eight with syntax

bound2eight

```
b8 = bound2eight(b)
```

removes from b pixels that are necessary for 4-connectedness but not necessary for 8-connectedness, leaving a boundary whose pixels are only 8-connected. Input b must be an $np \times 2$ matrix, each row of which contains the (x, y) coordinates of a boundary pixel. It is required that b be a closed, connected set of pixels ordered sequentially in the clockwise or counterclockwise direction. The same conditions apply to function bound2four:

bound2four

```
b4 = bound2four(b)
```

This function inserts new boundary pixels wherever there is a diagonal connection, thus producing an output boundary in which pixels are only 4-connected. Code listings for both functions can be found in Appendix C.

Function

```
g = bound2im(b, M, N, x0, y0)
```

bound2im

generates a binary image, g, of size $M \times N$, with 1s for boundary points and a background of 0s. Parameters x0 and y0 determine the location of the minimum x- and y-coordinates of b in the image. Boundary b must be an $np \times 2$ (or $2 \times np$) array of coordinates, where, as mentioned earlier, np is the number of points. If x0 and y0 are omitted, the boundary is centered approximately in the $M \times N$ array. If, in addition, M and N are omitted, the vertical and horizontal dimensions of the image are equal to the height and width of boundary b. If function boundaries finds multiple boundaries, we can get all the coordinates for use in function bound2im by concatenating the various elements of cell array B:

See Section 6.1.1 for an explanation of the cat *operator. See also Example 11.13.*

```
b = cat(1, B{:})
```

where the 1 indicates concatenation along the first (vertical) dimension of the array.

Function

```
[s, su] = bsubsamp(b, gridsep)
```

bsubsamp

subsamples a (single) boundary b onto a grid whose lines are separated by gridsep pixels. The output s is a boundary with fewer points than b, the number of such points being determined by the value of gridsep, and su is the set of boundary points scaled so that transitions in their coordinates are unity. This is useful for coding the boundary using chain codes, as discussed in Section 11.1.2. It is required that the points in b be ordered in a clockwise or counterclockwise direction.

When a boundary is subsampled using bsubsamp, its points cease to be connected. They can be reconnected by using

```
z = connectpoly(s(:, 1), s(:, 2))
```

connectpoly

where the rows of s are the coordinates of a subsampled boundary. It is required that the points in s be ordered, either in a clockwise or counterclockwise direction. The rows of output z are the coordinates of a connected boundary formed by connecting the points in s with the shortest possible path consisting of 4- or 8-connected straight segments. This function is useful for producing a polygonal, fully connected boundary that is generally smoother (and simpler) than the original boundary, b, from which s was obtained. Function connectpoly also is quite useful when working with functions that generate only the vertices of a polygon, such as minperpoly, discussed in Section 11.2.3.

Computing the integer coordinates of a straight line joining two points is a basic tool when working with boundaries (for example, function `connectpoly` requires a subfunction that does this). IPT function `intline` is well suited for this purpose. Its syntax is

```
[x, y] = intline(x1, x2, y1, y2)
```

`intline` is an undocumented IPT utility function. Its code is included in Appendix C.

where `(x1, y1)` and `(x2, y2)` are the integer coordinates of the two points to be connected. The outputs `x` and `y` are column vectors containing the integer *x*- and *y*-coordinates of the straight line joining the two points.

11.2 Representation

As noted at the beginning of this chapter, the segmentation techniques discussed in Chapter 10 yield raw data in the form of pixels along a boundary or pixels contained in a region. Although these data sometimes are used directly to obtain descriptors (as in determining the texture of a region), standard practice is to use schemes that compact the data into representations that are considerably more useful in the computation of descriptors. In this section we discuss the implementation of various representation approaches.

11.2.1 Chain Codes

Chain codes are used to represent a boundary by a connected sequence of straight-line segments of specified length and direction. Typically, this representation is based on 4- or 8-connectivity of the segments. The direction of each segment is coded by using a numbering scheme such as the ones shown in Figs. 11.1(a) and (b). Chain codes based on this scheme are referred to as *Freeman chain codes.*

The chain code of a boundary depends on the starting point. However, the code can be normalized with respect to the starting point by treating it as a circular sequence of direction numbers and redefining the starting point so that the resulting sequence of numbers forms an integer of minimum magnitude. We can normalize for rotation [in increments of 90° or 45°, as shown in Figs. 11.1(a) and (b)] by using the *first difference* of the chain code instead of the code itself. This difference is obtained by counting the number of direction changes (in a coun-

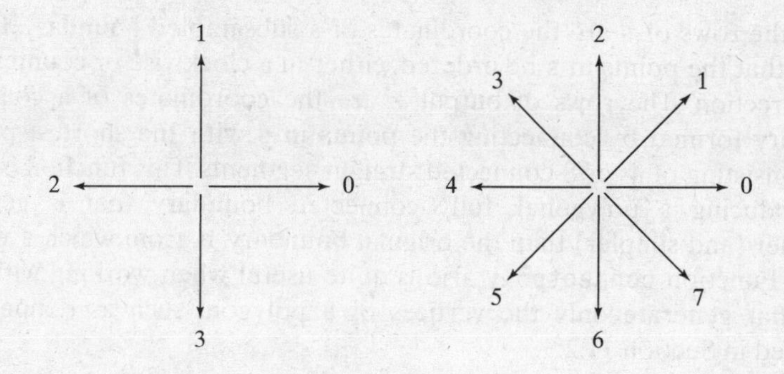

a b

FIGURE 11.1 (a) Direction numbers for (a) a 4-directional chain code, and (b) an 8-directional chain code.

terclockwise direction in Fig. 11.1) that separate two adjacent elements of the code. For instance, the first difference of the 4-direction chain code 10103322 is 3133030. If we elect to treat the code as a circular sequence, then the first element of the difference is computed by using the transition between the last and first components of the chain. Here, the result is 33133030. Normalization with respect to arbitrary rotational angles is achieved by orienting the boundary with respect to some dominant feature, such as its major axis, as discussed in Section 11.3.2.

Function `fchcode`, with syntax

fchcode

```
c = fchcode(b, conn, dir)
```

computes the Freeman chain code of an $np \times 2$ set of ordered boundary points stored in array `b`. The output `c` is a structure with the following fields, where the numbers inside the parentheses indicate array size:

`c.fcc` = Freeman chain code ($1 \times np$)
`c.diff` = First difference of code `c.fcc` ($1 \times np$)
`c.mm` = Integer of minimum magnitude ($1 \times np$)
`c.diffmm` = First difference of code `c.mm` ($1 \times np$)
`c.x0y0` = Coordinates where the code starts (1×2)

Parameter `conn` specifies the connectivity of the code; its value can be 4 or 8 (the default). A value of 4 is valid only when the boundary contains no diagonal transitions.

Parameter `dir` specifies the direction of the output code: If `'same'` is specified, the code is in the same direction as the points in `b`. Using `'reverse'` causes the code to be in the opposite direction. The default is `'same'`. Thus, writing `c = fchcode(b, conn)` uses the default direction, and `c = fchcode(b)` uses the default connectivity and direction.

EXAMPLE 11.3: Freeman chain code and some of its variations.

■ Figure 11.2(a) shows an image, `f`, of a circular stroke embedded in specular noise. The objective of this example is to obtain the chain code and first difference of the object's boundary. It is obvious by looking at Fig. 11.2(a) that the noise fragments attached to the object would result in a very irregular boundary, not truly descriptive of the general shape of the object. Smoothing is a routine process when working with noisy boundaries. Figure 11.2(b) shows the result, `g`, of using a 9×9 averaging mask:

```
>> h = fspecial('average', 9);
>> g = imfilter(f, h, 'replicate');
```

The binary image in Fig. 11.2(c) was then obtained by thresholding:

```
>> g = im2bw(g, 0.5);
```

The boundary of this image was computed using function `boundaries` discussed in the previous section:

```
>> B = boundaries(g);
```

As in the illustration in Section 11.1.3, we are interested in the longest boundary:

```
>> d = cellfun('length', B);
>> [max_d, k] = max(d);
>> b = B{1};
```

The boundary image in Fig. 11.2(d) was generated using the commands:

```
>> [M N] = size(g);
>> g = bound2im(b, M, N, min(b(:, 1)), min(b(:, 2)));
```

Obtaining the chain code of b directly would result in a long sequence with small variations that are not necessarily representative of the general shape of the image. Thus, as is typical in chain-code processing, we subsample the boundary using function bsubsamp discussed in the previous section:

```
>> [s, su] = bsubsamp(b, 50);
```

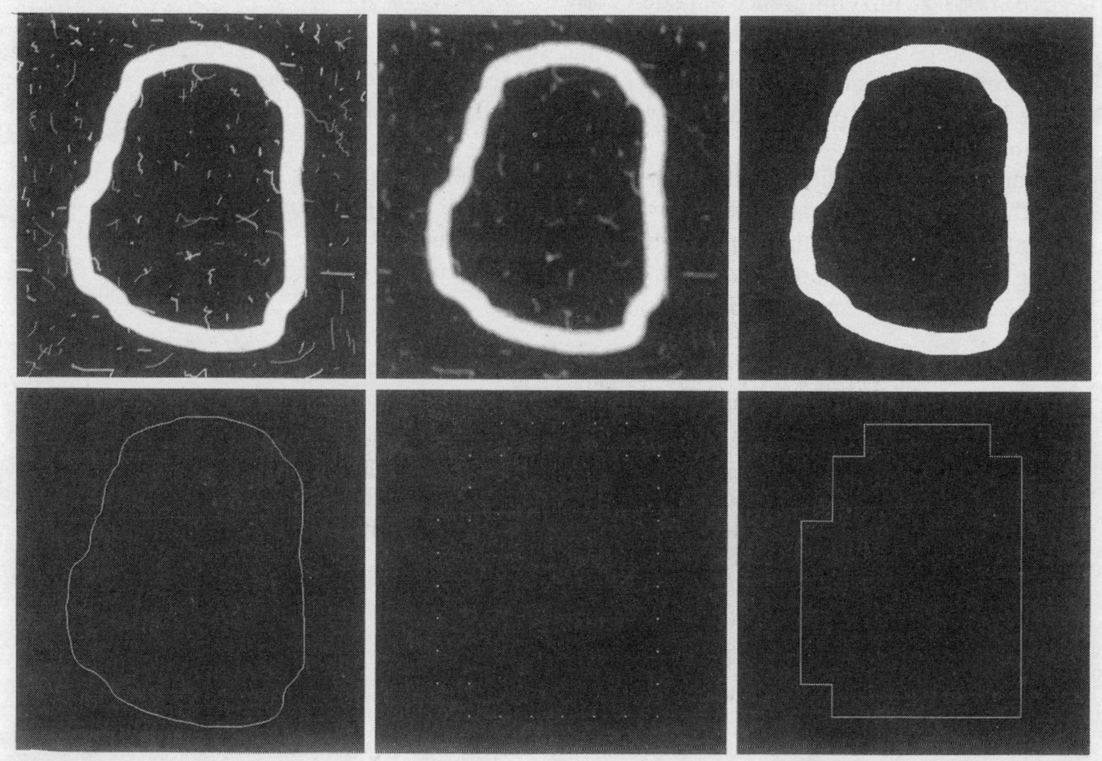

a b c
d e f

FIGURE 11.2 (a) Noisy image. (b) Image smoothed with a 9×9 averaging mask. (c) Thresholded image. (d) Boundary of binary image. (e) Subsampled boundary. (f) Connected points from (e).

Here, we used a grid separation equal to approximately 10% the width of the image, which in this case was of size 570 × 570 pixels. The resulting points can be displayed as an image [Fig. 11.2(e)]:

```
>> g2 = bound2im(s, M, N, min(s(:, 1)), min(s(:, 2)));
```

or as a connected sequence [Fig. 11.2(f)] by using the commands

```
>> cn = connectpoly(s(:, 1), s(:, 2));
>> g2 = bound2im(cn, M, N, min(cn(:, 1)), min(cn(:, 2)));
```

The advantage of using this representation, as opposed to Fig. 11.2(d), for chain-coding purposes is evident by comparing this figure with Fig. 11.2(f). The chain code is obtained from the scaled sequence su:

```
>> c = fchcode(su);
```

This command resulted in the following outputs:

```
>> c.x0y0
ans =
  7    3
>> c.fcc
ans =
2 2 0 2 2 0 2 0 0 0 0 6 0 6 6 6 6 6 6 6 6 4 4 4 4 4 4 2 4 2 2 2
>> c.mm
ans =
0 0 0 0 6 0 6 6 6 6 6 6 6 6 4 4 4 4 4 4 2 4 2 2 2 2 2 0 2 2 0 2
>> c.diff
ans =
0 6 2 0 6 2 6 0 0 0 6 2 6 0 0 0 0 0 0 0 6 0 0 0 0 0 6 2 6 0 0 0
>> c.diffmm
ans =
0 0 0 6 2 6 0 0 0 0 0 0 0 6 0 0 0 0 0 6 2 6 0 0 0 0 0 6 2 0 6 2 6
```

By examining c.fcc , Fig. 11.2(f), and c.x0y0 we see that the code starts on the left of the figure and proceeds in the clockwise direction, which is the same direction as the coordinates of the boundary. ■

11.2.2 Polygonal Approximations Using Minimum-Perimeter Polygons

A digital boundary can be approximated with arbitrary accuracy by a polygon. For a closed curve, the approximation is exact when the number of segments in the polygon is equal to the number of points in the boundary, so that each pair of adjacent points defines an edge of the polygon. In practice, the goal of a polygonal approximation is to use the fewest vertices possible to capture the "essence" of the boundary shape.

A particularly attractive approach to polygonal approximation is to find the *minimum-perimeter polygon* (MPP) of a region or boundary. The theoretical underpinnings and an algorithm for finding MPPs are discussed in the classic paper by Sklansky et al. [1972] (see also Kim and Sklansky [1982]). In this section we present the fundamentals of the algorithm and give an M-function implementation of the procedure. The method is restricted to *simple* polygons (i.e., polygons with no self-intersections). Also, regions with peninsular protrusions that are one pixel thick are excluded. Such protrusions can be extracted using morphological methods and then reappended after the polygonal approximation has been computed.

Foundation

We begin with a simple example to fix ideas. Suppose that we enclose a boundary by a set of concatenated cells, as shown in Fig. 11.3(a). It helps to visualize this enclosure as two walls corresponding to the outside and inside boundaries of the strip of cells and think of the object boundary as a rubber band contained within the two walls. If the rubber band is allowed to shrink, it takes the shape shown in Fig. 11.3(b), producing a polygon of minimum perimeter that fits the geometry established by the cell strip.

Sklansky's approach uses a so-called *cellular complex* or *cellular mosaic*, which, for our purposes, is the set of *square* cells used to enclose a boundary, as in Fig. 11.3(a). Figure 11.4(a) shows the region (shaded) enclosed by the cellular complex. Note that the boundary of this region forms a 4-connected path. As we traverse this path in a clockwise direction, we assign a black dot (•) to the convex corners (those with interior angles equal to 90°) and a white dot (∘) to the concave corners (those with interior angles equal to 270°). As Fig. 11.4(b) shows, the black dots are placed on the convex corners themselves. The white dots are placed diagonally opposite their corresponding concave corners. This corresponds to the cellular complex and vertex definitions of the algorithm.

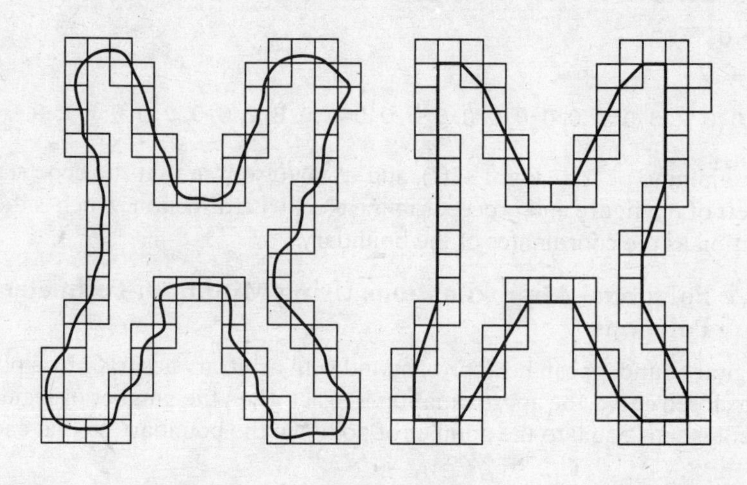

a b

FIGURE 11.3 (a) Object boundary enclosed by cells. (b) Minimum-perimeter polygon.

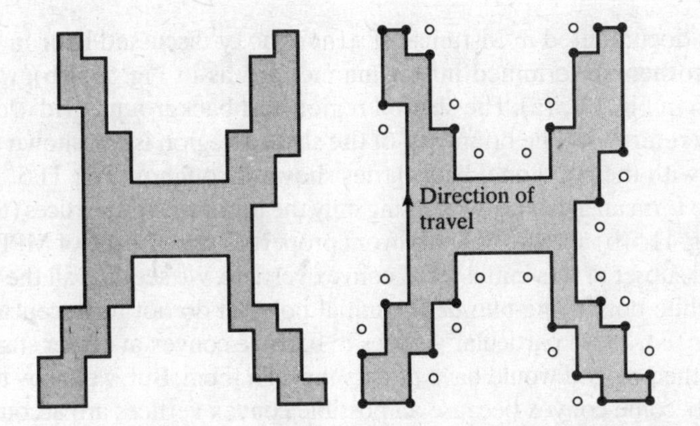

a b

FIGURE 11.4 (a) Region enclosed by the inner wall of the cellular complex in Fig. 11.3(a). (b) Convex (•) and concave (∘) corner markers for the boundary of the region in (a). Note that concave markers are placed diagonally opposite their corresponding corners.

The following properties are basic in formulating an approach for finding MPPs:

1. The MPP corresponding to a simply connected cellular complex is not self-intersecting. Let P denote this MPP.
2. Every *convex* vertex of P coincides with a • (but not every • is a vertex of P).
3. Every *concave* vertex of P coincides with a ∘ (but not every ∘ is a vertex of P).
4. If a • in fact is part of P, but it is not a convex vertex of P, then it lies on the edge of P.

In our discussion, a vertex of a polygon is defined to be *convex* if its *interior* angle is in the range $0° < \theta < 180°$; otherwise the vertex is *concave*. As in the previous paragraph, convexity is measured with respect to the interior region as we travel in a clockwise direction.

The condition $\theta = 0°$ is not allowed, and $\theta = 180°$ is treated as a special case.

An Algorithm for Finding MPPs

Properties 1 through 4 are the basis for finding the vertices of an MPP. There are various ways to do this (e.g., see Sklansky et al. [1972], and Kim and Sklansky [1982]). The approach we follow here is designed to take advantage of two basic IPT/MATLAB functions. The first is `qtdecomp`, which performs quadtree decompositions that lead to the cellular wall enclosing the data of interest. The second is function `inpolygon`, used to determine which points lie outside, on, or inside the boundary of a polygon defined by a given set of vertices.

It will be helpful to develop the procedure for finding MPPs in the context of an illustration. We use Figs. 11.3 and 11.4 again for this purpose. An approach for finding the 4-connected boundary of the shaded inner region in Fig. 11.4(a) is discussed later in this section. After the boundary has been obtained, the next step is to find its corners, which we do by obtaining its Freeman chain code. Changes in code direction indicate a corner in the boundary. By analyzing direction changes as we travel in a clockwise direction through the boundary, it becomes a fairly easy task to determine and mark the convex and concave corners, as in Fig. 11.4(b). The specific approach for obtaining the markers is documented in M-function `minperpoly` discussed later in this section. The corners determined in this manner are as in Fig. 11.4(b), which we show again in Fig. 11.5(a). The shaded region and background grid are included for easy reference. The boundary of the shaded region is not shown to avoid confusion with the polygonal boundaries shown throughout Fig. 11.5.

Next, we form an initial polygon using only the initial convex vertices (the black dots), as Fig. 11.5(b) shows. We know from property 2 that the set of MPP convex vertices is a subset of this initial set of convex vertices. We see that all the concave vertices (white dots) lying *outside* the initial polygon do not form concavities in the polygon. For those particular vertices to become convex at a later stage in the algorithm, the polygon would have to pass through them. But, we know that they can never become convex because all possible convex vertices are accounted for at this point (it is possible that their angle could become 180° later, but that would have no effect on the shape of the polygon). Thus, the white dots outside the initial polygon can be eliminated from further analysis, as Fig. 11.5(c) shows.

The concave vertices (white dots) inside the polygon are associated with concavities in the boundary that were ignored in the first pass. Thus, these vertices must be incorporated into the polygon, as shown in Fig. 11.5(d). At this point generally there are vertices that are black dots but that have ceased to be convex in the new polygon [see the black dots marked with arrows in Fig. 11.5(d)]. There are two possible reasons for this. The first reason may be that these vertices are part of the starting polygon in Fig. 11.5(b), which includes *all* convex (black) vertices. The second reason could be that they have become convex as a result of our having incorporated additional (white) vertices into the polygon as in Fig. 11.5(d). Therefore, all black dots in the polygon must be tested to see if any of the vertex angles at those points now exceed 180°. All those that do are deleted. The procedure in then repeated.

Figure 11.5(e) shows only one new black vertex that has become concave during the second pass through the data. The procedure terminates when no further vertex changes take place, at which time all vertices with angles of 180° are deleted because they are on an edge, and thus do not affect the shape of the final polygon. The boundary in Fig. 11.5(f) is the MPP for our example. This polygon is the same as the polygon in Fig. 11.3(b). Finally, Fig. 11.4(g) shows the original cellular complex superimposed on the MPP.

The preceding discussion is summarized in the following steps for finding the MPP of a region:

1. Obtain the cellular complex (the approach is discussed later in this section).
2. Obtain the region internal to the cellular complex.
3. Use function `boundaries` to obtain the boundary of the region in step 2 as a 4-connected, clockwise sequence of *coordinates*.
4. Obtain the Freeman chain code of this 4-connected sequence using function `fchcode`.
5. Obtain the convex (black dots) and concave (white dots) vertices from the chain code.
6. Form an initial polygon using the black dots as vertices, and delete from further analysis any white dots that are outside this polygon (white dots on the polygon boundary are kept).

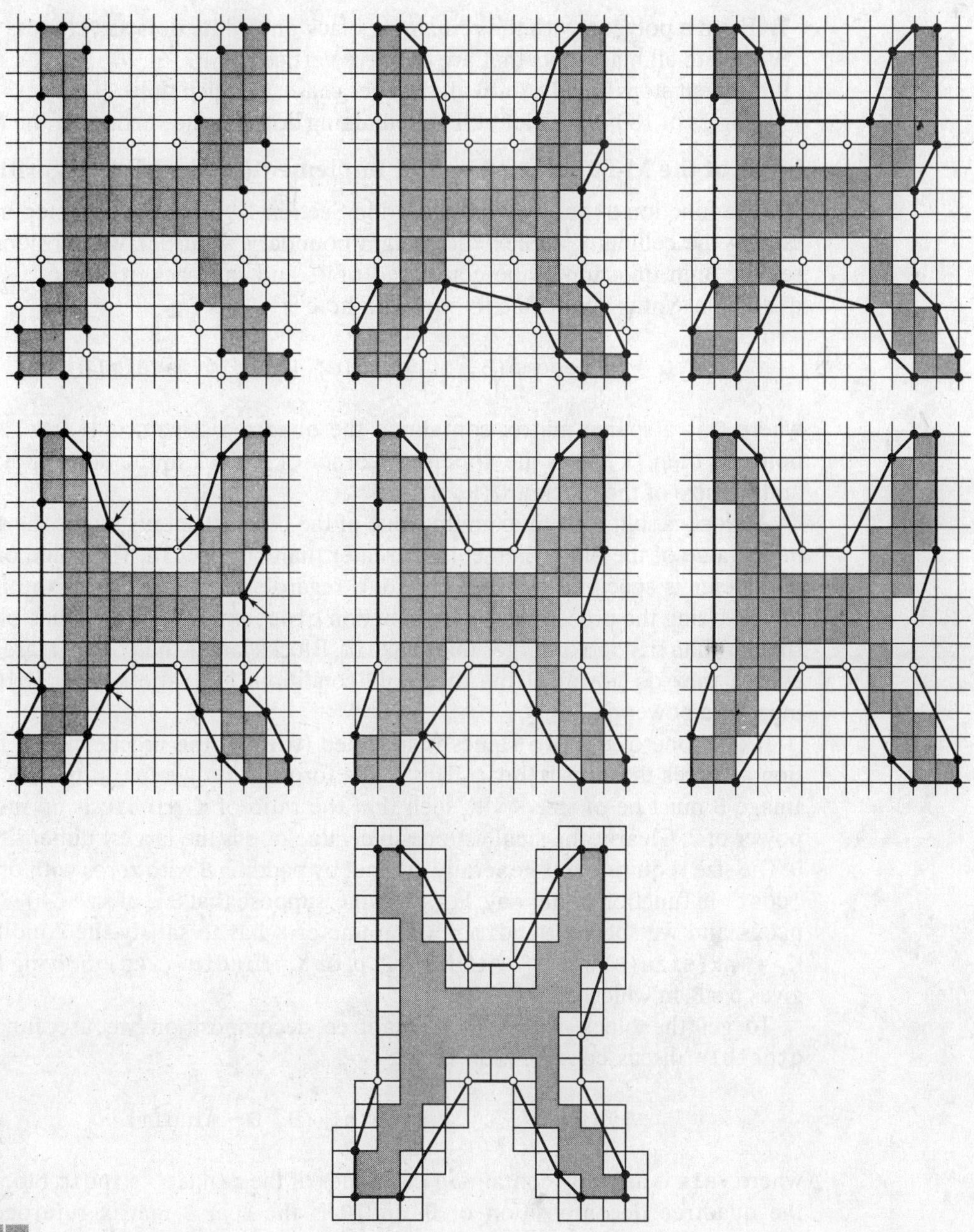

FIGURE 11.5 (a) Convex (black) and concave (white) vertices of the boundary in Fig. 11.4(a). (b) Initial polygon joining all convex vertices. (c) Result after deleting concave vertices outside of the polygon. (d) Result of incorporating the remaining concave vertices into the polygon (the arrows indicate black vertices that have become concave and will be deleted). (e) Result of deleting concave black vertices (the arrow indicates a black vertex that now has become concave). (f) Final result showing the MPP. (g) MPP with boundary cells superimposed.

7. Form a polygon with the remaining black and white dots as vertices.
8. Delete all black dots that are concave vertices.
9. Repeat steps 7 and 8 until all changes cease, at which time all vertices with angles of 180° are deleted. The remaining dots are the vertices of the MPP.

Some of the M-Functions Used in Implementing the MPP Algorithm

We use function qtdecomp introduced in Section 10.4.2 as the first step in obtaining the cellular complex enclosing a boundary. As usual, we consider the region, B, in question to be composed of 1s and the background of 0s. The qtdecomp syntax applicable to our work here is

```
Q = qtdecomp(B, threshold, [mindim maxdim])
```

where Q is a sparse matrix containing the quadtree structure. If Q(k, m) is nonzero, then (k, m) is the upper-left corner of a block in the decomposition and the size of the block is Q(k, m).

A block is split if the maximum value of the block elements minus the minimum value of the block elements is greater than threshold. The value of this parameter is specified between 0 and 1, regardless of the class of the input image. Using the preceding syntax, function qtdecomp will not produce blocks smaller than mindim or larger than maxdim. Blocks larger than maxdim are split even if they do not meet the threshold condition. The ratio maxdim/mindim must be a power of 2.

If only one of the two values is specified (without the brackets), the function assumes that it is mindim. This is the formulation we use in this section. Image B must be of size K × K, such that the ratio of K/mindim is an integer power of 2. Clearly, the smallest possible value of K is the largest dimension of B. The size requirements generally are met by padding B with zeros with option 'post' in function padarray. For example, suppose that B is of size 640 × 480 pixels, and we specify mindim = 3. Parameter K has to satisfy the conditions K >= max(size(B)) and K/mindim = 2^p, or K = mindim*(2^p). Solving for p gives p = 8, in which case K = 768.

To get the block values in a quadtree decomposition we use function qtgetblk, discussed in Section 10.4.2:

```
[vals, r, c] = qtgetblk(B, Q, mindim)
```

where vals is an array containing the values of the mindim × mindim blocks in the quadtree decomposition of B, and Q is the sparse matrix returned by qtdecomp. Parameters r and c are vectors containing the row and column coordinates of the upper-left corners of the blocks.

EXAMPLE 11.4: Obtaining the cellular wall of the boundary of a region.

■ To see how steps 1 through 4 of the MPP algorithm are implemented, consider the image in Fig. 11.6(a), and suppose that we specify mindim = 2. We show individual pixels as small squares to facilitate explanation of function qtdecomp. The image is of size 32 × 32, and it is easily verified that no addi

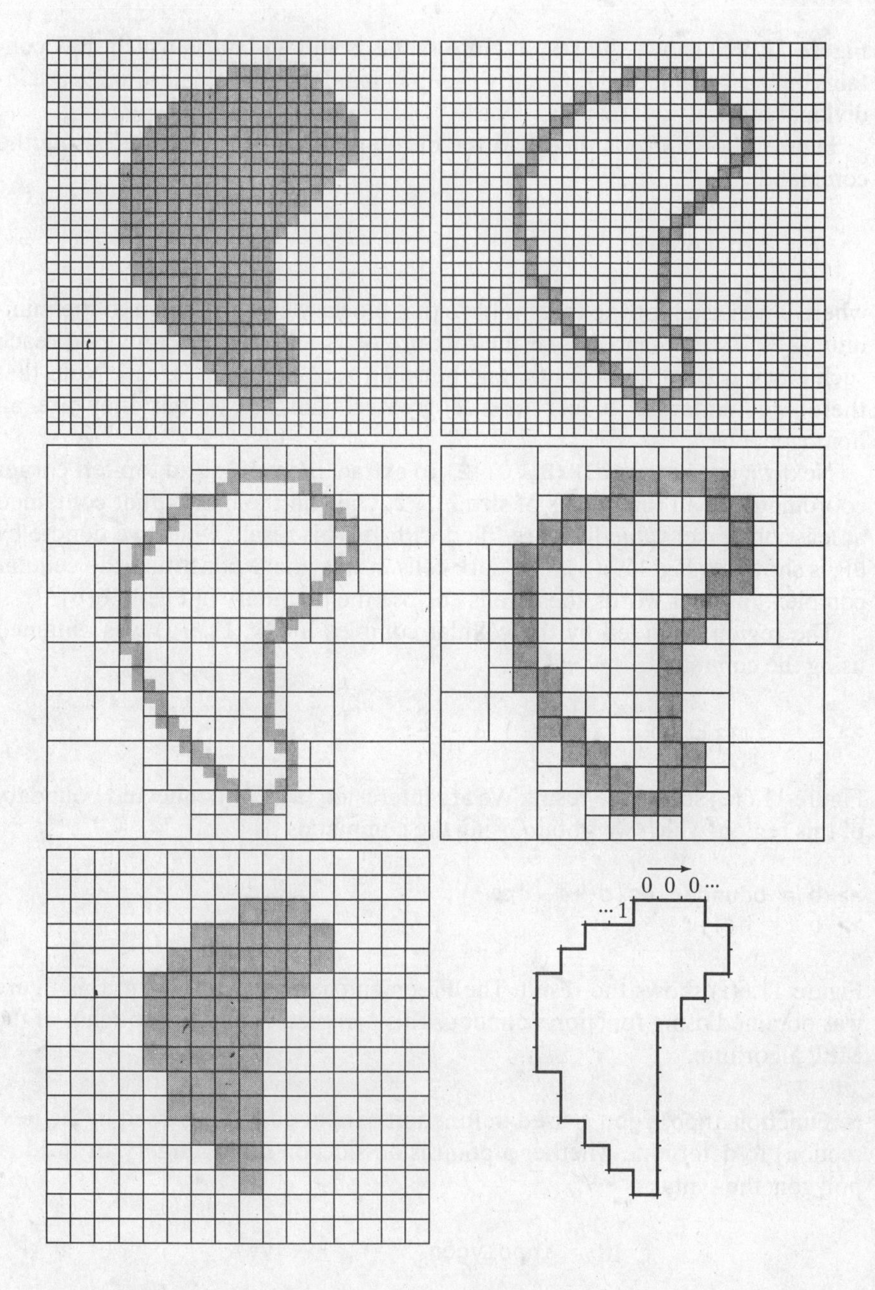

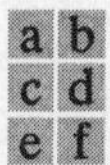

FIGURE 11.6 (a) Original image, where the small squares denote individual pixels. (b) 4-connected boundary. (c) Quadtree decomposition using blocks of minimum size 2 pixels on the side. (d) Result of filling with 1s all blocks of size 2×2 that contained at least one element valued 1. This is the cellular complex. (e) Inner region of (d). (f) 4-connected boundary points obtained using function `boundaries`. The chain code was obtained using function `fchcode`.

tional padding is required for the specified value of `mindim`. The 4-connected boundary of the region is obtained using the following command (the margin note in the next page explains why 8 is used in the function call):

```
>> B = bwperim(B, 8);
```

`bwperim`

The syntax for `bwperim` *is* `g = bwperim(f, conn)` *where* `conn` *identifies the desired connectivity: 4 (the default) or 8. The connectivity is with respect to the background pixels. Thus, to obtain 4-connected object boundaries we specify 8 for* `conn`. *Conversely, 8-connected boundaries result from specifying a value of 4 for* `conn`. *Output* `g` *is a binary image containing the boundaries of the objects in* `f`. *This function is discussed in detail in Section 11.3.1.*

Figure 11.6(b) shows the result. Note that `B` is still an image, which now contains only a 4-connected boundary (keep in mind that the small squares are individual pixels).

Figure 11.6(c) shows the quadtree decomposition of `B`, obtained using the command

```
>> Q = qtdecomp(B, 0, 2);
```

where 0 was used for the threshold so that blocks were split down to the minimum 2 × 2 size, regardless of the mixture of 1s and 0s they contained (each such block is capable of containing between zero and four pixels). Note that there are numerous blocks of size greater than 2 × 2, but they are all homogeneous.

Next we used `qtgetblk(B, Q, 2)` to extract the values and top-left corner coordinates of all the blocks of size 2 × 2. Then all the blocks that contained at least one pixel valued 1 were filled with 1s. This result, which we denote by `BF`, is shown in Fig. 11.6(d). The dark cells in this image constitute the cellular complex. In other words, these cells enclose the boundary in Fig. 11.6(b).

The region bounded by the cellular complex in Fig. 11.6(d) was obtained using the command

```
>> R = imfill(BF, 'holes') & ~BF;
```

Figure 11.6(e) shows the result. We are interested in the 4-connected boundary of this region, which we obtain using the commands

```
>> b = boundaries(b, 4, 'cw');
>> b  = b{1};
```

Figure 11.6(f) shows the result. The Freeman chain code shown in this figure was obtained using function `fchcode`. This completes steps 1 through 4 of the MPP algorithm. ■

Function `inpolygon` is used in function `minperpoly` (discussed in the next section) to determine whether a point is outside, on the boundary, or inside a polygon; the syntax is

`inpolygon`

```
IN = inpolygon(X, Y, xv, yv)
```

where `X` and `Y` are vectors containing the x- and y-coordinates of the points to be tested, and `xv` and `yv` are vectors containing the the x- and y-coordinates of the polygon vertices, arranged in a clockwise or counterclockwise sequence. Array `IN` is a vector whose length is equal to the number of points being tested. Its values are 1 for points inside or on the boundary of the polygon, and 0 for points outside the boundary.

An M-Function for Computing MPPs

Steps 1 through 9 of the MPP algorithm are implemented in function `minperpoly`, whose listing is included in Appendix C. The syntax is

```
[x, y] = minperpoly(B, cellsize)
```

minperpoly

where `B` is an input binary image containing a single region or boundary, and `cellsize` is the size of the square cells in the cellular complex used to enclose the boundary. Column vectors `x` and `y` contain the x- and y-coordinates of the MPP vertices.

EXAMPLE 11.5: Using function `minperpoly`.

■ Figure 11.7(a) shows an image, `B`, of a maple leaf, and Fig. 11.7(b) is the boundary obtained using the commands

```
>> b = boundaries(B, 4, 'cw');
>> b = b{1};
>> [M, N] = size(B);
>> xmin = min(b(:, 1));
>> ymin = min(b(:, 2));
>> bim = bound2im(b, M, N, xmin, ymin);
>> imshow(bim)
```

This is the reference boundary against which various MMPs are compared in this example. Figure 11.7(c) is the result of using the commands

```
>> [x, y] = minperpoly(B, 2);
>> b2 = connectpoly(x, y);
>> B2 = bound2im(b2, M, N, xmin, ymin);
>> imshow(B2)
```

Similarly, Figs. 11.7(d) through (f) show the MPPs obtained using square cells of sizes 3, 4, and 8. The thin stem is lost with cells larger than 2×2 due to a loss of resolution. The second major shape characteristic of the leaf is its set of three main lobes. These are preserved reasonably well even for cells of size 8, as Fig. 11.7(f) shows. Further increases in the size of the cells to 10 and even to 16 still preserve this feature, as Figs. 11.8(a) and (b) show. However, as shown in Figs. 11.8(c) and (d), values of 20 and higher cause this characteristic to be lost.

The arrows in Figs. 11.7(c) and (e) point to nodes formed by self-intersecting lines. These nodes can arise if the size of the indentation in the boundary with respect to the cell size is such that when the concave vertices are created, their positions "cross" each other, altering the clockwise sequence of the vertices. One approach for solving this problem is to delete one of the vertices. The other is to increase or decrease the cell size. For example, Fig. 11.7(d), which corresponds to a cell size of 3, does not have the problem exhibited by the vertices generated with cells of sizes 2 and 4. ■

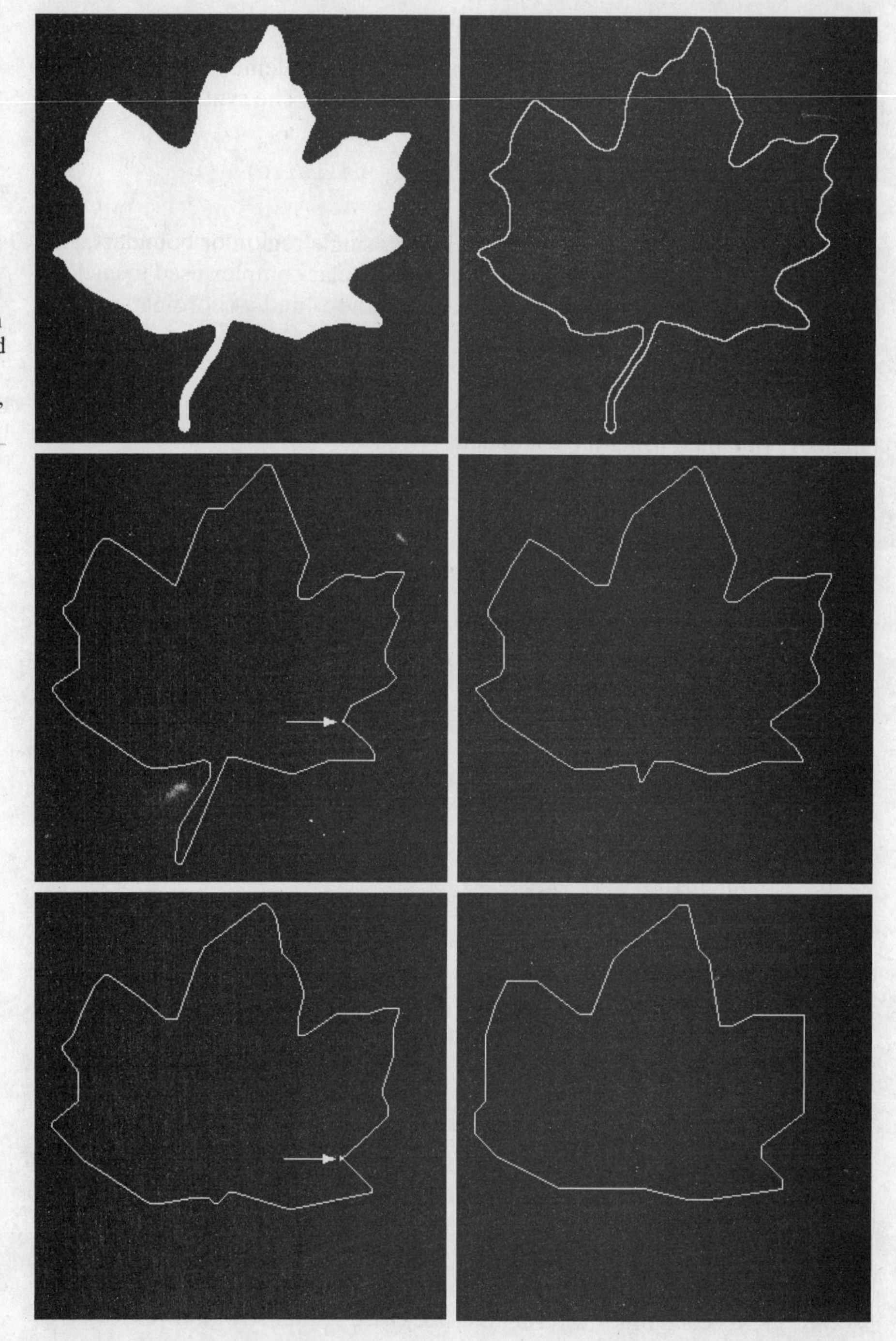

a	b
c	d
e	f

FIGURE 11.7 (a) Original image. (b) 4-connected boundary. (c) MPP obtained using square bounding cells of size 2. (d) through (f) MPPs obtained using square cells of sizes 3, 4, and 8, respectively.

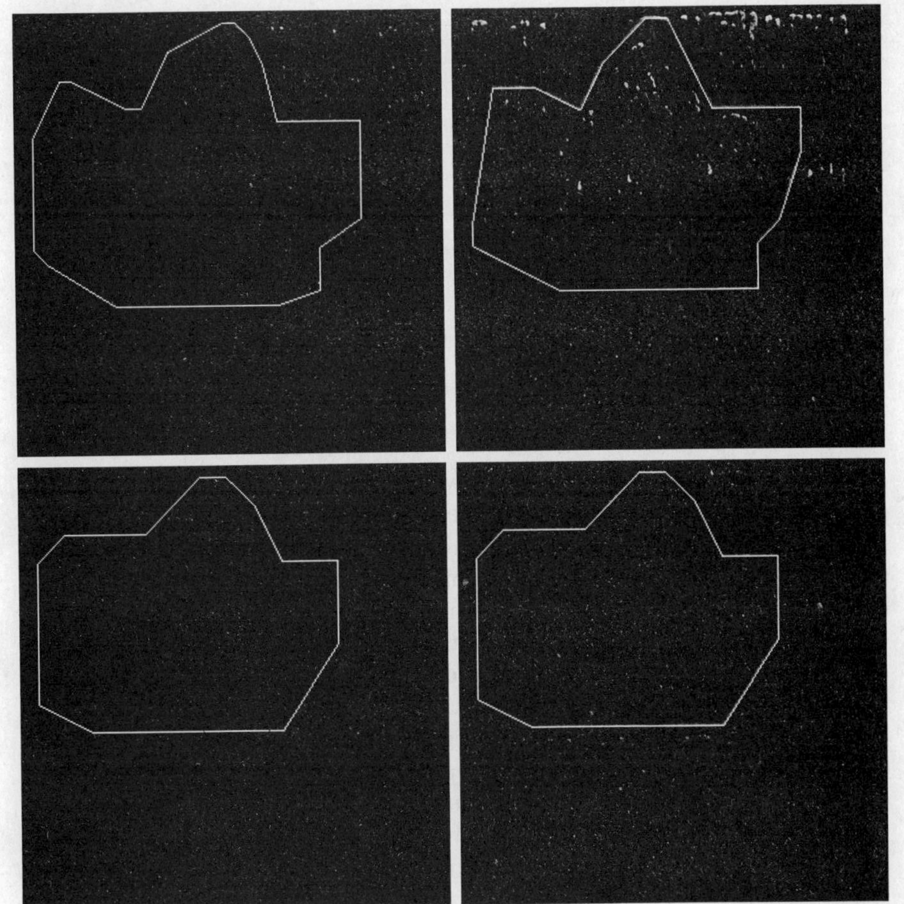

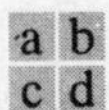

FIGURE 11.8 MPPs obtained with even larger bounding square cells of sizes (a) 10, (b) 16, (c) 20, and (d) 32.

11.2.3 Signatures

A *signature* is a 1-D functional representation of a boundary and may be generated in various ways. One of the simplest is to plot the distance from an interior point (e.g., the centroid) to the boundary as a function of angle, as illustrated in Fig. 11.9. Regardless of how a signature is generated, however, the basic idea is to reduce the boundary representation to a 1-D function, which presumably is easier to describe than the original 2-D boundary. Keep in mind that it makes sense to consider using signatures only when it can be guaranteed that the vector extending from its origin to the boundary intersects the boundary only once, thus yielding a single-valued function of increasing angle. This excludes boundaries with self-intersections, and it also typically excludes boundaries with deep, narrow concavities or thin, long protrusions.

Signatures generated by the approach just described are invariant to translation, but they do depend on rotation and scaling. Normalization with respect to

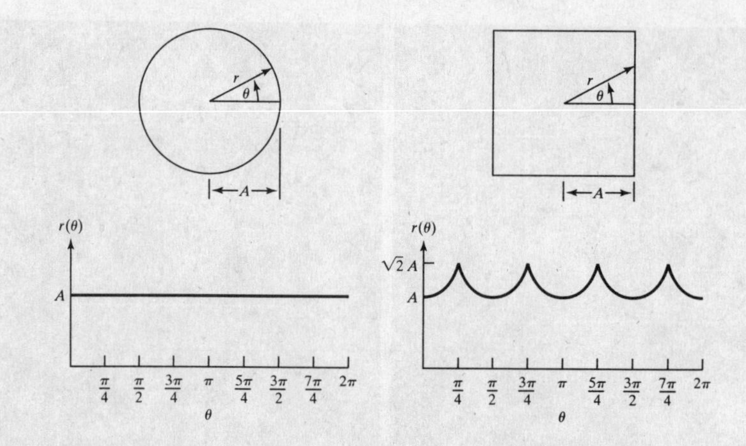

a b
c d

FIGURE 11.9 (a) and (b) Circular and square objects. (c) and (d) Corresponding distance versus angle signatures.

rotation can be achieved by finding a way to select the same starting point to generate the signature, regardless of the shape's orientation. One way to do so is to select the starting point as the point farthest from the origin of the vector (see Section 11.3.1), if this point happens to be unique and independent of rotational aberrations for each shape of interest.

Another way is to select a point on the major eigen axis (see Section 11.5). This method requires more computation but is more rugged because the direction of the eigen axes is determined by using all contour points. Yet another way is to obtain the chain code of the boundary and then use the approach discussed in Section 11.1.2, assuming that the rotation can be approximated by the discrete angles in the code directions defined in Fig. 11.1.

Based on the assumptions of uniformity in scaling with respect to both axes, and that sampling is taken at equal intervals of θ, changes in size of a shape result in changes in the amplitude values of the corresponding signature. One way to normalize for this dependence is to scale all functions so that they always span the same range of values, say, [0, 1]. The main advantage of this method is simplicity, but it has the potentially serious disadvantage that scaling of the entire function is based on only two values: the minimum and maximum. If the shapes are noisy, this can be a source of error from object to object. A more rugged approach is to divide each sample by the variance of the signature, assuming that the variance is not zero—as in the case of Fig. 11.9(a)—or so small that it creates computational difficulties. Use of the variance yields a variable scaling factor that is inversely proportional to changes in size and works much as automatic gain control does. Whatever the method used, keep in mind that the basic idea is to remove dependency on size while preserving the fundamental shape of the waveforms.

Function `signature`, included in Appendix C, finds the signature of a given boundary. Its syntax is

signature

```
[st, angle, x0, y0] = signature(b, x0, y0)
```

where `b` is an $np \times 2$ array containing the xy-coordinates of a boundary ordered in a clockwise or counterclockwise direction. The amplitude of the

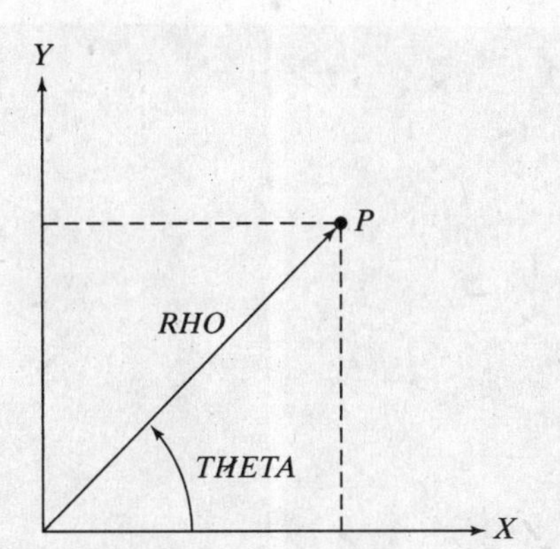

FIGURE 11.10 Axis convention used by MATLAB for performing conversions between polar and Cartesian coordinates, and vice versa.

signature as a function of increasing `angle` is output in `st`. Coordinates `(x0, y0)` in the input are the coordinates of the origin of the vector extending to the boundary. If these coordinates are not included in the argument, the function uses the coordinates of the centroid of the boundary by default. In either case, the values of `(x0, y0)` used by the function are included in the output. The size of arrays `st` and `angle` is 360×1, indicating a resolution of one degree. The input must be a one-pixel-thick boundary obtained, for example, using function `boundaries` (see Section 11.1.3). As before, we assume that a boundary is a closed curve.

Function `signature` utilizes MATLAB's function `cart2pol` to convert Cartesian to polar coordinates. The syntax is

```
[THETA, RHO] = cart2pol(X, Y)
```

where `X` and `Y` are vectors containing the coordinates of the Cartesian points. The vectors `THETA` and `RHO` contain the corresponding angle and length of the polar coordinates. If `X` and `Y` are row vectors, so are `THETA` and `RHO`, and similarly in the case of columns. Figure 11.10 shows the convention used by MATLAB for coordinate conversions. Note that the MATLAB coordinates `(X, Y)` in this situation are related to our image coordinates `(x, y)` as `X = y` and `Y = -x` [see Fig. 2.1(a)]. Function `pol2cart` is used for converting back to Cartesian coordinates:

```
[X, Y] = pol2cart(THETA, RHO)
```

EXAMPLE 11.6: Signatures.

■ Figures 11.11(a) and (b) show the boundaries, `bs` and `bt`, of an irregular square and triangle, respectively, embedded in arrays of size 674×674 pixels. Figure 11.11(c) shows the signature of the square, obtained using the commands

```
>> [st, angle, x0, y0] = signature(bs);
>> plot(angle, st)
```

The values of `x0` and `y0` obtained in the preceding command were [342, 326]. A similar pair of commands yielded the plot in Fig. 11.11(d), whose centroid is

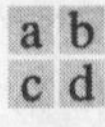
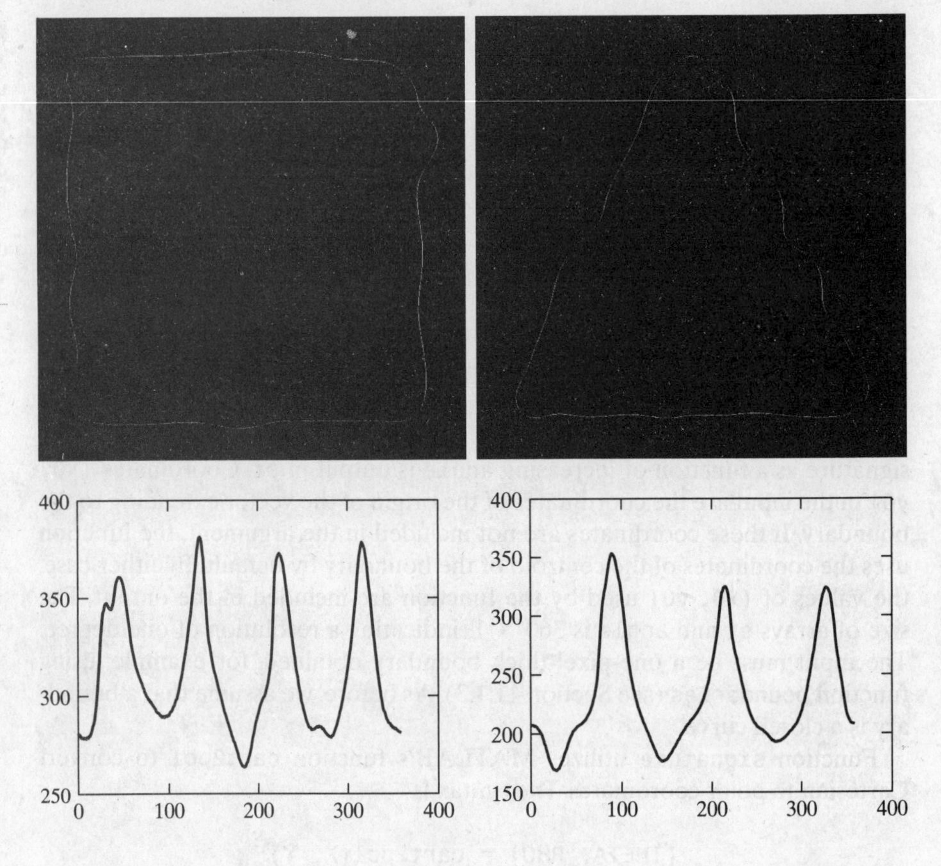

FIGURE 11.11 (a) and (b) Boundaries of an irregular square and triangle. (c) and (d) Corresponding signatures.

located at [416, 335]. Note that simply counting the number of prominent peaks in the two signatures is sufficient to differentiate between the two boundaries. ■

11.2.4 Boundary Segments

Decomposing a boundary into segments often is useful. Decomposition reduces the boundary's complexity and thus simplifies the description process. This approach is particularly attractive when the boundary contains one or more significant concavities that carry shape information. In this case use of the convex hull of the region enclosed by the boundary is a powerful tool for robust decomposition of the boundary.

The *convex hull* H of an arbitrary set S is the smallest convex set containing S. The set difference $H - S$ is called the *convex deficiency, D*, of the set S. To see how these concepts might be used to partition a boundary into meaningful segments, consider Fig. 11.12(a), which shows an object (set S) and its convex deficiency (shaded regions). The region boundary can be partitioned by following the contour of S and marking the points at which a transition is made into or out of a component of the convex deficiency. Figure 11.12(b) shows the result in this case. In principle, this scheme is independent of region size and

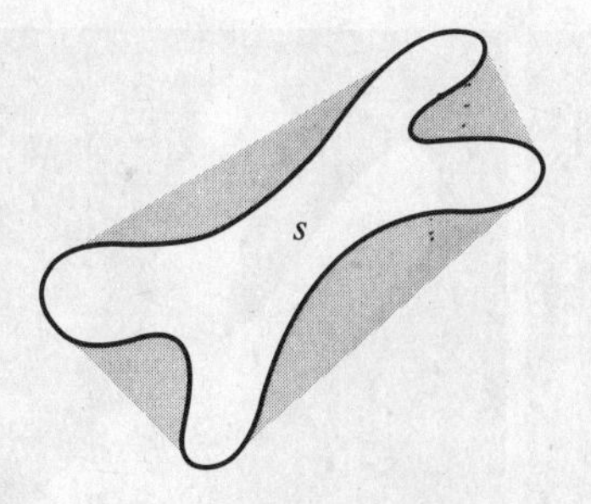

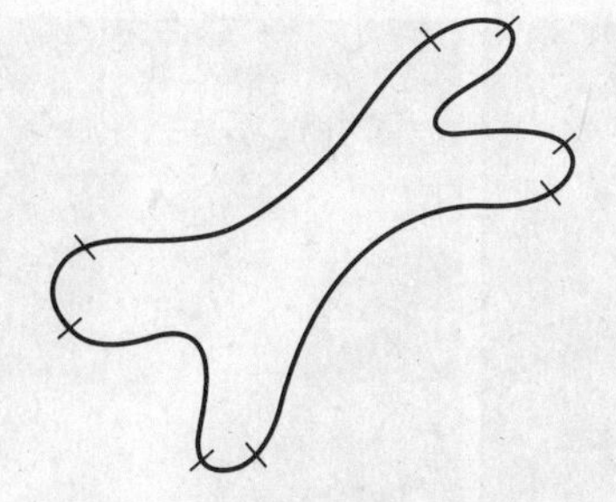

a b

FIGURE 11.12 (a) A region S and its convex deficiency (shaded). (b) Partitioned boundary.

orientation. In practice, this type of processing is preceded typically by aggressive image smoothing to reduce the number of "insignificant" concavities. The MATLAB tools necessary to implement boundary decomposition in the manner just described are contained in function `regionprops`, which is discussed in Section 11.4.1.

11.2.5 Skeletons

An important approach for representing the structural shape of a plane region is to reduce it to a graph. This reduction may be accomplished by obtaining the *skeleton* of the region via a thinning (also called *skeletonizing*) algorithm.

The skeleton of a region may be defined via the medial axis transformation (MAT). The MAT of a region R with border b is as follows. For each point p in R, we find its closest neighbor in b. If p has more than one such neighbor, it is said to belong to the *medial axis* (skeleton) of R.

Although the MAT of a region is an intuitive concept, direct implementation of this definition is expensive computationally, as it involves calculating the distance from every interior point to every point on the boundary of a region. Numerous algorithms have been proposed for improving computational efficiency while at the same time attempting to approximate the medial axis representation of a region.

As noted in Section 9.3.4, IPT generates the skeleton of all regions contained in a binary image `B` via function `bwmorph`, using the following syntax:

```
S = bwmorph(B, 'skel', Inf)
```

This function removes pixels on the boundaries of objects but does not allow objects to break apart. The pixels remaining make up the image skeleton. This option preserves the Euler number (defined in Table 11.1).

EXAMPLE 11.7: Computing the skeleton of a region.

■ Figure 11.13(a) shows an image, `f`, representative of what a human chromosome looks like after it has been segmented out of an electron microscope image with magnification on the order of 30,000X. The objective of this example is to compute the skeleton of the chromosome.

Clearly, the first step in the process must be to isolate the chromosome from the background of irrelevant detail. One approach is to smooth the image and then threshold it. Figure 11.13(b) shows the result of smoothing `f` with a 25×25 Gaussian spatial mask with `sig = 15`:

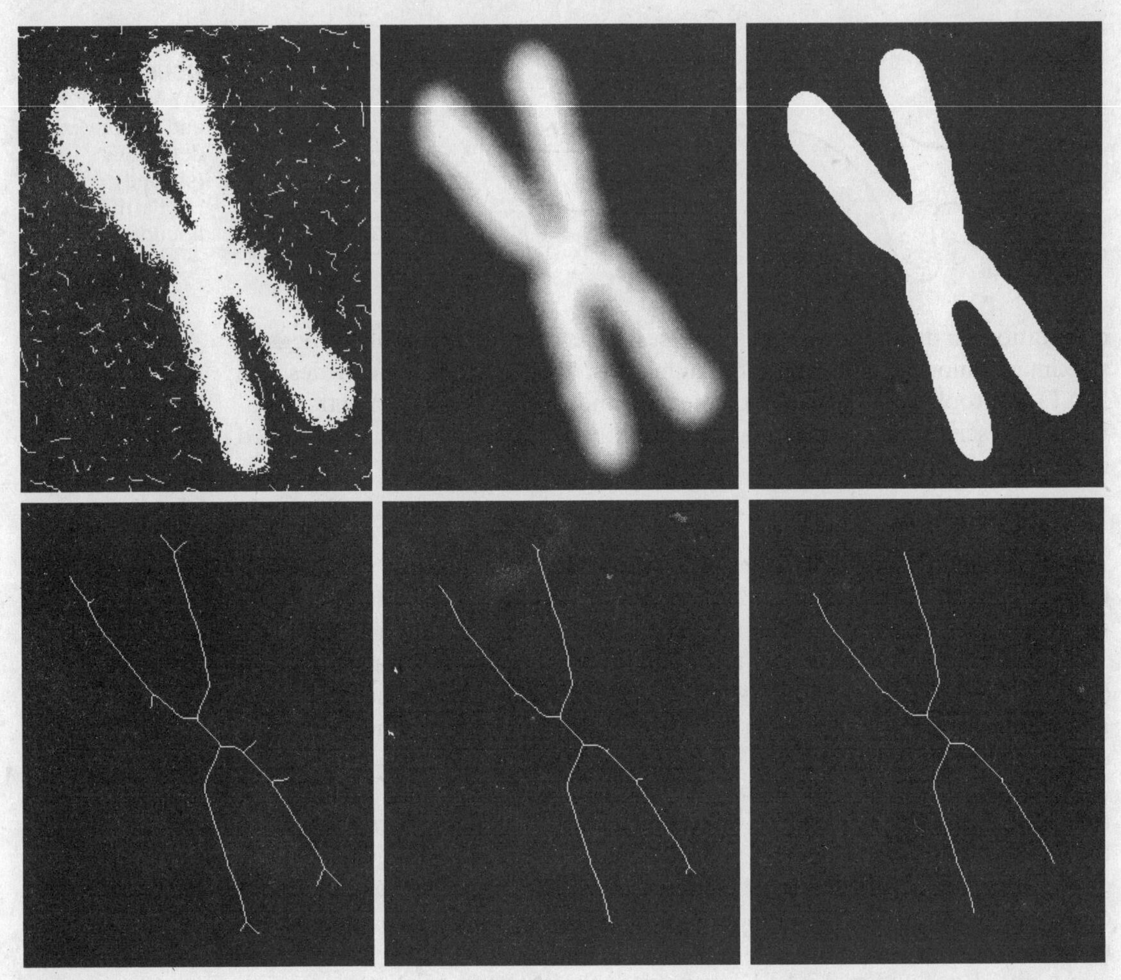

FIGURE 11.13 (a) Segmented human chromosome. (b) Image smoothed using a 25 × 25 Gaussian averaging mask with `sig = 15`. (c) Thresholded image. (d) Skeleton. (e) Skeleton after 8 applications of spur removal. (f) Result of 7 additional applications of spur removal.

```
>> f = im2double(f);
>> h = fspecial('gaussian', 25, 15);
>> g = imfilter(f, h, 'replicate');
>> imshow(g) % Fig. 11.13(b)
```

Next, we threshold the smoothed image:

```
>> g = im2bw(g, 1.5*graythresh(g));
>> figure, imshow(g) % Fig. 11.13(c)
```

where the automatically determined threshold, `graythresh(g)`, was multiplied by 1.5 to increase by 50% the amount of thresholding. The reasoning for this is that increasing the threshold value increases the amount of data removed from the boundary, thus achieving additional smoothing. The skeleton of Fig. 11.13(d) was obtained using the command

```
>> s = bwmorph(g, 'skel', Inf); % Fig. 11.13(d)
```

The spurs in the skeleton were reduced using the command

```
>> s1 = bwmorph(s, 'spur', 8); % Fig. 11.13(e)
```

where we repeated the operation 8 times, which in this case is equal to the approximately one-half the value of `sig` in the smoothing filter. Several small spurs still remain in the skeleton. However, applying the previous function an additional 7 times (to complete the value of `sig`) yielded the result in Fig. 11.13(f), which is a reasonable skeleton representation of the input. As a rule of thumb, the value of `sig` of a Gaussian smoothing mask is a good guideline for the selection of the number of times a spur removal algorithm is applied. ■

11.3 Boundary Descriptors

In this section we discuss a number of descriptors that are useful when working with region boundaries. As will become evident shortly, many of these descriptors can be used for boundaries and/or regions, and the grouping of these descriptors in IPT does not make a distinction regarding their applicability. Therefore, some of the concepts introduced here are mentioned again in Section 11.4 when we discuss regional descriptors.

11.3.1 Some Simple Descriptors

The *length* of a boundary is one of its simplest descriptors. The length of a 4-connected boundary is simply the number of pixels in the boundary, minus 1. If the boundary is 8-connected, we count vertical and horizontal transitions as 1, and diagonal transitions as $\sqrt{2}$.

We extract the boundary of objects contained in image `f` using function `bwperim`, introduced in Section 11.2.2:

```
g = bwperim(f, conn)
```

where `g` is a binary image containing the boundaries of the objects in `f`. For 2-D connectivity, which is our focus, `conn` can have the values 4 or 8, depending on whether 4- or 8-connectivity (the default) is desired (see the margin note in Example 11.4 concerning the interpretation of these connectivity values). The objects in `f` can have any pixel values consistent with the image class, but all background pixels have to be 0. By definition, the perimeter pixels are nonzero and are connected to at least one other nonzero pixel.

Connectivity can be defined in a more general way in IPT by using a 3×3 matrix of 0s and 1s for `conn`. The 1-valued elements define neighborhood locations relative to the center element of `conn`. For example, `conn = ones(3)` defines 8-connectivity. Array `conn` must be symmetric about its center element. The input image can be of any class. The output image containing the boundary of each object in the input is of class `logical`.

The *diameter* of a boundary is defined as the Euclidean distance between the two farthest points on the boundary. These points are not always unique, as in a circle or square, but generally the assumption is that if the diameter is to be a useful descriptor, it is best applied to boundaries with a single pair of farthest points.[†] The line segment connecting these points is called the *major axis* of the boundary. The *minor axis* of a boundary is defined as the line perpendicular to the major axis and of such length that a box passing through the outer four points of intersection of the boundary with the two axes completely encloses the boundary. This box is called the *basic rectangle*, and the ratio of the major to the minor axis is called the *eccentricity* of the boundary.

Function `diameter` (see Appendix C for a listing) computes the diameter, major axis, minor axis, and basic rectangle of a boundary or region. Its syntax is

diameter

```
s = diameter(L)
```

where `L` is a label matrix (Section 9.4) and `s` is a structure with the following fields:

`s.Diameter`	A scalar, the maximum distance between any two pixels in the corresponding region.
`s.MajorAxis`	A 2×2 matrix. The rows contain the row and column coordinates for the endpoints of the major axis of the corresponding region.
`s.MinorAxis`	A 2×2 matrix. The rows contain the row and column coordinates for the endpoints of the minor axis of the corresponding region.
`s.BasicRectangle`	A 4×2 matrix. Each row contains the row and column coordinates of a corner of the basic rectangle.

11.3.2 Shape Numbers

The *shape number* of a boundary, generally based on 4-directional Freeman chain codes, is defined as the first difference of smallest magnitude (Bribiesca and Guzman [1980], Bribiesca [1981]). The *order* of a shape number is defined at the number of digits in its representation. Thus, the shape number of a boundary is given by parameter `c.diffmm` in function `fchcode` discussed in Section 11.2.1, and the order of the shape number is computed as `length(c.diffmm)`.

As noted in Section 11.2.1, 4-directional Freeman chain codes can be made insensitive to the starting point by using the integer of minimum magnitude, and made insensitive to rotations that are multiples of 90° by using the first

---

[†]When more than one pair of farthest points exist, they should be near each other and be dominant factors in determining boundary shape.

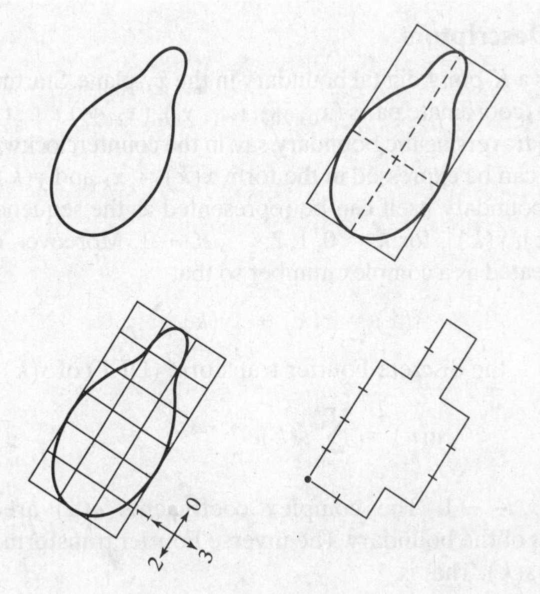

a b
c d

FIGURE 11.14 Steps in the generation of a shape number.

difference of the code. Thus, shape numbers are insensitive to the starting point and to rotations that are multiples of 90°. An approach used frequently to normalize for arbitrary rotations is to align one of the coordinate axes with the major axis and then extract the 4-code based on the rotated figure. The procedure is illustrated in Fig. 11.14.

The tools required to implement an M-function that calculates shape numbers have been developed already. They consist of function `boundaries` to extract the boundary, function `diameter` to find the major axis, function `bsubsamp` to reduce the resolution of the sampling grid, and function `fchcode` to extract the shape number. Keep in mind when using function `boundaries` to extract 4-connected boundaries that the input image must be labeled using `bwlabel` with 4-connectivity specified. As indicated in Fig. 11.14, compensation for rotation is based on aligning one of the coordinate axes with the major axis. The x-axis can be aligned with the major axis of a region or boundary by using function `x2majoraxis`. The syntax of this function follows; the code is included in Appendix C:

```
[B, theta] = x2majoraxis(A, B)
```

`x2majoraxis`

Here, `A = s.MajorAxis` from function `diameter`, and `B` is an input (binary) image or boundary list. (As before, we assume that a boundary is a connected, closed curve.) On the output, `B` has the same form as the input (i.e., a binary image or a coordinate sequence. Because of possible round-off error, rotations can result in a disconnected boundary sequence, so postprocessing to relink the points (using, for example, `bwmorph`) may be required.

11.3.3 Fourier Descriptors

Figure 11.15 shows a K-point digital boundary in the xy-plane. Starting at an arbitrary point (x_0, y_0), coordinate pairs $(x_0, y_0), (x_1, y_1), (x_2, y_2), \ldots, (x_{K-1}, y_{K-1})$ are encountered in traversing the boundary, say, in the counterclockwise direction. These coordinates can be expressed in the form $x(k) = x_k$ and $y(k) = y_k$. With this notation, the boundary itself can be represented as the sequence of coordinates $s(k) = [x(k), y(k)]$, for $k = 0, 1, 2, \ldots, K - 1$. Moreover, each coordinate pair can be treated as a complex number so that

$$s(k) = x(k) + jy(k)$$

From Section 4.1, the discrete Fourier transform (DFT) of $s(k)$ is

$$a(u) = \sum_{k=0}^{K-1} s(k)e^{-j2\pi uk/K}$$

for $u = 0, 1, 2, \ldots, K - 1$. The complex coefficients $a(u)$ are called the *Fourier descriptors* of the boundary. The inverse Fourier transform of these coefficients restores $s(k)$. That is,

$$s(k) = \frac{1}{K}\sum_{u=0}^{K-1} a(u)e^{j2\pi uk/K}$$

for $k = 0, 1, 2, \ldots, K - 1$. Suppose, however, that instead of all the Fourier coefficients, only the first P coefficients are used. This is equivalent to setting $a(u) = 0$ for $u > P - 1$ in the preceding equation for $a(u)$. The result is the following *approximation* to $s(k)$:

$$\hat{s}(k) = \frac{1}{P}\sum_{u=0}^{P-1} a(u)e^{j2\pi uk/K}$$

for $k = 0, 1, 2, \ldots, K - 1$. Although only P terms are used to obtain each component of $\hat{s}(k)$, k still ranges from 0 to $K - 1$. That is, the *same* number of points exists in the approximate boundary, but not as many terms are used in the reconstruction of each point. Recall from Chapter 4 that high-frequency components account for fine detail, and low-frequency components determine global shape. Thus, loss of detail in the boundary increases as P decreases.

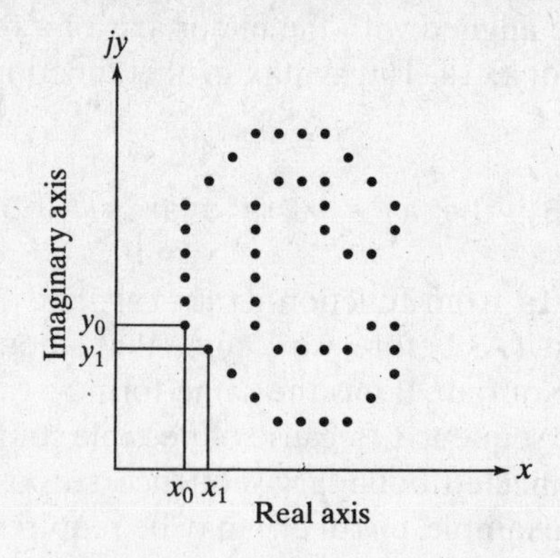

FIGURE 11.15 A digital boundary and its representation as a complex sequence. The points (x_0, y_0) and (x_1, y_1) are (arbitrarily) the first two points in the sequence.

The following function, frdescp, computes the Fourier descriptors of a boundary, s. Similarly, given a set of Fourier descriptors, function ifrdescp computes the inverse using a specified number of descriptor, to yield a closed spatial curve. The documentation section of each function explains its syntax.

frdescp

```
function z = frdescp(s)
%FRDESCP Computes Fourier descriptors.
%   Z = FRDESCP(S) computes the Fourier descriptors of S, which is an
%   np-by-2 sequence of image coordinates describing a boundary.
%
%   Due to symmetry considerations when working with inverse Fourier
%   descriptors based on fewer than np terms, the number of
%   points in S when computing the descriptors must be even. If the
%   number of points is odd, FRDESCP duplicates the end point and
%   adds it at the end of the sequence. If a different treatment is
%   desired, the sequence must be processed externally so that it has
%   an even number of points.
%
%   See function IFRDESCP for computing the inverse descriptors.

% Preliminaries
[np, nc] = size(s);
if nc ~= 2
   error('S must be of size np-by-2.');
end
if np/2 ~= round(np/2);
   s(end + 1, :) = s(end, :);
   np = np + 1;
end

% Create an alternating sequence of 1s and -1s for use in centering
% the transform.
x = 0:(np - 1);
m = ((-1) .^ x)';

% Multiply the input sequence by alternating 1s and -1s to
% center the transform.
s(:, 1) = m .* s(:, 1);
s(:, 2) = m .* s(:, 2);
% Convert coordinates to complex numbers.
s = s(:, 1) + i*s(:, 2);
% Compute the descriptors.
z = fft(s);
```

Function ifrdescp is as follows:

ifrdescp

```
function s = ifrdescp(z, nd)
%IFRDESCP Computes inverse Fourier descriptors.
%   I = IFRDESCP(Z, ND) computes the inverse Fourier descriptors of
%   of Z, which is a sequence of Fourier descriptor obtained, for
%   example, by using function FRDESCP. ND is the number of
%   descriptors used to computing the inverse; ND must be an even
```

```
%    integer no greater than length(Z). If ND is omitted, it defaults
%    to length(Z). The output, S, is an ND-by-2 matrix containing the
%    coordinates of a closed boundary.

% Preliminaries.
np = length(z);
% Check inputs.
if nargin == 1 | nd > np
   nd = np;
end

% Create an alternating sequence of 1s and -1s for use in centering
% the transform.
x = 0:(np - 1);
m = ((-1) .^ x)';

% Use only nd descriptors in the inverse. Since the
% descriptors are centered, (np - nd)/2 terms from each end of
% the sequence are set to 0.
d = round((np - nd)/2); % Round in case nd is odd.
z(1:d) = 0;
z(np - d + 1:np) = 0;
% Compute the inverse and convert back to coordinates.
zz = ifft(z);
s(:, 1) = real(zz);
s(:, 2) = imag(zz);
% Multiply by alternating 1 and -1s to undo the earlier
% centering.
s(:, 1) = m.*s(:, 1);
s(:, 2) = m.*s(:, 2);
```

EXAMPLE 11.8: Fourier descriptors.

■ Figure 11.16(a) shows a binary image, `f`, similar to the one in Fig. 11.13(c), but obtained using a Gaussian mask of size 15×15 with `sigma = 9`, and thresholded at 0.7. The purpose was to generate an image that was not overly

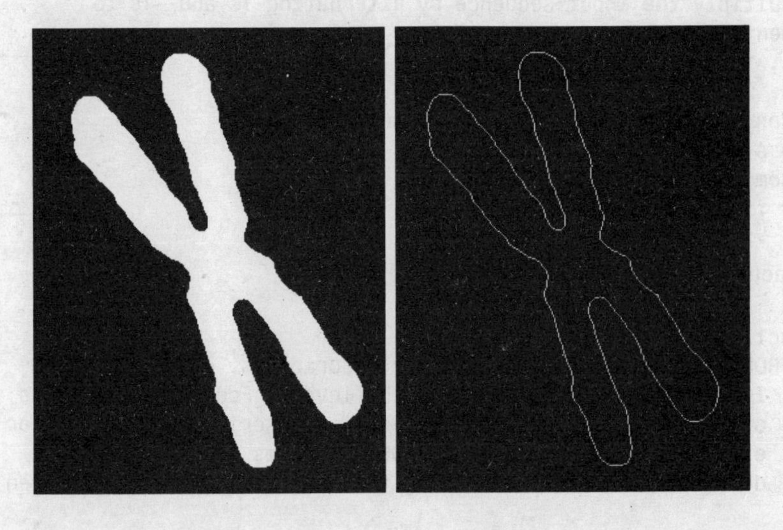

a b

FIGURE 11.16 (a) Binary image. (b) Boundary extracted using function `boundaries`. The boundary has 1090 points.

smooth in order to illustrate the effect that reducing the number of descriptors has on the shape of a boundary. The image in Fig. 11.16(b) was generated using the commands

```
>> b = boundaries(f);
>> b = b{1};
>> bim = bound2im(b, 344, 270);
```

where the dimensions shown are the dimensions of `f`. Figure 11.16(b) shows image `bim`. The boundary shown has 1090 points. Next, we computed the Fourier descriptors,

```
>> z = frdescp(b);
```

and obtained the inverse using approximately 50% of the possible 1090 descriptors:

```
>> z546 = ifrdescp(z, 546);
>> z546im = bound2im(z546, 344, 270);
```

Image `z546im` [Fig. 11.17(a)] shows close correspondence with the original boundary in Fig. 11.16(b). Some subtle details, like a 1-pixel bay in the bottom-facing cusp in the original boundary, were lost, but, for all practical purposes, the two boundaries are identical. Figures 11.17(b) through (f) show the results obtained using 110, 56, 28, 14, and 8 descriptors, which are approximately 10%, 5%, 2.5%, 1.25% and 0.7%, of the possible 1090 descriptors. The result obtained using 110 descriptors [Fig. 11.17(c)] shows slight further smoothing of the boundary, but, again, the general shape is quite close to the original. Figure 11.17(e) shows that even the result with 14 descriptors, a mere 1.25% of the total, retained the principal features of the boundary. Figure 11.17(f) shows distortion that is unacceptable because the main feature of the boundary (the four long protrusions) was lost. Further reduction to 4 and 2 descriptors would result in an ellipse and, finally, a circle.

Some of the boundaries in Fig. 11.17 have one-pixel gaps due to round off in pixel values. These small gaps, common with Fourier descriptors, can be repaired with function `bwmorph` using the `'bridge'` option. ■

As mentioned earlier, descriptors should be as insensitive as possible to translation, rotation, and scale changes. In cases where results depend on the order in which points are processed, an additional constraint is that descriptors should be insensitive to starting point. Fourier descriptors are not directly insensitive to these geometric changes, but the changes in these parameters can be related to simple transformations on the descriptors (see Gonzalez and Woods [2002]).

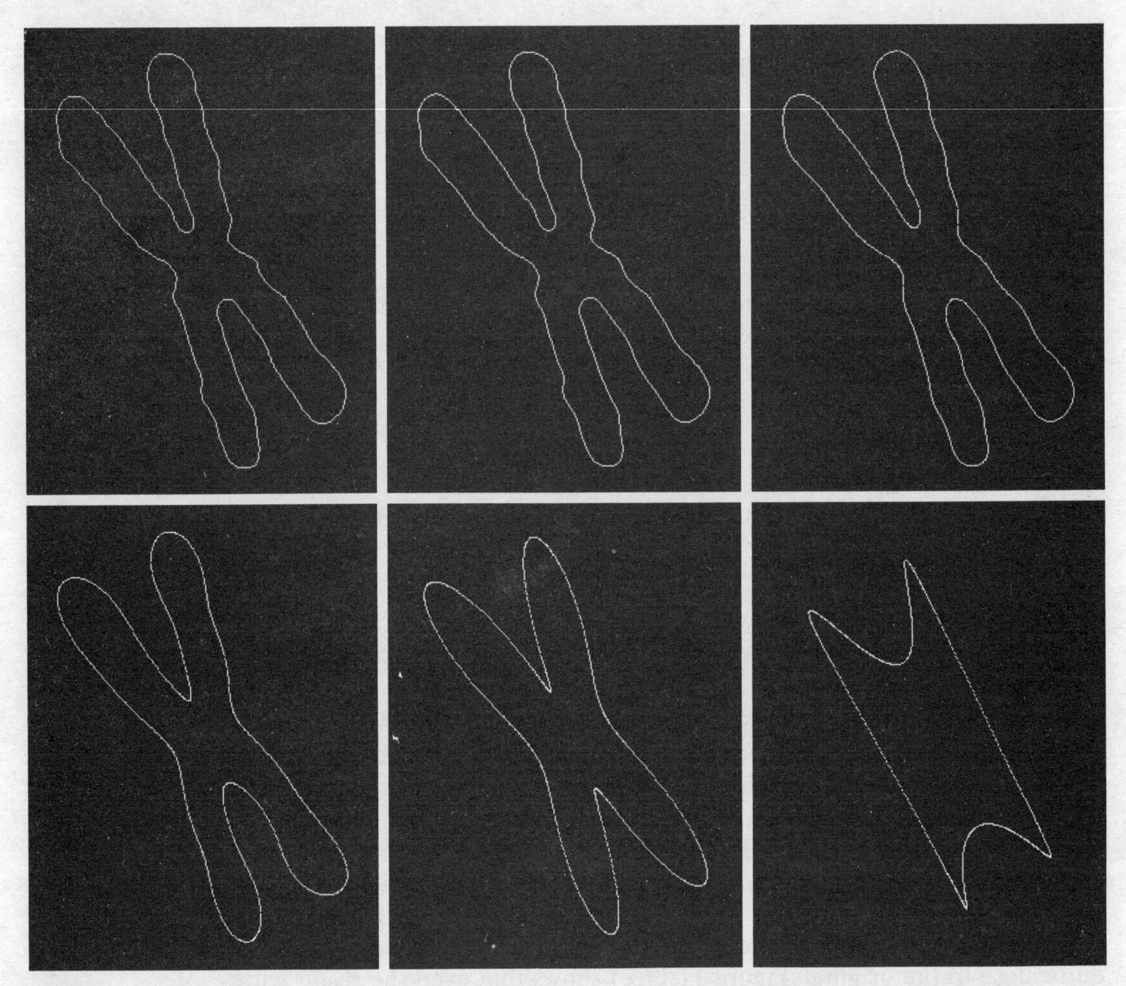

a b c
d e f

FIGURE 11.17 (a)–(f) Boundary reconstructed using 546, 110, 56, 28, 14, and 8 Fourier descriptors out of a possible 1090 descriptors.

11.3.4 Statistical Moments

The shape of 1-D boundary representations (e.g., boundary segments and signature waveforms) can be described quantitatively by using statistical moments, such as the mean, variance, and higher-order moments. Consider Fig. 11.18(a), which shows a boundary segment, and Fig. 11.18(b), which shows the segment represented as a 1-D function, $g(r)$, of an arbitrary variable r. This function was obtained by connecting the two end points of the segment to form a "major" axis and then using function `x2majoraxis` discussed in Section 11.3.2 to align the major axis with the x-axis.

One approach for describing the shape of $g(r)$ is to normalize it to unit area and treat it as a histogram. In other words, $g(r_i)$ is treated as the probability of value r_i occurring. In this case, r is considered a random variable and the moments are

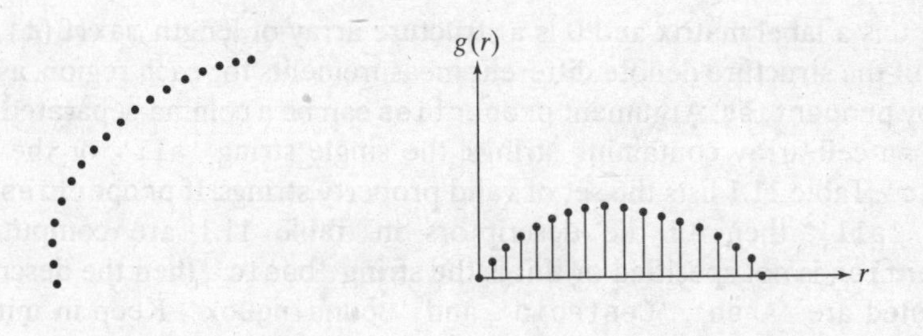

a b

FIGURE 11.18 (a) Boundary segment. (b) Representation as a 1-D function.

$$\mu_n = \sum_{i=0}^{K-1} (r_i - m)^n g(r_i)$$

where

$$m = \sum_{i=0}^{K-1} r_i g(r_i)$$

In this notation, K is the number of points on the boundary, and $\mu_n(r)$ is directly related to the shape of $g(r)$. For example, the second moment $\mu_2(r)$ measures the spread of the curve about the mean value of r and the third moment, $\mu_3(r)$, measures its symmetry with reference to the mean. Statistical moments are computed with function `statmoments`, discussed in Section 5.2.4.

What we have accomplished is to reduce the description task to 1-D functions. Although moments are a popular approach, they are not the only descriptors that could be used for this purpose. For instance, another method involves computing the 1-D discrete Fourier transform, obtaining its spectrum, and using the first q components of the spectrum to describe $g(r)$. The advantage of moments over other techniques is that implementation of moments is straightforward, and moments also carry a "physical" interpretation of boundary shape. The insensitivity of this approach to rotation is clear from Fig. 11.18. Size normalization, if desired, can be achieved by scaling the range of values of g and r.

11.4 Regional Descriptors

In this section we discuss a number of IPT functions for region processing and introduce several additional functions for computing texture, moment invariants, and several other regional descriptors. Keep in mind that function `bwmorph` discussed in Section 9.3.4 is used frequently for the type of processing we outline in this section. Function `roipoly` (Section 5.2.4) also is used frequently in this context.

11.4.1 Function regionprops

Function `regionprops` is IPT's principal tool for computing region descriptors. This function has the syntax

```
D = regionprops(L, properties)
```

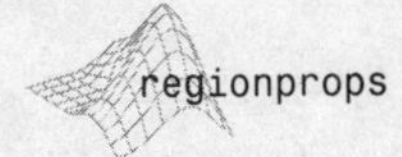

where L is a label matrix and D is a structure array of length max(L(:)). The fields of the structure denote different measurements for each region, as specified by properties. Argument properties can be a comma-separated list of strings, a cell array containing strings, the single string 'all', or the string 'basic'. Table 11.1 lists the set of valid property strings. If properties is the string 'all', then all the descriptors in Table 11.1 are computed. If properties is not specified or if it is the string 'basic', then the descriptors computed are 'Area', 'Centroid', and 'BoundingBox'. Keep in mind (as discussed in Section 2.1.1) that IPT uses x and y to indicate horizontal and vertical coordinates, respectively, with the origin being located in the top, left. Coordinates x and y increase to the right and downward from the origin, respectively. For the purposes of our discussion, on pixels are valued 1 while off pixels are valued 0.

EXAMPLE 11.9: Using function regionprops.

■ As a simple illustration, suppose that we want to obtain the area and the bounding box for each region in an image B. We write

```
>> B = bwlabel(B); % Convert B to a label matrix.
>> D = regionprops(B, 'area', 'boundingbox');
```

To extract the areas and number of regions we write

```
>> w = [D.Area];
>> NR = length(w);
```

where the elements of vector w are the areas of the regions and NR is the number of regions. Similarly, we can obtain a single matrix whose rows are the bounding boxes of each region using the statement

```
V = cat(1, D.BoundingBox);
```

This array is of dimension NR $\times$ 4. The cat operator is explained in Section 6.1.1. ■

11.4.2 Texture

An important approach for describing a region is to quantify its texture content. In this section we illustrate the use of two new functions for computing texture based on statistical and spectral measures.

Statistical Approaches

A frequently used approach for texture analysis is based on statistical properties of the intensity histogram. One class of such measures is based on statistical moments. As discussed in Section 5.2.4, the expression for the nth moment about the mean is given by

$$\mu_n = \sum_{i=0}^{L-1} (z_i - m)^n p(z_i)$$

TABLE 11.1 Regional descriptors computed by function `regionprops`.

Valid Strings for `properties`	Explanation
`'Area'`	The number of pixels in a region.
`'BoundingBox'`	1 × 4 vector defining the smallest rectangle containing a region. `BoundingBox` is defined by `[ul_corner width]`, where `ul_corner` is in the form `[x y]` and specifies the upper-left corner of the bounding box, and `width` is in the form `[x_width y_width]` and specifies the width of the bounding box along each dimension. Note that the `BoundingBox` is aligned with the coordinate axes and, in that sense, is a special case of the basic rectangle discussed in Section 11.3.1.
`'Centroid'`	1 × 2 vector; the center of mass of the region. The first element of `Centroid` is the horizontal coordinate (or x-coordinate) of the center of mass, and the second element is the vertical coordinate (or y-coordinate).
`'ConvexArea'`	Scalar; the number of pixels in `'ConvexImage'`.
`'ConvexHull'`	p × 2 matrix; the smallest convex polygon that can contain the region. Each row of the matrix contains the x- and y-coordinates of one of the p vertices of the polygon.
`'ConvexImage'`	Binary image; the convex hull, with all pixels within the hull filled in (i.e., set to `on`). (For pixels that the boundary of the hull passes through, `regionprops` uses the same logic as `roipoly` to determine whether the pixel is inside or outside the hull.) The image is the size of the bounding box of the region.
`'Eccentricity'`	Scalar; the eccentricity of the ellipse that has the same second moments as the region. The eccentricity is the ratio of the distance between the foci of the ellipse and its major axis length. The value is between 0 and 1, with 0 and 1 being degenerate cases (an ellipse whose eccentricity is 0 is a circle, while an ellipse with an eccentricity of 1 is a line segment).
`'EquivDiameter'`	Scalar; the diameter of a circle with the same area as the region. Computed as `sqrt(4*Area/pi)`.
`'EulerNumber'`	Scalar; equal to the number of objects in the region minus the number of holes in those objects.
`'Extent'`	Scalar; the proportion of the pixels in the bounding box that are also in the region. Computed as `Area` divided by the area of the bounding box.
`'Extrema'`	8 × 2 matrix; the extremal points in the region. Each row of the matrix contains the x- and y-coordinates of one of the points. The format of the vector is `[top-left, top-right, right-top, right-bottom, bottom-right, bottom-left, left-bottom, left-top]`.
`'FilledArea'`	The number of on pixels in `FilledImage`.
`'FilledImage'`	Binary image of the same size as the bounding box of the region. The `on` pixels correspond to the region, with all holes filled.
`'Image'`	Binary image of the same size as the bounding box of the region; the `on` pixels correspond to the region, and all other pixels are `off`.
`'MajorAxisLength'`	The length (in pixels) of the major axis† of the ellipse that has the same second moments as the region.
`'MinorAxisLength'`	The length (in pixels) of the minor axis† of the ellipse that has the same second moments as the region.
`'Orientation'`	The angle (in degrees) between the x-axis and the major axis† of the ellipse that has the same second moments as the region.
`'PixelList'`	A matrix whose rows are the `[x, y]` coordinates of the actual pixels in the region.
`'Solidity'`	Scalar; the proportion of the pixels in the convex hull that are also in the region. Computed as `Area/ConvexArea`.

† Note that the use of major and minor axis in this context is different from the major and minor axes of the basic rectangle discussed in Section 11.3.1. For a discussion of moments of an ellipse, see Haralick and Shapiro [1992].

TABLE 11.2 Some descriptors of texture based on the intensity histogram of a region.

Moment	Expression	Measure of Texture
Mean	$m = \sum_{i=0}^{L-1} z_i p(z_i)$	A measure of average intensity.
Standard deviation	$\sigma = \sqrt{\mu_2(z)} = \sqrt{\sigma^2}$	A measure of average contrast.
Smoothness	$R = 1 - 1/(1 + \sigma^2)$	Measures the relative smoothness of the intensity in a region. R is 0 for a region of constant intensity and approaches 1 for regions with large excursions in the values of its intensity levels. In practice, the variance used in this measure is normalized to the range [0, 1] by dividing it by $(L - 1)^2$.
Third moment	$\mu_3 = \sum_{i=0}^{L-1} (z_i - m)^3 p(z_i)$	Measures the skewness of a histogram. This measure is 0 for symmetric histograms, positive by histograms skewed to the right (about the mean) and negative for histograms skewed to the left. Values of this measure are brought into a range of values comparable to the other five measures by dividing μ_3 by $(L - 1)^2$ also, which is the same divisor we used to normalize the variance.
Uniformity	$U = \sum_{i=0}^{L-1} p^2(z_i)$	Measures uniformity. This measure is maximum when all gray levels are equal (maximally uniform) and decreases from there.
Entropy	$e = -\sum_{i=0}^{L-1} p(z_i) \log_2 p(z_i)$	A measure of randomness.

where z_i is a random variable indicating intensity, $p(z)$ is the histogram of the intensity levels in a region, L is the number of possible intensity levels, and

$$m = \sum_{i=0}^{L-1} z_i p(z_i)$$

is the mean (average) intensity. These moments can be computed with function `statmoments` discussed in Section 5.2.4. Table 11.2 lists some common descriptors based on statistical moments and also on uniformity and entropy. Keep in mind that the second moment, $\mu_2(z)$, is the variance, σ^2.

Writing an M-function to compute the texture measures in Table 11.3 is straightforward. Function `statxture`, written for this purpose, is included in Appendix C. The syntax of this function is

TABLE 11.3 Texture measures for the regions shown in Fig. 11.19.

Texture	Average Intensity	Average Contrast	R	Third Moment	Uniformity	Entropy
Smooth	87.02	11.17	0.002	−0.011	0.028	5.367
Coarse	119.93	73.89	0.078	2.074	0.005	7.842
Periodic	98.48	33.50	0.017	0.557	0.014	6.517

```
t = statxture(f, scale)
```

statxture

where `f` is an input image (or subimage) and `t` is a 6-element row vector whose components are the descriptors in Table 11.2, arranged in the same order. Parameter `scale` also is a 6-element row vector, whose components multiply the corresponding elements of `t` for scaling purposes. If omitted, `scale` defaults to all 1s.

EXAMPLE 11.10: Statistical texture measures.

■ The three regions outlined by the white boxes in Fig. 11.19 represent, from left to right, examples of smooth, coarse and periodic texture. The histograms of these regions, obtained using function `imhist`, are shown in Fig. 11.20. The entries in Table 11.3 were obtained by applying function `statxture` to each of the subimages in Fig. 11.19. These results are in general agreement with the texture content of the respective subimages. For example, the entropy of the coarse region [Fig. 11.19(b)] is higher than the others because the values of the pixels in that region are more random than the values in the other

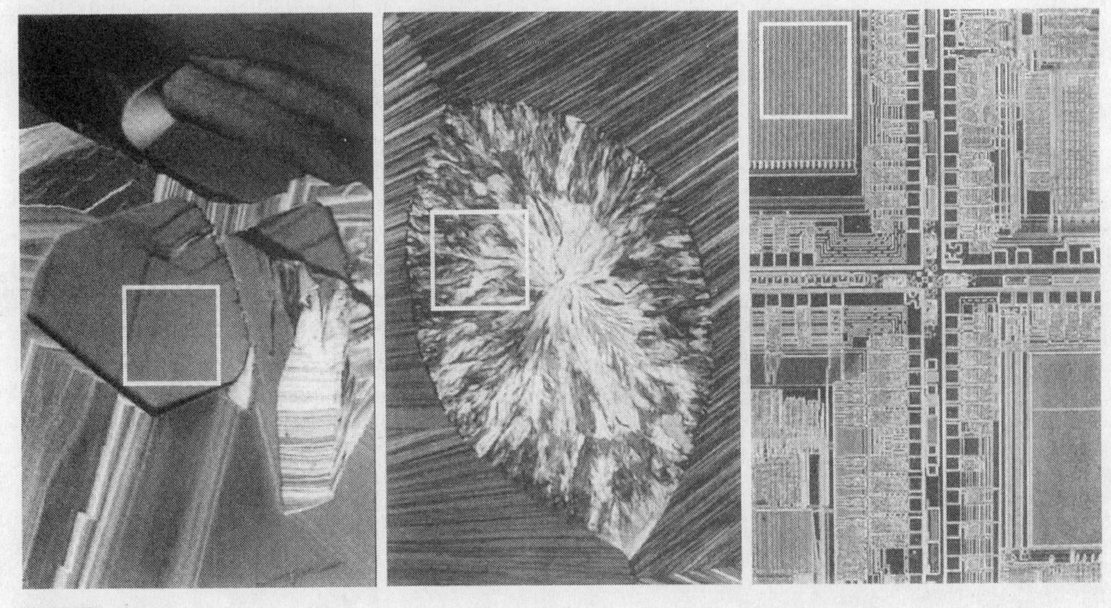

a b c

FIGURE 11.19 The subimages shown represent, from left to right, smooth, coarse, and periodic texture. These are optical microscope images of a superconductor, human cholesterol, and a microprocessor. (Original images courtesy of Dr. Michael W. Davidson, Florida State University.)

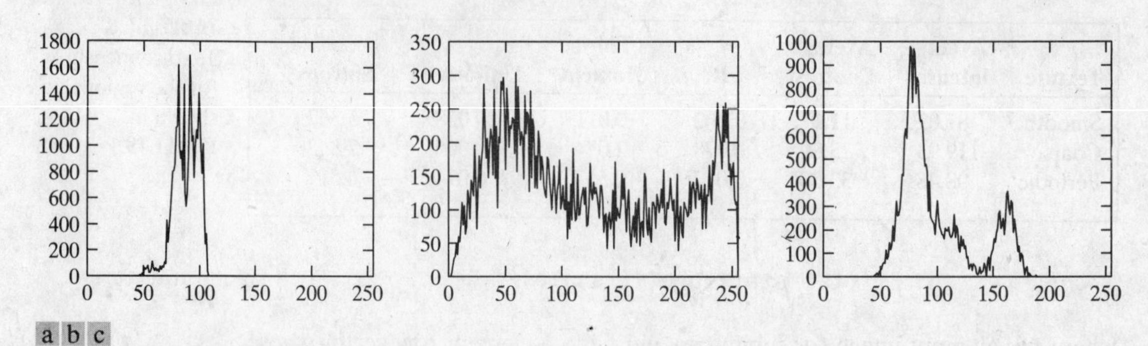

a b c

FIGURE 11.20 Histograms corresponding to the subimages in Fig. 11.19.

regions. This also is true for the contrast and for the average intensity in this case. On the other hand, this region is the least smooth and the least uniform, as revealed by the values of R and the uniformity measure. The histogram of the coarse region also shows the greatest lack of symmetry with respect to the location of the mean value, as is evident in Fig. 11.20(b), and also by the largest value of the third moment shown in Table 11.3. ■

Spectral Measures of Texture

Spectral measures of texture are based on the Fourier spectrum, which is ideally suited for describing the directionality of periodic or almost periodic 2-D patterns in an image. These global texture patterns, easily distinguishable as concentrations of high-energy bursts in the spectrum, generally are quite difficult to detect with spatial methods because of the local nature of these techniques. Thus spectral texture is useful for discriminating between periodic and nonperiodic texture patterns, and, further, for quantifying differences between periodic patterns.

Interpretation of spectrum features is simplified by expressing the spectrum in polar coordinates to yield a function $S(r, \theta)$, where S is the spectrum function and r and θ are the variables in this coordinate system. For each direction θ, $S(r, \theta)$ may be considered a 1-D function, $S_\theta(r)$. Similarly, for each frequency r, $S_r(\theta)$ is a 1-D function. Analyzing $S_\theta(r)$ for a fixed value of θ yields the behavior of the spectrum (such as the presence of peaks) along a radial direction from the origin, whereas analyzing $S_r(\theta)$ for a fixed value of r yields the behavior along a circle centered on the origin.

A global description is obtained by integrating (summing for discrete variables) these functions:

$$S(r) = \sum_{\theta=0}^{\pi} S_\theta(r)$$

and

$$S(\theta) = \sum_{r=1}^{R_0} S_r(\theta)$$

where R_0 is the radius of a circle centered at the origin.

The results of these two equations constitute a pair of values $[S(r), S(\theta)]$ for *each* pair of coordinates (r, θ). By varying these coordinates we can generate two 1-D functions, $S(r)$ and $S(\theta)$, that constitute a spectral-energy description of texture for an entire image or region under consideration. Furthermore, descriptors of these functions themselves can be computed in order to characterize their behavior quantitatively. Descriptors typically used for this purpose are the location of the highest value, the mean and variance of both the amplitude and axial variations, and the distance between the mean and the highest value of the function.

Function specxture (see Appendix C for a listing) can be used to compute the two preceding texture measures. The syntax is

```
[srad, sang, S] = specxture(f)
```

specxture

where srad is $S(r)$, sang is $S(\theta)$, and S is the spectrum image (displayed using the log, as explained in Chapter 4).

EXAMPLE 11.11: Computing spectral texture.

■ Figure 11.21(a) shows an image with randomly distributed objects and Fig. 11.22(b) shows an image containing the same objects, but arranged periodically. The corresponding Fourier spectra, computed using function specxture, are shown in Figs. 11.21(c) and (d). The periodic bursts of energy extending quadrilaterally in two dimensions in the Fourier spectra are due to the periodic texture of the coarse background material on which the matches rest. The other components of the spectra in Fig. 11.21(c) are clearly caused by the random orientation of the strong edges in Fig. 11.21(a). By contrast, the main energy in Fig. 11.21(d) not associated with the background is along the horizontal axis, corresponding to the strong vertical edges in Fig. 11.21(b).

Figures 11.22(a) and (b) are plots of $S(r)$ and $S(\theta)$ for the random matches, and similarly in (c) and (d) for the ordered matches, all computed with function specxture. The plots were obtained with the commands plot(srad) and plot(sang). The axes in Figs. 11.22(a) and (c) were scaled using

```
>> axis([horzmin horzmax vertmin vertmax])
```

discussed in Section 3.3.1, with the maximum and minimum values obtained from Fig. 11.22(a).

The plot of $S(r)$ corresponding to the randomly-arranged matches shows no strong periodic components (i.e., there are no peaks in the spectrum besides the peak at the origin, which is the DC component). On the other hand, the plot of $S(r)$ corresponding to the ordered matches shows a strong peak near $r = 15$ and a smaller one near $r = 25$. Similarly, the random nature of the energy bursts in Fig. 11.21(c) is quite apparent in the plot of $S(\theta)$ in Fig. 11.22(b). By contrast, the plot in Fig. 11.22(d) shows strong energy components in the region near the origin and at 90° and 180°. This is consistent with the energy distribution in Fig. 11.21(d). ■

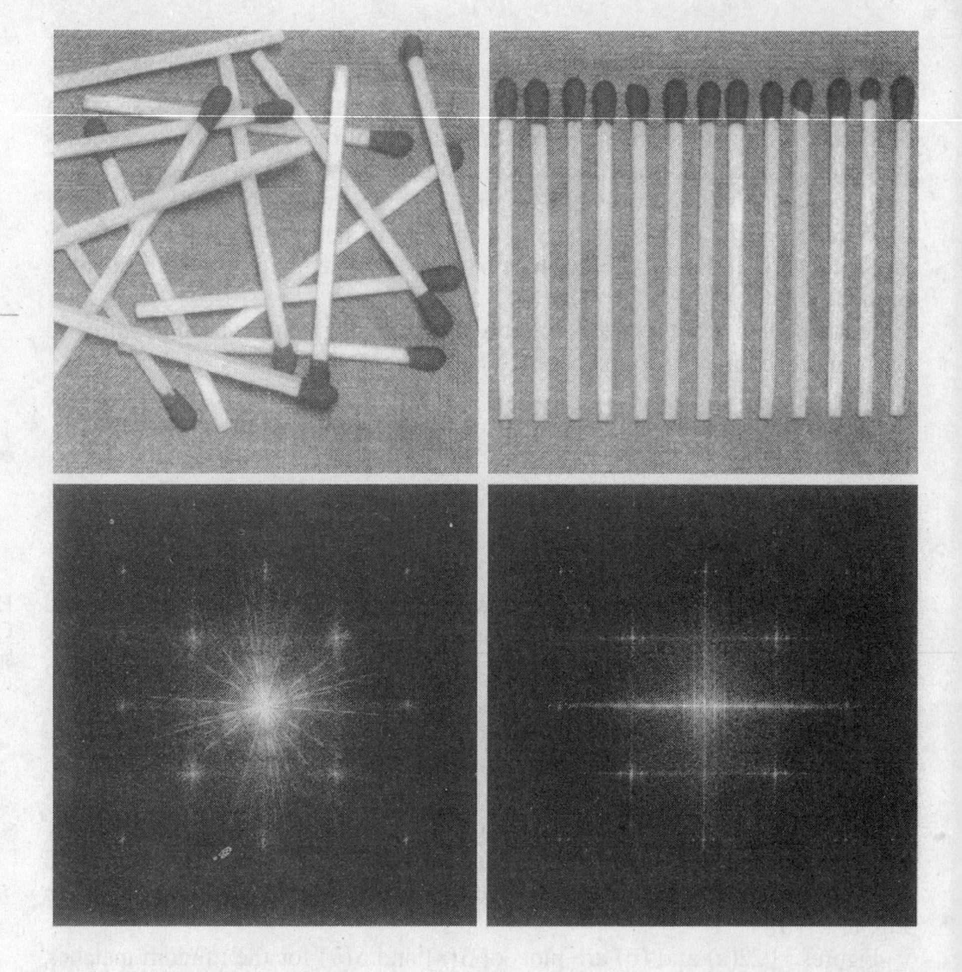

a b
c d

FIGURE 11.21 (a) and (b) Images of unordered and ordered objects. (c) and (d) Corresponding spectra.

11.4.3 Moment Invariants

The 2-D *moment* of order $(p + q)$ of a digital image $f(x, y)$ is defined as

$$m_{pq} = \sum_x \sum_y x^p y^q f(x, y)$$

for $p, q = 0, 1, 2, \ldots$, where the summations are over the values of the spatial coordinates x and y spanning the image. The corresponding *central moment* is defined as

$$\mu_{pq} = \sum_x \sum_y (x - \overline{x})^p (y - \overline{y})^q f(x, y)$$

where

$$\overline{x} = \frac{m_{10}}{m_{00}} \quad \text{and} \quad \overline{y} = \frac{m_{01}}{m_{00}}$$

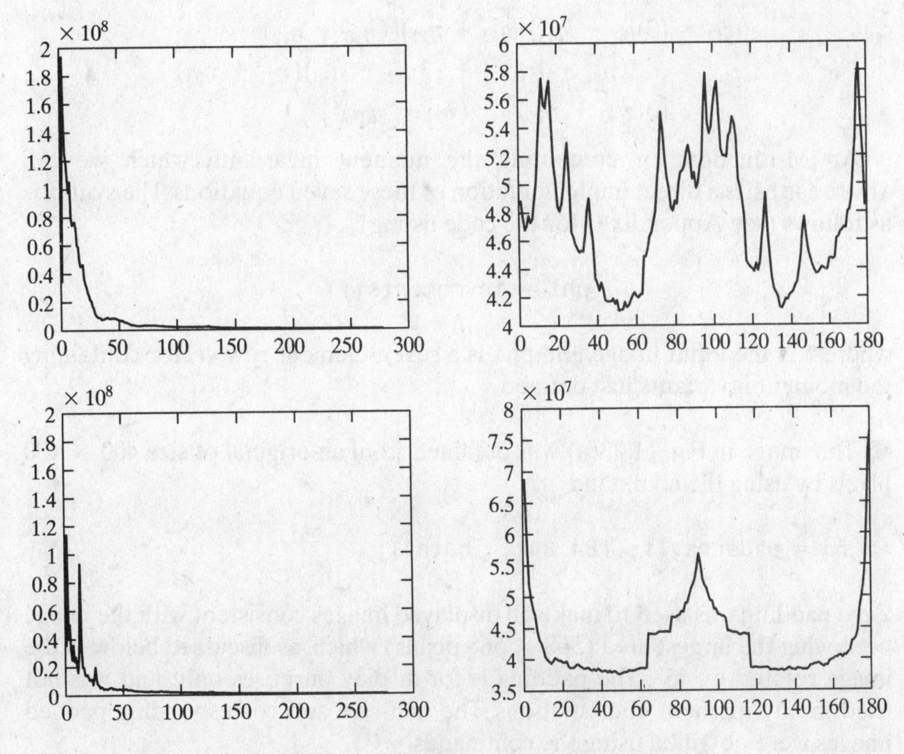

a b
c d

FIGURE 11.22 Plots of (a) $S(r)$ and (b) $S(\theta)$ for the random image. (c) and (d) are plots of $S(r)$ and $S(\theta)$ for the ordered image.

The *normalized central moment* of order $(p + q)$ is defined as

$$\eta_{pq} = \frac{\mu_{pq}}{\mu_{00}^{\gamma}}$$

for $p, q = 0, 1, 2, \ldots$, where

$$\gamma = \frac{p + q}{2} + 1$$

for $p + q = 2, 3, \ldots$.

A set of seven 2-D *moment invariants* that are insensitive to translation, scale change, mirroring, and rotation can be derived from these equations. They are

$$\phi_1 = \eta_{20} + \eta_{02}$$
$$\phi_2 = (\eta_{20} - \eta_{02})^2 + 4\eta_{11}^2$$
$$\phi_3 = (\eta_{30} - 3\eta_{12})^2 + (3\eta_{21} - \eta_{03})^2$$
$$\phi_4 = (\eta_{30} + \eta_{12})^2 + (\eta_{21} + \eta_{03})^2$$
$$\begin{aligned}\phi_5 = {} & (\eta_{30} - 3\eta_{12})(\eta_{30} + \eta_{12})[(\eta_{30} + \eta_{12})^2 \\ & -3(\eta_{21} + \eta_{03})^2] + (3\eta_{21} - \eta_{03})(\eta_{21} + \eta_{03}) \\ & [3(\eta_{30} + \eta_{12})^2 - (\eta_{21} + \eta_{03})^2]\end{aligned}$$
$$\begin{aligned}\phi_6 = {} & (\eta_{20} - \eta_{02})[(\eta_{30} + \eta_{12})^2 - (\eta_{21} + \eta_{03})^2] \\ & + 4\eta_{11}(\eta_{30} + \eta_{12})(\eta_{21} + \eta_{03})\end{aligned}$$

$$\phi_7 = (3\eta_{21} - \eta_{03})(\eta_{30} + \eta_{12})[(\eta_{30} + \eta_{12})^2 - 3(\eta_{21} + \eta_{03})^2] + (3\eta_{12} - \eta_{30})(\eta_{21} + \eta_{03})[3(\eta_{30} + \eta_{12})^2 - (\eta_{21} + \eta_{03})^2]$$

An M-function for computing the moment invariants, which we call `invmoments`, is a direct implementation of these seven equations. The syntax is as follows (see Appendix C for the code listing):

invmoments

```
phi = invmoments(f)
```

where `f` is the input image and `phi` is a seven-element row vector containing the moment invariants just defined.

EXAMPLE 11.12: Moment invariants.

■ The image in Fig. 11.23(a) was obtained from an original of size 400 × 400 pixels by using the command

```
>> fp = padarray(f, [84 84], 'both');
```

Zero padding was used to make all displayed images consistent with the image occupying the largest area (568 × 568 pixels) which, as discussed below, is the image rotated by 45°. The padding is for display purposes only, and was not used in any moment computations. The half-size and corresponding padded images were obtained using the commands

`B = fliplr(A)` *returns* `A` *with the columns flipped about the vertical axis, and* `B = flipud(A)` *returns* `A` *with the rows flipped about the horizontal axis.*

```
>> fhs = f(1:2:end, 1:2:end);
>> fhsp = padarray(fhs, [184 184], 'both');
```

The mirrored image was obtained using MATLAB function `fliplr`:

```
>> fm = fliplr(f);
>> fmp = padarray(fm, [84 84], 'both');
```

To rotate an image we use function `imrotate`:

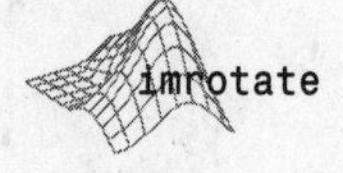


```
g = imrotate(f, angle, method, 'crop')
```

which rotates `f` by `angle` degrees in the counterclockwise direction. Parameter `method` can be one of the following:

`'nearest'` uses nearest neighbor interpolation;
`'bilinear'` uses bilinear interpolation (typically a good choice); and
`'bicubic'` uses bicubic interpolation.

The image size is increased automatically by padding to fit the rotation. If `'crop'` is included in the argument, the central part of the rotated image is cropped to the same size as the original. The default is to specify `angle` only, in which case `'nearest'` interpolation is used and no cropping takes place.

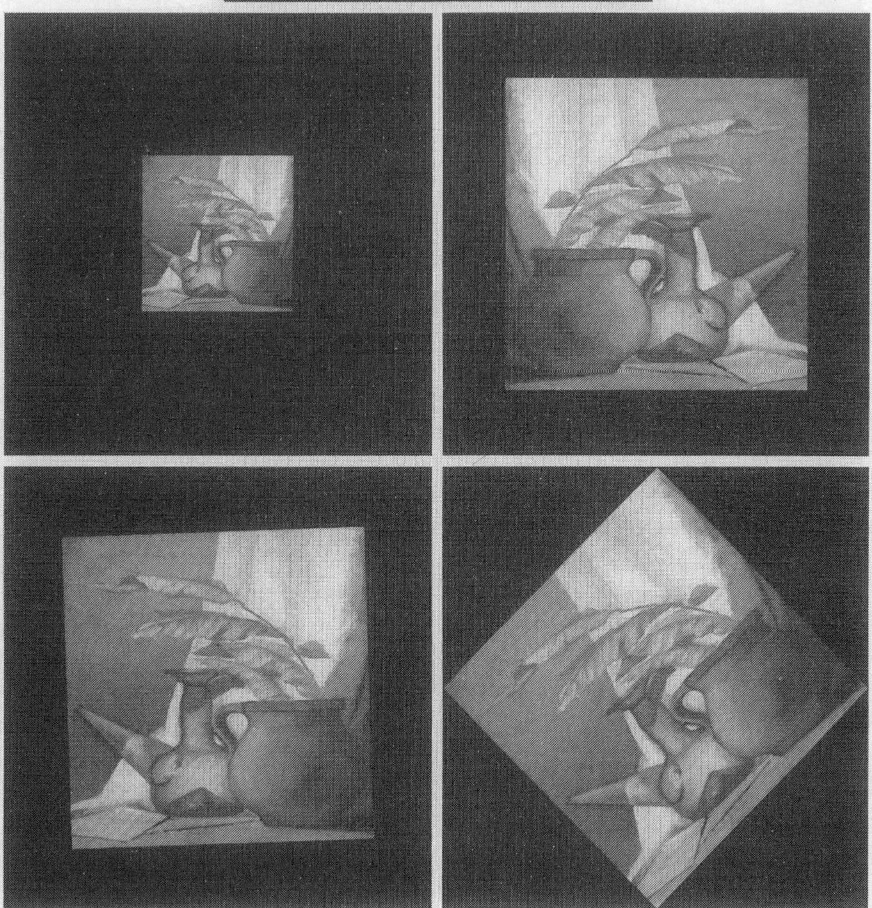

a
b c
d e

FIGURE 11.23 (a) Original, padded image. (b) Half size image. (c) Mirrored image. (d) Image rotated by 2°. (e) Image rotated 45°. The zero padding in (a) through (d) was done to make the images consistent in size with (e) for viewing purposes only.

TABLE 11.4 The seven moment invariants of the images in Figs. 11.23(a) through (e). Note the use of the magnitude of the log in the first column.

Invariant (|log|)	Original	Half Size	Mirrored	Rotated 2°	Rotated 45°
ϕ_1	6.600	6.600	6.600	6.600	6.600
ϕ_2	16.410	16.408	16.410	16.410	16.410
ϕ_3	23.972	23.958	23.972	23.978	23.973
ϕ_4	23.888	23.882	23.888	23.888	23.888
ϕ_5	49.200	49.258	49.200	49.200	49.198
ϕ_6	32.102	32.094	32.102	32.102	32.102
ϕ_7	47.953	47.933	47.850	47.953	47.954

The rotated images for our example were generated as follows:

```
>> fr2 = imrotate(f, 2, 'bilinear');
>> fr2p = padarray(fr2, [76 76], 'both');
>> fr45 = imrotate(f, 45, 'bilinear');
```

Note that no padding was required in the last image because it is the largest image in the set. The 0s in both rotated images were generated by IPT in the process of rotation.

The seven moment invariants of the five images just discussed were generated using the commands

```
>> phiorig = abs(log(invmoments(f)));
>> phihalf = abs(log(invmoments(fhs)));
>> phimirror = abs(log(invmoments(fm)));
>> phirot2 = abs(log(invmoments(fr2)));
>> phirot45 = abs(log(invmoments(fr45)));
```

Note that the absolute value of the `log` was used instead of the moment invariant values themselves. Use of the `log` reduces dynamic range, and the absolute value avoids having to deal with the complex numbers that result when computing the `log` of negative moment invariants. Because interest generally lies on the invariance of the moments, and not on their sign, use of the absolute value is common practice.

The seven moments of the original, half-size, mirrored, and rotated images are summarized in Table 11.4. Note how close the numbers are, indicating a high degree of invariance to the changes just mentioned. Results like these are the reason why moment invariants have been a basic staple in image description for more than four decades. ■

11.5 Using Principal Components for Description

Suppose that we have n registered images, "stacked" in the arrangement shown in Fig. 11.24. There are n pixels for any given pair of coordinates (i, j), one pixel at that location for each image. These pixels may be arranged in the form of a column vector

$$\mathbf{x} = \begin{bmatrix} x_1 \\ x_2 \\ \vdots \\ x_n \end{bmatrix}$$

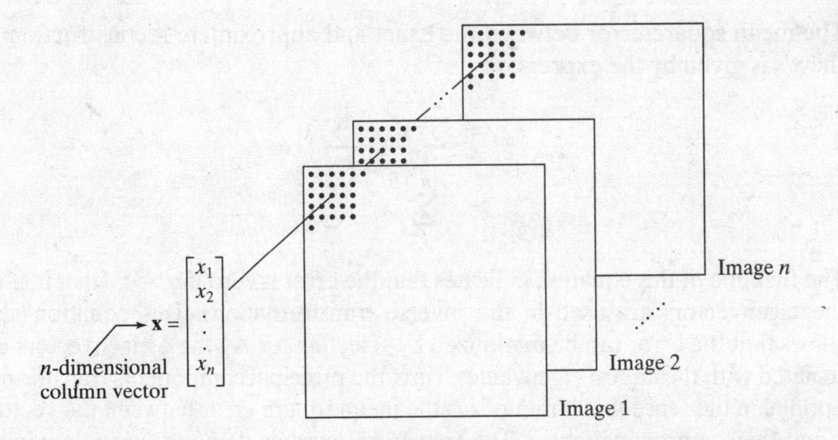

FIGURE 11.24 Forming a vector from corresponding pixels in a stack of images of the same size.

If the images are of size $M \times N$, there will be total of MN such n-dimensional vectors comprising all pixels in the n images.

The mean vector, $\mathbf{m_x}$, of a vector population can be approximated by the sample average:

$$\mathbf{m_x} = \frac{1}{K}\sum_{k=1}^{K}\mathbf{x}_k$$

with $K = MN$. Similarly, the $n \times n$ covariance matrix, $\mathbf{C_x}$, of the population can be approximated by

$$\mathbf{C_x} = \frac{1}{K-1}\sum_{k=1}^{K}(\mathbf{x}_k - \mathbf{m_x})(\mathbf{x}_k - \mathbf{m_x})^T$$

where $K - 1$ instead of K is used to obtain an unbiased estimate of $\mathbf{C_x}$ from the samples. Because $\mathbf{C_x}$ is real and symmetric, finding a set of n orthonormal eigenvectors always is possible.

The *principal components transform* (also called the *Hotelling transform*) is given by

$$\mathbf{y} = \mathbf{A}(\mathbf{x} - \mathbf{m_x})$$

It is not difficult to show that the elements of vector $\mathbf{y}$ are uncorrelated. Thus, the covariance matrix $\mathbf{C_y}$ is diagonal. The rows of matrix $\mathbf{A}$ are the normalized eigenvectors of $\mathbf{C_x}$. Because $\mathbf{C_x}$ is real and symmetric, these vectors form an orthonormal set, and it follows that the elements along the main diagonal of $\mathbf{C_y}$ are the eigenvalues of $\mathbf{C_x}$. The main diagonal element in the ith row of $\mathbf{C_y}$ is the variance of vector element y_i.

Because the rows of $\mathbf{A}$ are orthonormal, its inverse equals its transpose. Thus, we can recover the $\mathbf{x}$'s by performing the inverse transformation

$$\mathbf{x} = \mathbf{A}^T\mathbf{y} + \mathbf{m_x}$$

The importance of the principal components transform becomes evident when only q eigenvectors are used, in which case $\mathbf{A}$ becomes a $q \times n$ matrix, $\mathbf{A}_q$. Now the reconstruction is an approximation:

$$\hat{\mathbf{x}} = \mathbf{A}_q^T\mathbf{y} + \mathbf{m_x}$$

The mean square error between the exact and approximate reconstruction of the **x**'s is given by the expression

$$\begin{aligned} e_{\text{ms}} &= \sum_{j=1}^{n} \lambda_j - \sum_{j=1}^{q} \lambda_j \\ &= \sum_{j=q+1}^{n} \lambda_j \end{aligned}$$

The first line of this equation indicates that the error is zero if $q = n$ (that is, if all the eigenvectors are used in the inverse transformation). This equation also shows that the error can be minimized by selecting for $\mathbf{A}_q$ the q eigenvectors associated with the largest eigenvalues. Thus, the principal components transform is optimal in the sense that it minimizes the mean square error between the vectors $\mathbf{x}$ and their approximations $\hat{\mathbf{x}}$. The transform owes its name to using the eigenvectors corresponding to the largest (principal) eigenvalues of the covariance matrix. The example given later in this section further clarifies this concept.

A set of n registered images (each of size $M \times N$) is converted to a stack of the form shown in Fig. 11.24 by using the command:

```
>> S = cat(3, f1, f2,..., fn);
```

This image stack array, which is of size $M \times N \times n$, is converted to an array whose *rows* are n-dimensional vectors by using function `imstack2vectors` (see Appendix C for the code), which has the syntax

imstack2vectors

```
[X, R] = imstack2vectors(S, MASK)
```

where `S` is the image stack and `X` is the array of vectors extracted from `S` using the approach shown in Fig. 11.24. Input `MASK` is an $M \times N$ logical or numeric image with nonzero elements in the locations where elements of `S` are to be used in forming `X` and 0s in locations to be ignored. For example, if we wanted to use only vectors in the right, upper quadrant of the images in the stack, then `MASK` would contain 1s in that quadrant and 0s elsewhere. If `MASK` is not included in the argument, then all image locations are used in forming `X`. Finally, parameter `R` is an array whose rows are the 2-D coordinates corresponding to the location of the vectors used to form `X`. We show how to use `MASK` in Example 12.2. In the present discussion we use the default.

The following M-function, `covmatrix`, computes the mean vector and covariance matrix of the vectors in `X`.

covmatrix

```
function [C, m] = covmatrix(X)
%COVMATRIX Computes the covariance matrix of a vector population.
%   [C, M] = COVMATRIX(X) computes the covariance matrix C and the
%   mean vector M of a vector population organized as the rows of
%   matrix X. C is of size N-by-N and M is of size N-by-1, where N is
%   the dimension of the vectors (the number of columns of X).

[K, n] = size(X);
X = double(X);
```

```
if n == 1 % Handle special case.
   C = 0;
   m = X;
else
   % Compute an unbiased estimate of m.
   m = sum(X, 1)/K;
   % Subtract the mean from each row of X.
   X = X - m(ones(K, 1), :);
   % Compute an unbiased estimate of C. Note that the product is
   % X'*X because the vectors are rows of X.
   C = (X'*X)/(K - 1);
   m = m'; % Convert to a column vector.
end
```

The following function implements the concepts developed in this section. Note the use of structures to simplify the output arguments.

princomp

```
function P = princomp(X, q)
%PRINCOMP Obtain principal-component vectors and related quantities.
%   P = PRINCOMP(X, Q) Computes the principal-component vectors of
%   the vector population contained in the rows of X, a matrix of
%   size K-by-n where K is the number of vectors and n is their
%   dimensionality. Q, with values in the range [0, n], is the number
%   of eigenvectors used in constructing the principal-components
%   transformation matrix. P is a structure with the following
%   fields:
%
%      P.Y      K-by-Q matrix whose columns are the principal-
%               component vectors.
%      P.A      Q-by-n principal components transformation matrix
%               whose rows are the Q eigenvectors of Cx corresponding
%               to the Q largest eigenvalues.
%      P.X      K-by-n matrix whose rows are the vectors reconstructed
%               from the principal-component vectors. P.X and P.Y are
%               identical if Q = n.
%     P.ems     The mean square error incurred in using only the Q
%               eigenvectors corresponding to the largest
%               eigenvalues. P.ems is 0 if Q = n.
%     P.Cx      The n-by-n covariance matrix of the population in X.
%     P.mx      The n-by-1 mean vector of the population in X.
%     P.Cy      The Q-by-Q covariance matrix of the population in
%               Y. The main diagonal contains the eigenvalues (in
%               descending order) corresponding to the Q eigenvectors.

[K, n] = size(X);
X = double(X);

% Obtain the mean vector and covariance matrix of the vectors in X.
[P.Cx, P.mx] = covmatrix(X);
P.mx = P.mx'; % Convert mean vector to a row vector.

% Obtain the eigenvectors and corresponding eigenvalues of Cx. The
% eigenvectors are the columns of n-by-n matrix V. D is an n-by-n
```

`[V, D] = eig(A)` *returns the eigenvectors of* `A` *as the columns of matrix* `V`*, and the corresponding eigenvalues along the main diagonal of diagonal matrix* `D`.

```
% diagonal matrix whose elements along the main diagonal are the
% eigenvalues corresponding to the eigenvectors in V, so that X*V =
% D*V.
[V, D] = eig(P.Cx);

% Sort the eigenvalues in decreasing order. Rearrange the
% eigenvectors to match.
d = diag(D);
[d, idx] = sort(d);
d = flipud(d);
idx = flipud(idx);
D = diag(d);
V = V(:, idx);

% Now form the q rows of A from first q columns of V.
P.A = V(:, 1:q)';

% Compute the principal component vectors.
Mx = repmat(P.mx, K, 1); % M-by-n matrix. Each row = P.mx.
P.Y = P.A*(X - Mx)'; % q-by-K matrix.

% Obtain the reconstructed vectors.
P.X = (P.A'*P.Y)' + Mx;

% Convert P.Y to K-by-q array and P.mx to n-by-1 vector.
P.Y = P.Y';
P.mx = P.mx';

% The mean square error is given by the sum of all the
% eigenvalues minus the sum of the q largest eigenvalues.
d = diag(D);
P.ems = sum(d(q + 1:end));

% Covariance matrix of the Y's:
P.Cy = P.A*P.Cx*P.A';
```

EXAMPLE 11.13: Principal components.

■ Figure 11.25 shows six satellite images of size 512 × 512, corresponding to six spectral bands: visible blue (450–520 nm), visible green (520–600 nm), visible red (630–690 nm), near infrared (760–900 nm), middle infrared (1550–1750 nm), and thermal infrared (10,400–12,500 nm). The objective of this example is to illustrate the use of function `princomp` for principal-components work. The first step is to organize the elements of the six images in a stack of size 512 × 512 × 6, as discussed earlier:

```
>> S = cat(3, f1, f2, f3, f4, f5, f6);
```

where the `f`'s correspond to the six multispectral images just discussed. Then we organize the stack into array `X`:

```
>> [X, R] = imstack2vectors(S);
```

Next, we obtain the six principal-component images by using `q = 6` in function `princomp`:

```
>> P = princomp(X, 6);
```

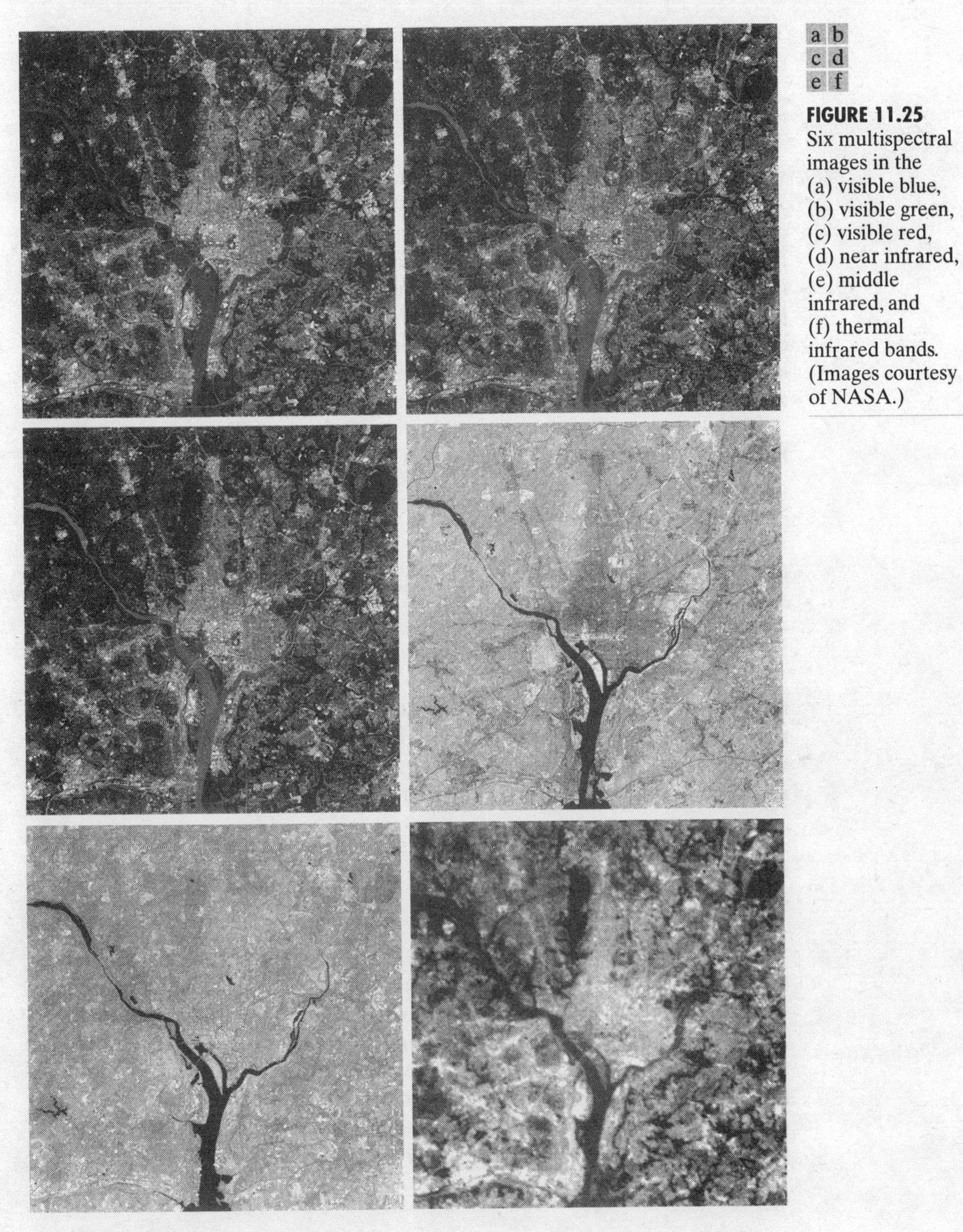

a b
c d
e f

FIGURE 11.25 Six multispectral images in the (a) visible blue, (b) visible green, (c) visible red, (d) near infrared, (e) middle infrared, and (f) thermal infrared bands. (Images courtesy of NASA.)

The first component image is generated and displayed with the commands

```
>> g1 = P.Y(:, 1);
>> g1 = reshape(g1, 512, 512);
>> imshow(g1, [ ])
```

The other five images are obtained and displayed in the same manner. The eigenvalues are along the main diagonal of `P.Cy`, so we use

```
>> d = diag(P.Cy);
```

where `d` is a 6-dimensional column vector because we used `q = 6` in the function.

Figure 11.26 shows the six principal-component images just computed. The most obvious feature is that a significant portion of the contrast detail is contained in the first two images, and it decreases rapidly from there. The reason is easily explained by looking at the eigenvalues. As Table 11.5 shows, the first two eigenvalues are quite large in comparison with the others. Because the eigenvalues are the variances of the elements of the **y** vectors, and variance is a measure of contrast, it is not unexpected that the images corresponding to the dominant eigenvalues would exhibit significantly higher contrast.

Suppose that we use a smaller value of `q`, say `q = 2`. Then reconstruction is based only on two principal component images. Using

```
>> P = princomp(X, 2);
```

and statements of the form

```
>> h1 = P.X(:, 1);
>> h1 = reshape(h1, 512, 512);
```

for each image resulted in the reconstructed images in Fig. 11.27. Visually, these images are quite close to the originals in Fig. 11.25. In fact, even the difference images show little degradation. For instance, to compare the original and reconstructed band 1 images, we write

```
>> D1 = double(f1) - double(h1);
>> D1 = gscale(D1);
>> imshow(D1)
```

Figure 11.28(a) shows the result. The low contrast in this image is an indication that little visual data was lost when only two principal component images were used to reconstruct the original image. Figure 11.28(b) shows the difference of the band 6 images. The difference here is more pronounced because the original band 6 image is actually blurry. But the two principal-component images used in the reconstruction are sharp, and they have the strongest influence on the reconstruction. The mean square error incurred in using only two principal component images is given by

```
P.ems

ans =

    1.7311e+003
```

which is the sum of the four smaller eigenvalues in Table 11.5. ■

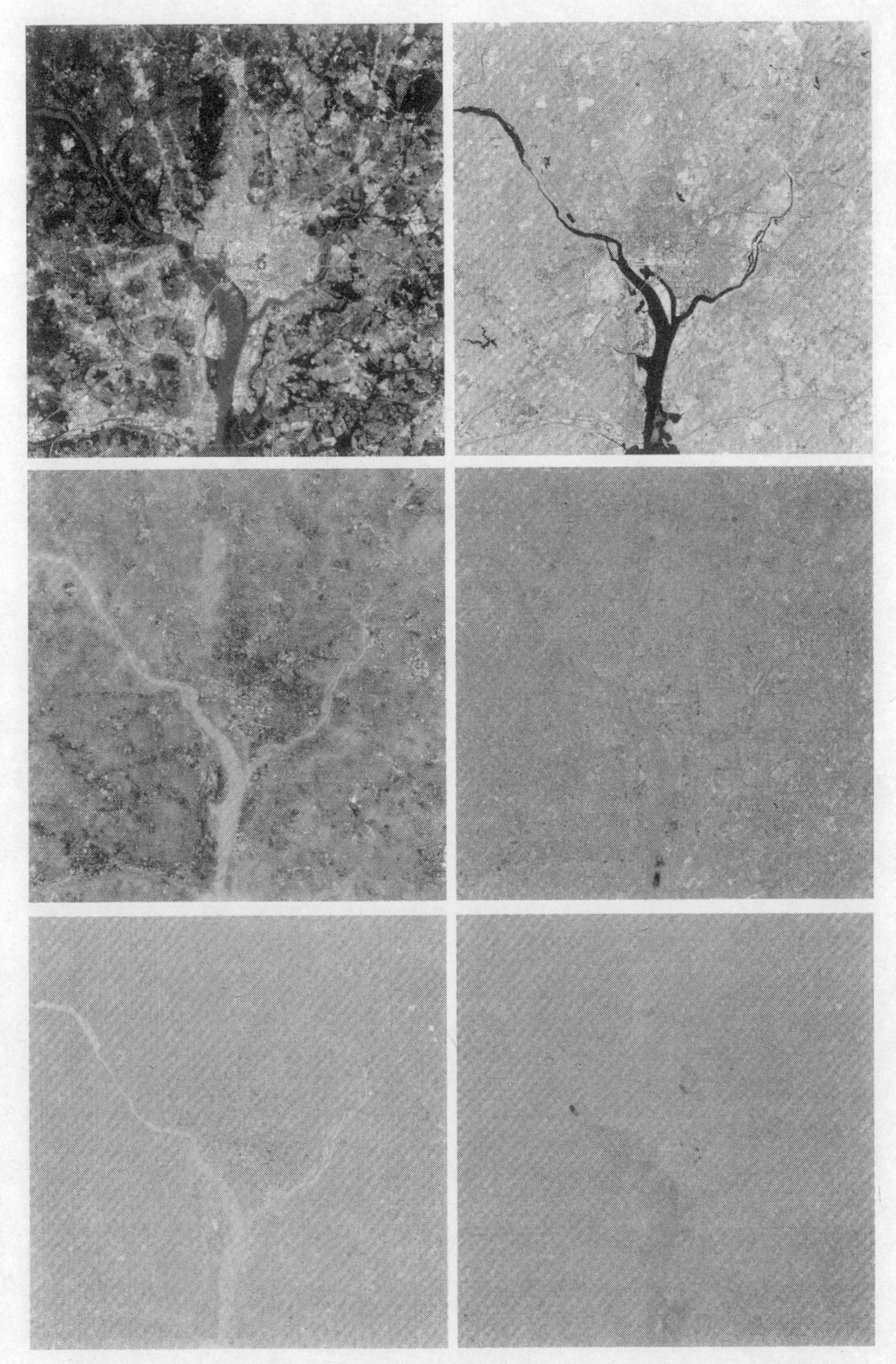

a b
c d
e f

FIGURE 11.26 Principal-component images corresponding to the images in Fig. 11.25.

λ_1	λ_2	λ_3	λ_4	λ_5	λ_6
10352	2959	1403	203	94	31

TABLE 11.5 Eigenvalues of P.Cy when q = 6.

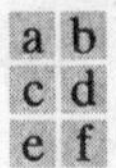

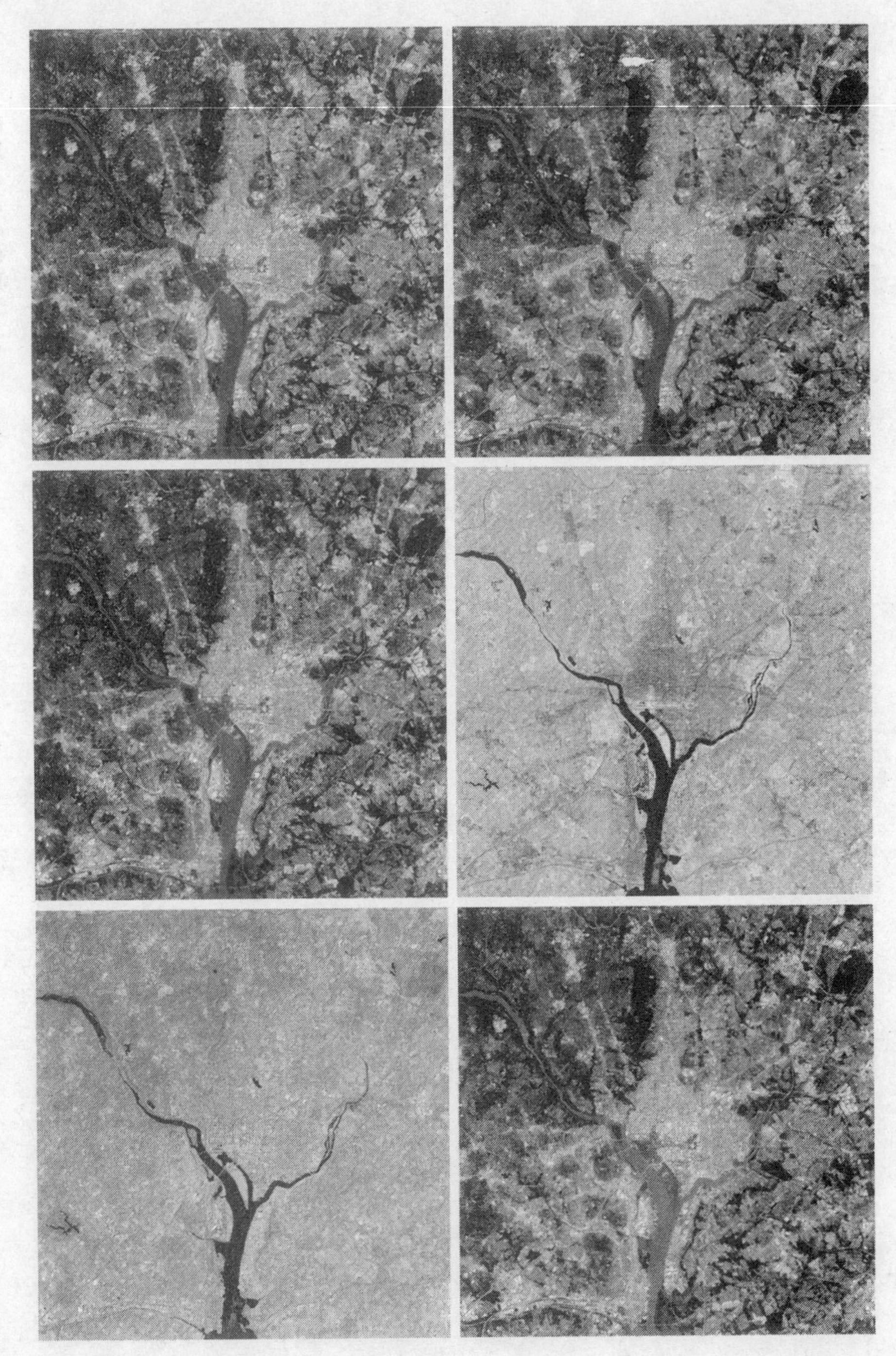

FIGURE 11.27 Multispectral images reconstructed using only the two principal-component images with the largest variance. Compare with the originals in Fig. 11.25.

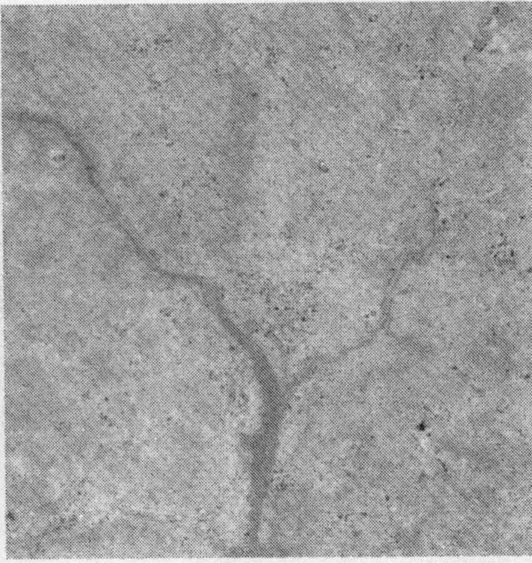

a b

FIGURE 11.28 (a) Difference between Figs. 11.27(a) and 11.25(a). (b) Difference between Figs. 11.27(f) and 11.25(f). Both images are scaled to the full [0, 255] 8-bit intensity scale.

Before leaving this section we point out that function `princomp` can be used to align objects (regions or boundaries) with the eigenvectors of the objects. The coordinates of the objects are arranged as the columns of `X`, and we use `q = 2`. The transformed data, aligned with the eigenvectors, is contained in `P.Y`. This is a rugged alignment procedure that uses all coordinates to compute the transformation matrix and aligns the data in the direction of its principal spread.

Summary

The representation of objects or regions that have been segmented out of an image is an early step in the preparation of image data for subsequent use in automation. For example, descriptors such as the ones just covered constitute the input to the object recognition algorithms developed in the next chapter. The M-functions developed in the preceding sections of this chapter are a significant extension to the power of standard IPT functions for image representation and description. It is undoubtedly clear by now that the choice of one type of descriptor over another is dictated to a large degree by the problem at hand. This is one of the principal reasons why the solution of image processing problems is aided significantly by having a flexible prototyping environment in which existing functions can be integrated with new code to gain flexibility and reduce development time. The material in this chapter is a good example of how to construct the basis for such an environment.

12 Object Recognition

Preview

We conclude the book with a discussion and development of several M-functions for region and/or boundary recognition, which in this chapter we call *objects* or *patterns*. Approaches to computerized pattern recognition may be divided into two principal areas: decision-theoretic and structural. The first category deals with patterns described using quantitative descriptors, such as length, area, texture, and many of the other descriptors discussed in Chapter 11. The second category deals with patterns best represented by symbolic information, such as strings, and described by the properties and relationships between those symbols, as explained in Section 12.4. Central to the theme of recognition is the concept of "learning" from sample patterns. Learning techniques for both decision-theoretic and structural approaches are implemented and illustrated in the material that follows.

12.1 Background

A *pattern* is an *arrangement of descriptors*, such as those discussed in Chapter 11. The name *feature* is used often in the pattern recognition literature to denote a descriptor. A *pattern class* is a family of patterns that share a set of common properties. Pattern classes are denoted $\omega_1, \omega_2, \ldots, \omega_W$, where W is the number of classes. Pattern recognition by machine involves techniques for assigning patterns to their respective classes—automatically and with as little human intervention as possible.

The two principal pattern arrangements used in practice are vectors (for quantitative descriptions) and strings (for structural descriptions). Pattern vectors are represented by bold lowercase letters, such as $\mathbf{x}$, $\mathbf{y}$, and $\mathbf{z}$, and have the $n \times 1$ vector form

$$\mathbf{x} = \begin{bmatrix} x_1 \\ x_2 \\ \vdots \\ x_n \end{bmatrix}$$

where each component, x_i, represents the ith descriptor and n is the total number of such descriptors associated with the pattern. Sometimes it is necessary in computations to use row vectors of dimension $1 \times n$, obtained simply by forming the transpose, $\mathbf{x}^T$, of the preceding column vector.

The nature of the components of a pattern vector **x** depends on the approach used to describe the physical pattern itself. For example, consider the problem of automatically classifying alphanumeric characters. Descriptors suitable for a decision-theoretic approach might include measures such 2-D moment invariants or a set of Fourier coefficients describing the outer boundary of the characters.

In some applications, pattern characteristics are best described by structural relationships. For example, fingerprint recognition is based on the interrelationships of print features called *minutiae*. Together with their relative sizes and locations, these features are primitive components that describe fingerprint ridge properties, such as abrupt endings, branching, merging, and disconnected segments. Recognition problems of this type, in which not only quantitative measures about each feature but also the spatial relationships between the features determine class membership, generally are best solved by structural approaches.

The material in the following sections is representative of techniques for implementing pattern recognition solutions in MATLAB. A basic concept in recognition, especially in decision-theoretic applications, is the idea of pattern matching based on measures of distance between pattern vectors. Therefore, we begin the discussion with various approaches for the efficient computation of distance measures in MATLAB.

12.2 Computing Distance Measures in MATLAB

The material in this section deals with vectorizing distance computations that otherwise would involve `for` or `while` loops. Some of the vectorized expressions presented here are considerably more subtle than most of the examples in the previous chapters, so the reader is encouraged to study them in detail. The following formulations are based on a summary of similar expressions compiled by Acklam [2002].

The *Euclidean distance* between two n-dimensional (row or column) vectors **x** and **y** is defined as the scalar

$$d(\mathbf{x}, \mathbf{y}) = \|\mathbf{x} - \mathbf{y}\| = \|\mathbf{y} - \mathbf{x}\| = [(x_1 - y_1)^2 + \cdots + (x_n - y_n)^2]^{1/2}$$

This expression is simply the norm of the difference between the two vectors, so we compute it using MATLAB's function `norm`:

```
d = norm(x - y)
```

where x and y are vectors corresponding to $\mathbf{x}$ and $\mathbf{y}$ in the preceding equation for $d(\mathbf{x}, \mathbf{y})$. Often, it is necessary to compute a set of Euclidean distances between a vector $\mathbf{y}$ and each member of a vector *population* consisting of p, n-dimensional vectors arranged as the rows of a $p \times n$ matrix $\mathbf{X}$. For the dimensions to line up properly, $\mathbf{y}$ has to be of dimension $1 \times n$. Then the distance between $\mathbf{y}$ and each element of $\mathbf{X}$ is contained in the $p \times 1$ vector

```
d = sqrt(sum(abs(X – repmat(y, p, 1)).^2, 2))
```

where d(i) is the Euclidean distance between $\mathbf{y}$ and the ith row of $\mathbf{X}$ [i.e., X(i, :)]. Note the use of function repmat to duplicate row vector $\mathbf{y}$ p times and thus form a $p \times n$ matrix to match the dimensions of $\mathbf{X}$. The last 2 on the right of the preceding line of code indicates that sum is to operate along dimension 2; that is, to sum the elements along the horizontal dimension.

Suppose next that we have two vector populations $\mathbf{X}$, of dimension $p \times n$, and $\mathbf{Y}$, of dimension $q \times n$. The matrix containing the distances between rows of these two populations can be obtained using the expression

```
D = sqrt(sum(abs(repmat(permute(X, [1 3 2]), [1 q 1]) ...
        – repmat(permute(Y, [3 1 2]), [p 1 1])).^2, 3))
```

where D(i,j) is the Euclidean distance between the ith and jth rows of the populations; that is, the distance between X(i,:) and Y(j,:).

The syntax for function permute in the preceding expression is

```
B = permute(A, order)
```

This function reorders the dimensions of A according to the elements of the vector order (the elements of this vector must be unique). For example, if A is a 2-D array, the statement B = permute(A,[2 1]) simply interchanges the rows and columns of A, which is equivalent to letting B equal the transpose of A. If the length of vector order is greater than the number of dimensions of A, MATLAB processes the components of the vector from left to right, until all elements are used. In the preceding expression for D, permute(X, [1 3 2]) creates arrays in the third dimension, each being a column (dimension 1) of X. Since there are n columns in X, n such arrays are created, with each array being of dimension $p \times 1$. Therefore, the command permute(X, [1 3 2]) creates an array of dimension $p \times 1 \times n$. Similarly, the command permute(Y, [3 1 2]) creates an array of dimension $1 \times q \times n$. Finally, the command repmat(permute(X, [1 3 2]), [1 q 1]) duplicates q times each of the n columns produced by the permute function, thus creating an array of dimension $p \times q \times n$. Similar comments hold for the other command involving Y. Basically, the preceding expression for D is simply a vectorization of the expressions that would be written using for or while loops.

In addition to the expressions just discussed, we use in this chapter a distance measure from a vector $\mathbf{y}$ to the mean $\mathbf{m_x}$ of a vector population, weighted inversely by the covariance matrix, $\mathbf{C_x}$, of the population. This metric, called the *Mahalanobis distance*, is defined as

$$d(\mathbf{y}, \mathbf{m_x}) = (\mathbf{y} - \mathbf{m_x})^T \mathbf{C_x}^{-1} (\mathbf{y} - \mathbf{m_x})$$

The inverse matrix operation is the most time-consuming computational task required to implement the Mahalanobis distance. This operation can be optimized significantly by using MATLAB's matrix right division operator (/) introduced in Table 2.4 (see also the margin note in the following page). Expressions for $\mathbf{m_x}$ and $\mathbf{C_x}$ are given in Section 11.5.

Let $\mathbf{X}$ denote a population of p, n-dimensional vectors, and let $\mathbf{Y}$ denote a population of q, n-dimensional vectors, such that the vectors in both $\mathbf{X}$ and $\mathbf{Y}$ are the rows of these arrays. The objective of the following M-function is to compute the Mahalanobis distance between every vector in $\mathbf{Y}$ and the mean, $\mathbf{m_x}$:

mahalanobis

```
function d = mahalanobis(varargin)
%MAHALANOBIS Computes the Mahalanobis distance.
%   D = MAHALANOBIS(Y, X) computes the Mahalanobis distance between
%   each vector in Y to the mean (centroid) of the vectors in X, and
%   outputs the result in vector D, whose length is size(Y, 1). The
%   vectors in X and Y are assumed to be organized as rows. The
%   input data can be real of complex. The outputs are real
%   quantities.
%
%   D = MAHALANOBIS(Y, CX, MX) computes the Mahalanobis distance
%   between each vector in Y and the given mean vector, MX. The
%   results are output in vector D, whose length is size(Y, 1). The
%   vectors in Y are assumed to be organized as the rows of this
%   array. The input data can be real or complex. The outputs are
%   real quantities. In addition to the mean vector MX, the
%   covariance matrix CX of a population of vectors X also must be
%   provided. Use function COVMATRIX (Section 11.5) to compute MX and
%   CX.

% Reference: Acklam, P. J. [2002]. "MATLAB Array Manipulation Tips
% and Tricks." Available at
%    home.online.no/~pjacklam/matlab/doc/mtt/index.html
% or at
%    www.prenhall.com/gonzalezwoodseddins

param = varargin; % Keep in mind that param is a cell array.
Y = param{1};
ny = size(Y, 1); % Number of vectors in Y.

if length(param) == 2
   X = param{2};
   % Compute the mean vector and covariance matrix of the vectors
   % in X.
   [Cx, mx] = covmatrix(X);
elseif length(param) == 3 % Cov. matrix and mean vector provided.
   Cx = param{2};
   mx = param{3};
else
   error('Wrong number of inputs.')
end
```

```
mx = mx(:)'; % Make sure that mx is a row vector.

% Subtract the mean vector from each vector in Y.
Yc = Y - mx(ones(ny, 1), :);

% Compute the Mahalanobis distances.
d = real(sum(Yc/Cx.*conj(Yc), 2));
```

With `A` *a square matrix, the MATLAB matrix operation* `A/B` *is a more accurate (and generally faster) implementation of the operation* `B*inv(A)`. *Similarly,* `A\B` *is a preferred implementation of the operation* `inv(A)*B`. *See Table 2.4.*

The call to `real` in the last line of code is to remove "numeric noise," as we did in Chapter 4 after filtering an image. If the data are known to always be real, the code can be simplified by removing functions `real` and `conj`.

12.3 Recognition Based on Decision-Theoretic Methods

Decision-theoretic approaches to recognition are based on the use of *decision* (also called *discriminant*) *functions*. Let $\mathbf{x} = (x_1, x_2, \ldots, x_n)^T$ represent an n-dimensional pattern vector, as discussed in Section 12.1. For W pattern classes, $\omega_1, \omega_2, \ldots, \omega_W$, the basic problem in decision-theoretic pattern recognition is to find W decision functions $d_1(\mathbf{x}), d_2(\mathbf{x}), \ldots, d_W(\mathbf{x})$ with the property that if a pattern $\mathbf{x}$ belongs to class ω_i, then

$$d_i(\mathbf{x}) > d_j(\mathbf{x}) \qquad j = 1, 2, \ldots, W; j \neq i$$

In other words, an unknown pattern $\mathbf{x}$ is said to belong to the ith pattern class if, upon substitution of $\mathbf{x}$ into all decision functions, $d_i(\mathbf{x})$ yields the largest numerical value. Ties are resolved arbitrarily.

The *decision boundary* separating class ω_i from ω_j is given by values of $\mathbf{x}$ for which $d_i(\mathbf{x}) = d_j(\mathbf{x})$ or, equivalently, by values of $\mathbf{x}$ for which

$$d_i(\mathbf{x}) - d_j(\mathbf{x}) = 0$$

Common practice is to express the decision boundary between two classes by the single function $d_{ij}(\mathbf{x}) = d_i(\mathbf{x}) - d_j(\mathbf{x}) = 0$. Thus $d_{ij}(\mathbf{x}) > 0$ for patterns of class ω_i and $d_{ij}(\mathbf{x}) < 0$ for patterns of class ω_j.

As will become clear in the following sections, finding decision functions entails estimating parameters from patterns that are representative of the classes of interest. Patterns used for parameter estimation are called *training patterns*, or *training sets*. Sets of patterns of known classes that are not used for training but are used instead to test the performance of a particular recognition approach are referred to as *test* or *independent* patterns or sets. The principal objective of Sections 12.3.2 and 12.3.4 is to develop various approaches for finding decision functions via the use of parameter estimation from training sets. Section 12.3.3 deals with matching by correlation, an approach that could be expressed in the form of decision functions but is traditionally presented in the form of direct image matching instead.

12.3.1 Forming Pattern Vectors

As noted at the beginning of this chapter, pattern vectors can be formed from quantitative descriptors, such as those discussed in Chapter 11 for regions and/or boundaries. For example, suppose that we describe a boundary by using

Fourier descriptors. The value of the ith descriptor becomes the value of x_i, the ith component of a pattern vector. In addition, we could append other components to pattern vectors. For instance, we could incorporate six additional components to the Fourier-descriptor by appending to each vector the six measures of texture in Table 11.2.

Another approach used quite frequently when dealing with (registered) multispectral images is to stack the images and then form vectors from corresponding pixels in the images, as illustrated in Fig. 11.24. The images are stacked by using function `cat`:

```
S = cat(3, f1, f2, ..., fn)
```

where `S` is the stack and `f1, f2, ..., fn` are the images from which the stack is formed. The vectors then are generated by using function `imstack2vectors` discussed in Section 11.5. See Example 12.2 for an illustration.

12.3.2 Pattern Matching Using Minimum-Distance Classifiers

Suppose that each pattern class, ω_j, is characterized by a mean vector $\mathbf{m}_j$. That is, we use the mean vector of each population of training vectors as being representative of that class of vectors:

$$\mathbf{m}_j = \frac{1}{N_j} \sum_{\mathbf{x} \in \omega_j} \mathbf{x} \qquad j = 1, 2, \ldots, W$$

where N_j is the number of training pattern vectors from class ω_j and the summation is taken over these vectors. As before, W is the number of pattern classes. One way to determine the class membership of an *unknown* pattern vector $\mathbf{x}$ is to assign it to the class of its closest prototype. Using the Euclidean distance as a measure of closeness (i.e., similarity) reduces the problem to computing the distance measures:

$$D_j(\mathbf{x}) = \|\mathbf{x} - \mathbf{m}_j\| \qquad j = 1, 2, \ldots, W$$

We then assign $\mathbf{x}$ to class ω_i if $D_i(\mathbf{x})$ is the smallest distance. That is, the smallest distance implies the best match in this formulation.

Suppose that all the mean vectors are organized as rows of a matrix $\mathbf{M}$. Then computing the distances from an arbitrary pattern $\mathbf{x}$ to all the mean vectors is accomplished by using the expression discussed in Section 12.2:

```
d = sqrt(sum(abs(M - repmat(x, W, 1)).^2, 2))
```

Because all distances are positive, this statement can be simplified by ignoring the `sqrt` operation. The minimum of `d` determines the class membership of pattern vector `x`:

```
>> class = find(d == min(d));
```

In other words, if the minimum of `d` is in its kth position (i.e., `x` belongs to the kth pattern class), then scalar `class` will equal `k`. If more than one minimum exists, `class` would equal a vector, with each of its elements pointing to a different location of the minimum.

If, instead of a single pattern, we have a set of patterns arranged as the rows of a matrix, `X`, then we use an expression similar to the longer expression in Section 12.2 to obtain a matrix `D`, whose element `D(I, J)` is the Euclidean distance between the ith pattern vector in `X` and the jth mean vector in `M`. Thus, to find the class membership of, say, the ith pattern in `X`, we find the column location in row `i` of `D` that yields the smallest value. Multiple minima yield multiple values, as in the single-vector case discussed in the last paragraph.

It is not difficult to show that selecting the smallest distance is equivalent to evaluating the functions

$$d_j(\mathbf{x}) = \mathbf{x}^T\mathbf{m}_j - \frac{1}{2}\mathbf{m}_j^T\mathbf{m}_j \qquad j = 1, 2, \ldots, W$$

and assigning $\mathbf{x}$ to class ω_i if $d_i(\mathbf{x})$ yields the *largest* numerical value. This formulation agrees with the concept of a decision function defined earlier.

The decision boundary between classes ω_i and ω_j for a minimum distance classifier is

$$\begin{aligned} d_{ij}(\mathbf{x}) &= d_i(\mathbf{x}) - d_j(\mathbf{x}) \\ &= \mathbf{x}^T(\mathbf{m}_i - \mathbf{m}_j) - \frac{1}{2}(\mathbf{m}_i - \mathbf{m}_j)^T(\mathbf{m}_i + \mathbf{m}_j) = 0 \end{aligned}$$

The surface given by this equation is the perpendicular bisector of the line segment joining $\mathbf{m}_i$ and $\mathbf{m}_j$. For $n = 2$, the perpendicular bisector is a line, for $n = 3$ it is a plane, and for $n > 3$ it is called a *hyperplane.*

12.3.3 Matching by Correlation

Correlation is quite simple in principle. Given an image $f(x, y)$, the correlation problem is to find all places in the image that match a given subimage (also called a *mask* or *template*) $w(x, y)$. Typically, $w(x, y)$ is much smaller than $f(x, y)$. One approach for finding matches is to treat $w(x, y)$ as a spatial filter and compute the sum of products (or a normalized version of it) for each location of w in f, in exactly the same manner explained in Section 3.4.1. Then the best match (matches) of $w(x, y)$ in $f(x, y)$ is (are) the location(s) of the maximum value(s) in the resulting correlation image. Unless $w(x, y)$ is small, the approach just described generally becomes computationally intensive. For this reason, practical implementations of spatial correlation typically rely on hardware-oriented solutions.

For prototyping, an alternative approach is to implement correlation in the frequency domain, making use of the correlation theorem, which, like the convolution theorem discussed in Chapter 4, relates spatial correlation to the product of the image transforms. Letting "$\circ$" denote correlation and "$*$" the complex conjugate, the correlation theorem states that

$$f(x, y) \circ w(x, y) \Leftrightarrow F(u, v)H^*(u, v)$$

In other words, spatial correlation can be obtained as the inverse Fourier transform of the product of the transform of one function times the conjugate of the transform of the other. Conversely, it follows that

$$f(x, y)w^*(x, y) \Leftrightarrow F(u, v) \circ H(u, v)$$

This second aspect of the correlation theorem is included for completeness. It is not used in this chapter.

Implementation of the first correlation result in the form of an M-function is straightforward, as the following code shows.

dftcorr

```
function g = dftcorr(f, w)
%DFTCORR 2-D correlation in the frequency domain.
%   G = DFTCORR(F, W) performs the correlation of a mask, W, with
%   image F. The output, G, is the correlation image, of class
%   double. The output is of the same size as F. When, as is
%   generally true in practice, the mask image is much smaller than
%   G, wraparound error is negligible if W is padded to size(F).

[M, N] = size(f);
f = fft2(f);
w = conj(fft2(w, M, N));
g = real(ifft2(w.*f));
```

EXAMPLE 12.1: Using correlation for image matching.

■ Figure 12.1(a) shows an image of Hurricane Andrew, in which the eye of the storm is clearly visible. As an example of correlation, we wish to find the location of the best match in (a) of the eye image in Fig. 12.1(b). The image is of size 912 × 912 pixels; the mask is of size 32 × 32 pixels. Figure 12.1(c) is the result of the following commands:

```
>> g = dftcorr(f, w);
>> gs = gscale(g);
>> imshow(gs)
```

The blurring evident in the correlation image of Fig. 12.1(c) should not be a surprise because the image in 12.1(b) has two dominant, nearly constant regions, and thus behaves similarly to a lowpass filter.

The feature of interest is the location of the best match, which, for correlation, implies finding the location(s) of the highest value in the correlation image:

```
>> [I, J] = find(g == max(g(:)))
I =

   554

J =

   203
```

In this case the highest value is unique. As explained in Section 3.4.1, the coordinates of the correlation image correspond to displacements of the template, so coordinates `[I, J]` correspond to the location of the bottom, left corner of

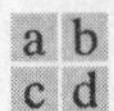

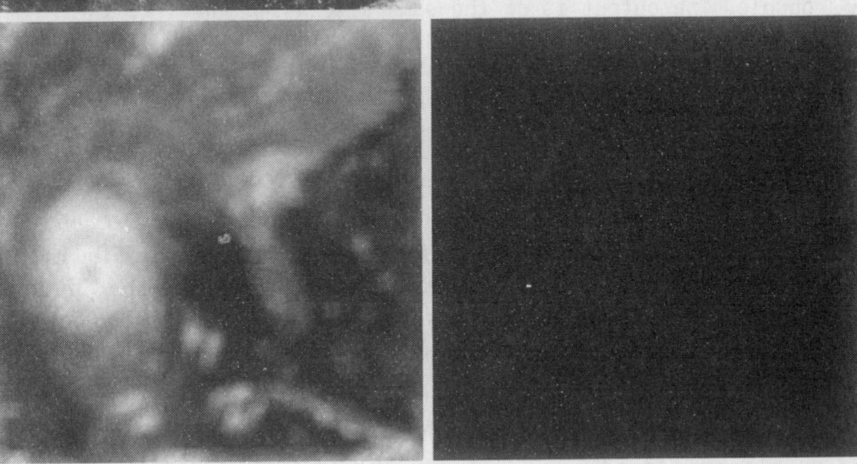

FIGURE 12.1 (a) Multispectral image of Hurricane Andrew. (b) Template. (c) Correlation of image and template. (d) Location of the best match. (Original image courtesy of NOAA.)

See Fig. 3.14 for an explanation of the mechanics of correlation.

the template. If the template were so located on top of the image, we would find that the template aligns quite closely with the eye of the hurricane at those coordinates. Another approach for finding the locations of the matches is to threshold the correlation image near its maximum, or threshold its scaled version, gs, whose highest value is known to be 255. For example, the image in Fig. 12.1(d) was obtained using the command

```
>> imshow(gs > 254)
```

Aligning the bottom, left corner of the template with the small white dot in Fig. 12.1(d) again reveals that the best match is near the eye of the hurricane. ■

12.3.4 Optimum Statistical Classifiers

The well-known Bayes classifier for a 0-1 loss function (Gonzalez and Woods [2002]) has decision functions of the form

$$d_j(\mathbf{x}) = p(\mathbf{x}/\omega_j)P(\omega_j) \quad j = 1, 2, \ldots, W$$

where $p(\mathbf{x}/\omega_j)$ is the probability density function (PDF) of the pattern vectors of class ω_j, and $P(\omega_j)$ is the probability (a scalar) that class ω_j occurs. As before, given an unknown pattern vector, the process is to compute a total of W decision functions and then assign the pattern to the class whose decision function yielded the largest numerical value. Ties are resolved arbitrarily.

The case when the probability density functions are (or are assumed to be) Gaussian is of particular practical interest. The n-dimensional Gaussian PDF has the form

$$p(\mathbf{x}/\omega_j) = \frac{1}{(2\pi)^{n/2}|\mathbf{C}_j|^{1/2}} e^{-\frac{1}{2}[(\mathbf{x}-\mathbf{m}_j)^T\mathbf{C}_j^{-1}(\mathbf{x}-\mathbf{m}_j)]}$$

where $\mathbf{C}_j$ and $\mathbf{m}_j$ are the covariance matrix and mean vector of the pattern population of class ω_j, and $|\mathbf{C}_j|$ is the determinant of $\mathbf{C}_j$.

Because the logarithm is a monotonically increasing function, choosing the largest $d_j(\mathbf{x})$ to classify patterns is equivalent to choosing the largest $\ln[d_j(\mathbf{x})]$, so we can use instead decision functions of the form

$$\begin{aligned} d_j(\mathbf{x}) &= \ln[p(\mathbf{x}/\omega_j)P(\omega_j)] \\ &= \ln p(\mathbf{x}/\omega_j) + \ln P(\omega_j) \end{aligned}$$

where the logarithm is guaranteed to be real because $p(\mathbf{x}/\omega_j)$ and $P(\omega_j)$ are nonnegative. Substituting the expression for the Gaussian PDF gives the equation

$$d_j(\mathbf{x}) = \ln P(\omega_j) - \frac{n}{2}\ln 2\pi - \frac{1}{2}\ln|\mathbf{C}_j| - \frac{1}{2}[(\mathbf{x}-\mathbf{m}_j)^T\mathbf{C}_j^{-1}(\mathbf{x}-\mathbf{m}_j)]$$

The term $(n/2)\ln 2\pi$ is the same positive constant for all classes, so it can be deleted, yielding the decision functions

$$d_j(\mathbf{x}) = \ln P(\omega_j) - \frac{1}{2}\ln|\mathbf{C}_j| - \frac{1}{2}[(\mathbf{x}-\mathbf{m}_j)^T\mathbf{C}_j^{-1}(\mathbf{x}-\mathbf{m}_j)]$$

for $j = 1, 2, \ldots, W$. The term inside the brackets is recognized as the Mahalanobis distance discussed in Section 12.2, for which we have a vectorized implementation. We also have an efficient method for computing the mean and covariance matrix from Section 11.5, so implementing the Bayes classifier for the multivariate Gaussian case is straightforward, as the following function shows.

bayesgauss

```
function d = bayesgauss(X, CA, MA, P)
%BAYESGAUSS Bayes classifier for Gaussian patterns.
%   D = BAYESGAUSS(X, CA, MA, P) computes the Bayes decision
%   functions of the patterns in the rows of array X using the
%   covariance matrices and and mean vectors provided in the arrays
%   CA and MA. CA is an array of size n-by-n-by-W, where n is the
%   dimensionality of the patterns and W is the number of
%   classes. Array MA is of dimension n-by-W (i.e., the columns of MA
%   are the individual mean vectors). The location of the covariance
%   matrices and the mean vectors in their respective arrays must
%   correspond. There must be a covariance matrix and a mean vector
```

```
%   for each pattern class, even if some of the covariance matrices
%   and/or mean vectors are equal. X is an array of size K-by-n,
%   where K is the total number of patterns to be classified (i.e.,
%   the pattern vectors are rows of X). P is a 1-by-W array,
%   containing the probabilities of occurrence of each class. If
%   P is not included in the argument list, the classes are assumed
%   to be equally likely.
%
%   The output, D, is a column vector of length K. Its Ith element is
%   the class number assigned to the Ith vector in X during Bayes
%   classification.

d = [ ]; % Initialize d.
error(nargchk(3, 4, nargin)) % Verify correct no. of inputs.
n = size(CA, 1);             % Dimension of patterns.

% Protect against the possibility that the class number is
% included as an (n+1)th element of the vectors.
X = double(X(:, 1:n));
W = size(CA, 3); % Number of pattern classes.
K = size(X, 1);  % Number of patterns to classify.
if nargin == 3
   P(1:W) = 1/W; % Classes assumed equally likely.
else
   if sum(P) ~= 1
      error('Elements of P must sum to 1.');
   end
end

% Compute the determinants.
for J = 1:W
   DM(J) = det(CA(:, :, J));
end

% Compute inverses, using right division (IM/CA), where IM =
% eye(size(CA, 1)) is the n-by-n identity matrix. Reuse CA to
% conserve memory.
IM = eye(size(CA,1));
for J = 1:W
   CA(:, :, J) = IM/CA(:, :, J);
end

% Evaluate the decision functions. The sum terms are the
% Mahalanobis distances discussed in Section 12.2.
MA = MA'; % Organize the mean vectors as rows.
for I = 1:K
   for J = 1:W
      m = MA(J, :);
      Y = X - m(ones(size(X, 1), 1), :);
      if P(J) == 0
         D(I, J) = -Inf;
      else
         D(I, J) = log(P(J)) - 0.5*log(DM(J)) ...
                   - 0.5*sum(Y(I, :)*(CA(:, :, J)*Y(I, :)'));
```

`eye(n)` *returns the* `n` × `n` *identity matrix;* `eye(m, n)` *or* `eye([m n])` *returns an* `m` × `n` *matrix with 1s along the diagonal and 0s elsewhere. The syntax* `eye(size(A))` *gives the same result as the previous format, with* `m` *and* `n` *being the number of rows and columns in* `A`*, respectively.*

```
        end
    end
end

% Find the maximum in each row of D. These maxima
% give the class of each pattern:
for I = 1:K
    J = find(D(I, :) == max(D(I, :)));
    d(I, :) = J(:);
end

% When there are multiple maxima the decision is
% arbitrary. Pick the first one.
d = d(:, 1);
```

EXAMPLE 12.2: Bayes classification of multispectral data.

■ Bayes recognition is used frequently for automatically classifying regions in multispectral imagery. Figure 12.2 shows the first four images from Fig. 11.25 (three visual bands and one infrared band). As a simple illustration, we apply the Bayes classification approach to three types (classes) of regions in these images: water, urban, and vegetation. The pattern vectors in this example are formed by the method discussed in Sections 11.5 and 12.3.1, in which corresponding pixels in the images are organized as vectors. We are dealing with four images, so the pattern vectors are four dimensional.

To obtain the mean vectors and covariance matrices, we need samples representative of each pattern class. A simple way to obtain such samples interactively is to use function `roipoly` (see Section 5.2.4) with the statement

```
>> B = roipoly(f);
```

where `f` is any of the multispectral images and `B` is a binary mask image. With this format, image `B` is generated interactively on the screen. Figure 12.2(e) shows a composite of three mask images, `B1`, `B2`, and `B3`, generated using this method. The numbers 1, 2, and 3 identify regions containing samples representative of water, urban development, and vegetation, respectively.

Next we obtain the vectors corresponding to each region. The four images already are registered spatially, so they simply are concatenated along the third dimension to obtain an image stack:

```
>> stack = cat(3, f1, f2, f3, f4);
```

where `f1` thorough `f4` are the four images in Figs. 12.2(a) through (d). Any point, when viewed through these four images, corresponds to a four-dimensional pattern vector (see Fig. 11.24). We are interested in the vectors contained in the three regions shown in Fig. 12.2(e), which we obtain by using function `imstack2vectors` discussed in Section 11.5:

```
>> [X, R] = imstack2vectors(stack, B);
```

where `X` is an array whose rows are the vectors, and `R` is an array whose rows are the locations (2-D region coordinates) corresponding to the vectors in `X`.

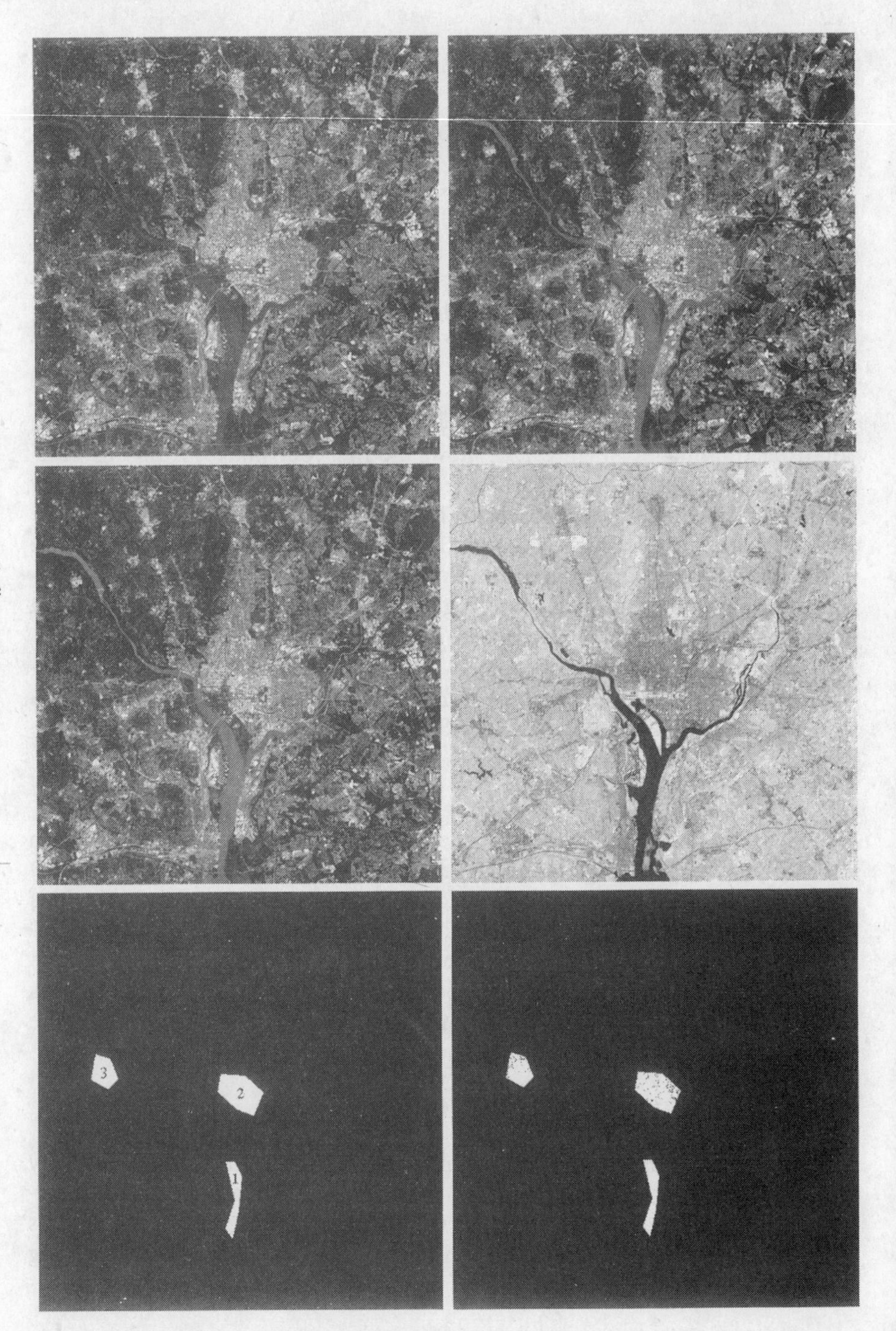

a b
c d
e f

FIGURE 12.2 Bayes classification of multispectral data. (a)–(c) Images in the blue, green, and red visible wavelengths. (d) Infrared image. (e) Mask showing sample regions of water (1), urban development (2), and vegetation (3). (f) Results of classification. The black dots denote points classified incorrectly. The other (white) points in the regions were classified correctly. (Original images courtesy of NASA.)

Using `imstack2vectors` with the three masks B1, B2, and B3 yielded three vector sets, X1, X2, and X3, and three sets of coordinates, R1, R2, and R3. Then three subsets Y1, Y2, and Y3 were extracted from the X's to use as training samples to estimate the covariance matrices and mean vectors. The Y's were generated by skipping every other row of X1, X2, and X3. The covariance matrix and mean vector of the vectors in Y1 were obtained with the command

```
>> [C1, m1] = covmatrix(Y1);
```

and similarly for the other two classes. Then we formed arrays CA and MA for use in `bayesgauss` as follows:

```
>> CA = cat(3, C1, C2, C3);
>> MA = cat(2, m1, m2, m3);
```

The performance of the classifier with the training patterns was determined by classifying the training sets:

```
>> dY1 = bayesgauss(Y1, CA, MA);
```

and similarly for the other two classes. The number of misclassified patterns of class 1 was obtained by writing

```
>> IY1 = find(dY1 ~= 1);
```

Finding the class into which the patterns were misclassified is straightforward. For instance, `length(find(dY1 == 2))` gives the number of patterns from class 1 that were misclassified into class 2. The other pattern sets were handled in a similar manner.

Table 12.1 summarizes the recognition results obtained with the training and independent pattern sets. The percentage of training and independent patterns recognized correctly was about the same with both sets, indicating stability in the parameter estimates. The largest error in both cases was with patterns from the urban area. This is not unexpected, as vegetation is present there also (note that no patterns in the urban or vegetation areas were misclassified as water). Figure 12.2(f) shows as black dots the points that were misclassified and as white dots the points that were classified correctly in each region. No black dots are readily visible in region 1 because the 7 misclassified points are very close to, or on, the boundary of the white region.

Additional work would be required to design an operable recognition system for multispectral classification. However, the important point of this example is the ease with which such a system could be prototyped using MATLAB and IPT functions, complemented by some of the functions developed thus far in the book. ■

TABLE 12.1 Bayes classification of multispectral image data.

Training Patterns						Independent Patterns					
Class	No. of Samples	Classified into Class 1	2	3	% Correct	Class	No. of Samples	Classified into Class 1	2	3	% Correct
1	484	482	2	0	99.6	1	483	478	3	2	98.9
2	933	0	885	48	94.9	2	932	0	880	52	94.4
3	483	0	19	464	96.1	3	482	0	16	466	96.7

12.3.5 Adaptive Learning Systems

The approaches discussed in Sections 12.3.1 and 12.3.3 are based on the use of sample patterns to estimate the statistical parameters of each pattern class. The minimum-distance classifier is specified completely by the mean vector of each class. Similarly, the Bayes classifier for Gaussian populations is specified completely by the mean vector and covariance matrix of each class of patterns.

In these two approaches, training is a simple matter. The training patterns of each class are used to compute the parameters of the decision function corresponding to that class. After the parameters in question have been estimated, the structure of the classifier is fixed, and its eventual performance will depend on how well the actual pattern populations satisfy the underlying statistical assumptions made in the derivation of the classification method being used.

As long as the pattern classes are characterized, at least approximately, by Gaussian probability density functions, the methods just discussed can be quite effective. However, when this assumption is not valid, designing a statistical classifier becomes a much more difficult task because estimating multivariate probability density functions is not a trivial endeavor. In practice, such decision-theoretic problems are best handled by methods that yield the required decision functions directly via training. Then making assumptions regarding the underlying probability density functions or other probabilistic information about the pattern classes under consideration is unnecessary.

The principal approach in use today for this type of classification is based on neural networks (Gonzalez and Woods [2002]). The scope of implementing neural networks suitable for image-processing applications is not beyond the capabilities of the functions available to us in MATLAB and IPT. However, this effort would be unwarranted in the present context because a comprehensive neural-networks toolbox has been available from The MathWorks for several years.

12.4 Structural Recognition

Structural recognition techniques are based generally on representing objects of interest as strings, trees, or graphs and then defining descriptors and recognition rules based on those representations. The key difference between decision-theoretic and structural methods is that the former uses quantitative descriptors expressed in the form of numeric vectors. Structural techniques, on the other hand, deal principally with symbolic information. For instance, suppose that object boundaries in a given application are represented by minimum-perimeter polygons. A decision-theoretic approach might be based on forming vectors whose elements are the numeric values of the interior angles of the polygons, while a structural approach might be based on defining symbols for *ranges* of angle values and then forming a string of such symbols to describe the patterns.

Strings are by far the most common representation used in structural recognition, so we focus on this approach in this section. As will become evident shortly, MATLAB has an extensive set of functions specialized for string manipulation.

12.4.1 Working with Strings in MATLAB

In MATLAB, a string is a one-dimensional array whose components are the numeric codes for the characters in the string. The characters displayed depend on the character set used in encoding a given font. The *length* of a string is the number of characters in the string, including spaces. It is obtained using the familiar function `length`. A string is defined by enclosing its characters in single quotes (a textual quote within a string is indicated by two quotes).

Table 12.2 lists the principal MATLAB functions that deal with strings.[†] Considering first the general category, function `blanks` has the syntax:

```
s = blanks(n)
```

It generates a string consisting of `n` blanks. Function `cellstr` creates a cell array of strings from a character array. One of the principal advantages of storing strings in cell arrays is that it eliminates the need to pad strings with blanks to create character arrays with rows of equal length (e.g., to perform string comparisons). The syntax

```
c = cellstr(S)
```

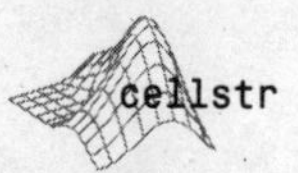

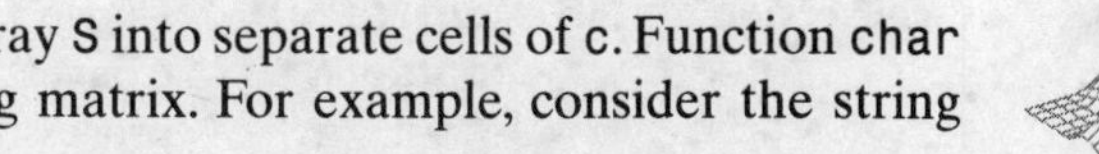

places the rows of the character array `S` into separate cells of `c`. Function `char` is used to convert back to a string matrix. For example, consider the string matrix

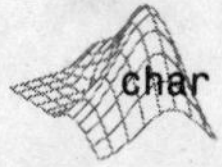


```
>> S = [' abc'; 'defg'; 'hi  ']   % Note the blanks.
S =
   abc
  defg
  hi
```

Typing `whos S` at the prompt displays the following information:

```
>> whos S
  Name      Size      Bytes     Class
  S         3x4       24        char array
```

[†]Some of the string functions discussed in this section were introduced in earlier chapters.

TABLE 12.2 MATLAB's string-manipulation functions.

Category	Function Name	Explanation
General	`blanks`	String of blanks.
	`cellstr`	Create cell array of strings from character array. Use function `char` to convert back to a character string.
	`char`	Create character array (string).
	`deblank`	Remove trailing blanks.
	`eval`	Execute string with MATLAB expression.
String tests	`iscellstr`	True for cell array of strings.
	`ischar`	True for character array.
	`isletter`	True for letters of the alphabet.
	`isspace`	True for whitespace characters.
String operations	`lower`	Convert string to lowercase.
	`regexp`	Match regular expression.
	`regexpi`	Match regular expression, ignoring case.
	`regexprep`	Replace string using regular expression.
	`strcat`	Concatenate strings.
	`strcmp`	Compare strings (see Section 2.10.5).
	`strcmpi`	Compare strings, ignoring case.
	`strfind`	Find one string within another.
	`strjust`	Justify string.
	`strmatch`	Find matches for string.
	`strncmp`	Compare first `n` characters of strings.
	`strncmpi`	Compare first `n` characters, ignoring case.
	`strread`	Read formatted data from a string. See Section 2.10.5 for a detailed explanation.
	`strrep`	Replace a string within another.
	`strtok`	Find token in string.
	`strvcat`	Concatenate strings vertically.
	`upper`	Convert string to uppercase.
String to number conversion	`double`	Convert string to numeric codes.
	`int2str`	Convert integer to string.
	`mat2str`	Convert matrix to a string suitable for processing with the `eval` function.
	`num2str`	Convert number to string.
	`sprintf`	Write formatted data to string.
	`str2double`	Convert string to double-precision value.
	`str2num`	Convert string to number (see Section 2.10.5).
	`sscanf`	Read string under format control.
Base number conversion	`base2dec`	Convert base B string to decimal integer.
	`bin2dec`	Convert binary string to decimal integer.
	`dec2base`	Convert decimal integer to base B string.
	`dec2bin`	Convert decimal integer to binary string.
	`dec2hex`	Convert decimal integer to hexadecimal string.
	`hex2dec`	Convert hexadecimal string to decimal integer.
	`hex2num`	Convert IEEE hexadecimal to double-precision number.

Note in the first command line that two of the three strings in S have trailing blanks because all rows in a string matrix must have the same number of characters. Note also that no quotes enclose the strings in the output because S is a character array. The following command returns a 3×1 cell array:

```
>> c = cellstr(S)
c =

    ' abc'
    'defg'
    'hi'
>> whos c
    Name        Size          Bytes         Class
    c           3x1           294           cell array
```

where, for example, `c(1) = ' abc'`. Note that quotes appear around the strings in the output, and that the strings have no trailing blanks. To convert back to a string matrix we let

```
Z = char(c)
Z =
  abc
 defg
 hi
```

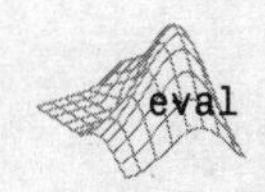

Function `eval` evaluates a string that contains a MATLAB expression. The call `eval(expression)` executes `expression`, a string containing any valid MATLAB expression. For example, if `t` is the character string `t = '3^2'`, typing `eval(t)` returns a 9.

The next category of functions deals with string tests. A 1 is returned if the funtion is `true`; otherwise the value returned is 0. Thus, in the preceding example, `iscellstr(c)` would return a 1 and `iscellstr(S)` would return a 0. Similar comments apply to the other functions in this category.

String operations are next. Functions `lower` (and `upper`) are self explanatory. They are discussed in Section 2.10.5. The next three functions deal with *regular expressions*,† which are sets of symbols and syntactic elements used commonly to match patterns of text. A simple example of the power of regular expressions is the use of the familiar wildcard symbol " * " in a file search. For instance, a search for image*.m in a typical search command window would return all the M-files that begin with the word "image." Another example of the use of regular expressions is in a search-and-replace function that searches for an instance of a given text string and replaces it with another. Regular expressions are formed using *metacharacters*, some of which are listed in Table 12.3.

---

†Regular expressions can be traced to the work of American mathematician Stephen Kleene, who developed regular expressions as a notation for describing what he called "the algebra of regular sets."

TABLE 12.3 Some of the metacharacters used in regular expressions for matching. See the `regular expressions` help page for a complete list.

Metacharacters	Usage
.	Matches any one character.
[ab...]	Matches any one of the characters, (a, b, . . .), contained within the brackets.
[^ab...]	Matches any character except those contained within the brackets.
?	Matches any character zero or one times.
*	Matches the preceding element zero or more times.
+	Matches the preceding element one or more times.
{num}	Matches the preceding element num times.
{min, max}	Matches the preceding element at least min times, but not more than max times.
|	Matches either the expression preceding or following the metacharacter |.
^chars	Matches when a string begins with chars.
chars$	Matches when a string ends with chars.
\<chars	Matches when a word begins with chars.
chars\>	Matches when a word ends with chars.
\<word\>	Exact word match.

In the context of this discussion, a "word" is a substring within a string, preceded by a space or the beginning of the string, and ending with a space or the end of the string. Several examples are given in the following paragraph.

Function regexp matches a regular expression. Using the basic syntax

```
idx = regexp(str, expr)
```

returns a row vector, idx, containing the indices (locations) of the substrings in str that match the regular expression string, expr. For example, suppose that expr = 'b.*a'. Then the expression idx = regexp(str, expr) would mean find matches in string str for any b that is followed by any character (as specified by the metacharacter ".") any number of times, including zero times (as specified by *), followed by an a. The indices of any locations in str meeting these conditions are stored in vector idx. If no such locations are found, then idx is returned as the empty matrix.

A few more examples of regular expressions for expr should clarify these concepts. The regular expression 'b. + a' would be as in the preceding example, except that "any number of times, including zero times" would be replaced by "one or more times." The expression 'b [0–9]' means any b followed by any number from 0 to 9; the expression 'b [0–9]*' means any b followed by any number from 0 to 9 any number of times; and 'b [0–9] +' means b followed by any number from 0 to 9 one or more times. For example, if str = 'b0123c234bcd', the preceding three instances of expr would give the following results: idx = 1; idx = [1 10]; and idx = 1.

As an example of the use of regular expressions for recognizing object characteristics, suppose that the boundary of an object has been coded with a four-directional Freeman chain code [see Fig. 11.1(a)], stored in string str, so that

```
>> str

str =

000300333222221111
```

Suppose also that we are interested in finding the locations in the string where the direction of travel turns from east (0) to south (3), and stays there for at least two increments, but no more than six increments. This is a "downward step" feature in the object, larger than a single transition, which may be due to noise. We can express these requirements in terms of the following regular expression:

```
>> expr = '0[3]{2, 6}';
```

Then

```
>> idx = regexp(str, expr)
idx =
    6
```

The value of idx identifies the point in this case where a 0 is followed by three 3s. More complex expressions are formed in a similar manner.

Function regexpi behaves in the manner just described for regexp, except that it ignores character (upper and lower) case. Function regexprep, with syntax

regexpi

```
s = regexprep(str, expr, replace)
```

regexprep

replaces with string replace all occurrences of the regular expression expr in string, str. The new string is returned. If no matches are found regexprep returns str, unchanged.

Function strcat has the syntax

```
C = strcat(S1, S2, S3, ...)
```

strcat

This function concatenates (horizontally) corresponding rows of the character arrays S1, S2, S3, and so on. All input arrays must have the same number of rows (or any can be a single string). When the inputs are all character arrays, the output is also a character array. If any of the inputs is a cell array of strings, strcat returns a cell array of strings formed by concatenating corresponding elements of S1, S2, S3, and so on. The inputs must all have the same size (or any can be a scalar). Any of the inputs can also be character arrays. Trailing spaces in character array inputs are ignored and do not appear in the output. This is not true for inputs that are cell arrays of strings. To preserve trailing spaces the familiar concatenation syntax based on square brackets, [S1 S2 S3 ...], should be used. For example,

```
>> a = 'hello '   % Note the trailing blank space.
>> b = 'goodbye'
>> strcat(a, b)
ans =
    hellogoodbye
[a b]
ans =
    hello goodbye
```

Function strvcat, with syntax

```
S = strvcat(t1, t2, t3, ...)
```

forms the character array S containing the text strings (or string matrices) t1,t2,t3, ... as rows. Blanks are appended to each string as necessary to form a valid matrix. Empty arguments are ignored. For example, using the strings a and b in the previous example,

```
>> strvcat(a, b)

ans =

    hello
    goodbye
```

Function strcmp, with syntax

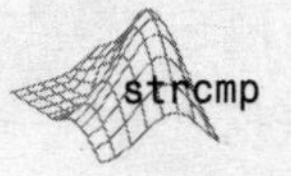


```
k = strcmp(str1, str2)
```

compares the two strings in the argument and returns 1 (true) if the strings are identical. Otherwise it returns a 0 (false). A more general syntax is

```
K = strcmp(S, T)
```

where either S or T is a cell array of strings, and K is an array (of the same size as S and T) containing 1s for the elements of S and T that match, and 0s for the ones that do not. S and T must be of the same size (or one can be a scalar cell). Either one can also be a character array with the proper number of rows.

strcmpi

Function strcmpi performs the same operation as strcmp, but it ignores character case.

Function strncmp, with syntax

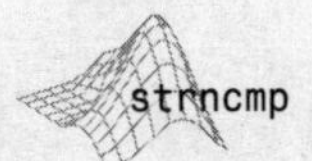


```
k = strncmp('str1', 'str2', n)
```

returns a logical true (1) if the first n characters of the strings str1 and str2 are the same, and returns a logical false (0) otherwise. Arguments str1 and str2 can be cell arrays of strings also. The syntax

```
R = strncmp(S, T, n)
```

where S and T can be cell arrays of strings, returns an array R the same size as S and T containing 1 for those elements of S and T that match (up to n characters), and 0 otherwise. S and T must be the same size (or one can be a scalar cell). Either one can also be a character array with the correct number of rows. The command strncmp is case sensitive. Any leading and trailing blanks in either of the strings are included in the comparison. Function strncmpi performs the same operation as strncmp, but ignores character case.

Function strfind, with syntax

```
I = strfind(str, pattern)
```

strfind

searches string str for occurrences of a *shorter* string, pattern, returning the starting index of each such occurrence in the double array, I. If pattern is not found in str, or if pattern is longer than str, then strfind returns the empty array, [ ].

Function strjust has the syntax

```
Q = strjust(A, direction)
```

where A is a character array, and direction can have the justification values 'right', 'left', and 'center'. The default justification is 'right'. The output array contains the same strings as A, but justified in the direction specified. Note that justification of a string implies the existence of leading and/or trailing blank characters to provide space for the specified operation. For instance, letting the symbol "□" represents a blank character, the string '□□abc' with two leading blank characters does not change under 'right' justification; becomes 'abc□□' with 'left' justification; and becomes the string '□abc□' with 'center' justification. Clearly, these operations have no effect on a string that does not contain any leading or trailing blanks.

Function strmatch, with syntax

```
m = strmatch('str', STRS)
```

looks through the rows of the character array or cell array of strings, STRS, to find strings that begin with string str, returning the matching row indices. The alternate syntax

```
m = strmatch('str', STRS, 'exact')
```

returns only the indices of the strings in STRS matching str exactly. For example, the statement

```
>> m = strmatch('max', strvcat('max', 'minimax', 'maximum'));
```

returns m = [1; 3] because rows 1 and 3 of the array formed by strvcat begin with 'max'. On the other hand, the statement

```
>> m = strmatch('max', strvcat('max', 'minimax', 'maximum'), 'exact');
```

returns m = 1, because only row 1 matches 'max' exactly.

Function strrep, with syntax

```
r = strrep('str1', 'str2', 'str3')
```

replaces all occurrences of the string str2 within string str1 with the string str3. If any of str1, str2, or str3 is a cell array of strings, this function returns a cell array the same size as str1, str2, and str3, obtained by performing a strrep using corresponding elements of the inputs. The inputs must all be of the same size (or any can be a scalar cell). Any one of the strings can also be a character array with the correct number of rows. For example,

```
>> s = 'Image processing and restoration.';
>> str = strrep(s, 'processing', 'enhancement')
str =

    Image enhancement and restoration.
```

Function strtok, with syntax

```
t = strtok('str', delim)
```

returns the first token in the text string str, that is, the first set of characters before a delimiter in delim is encountered. Parameter delim is a vector containing delimiters (e.g., blanks, other characters, strings). For example,

```
>> str = 'An image is an ordered set of pixels';
>> delim = ['x'];
>> t = strtok(str, delim)
t =

    An
```

Note that function strtok terminates after the first delimiter is encountered. (i.e., a blank character in the example just given). If we change delim to delim = ['x'], then the output becomes

```
>> t = strtok(str, delim)
t =

    An image is an ordered set of pi
```

The next set of functions in Table 12.2 deals with conversions between strings and numbers. Function int2str, with syntax

```
str = int2str(N)
```

converts an integer to a string with integer format. The input N can be a single integer or a vector or matrix of integers. Noninteger inputs are rounded before conversion. For example, int2str(2 + 3) is the *string* '5'. For matrix or vector inputs, int2str returns a string matrix:

```
>> str = int2str(eye(3))
ans =
    1  0  0
    0  1  0
    0  0  1
>> class(str)
ans =
    char
```

Function mat2str, with syntax

```
str = mat2str(A)
```

converts matrix A into a string, suitable for input to the eval function, using full precision. Using the syntax

```
str = mat2str(A, n)
```

converts matrix A using n digits of precision. For example, consider the matrix

```
>> A = [1 2;3 4]
A =
    1    2
    3    4
```

The statement

```
>> b = mat2str(A)
```

produces

```
b =
    [1 2;3 4]
```

where b is a string of 9 characters, including the square brackets, spaces, and a semicolon. The command

```
>> eval(mat2str(A))
```

reproduces A. The other functions in this category have similar interpretations.

The last category in Table 12.2 deals with base number conversions. For example, function dec2base, with syntax

```
str = dec2base(d, base)
```

converts the decimal integer d to the specified base, where d must be a nonnegative integer smaller than 2^52, and base must be an integer between 2 and 36. The returned argument str is a string. For example, the following command converts 23_{10} to base 2 and returns the result as a string:

```
>> str = dec2base(23, 2)
str =
   10111
>> class(str)
ans =
   char
```

Using the syntax

```
str = dec2base(d, base, n)
```

produces a representation with at least n digits.

12.4.2 String Matching

In addition to the string matching and comparing functions in Table 12.2, it is often useful to have available measures of similarity that behave much like the distance measures discussed in Section 12.2. We illustrate this approach using a measure defined as follows.

Suppose that two region boundaries, a and b, are coded into strings $a_1a_2 \ldots a_m$ and $b_1b_2 \ldots b_n$, respectively. Let α denote the number of matches between these two strings, where a match is said to occur in the kth position if $a_k = b_k$. The number of symbols that do not match is

$$\beta = \max(|a|, |b|) - \alpha$$

where $|\text{arg}|$ is the length (number of symbols) of the string in the argument. It can be shown that $\beta = 0$ if and only if a and b are identical strings.

A simple measure of similarity between a and b is the ratio

$$R = \frac{\alpha}{\beta} = \frac{\alpha}{\max(|a|, |b|) - \alpha}$$

This measure, proposed by Sze and Yang [1981], is infinite for a perfect match and 0 when none of the corresponding symbols in a and b match (α is 0 in this case).

Because matching is performed between corresponding symbols, it is required that all strings be "registered" in some position-independent manner in order for this method to make sense. One way to register two strings is to shift one string with respect to the other until a maximum value of R is obtained. This and other similar matching strategies can be developed using some of the string operations detailed in Table 12.2. Typically, a more efficient approach is to define the same starting point for all strings based on normalizing the boundaries with respect to size and orientation before their string representation is extracted. This approach is illustrated in Example 12.3.

The following M-function computes the preceding measure of similarity for two character strings.

strsimilarity

```
function R = strsimilarity(a, b)
%STRSIMILARITY Computes a similarity measure between two strings.
%   R = STRSIMILARITY(A, B) computes the similarity measure, R,
%   defined in Section 12.4.2 for strings A and B. The strings do not
%   have to be of the same length, but if one is shorter than other,
%   then it is assumed that the shorter string has been padded with
%   leading blanks so that it is brought into the necessary
%   registration prior to using this function. Only one of the
%   strings can have blanks, and these must be leading and/or
%   trailing blanks. Blanks are not counted when computing the length
%   of the strings for use in the similarity measure.

% Verify that a and b are character strings.
if ~ischar(a) | ~ischar(b)
   error('Inputs must be character strings.')
end

% Find any blank spaces.
I = find(a == ' ');
J = find(b == ' ');
LI = length(I); LJ = length(J);
if LI ~= 0 & LJ ~= 0
   error('Only one of the strings can contain blanks.')
end

% Pad the end of the appropriate string. It is assumed
% that they are registered in terms of their beginning
% positions.
a = a(:); b = b(:);
La = length(a); Lb = length(b);
if LI == 0 & LJ == 0
   if La > Lb
      b = [b; blanks(La - Lb)'];
   else
      a = [a; blanks(Lb - La)'];
   end
elseif isempty(I)
   Lb = length(b) - length(J);
   b = [b; blanks(La - Lb - LJ)'];
```

```
else
   La = length(a) - length(I);
   a = [a; blanks(Lb - La - LI)'];
end
% Compute the similarity measure.
I = find(a == b);
alpha = length(I);
den = max(La, Lb) - alpha;
if den == 0
   R = Inf;
else
   R = alpha/den;
end
```

EXAMPLE 12.3: Object recognition based on string matching.

■ Figures 12.3(a) and (d) show silhouettes of two samples of container bottles whose principal shape difference is the curvature of their sides. For purposes of differentiation, objects with the curvature characteristics of Fig. 12.3(a) are said to be from class 1. Objects with straight sides are said to be from class 2. The images are of size 372 × 288 pixels.

To illustrate the effectiveness of measure R for differentiating between objects of classes 1 and 2, the boundaries of the objects were approximated by minimum-perimeter polygons using function `minperpoly` (see Section 11.2.2) with a cell size of 8. Figures 12.3(b) and (e) show the results. Then noise was added to the coordinates of each vertex of the polygons using function `randvertex` (the listing is included in Appendix C), which has the syntax

`randvertex`

```
[xn, yn] = randvertex(x, y, npix)
```

where `x` and `y` are column vectors containing the coordinates of the vertices of a polygon, `xn` and `yn` are the corresponding noisy coordinates, and `npix` is the maximum number of pixels by which a coordinate is allowed to be displaced in either direction. Five sets of noisy vertices were generated for each class using `npix = 5`. Figures 12.3(c) and (f) show typical results.

Strings of symbols were generated for each class by coding the interior angles of the polygons using function `polyangles` (see Appendix C for the code listing):

`polyangles`

```
>> angles = polyangles(x, y);
```

The `x` and `y` inputs to function `polyangles` are vectors containing the x- and y-coordinates of the vertices of a polygon, ordered in the clockwise direction. The output is a vector containing the corresponding interior angles, in degrees.

Then a string, `s`, was generated from a given `angles` array by quantizing the angles into 45° increments, using the statement

```
>> s = floor(angles/45) + 1;
```

This yielded a string whose elements were numbers between 1 and 8, with 1 designating the range $0° \leq \theta < 45°$, 2 designating the range $45° \leq \theta < 90°$, and so forth, where θ denotes an interior angle.

Because the first vertex in the output of `minperpoly` is always the top, left vertex of the boundary of the input, `B`, the first element of string `s` corresponds

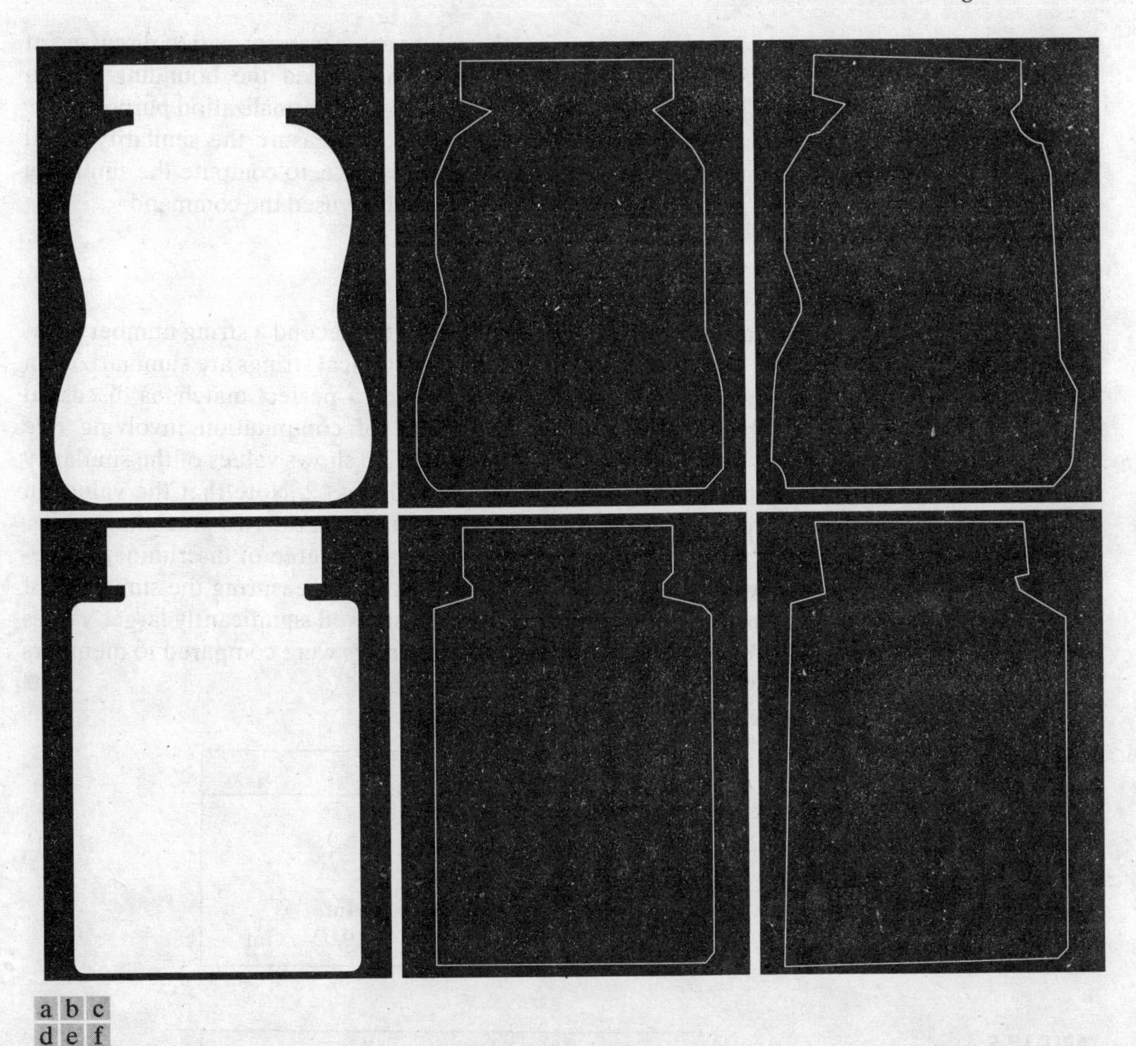

a b c
d e f

FIGURE 12.3 (a) An object. (b) Its minimum perimeter polygon obtained using function `minperpoly` with a cell size of 8. (c) A typical noisy boundary. (d)–(f) The same sequence for another object.

to the interior angle of that vertex. This automatically registers the strings (if the objects are not rotated) because they all start at the top, left vertex in all images. The direction of the vertices output by `minperpoly` is clockwise, so the elements of `s` also are in that direction. Finally, each `s` was converted from a string of integers to a character string using the command

```
>> s = int2str(s);
```

In this example the objects are of comparable size and they are all vertical, so normalization of neither size nor orientation was required. If the objects had been of arbitrary size and orientation, we could have aligned them along their principal directions by using the eigenvector transformation discussed at the end of Section 11.5. Then we could have used the bounding box in Section 11.4.1 to obtain the object dimensions for normalization purposes.

First, function `strsimilarity` was used to measure the similarity of all strings of class 1 between themselves. For instance, to compute the similarity between the first and second strings of class 1 we used the command

```
>> R = strsimilarity(s11, s12);
```

where the first subscript indicates class and the second a string number within that class. The results obtained using five typical strings are summarized in Table 12.4, where `Inf` indicates infinity (i.e., a perfect match, as discussed earlier). Table 12.5 shows the same type of computation involving five strings of class 2 against themselves. Table 12.6 shows values of the similarity measure between the strings of class 1 and class 2. Note that the values in this table are significantly lower than the entries in the two preceding tables, indicating that the R measure achieved a high degree of discrimination between the two classes of objects. In other words, measuring the similarity of strings against members of their own class showed significantly larger values of `R`, indicating a closer match than when strings were compared to members of the opposite class. ■

TABLE 12.4 Values of similarity measure, R, between the strings of class 1. (All values shown are ×10.)

R	s_{11}	s_{12}	s_{13}	s_{14}	s_{15}
s_{11}	Inf				
s_{12}	9.33	Inf			
s_{13}	26.25	12.31	Inf		
s_{14}	16.36	9.33	14.16	Inf	
s_{15}	22.22	14.17	14.01	19.02	Inf

TABLE 12.5 Values of similarity measure, R, between the strings of class 2. (All values shown are ×10.)

R	s_{21}	s_{22}	s_{23}	s_{24}	s_{25}
s_{21}	Inf				
s_{22}	10.00	Inf			
s_{23}	13.33	13.33	Inf		
s_{24}	7.50	13.31	18.00	Inf	
s_{25}	13.33	7.51	18.12	10.01	Inf

TABLE 12.6 Values of similarity measure, R, between the strings of classes 1 and 2. (All values shown are ×10.)

R	s_{11}	s_{12}	s_{13}	s_{14}	s_{15}
s_{21}	2.03	0.01	1.15	1.17	0.75
s_{22}	1.15	1.61	1.16	0.75	2.07
s_{23}	2.08	1.15	2.08	2.06	2.08
s_{24}	1.60	1.62	1.59	1.14	2.61
s_{25}	1.61	0.36	0.74	1.60	1.16

Summary

Starting with Chapter 9, our treatment of digital image processing began a transition from processes whose outputs are images to processes whose outputs are attributes about those images. Although the material in the present chapter is introductory in nature, the topics covered are fundamental to understanding the state of the art in object recognition. As mentioned in Section 1.2 at the onset of our journey, recognition of individual objects is a logical place at which to conclude this book.

Having finished study of the material in the preceding twelve chapters, the reader is now in the position of being able to master the fundamentals of how to prototype software solutions of image-processing problems using MATLAB and Image Processing Toolbox functions. What is even more important, the background and numerous new functions developed in the book constitute a basic blueprint on how to extend the power of MATLAB and IPT. Given the task-specific nature of most imaging problems, a clear understanding of this material enhances significantly the chances of arriving at successful solutions in a broad spectrum of image processing application areas.

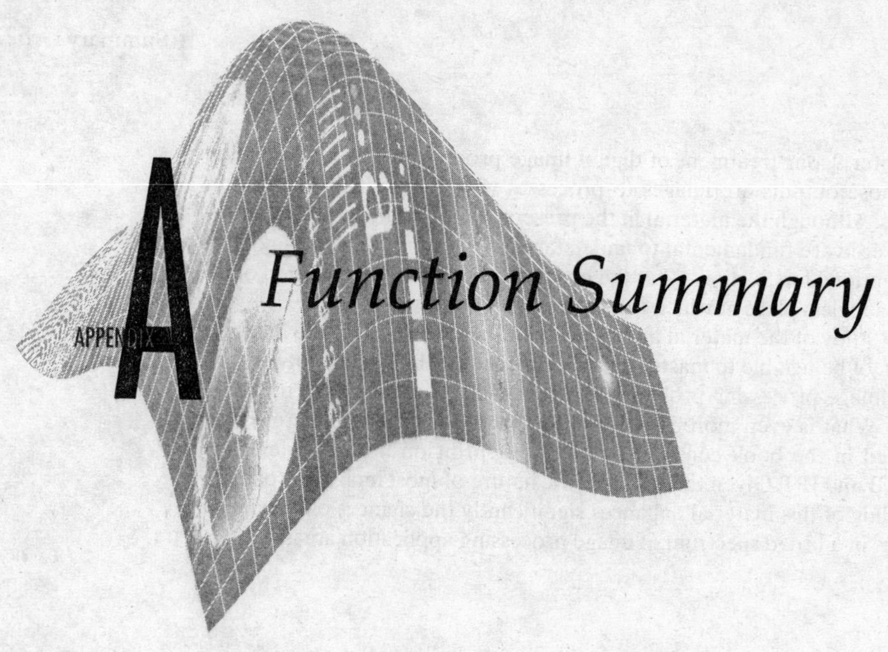

Preview

Section A.1 of this appendix contains a listing of all the functions in the Image Processing Toolbox, and all the new functions developed in the preceding chapters. The latter functions are referred to as *DIPUM* functions, a term derived from the first letter of the words in the title of the book. Section A.2 lists the MATLAB functions used throughout the book. All page numbers listed refer to pages in the book, indicating where a function is first used and illustrated. In some instances, more than one location is given, indicating that the function is explained in different ways, depending on the application. Some IPT functions were not used in our discussions. These are identified by a reference to online help instead of a page number. All MATLAB functions listed in Section A.2 are used in the book. Each page number in that section identifies the first use of the MATLAB function indicated.

A.1 IPT and DIPUM Functions

The following functions are loosely grouped in categories similar to those found in IPT documentation. A new category (e.g., wavelets) was created in cases where there are no existing IPT functions.

Function Category and Name	Description	Page or Other Location
Image Display		
`colorbar`	Display colorbar (MATLAB).	online
`getimage`	Get image data from axes.	online
`ice` (DIPUM)	Interactive color editing.	218
`image`	Create and display image object (MATLAB).	online
`imagesc`	Scale data and display as image (MATLAB).	online
`immovie`	Make movie from multiframe image.	online
`imshow`	Display image.	16
`imview`	Display image in Image Viewer.	online

montage	Display multiple image frames as rectangular montage.	online
movie	Play recorded movie frames (MATLAB).	online
rgbcube (DIPUM)	Display a color RGB cube.	195
subimage	Display multiple images in single figure.	online
truesize	Adjust display size of image.	online
warp	Display image as texture-mapped surface.	online

Image file I/O

dicominfo	Read metadata from a DICOM message.	online
dicomread	Read a DICOM image.	online
dicomwrite	Write a DICOM image.	online
dicom-dict.txt	Text file containing DICOM data dictionary.	online
dicomuid	Generate DICOM unique identifier.	online
imfinfo	Return information about image file (MATLAB).	19
imread	Read image file (MATLAB).	14
imwrite	Write image file (MATLAB).	18

Image arithmetic

imabsdiff	Compute absolute difference of two images.	42
imadd	Add two images, or add constant to image.	42
imcomplement	Complement image.	42, 67
imdivide	Divide two images, or divide image by constant.	42
imlincomb	Compute linear combination of images.	42, 159
immultiply	Multiply two images, or multiply image by constant.	42
imsubtract	Subtract two images, or subtract constant from image.	42

Geometric transformations

checkerboard	Create checkerboard image.	167
findbounds	Find output bounds for geometric transformation.	online
fliptform	Flip the input and output roles of a TFORM struct.	online
imcrop	Crop image.	online
imresize	Resize image.	online
imrotate	Rotate image.	472
imtransform	Apply geometric transformation to image.	188
intline	Integer-coordinate line drawing algorithm. (Undocumented IPT function).	43
makeresampler	Create resampler structure.	190
maketform	Create geometric transformation structure (TFORM).	183
pixeldup (DIPUM)	Duplicate pixels of an image in both directions.	168
tformarray	Apply geometric transformation to N-D array.	online
tformfwd	Apply forward geometric transformation.	184
tforminv	Apply inverse geometric transformation.	184
vistformfwd (DIPUM)	Visualize forward geometric transformation.	185

Image registration

cpstruct2pairs	Convert CPSTRUCT to valid pairs of control points.	online
cp2tform	Infer geometric transformation from control point pairs.	191
cpcorr	Tune control point locations using cross-correlation.	online
cpselect	Control point selection tool.	193
normxcorr2	Normalized two-dimensional cross-correlation.	online

Pixel values and statistics

`corr2`	Compute 2-D correlation coefficient.	online
`covmatrix` (DIPUM)	Compute the covariance matrix of a vector population.	476
`imcontour`	Create contour plot of image data.	online
`imhist`	Display histogram of image data.	77
`impixel`	Determine pixel color values.	online
`improfile`	Compute pixel-value cross-sections along line segments.	online
`mean2`	Compute mean of matrix elements.	75
`pixval`	Display information about image pixels.	17
`regionprops`	Measure properties of image regions.	463
`statmoments` (DIPUM)	Compute statistical central moments of an image histogram.	155
`std2`	Compute standard deviation of matrix elements.	415

Image analysis (includes segmentation, description, and recognition)

`bayesgauss` (DIPUM)	Bayes classifier for Gaussian patterns.	493
`bound2eight` (DIPUM)	Convert 4-connected boundary to 8-connected boundary.	434
`bound2four` (DIPUM)	Convert 8-connected boundary to 4-connected boundary.	434
`bwboundaries`	Trace region boundaries.	online
`bwtraceboundary`	Trace single boundary.	online
`bound2im` (DIPUM)	Convert a boundary to an image.	435
`boundaries` (DIPUM)	Trace region boundaries.	434
`bsubsamp` (DIPUM)	Subsample a boundary.	435
`colorgrad` (DIPUM)	Compute the vector gradient of an RGB image.	234
`colorseg` (DIPUM)	Segment a color image.	238
`connectpoly` (DIPUM)	Connect vertices of a polygon.	435
`diameter` (DIPUM)	Measure diameter of image regions.	456
`edge`	Find edges in an intensity image.	385
`fchcode` (DIPUM)	Compute the Freeman chain code of a boundary.	437
`frdescp` (DIPUM)	Compute Fourier descriptors.	459
`graythresh`	Compute global image threshold using Otsu's method.	406
`hough` (DIPUM)	Hough transform.	396
`houghlines` (DIPUM)	Extract line segments based on the Hough transform.	401
`houghpeaks` (DIPUM)	Detect peaks in Hough transform.	399
`houghpixels` (DIPUM)	Compute image pixels belonging to Hough transform bin.	401
`ifrdescp` (DIPUM)	Compute inverse Fourier descriptors.	459
`imstack2vectors` (DIPUM)	Extract vectors from an image stack.	476
`invmoments` (DIPUM)	Compute invariant moments of image.	472
`mahalanobis` (DIPUM)	Compute the Mahalanobis distance.	487
`minperpoly` (DIPUM)	Compute minimum perimeter polygon.	447
`polyangles` (DIPUM)	Compute internal polygon angles.	510
`princomp` (DIPUM)	Obtain principal-component vectors and related quantities.	477
`qtdecomp`	Perform quadtree decomposition.	413
`qtgetblk`	Get block values in quadtree decomposition.	413
`qtsetblk`	Set block values in quadtree decomposition.	online
`randvertex` (DIPUM)	Randomly displace polygon vertices.	510
`regiongrow` (DIPUM)	Perform segmentation by region growing.	409
`signature` (DIPUM)	Compute the signature of a boundary.	450
`specxture` (DIPUM)	Compute spectral texture of an image.	469
`splitmerge` (DIPUM)	Segment an image using a split-and-merge algorithm.	414
`statxture` (DIPUM)	Compute statistical measures of texture in an image.	467

strsimilarity (DIPUM)	Similarity measure between two strings.	509
x2majoraxis (DIPUM)	Align coordinate x with the major axis of a region.	457
Image Compression		
compare (DIPUM)	Compute and display the error between two matrices.	285
entropy (DIPUM)	Compute a first-order estimate of the entropy of a matrix.	288
huff2mat (DIPUM)	Decode a Huffman encoded matrix.	301
huffman (DIPUM)	Build a variable-length Huffman code for symbol source.	290
im2jpeg (DIPUM)	Compress an image using a JPEG approximation.	319
im2jpeg2k (DIPUM)	Compress an image using a JPEG 2000 approximation.	327
imratio (DIPUM)	Compute the ratio of the bytes in two images/variables.	283
jpeg2im (DIPUM)	Decode an IM2JPEG compressed image.	322
jpeg2k2im (DIPUM)	Decode an IM2JPEG2K compressed image.	330
lpc2mat (DIPUM)	Decompress a 1-D lossless predictive encoded matrix.	312
mat2huff (DIPUM)	Huffman encodes a matrix.	298
mat2lpc (DIPUM)	Compress a matrix using 1-D lossless predictive coding.	312
quantize (DIPUM)	Quantize the elements of a UINT8 matrix.	316
Image enhancement		
adapthisteq	Adaptive histogram equalization.	online
decorrstretch	Apply decorrelation stretch to multichannel image.	online
gscale (DIPUM)	Scale the intensity of the input image.	76
histeq	Enhance contrast using histogram equalization.	82
intrans (DIPUM)	Perform intensity transformations.	73
imadjust	Adjust image intensity values or colormap.	66
stretchlim	Find limits to contrast stretch an image.	online
Image noise		
imnoise	Add noise to an image.	106
imnoise2 (DIPUM)	Generate an array of random numbers with specified PDF.	148
imnoise3 (DIPUM)	Generate periodic noise.	152
Linear and nonlinear spatial filtering		
adpmedian (DIPUM)	Perform adaptive median filtering.	165
convmtx2	Compute 2-D convolution matrix.	online
dftcorr (DIPUM)	Perform frequency domain correlation.	491
dftfilt (DIPUM)	Perform frequency domain filtering.	122
fspecial	Create predefined filters.	99
medfilt2	Perform 2-D median filtering.	106
imfilter	Filter 2-D and N-D images.	92
ordfilt2	Perform 2-D order-statistic filtering.	105
spfilt (DIPUM)	Performs linear and nonlinear spatial filtering.	159
wiener2	Perform 2-D adaptive noise-removal filtering.	online
Linear 2-D filter design		
freqspace	Determine 2-D frequency response spacing (MATLAB).	online
freqz2	Compute 2-D frequency response.	123
fsamp2	Design 2-D FIR filter using frequency sampling.	online
ftrans2	Design 2-D FIR filter using frequency transformation.	online
fwind1	Design 2-D FIR filter using 1-D window method.	online
fwind2	Design 2-D FIR filter using 2-D window method.	online

`hpfilter` (DIPUM)	Computes frequency domain highpass filters.	136
`lpfilter` (DIPUM)	Computes frequency domain lowpass filters.	131

Image deblurring (restoration)

`deconvblind`	Deblur image using blind deconvolution.	180
`deconvlucy`	Deblur image using Lucy-Richardson method.	177
`deconvreg`	Deblur image using regularized filter.	175
`deconvwnr`	Deblur image using Wiener filter.	171
`edgetaper`	Taper edges using point-spread function.	172
`otf2psf`	Optical transfer function to point-spread function.	142
`psf2otf`	Point-spread function to optical transfer function.	142

Image transforms

`dct2`	2-D discrete cosine transform.	321
`dctmtx`	Discrete cosine transform matrix.	321
`fan2para`	Convert fan-beam projections to parallel-beam.	online
`fanbeam`	Compute fan-beam transform.	online
`fft2`	2-D fast Fourier transform (MATLAB).	112
`fftn`	N-D fast Fourier transform (MATLAB).	online
`fftshift`	Reverse quadrants of output of FFT (MATLAB).	112
`idct2`	2-D inverse discrete cosine transform.	online
`ifanbeam`	Compute inverse fan-beam transform.	online
`ifft2`	2-D inverse fast Fourier transform (MATLAB).	114
`ifftn`	N-D inverse fast Fourier transform (MATLAB).	online
`iradon`	Compute inverse Radon transform.	online
`para2fan`	Convert parallel-beam projections to fan-beam.	online
`phantom`	Generate a head phantom image.	online
`radon`	Compute Radon transform.	online

Wavelets

`wave2gray` (DIPUM)	Display wavelet decomposition coefficients.	267
`waveback` (DIPUM)	Perform a multi-level 2-dimensional inverse FWT.	272
`wavecopy` (DIPUM)	Fetch coefficients of wavelet decomposition structure.	265
`wavecut` (DIPUM)	Set to zero coefficients in a wavelet decomposition structure.	264
`wavefast` (DIPUM)	Perform a multilevel 2-dimensional fast wavelet transform.	255
`wavefilter` (DIPUM)	Create wavelet decomposition and reconstruction filters.	252
`wavepaste` (DIPUM)	Put coefficients in a wavelet decomposition structure.	265
`wavework` (DIPUM)	Edit wavelet decomposition structures.	262
`wavezero` (DIPUM)	Set wavelet detail coefficients to zero.	277

Neighborhood and block processing

`bestblk`	Choose block size for block processing.	online
`blkproc`	Implement distinct block processing for image.	321
`col2im`	Rearrange matrix columns into blocks.	322
`colfilt`	Columnwise neighborhood operations.	97
`im2col`	Rearrange image blocks into columns.	321
`nlfilter`	Perform general sliding-neighborhood operations.	96

Morphological operations (intensity and binary images)

`conndef`	Default connectivity.	online
`imbothat`	Perform bottom-hat filtering.	373
`imclearborder`	Suppress light structures connected to image border.	366

imclose	Close image.	348
imdilate	Dilate image.	340
imerode	Erode image.	347
imextendedmax	Extended-maxima transform.	online
imextendedmin	Extended-minima transform.	online
imfill	Fill image regions and holes.	366
imhmax	H-maxima transform.	online
imhmin	H-minima transform.	374
imimposemin	Impose minima.	424
imopen	Open image.	348
imreconstruct	Morphological reconstruction.	363
imregionalmax	Regional maxima.	online
imregionalmin	Regional minima.	422
imtophat	Perform tophat filtering.	373
watershed	Watershed transform.	420

Morphological operations (binary images)

applylut	Perform neighborhood operations using lookup tables.	353
bwarea	Compute area of objects in binary image.	online
bwareaopen	Binary area open (remove small objects).	online
bwdist	Compute distance transform of binary image.	418
bweuler	Compute Euler number of binary image.	online
bwhitmiss	Binary hit-miss operation.	352
bwlabel	Label connected components in 2-D binary image.	361
bwlabeln	Label connected components in N-D binary image.	online
bwmorph	Perform morphological operations on binary image.	356
bwpack	Pack binary image.	online
bwperim	Determine perimeter of objects in binary image.	445
bwselect	Select objects in binary image.	online
bwulterode	Ultimate erosion.	online
bwunpack	Unpack binary image.	online
endpoints (DIPUM)	Compute end points of a binary image.	354
makelut	Construct lookup table for use with applylut.	353

Structuring element (STREL) creation and manipulation

getheight	Get strel height.	online
getneighbors	Get offset location and height of strel neighbors.	online
getnhood	Get strel neighborhood.	online
getsequence	Get sequence of decomposed strels.	342
isflat	Return true for flat strels.	online
reflect	Reflect strel about its center.	online
strel	Create morphological structuring element.	341
translate	Translate strel.	online

Region-based processing

histroi (DIPUM)	Compute the histogram of an ROI in an image.	156
poly2mask	Convert ROI polygon to mask.	online
roicolor	Select region of interest, based on color.	online
roifill	Smoothly interpolate within arbitrary region.	online
roifilt2	Filter a region of interest.	online
roipoly	Select polygonal region of interest.	156

Colormap manipulation

`brighten`	Brighten or darken colormap (MATLAB).	online
`cmpermute`	Rearrange colors in colormap.	online
`cmunique`	Find unique colormap colors and corresponding image.	online
`colormap`	Set or get color lookup table (MATLAB).	132
`imapprox`	Approximate indexed image by one with fewer colors.	198
`rgbplot`	Plot RGB colormap components (MATLAB).	online

Color space conversions

`applycform`	Apply device-independent color space transformation.	online
`hsv2rgb`	Convert HSV values to RGB color space (MATLAB).	206
`iccread`	Read ICC color profile.	online
`lab2double`	Convert L*a*b* color values to class `double`.	online
`lab2uint16`	Convert L*a*b* color values to class `uint16`.	online
`lab2uint8`	Convert L*a*b* color values to class `uint8`.	online
`makecform`	Create device-independent color space transform structure.	online
`ntsc2rgb`	Convert NTSC values to RGB color space.	205
`rgb2hsv`	Convert RGB values to HSV color space (MATLAB).	206
`rgb2ntsc`	Convert RGB values to NTSC color space.	204
`rgb2ycbcr`	Convert RGB values to YCBCR color space.	205
`ycbcr2rgb`	Convert YCBCR values to RGB color space.	205
`rgb2hsi` (DIPUM)	Convert RGB values to HSI color space.	212
`hsi2rgb` (DIPUM)	Convert HSI values to RGB color space.	213
`whitepoint`	Returns XYZ values of standard illuminants.	online
`xyz2double`	Convert XYZ color values to class `double`.	online
`xyz2uint16`	Convert XYZ color values to class `uint16`.	online

Array operations

`circshift`	Shift array circularly (MATLAB).	433
`dftuv` (DIPUM)	Compute meshgrid arrays.	128
`padarray`	Pad array.	97
`paddedsize` (DIPUM)	Compute the minimum required pad size for use in FFTs.	117

Image types and type conversions

`changeclass`	Change the class of an image (undocumented IPT function).	72
`dither`	Convert image using dithering.	199
`gray2ind`	Convert intensity image to indexed image.	201
`grayslice`	Create indexed image from intensity image by thresholding.	201
`im2bw`	Convert image to binary image by thresholding.	26
`im2double`	Convert image array to double precision.	26
`im2java`	Convert image to Java image (MATLAB).	online
`im2java2d`	Convert image to Java buffered image object.	online
`im2uint8`	Convert image array to 8-bit unsigned integers.	26
`im2uint16`	Convert image array to 16-bit unsigned integers.	26
`ind2gray`	Convert indexed image to intensity image.	201
`ind2rgb`	Convert indexed image to RGB image (MATLAB).	202
`label2rgb`	Convert label matrix to RGB image.	online
`mat2gray`	Convert matrix to intensity image.	26
`rgb2gray`	Convert RGB image or colormap to grayscale.	202
`rgb2ind`	Convert RGB image to indexed image.	200

Miscellaneous

`conwaylaws` (DIPUM)	Apply Conway's genetic laws to a single pixel.	355
`manualhist` (DIPUM)	Generate a 2-mode histogram interactively.	87
`twomodegauss` (DIPUM)	Generate a 2-mode Gaussian function.	86
`uintlut`	Compute new array values based on lookup table.	online

Toolbox preferences

`iptgetpref`	Get value of Image Processing Toolbox preference.	online
`iptsetpref`	Set value of Image Processing Toolbox preference.	online

A.2 MATLAB Functions

The following MATLAB functions, listed alphabetically, are used in the book. See the pages indicated and/or online help for additional details.

MATLAB Function	Description	Pages
A		
`abs`	Absolute value and complex magnitude.	112
`all`	Test to determine if all elements are nonzero.	46
`ans`	The most recent answer.	48
`any`	Test for any nonzeros.	46
`axis`	Axis scaling and appearance.	78
B		
`bar`	Bar chart.	77
`bin2dec`	Binary to decimal number conversion.	300
`blanks`	A string of blanks.	499
`break`	Terminate execution of a `for` loop or `while` loop.	49
C		
`cart2pol`	Transform Cartesian coordinates to polar or cylindrical.	451
`cat`	Concatenate arrays.	195
`ceil`	Round toward infinity.	114
`cell`	Create cell array.	292
`celldisp`	Display cell array contents.	293, 428
`cellfun`	Apply a function to each element in a cell array.	428
`cellplot`	Graphically display the structure of cell arrays.	293
`cellstr`	Create cell array of strings from character array.	499
`char`	Create character array (string).	61, 499
`circshift`	Shift array circularly.	433
`colon`	Colon operator.	31, 41
`colormap`	Set and get the current colormap.	132, 199
`computer`	Identify information about computer on which MATLAB is running.	48
`continue`	Pass control to the next iteration of `for` or `while` loop.	49
`conv2`	Two-dimensional convolution.	257

`ctranspose`	Vector and matrix complex transpose. (See `transpose` for real data.)	41
`cumsum`	Cumulative sum.	82

D

`dec2base`	Decimal number to base conversion.	508
`dec2bin`	Decimal to binary number conversion.	298
`diag`	Diagonal matrices and diagonals of a matrix.	239
`diff`	Differences and approximate derivatives.	373
`dir`	Display directory listing.	284
`disp`	Display text or array.	59
`double`	Convert to double precision.	24

E

`edit`	Edit or create an M-file.	40
`eig`	Find eigenvalues and eigenvectors.	478
`end`	Terminate `for`, `while`, `switch`, `try`, and `if` statements or indicate last index.	31
`eps`	Floating-point relative accuracy.	48, 69
`error`	Display error message.	50
`eval`	Execute a string containing a MATLAB expression.	501
`eye`	Identity matrix.	494

F

`false`	Create false array. Shorthand for `logical(0)`.	38, 410
`feval`	Function evaluation.	415
`fft2`	Two-dimensional discrete Fourier transform.	112
`fftshift`	Shift zero-frequency component of DFT to center of spectrum.	112
`fieldnames`	Return field names of a structure, or property names of an object.	284
`figure`	Create a figure graphics object.	18
`find`	Find indices and values of nonzero elements.	147
`fliplr`	Flip matrices left-right.	472
`flipup`	Flip matrices up-down.	472
`floor`	Round towards minus infinity.	114
`for`	Repeat a group of statements a fixed number of times.	49
`full`	Convert sparse matrix to full matrix.	396

G

`gca`	Get current axes handle.	78
`get`	Get object properties.	218
`getfield`	Get field of structure array.	540
`global`	Define a global variable.	292
`grid`	Grid lines for two- and three-dimensional plots.	132
`guidata`	Store or retrieve application data.	539
`guide`	Start the GUI Layout Editor.	528

H

`help`	Display help for MATLAB functions in Command Window.	39
`hist`	Compute and/or display histogram.	150
`histc`	Histogram count.	299
`hold on`	Retain the current plot and certain axis properties.	81

`ndims` Number of array dimensions. 37
`nextpow2` Next power of two. 117
`norm` Vector and matrix norm. 485
`numel` Number of elements in an array. 51

O

`ones` Generate array of ones. 38

P

`patch` Create patch graphics object. 196
`permute` Rearrange the dimensions of a multidimensional array. 486
`persistent` Define persistent variable. 353
`pi` Ratio of a circle's circumference to its diameter. 48
`plot` Linear 2-D plot. 80
`plus` Array and matrix addition. 41
`pol2cart` Transform polar or cylindrical coordinates to Cartesian. 451
`pow2` Base 2 power and scale floating-point numbers. 300
`power` Array power. (See `mpower` for matrix power.) 41
`print` Print to file or to hardcopy device. 23
`prod` Product of array elements. 98

R

`rand` Uniformly distributed random numbers and arrays. 38, 145
`randn` Normally distributed random numbers and arrays. 38, 147
`rdivide` Array right division. (See `mrdivide` for matrix right division.) 41
`real` Real part of complex number. 115
`realmax` Largest floating-point number that your computer can represent. 48
`realmin` Smallest floating-point number that your computer can represent. 48
`regexp` Match regular expression. 502
`regexpi` Match regular expression, ignoring case. 503
`regexprep` Replace string using regular expression. 503
`rem` Remainder after division. 256
`repmat` Replicate and tile an array. 264
`reshape` Reshape array. 300
`return` Return to the invoking function. 49
`rot90` Rotate matrix multiples of 90 degrees. 94
`round` Round to nearest integer. 22

S

`save` Save workspace variables to disk. 301
`set` Set object properties. 78
`setfield` Set field of structure array. 546
`shading` Set color shading properties. We use the `interp` mode in the book. 135
`sign` Signum function. 326
`single` Convert to single precision. 24
`size` Return array dimensions. 15
`sort` Sort elements in ascending order. 293
`sortrows` Sort rows in ascending order. 433

`sparse`	Create sparse matrix.	395
`spline`	Cubic spline data interpolation.	218
`sprintf`	Write formatted data to a string.	52
`stem`	Plot discrete sequence data.	79
`str*`	String operations. See Table 12.2.	500
`str2num`	String to number conversion.	60
`strcat`	String concatenation.	503
`strcmp`	Compare strings.	62, 504
`strcmpi`	Compare strings ignoring case.	504
`strfind`	Find one string within another.	505
`strjust`	Justify a character array.	505
`strmatch`	Find possible matches for a string.	505
`strncmp`	Compare the first n characters of two strings.	504
`strncmpi`	Compare first n characters of strings ignoring case.	316, 505
`strread`	Read formatted data from a string.	61
`strrep`	String search and replace.	506
`strtok`	First token in string.	506
`strvcat`	Vertical concatenation of strings.	504
`subplot`	Subdivide figure window into array of axes or subplots.	249
`sum`	Sum of array elements.	35
`surf`	3-D shaded surface plot.	134
`switch`	Switch among several cases based on expression.	49

T

`text`	Create text object.	79
`tic, toc`	Stopwatch timer.	57
`times`	Array multiplication. (See `mtimes` for matrix multiplication.)	41
`title`	Add title to current graphic.	79
`transpose`	Matrix or vector transpose. (See `ctranspose` for complex data.)	30, 41
`true`	Create true array. Shorthand for `logical(1)`.	38, 410
`try...catch`	See Table 2.11.	49

U

`uicontrol`	Create user interface control object.	534
`uint16`	Convert to unsigned integer.	24
`uint32`	Convert to unsigned integer.	24
`uint8`	Convert to unsigned integer.	24
`uiresume`	Control program execution.	540
`uiwait`	Control program execution.	540
`uminus`	Unary minus.	41
`uplus`	Unary plus.	41
`unique`	Unique elements of a vector.	433
`upper`	Convert string to upper case.	62

V

`varargin`	Pass a variable number of arguments.	72
`vararout`	Return a variable number of arguments.	72
`version`	Get MATLAB version number.	48
`view`	Viewpoint specification.	132

W

`warning`	Display warning message.	159
`while`	Repeat statements an indefinite number of times.	49
`whitebg`	Change background color.	198
`whos`	List variables in the workspace.	16

X

`xlabel`	Label the x-axis.	79
`xlim`	Set or query x-axis limits.	80
`xor`	Exclusive or.	46
`xtick`	Set horizontal axis tick.	78

Y

`ylabel`	Label the y-axis.	79
`ylim`	Set or query y-axis limits.	80
`ytick`	Set vertical axis tick.	78

Z

`zeros`	Generate array of zeros.	38

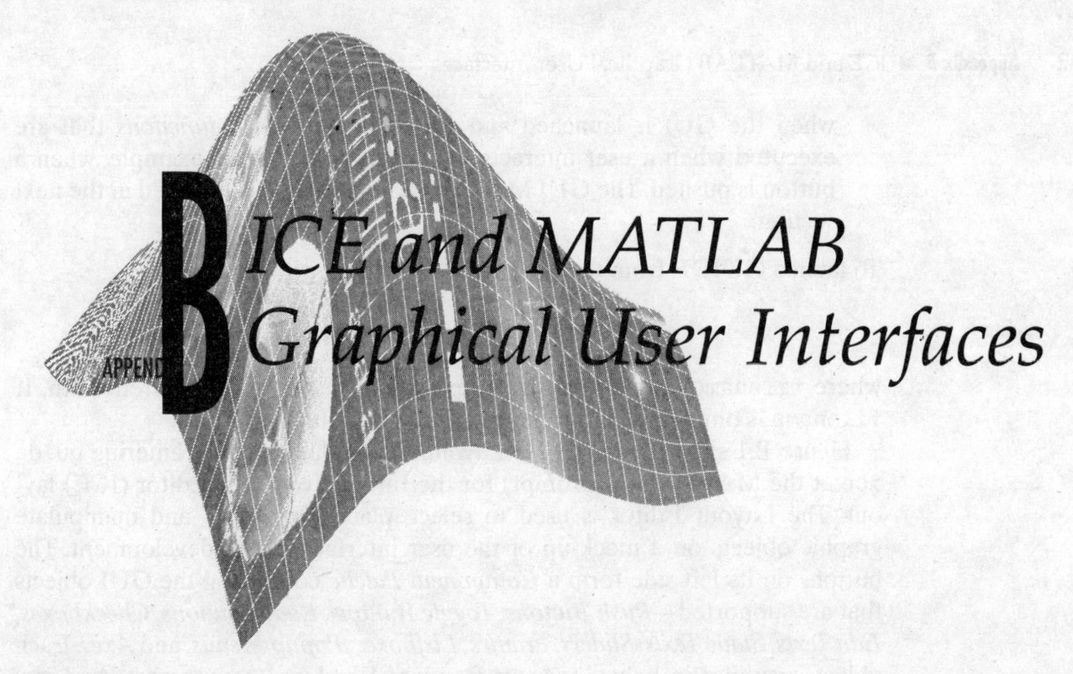

APPENDIX B ICE and MATLAB Graphical User Interfaces

Preview

In this appendix we develop the `ice` *interactive color editing* (ICE) function introduced in Chapter 6. The discussion assumes familiarity on the part of the reader with the material in Section 6.4. Section 6.4 provides many examples of using `ice` in both pseudo- and full-color image processing (Examples 6.3 through 6.7) and describes the `ice` calling syntax, input parameters, and graphical interface elements (they are summarized in Tables 6.4 through 6.6). The power of `ice` is its ability to let users generate color transformation curves interactively and graphically, while displaying the impact of the generated transformations on images in real or near real time.

B.1 Creating ICE's Graphical User Interface

MATLAB's *Graphical User Interface Development Environment* (GUIDE) provides a rich set of tools for incorporating *graphical user interfaces* (GUIs) in M-functions. Using GUIDE, the processes of (1) laying out a GUI (i.e., its buttons, pop-up menus, etc.) and (2) programming the operation of the GUI are divided conveniently into two easily managed and relatively independent tasks. The resulting graphical M-function is composed of two identically named (ignoring extensions) files:

1. A file with extension `.fig`, called a *FIG-file*, that contains a complete graphical description of all the function's GUI objects or elements and their spatial arrangement. A FIG-file contains binary data that does not need to be parsed when the associated GUI-based M-function is executed. The FIG-file for ICE (`ice.fig`) is described later in this section.
2. A file with extension `.m`, called a *GUI M-file*, which contains the code that controls the GUI operation. This file includes functions that are called

when the GUI is launched and exited, and *callback functions* that are executed when a user interacts with GUI objects—for example, when a button is pushed. The GUI M-file for ICE (`ice.m`) is described in the next section.

To launch GUIDE from the MATLAB command window, type

```
guide filename
```

where `filename` is the name of an existing FIG-file on the current path. If `filename` is omitted, GUIDE opens a new (i.e., blank) window.

Figure B.1 shows the GUIDE *Layout Editor* (launched by entering `guide ice` at the MATLAB >> prompt) for the Interactive Color Editor (ICE) layout. The Layout Editor is used to select, place, size, align, and manipulate graphic objects on a mock-up of the user interface under development. The buttons on its left side form a *Component Palette* containing the GUI objects that are supported—*Push Buttons*, *Toggle Buttons*, *Radio Buttons*, *Checkboxes*, *Edit Texts*, *Static Texts*, *Sliders*, *Frames*, *Listboxes*, *Popup Menus*, and *Axes*. Each object is similar in behavior to its standard Windows' counterpart. And any combination of objects can be added to the figure object in the layout area on the right side of the Layout Editor. Note that the ICE GUI includes checkboxes (`Smooth`, `Clamp Ends`, `Show PDF`, `Show CDF`, `Map Bars`, and `Map Image`), static text ("Component:", "Curve", ...), a frame outlining the curve controls, two

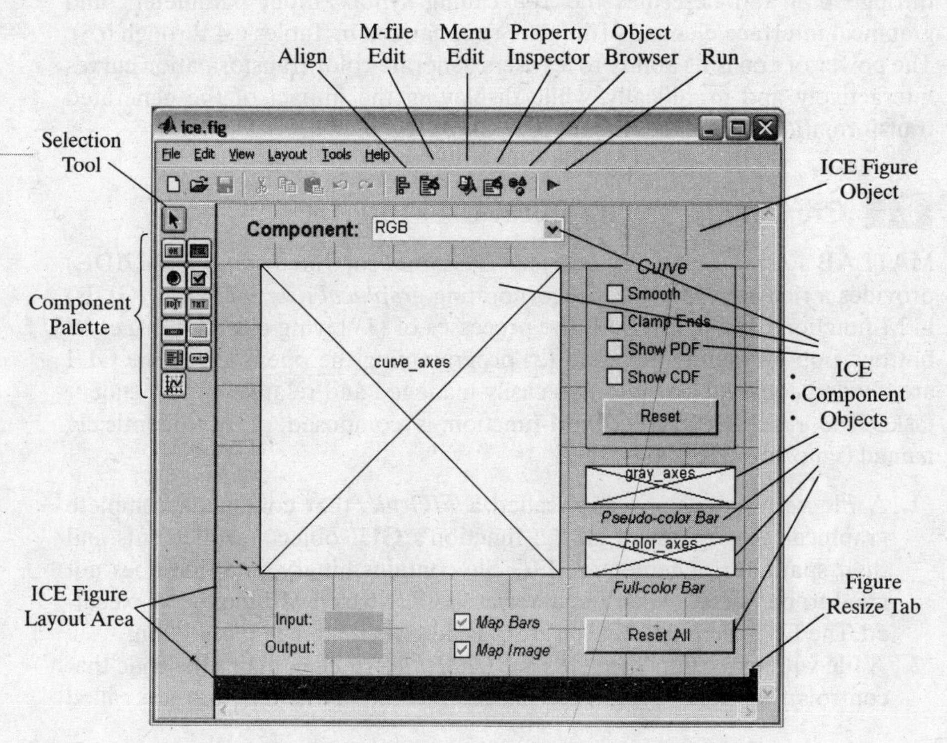

FIGURE B.1 The GUIDE Layout Editor mockup of the ICE GUI.

push buttons (`Reset` and `Reset All`), a popup menu for selecting a color transformation curve, and three `axes` objects for displaying the selected curve (with associated control points) and its effect on both a gray-scale wedge and hue wedge. A hierarchical list of the elements comprising ICE (obtained by clicking the *Object Browser* button in the task bar at the top of the Layout Editor) is shown in Fig. B.2(a). Note that each element has been given a unique name or tag. For example, the axes object for curve display (at the top of the list) is assigned the identifier `curve_axes` [the identifier is the first entry after the open parenthesis in Fig. B.2(a)].

Tags are one of several *properties* that are common to all GUI objects. A scrollable list of the properties characterizing a specific object can be obtained by selecting the object [in the Object Browser list of Fig. B.2(a) or layout area of Fig. B.1 using the *Selection Tool*] and clicking the *Property Inspector* button on the Layout Editor's task bar. Figure B.2(b) shows the list that is generated when the `figure` object of Fig. B.2(a) is selected. Note that the `figure` object's `Tag` property [highlighted in Fig. B.2(b)] is `ice`. This property is important because GUIDE uses it to automatically generate `figure` callback function names. Thus, for example, the `WindowButtonDownFcn` property at the bottom of the scrollable Property Inspector window, which is executed when a mouse button is pressed over the figure window, is assigned the name `ice_WindowButtonDownFcn`. Recall that callback functions are merely M-functions that are executed when a user interacts with a GUI object. Other

The GUIDE generated `figure` object is a container for all other objects in the interface.

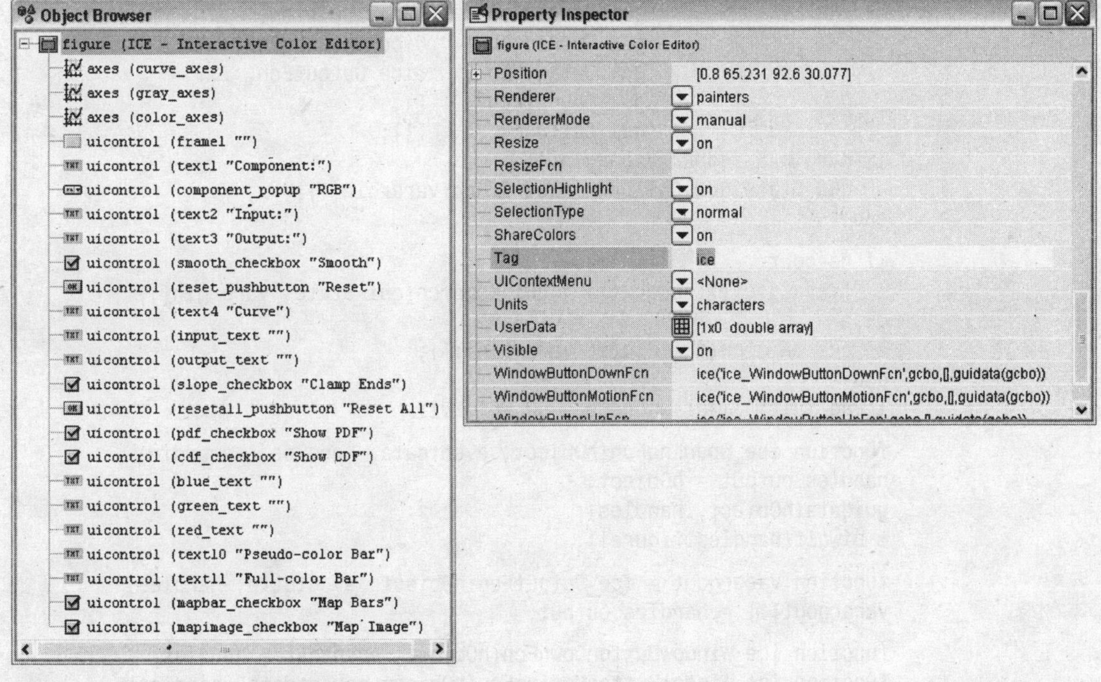

a b

FIGURE B.2 (a) The GUIDE Object Browser and (b) Property Inspector for the ICE "figure" object.

notable (and common to all GUI objects) properties include the Position and Units properties, which define the size and location of an object.

Finally, we note that some properties are unique to particular objects. A pushbutton object, for example, has a Callback property that defines the function that is executed when the button is pressed and the String property that determines the button's label. The Callback property of the ICE Reset button is reset_pushbutton_Callback [note the incorporation of its Tag property from Fig. B.2(a) in the callback function name]; its String property is "Reset". Note, however, that the Reset pushbutton does not have a WindowButtonMotionFcn property; it is specific to "figure" objects.

B.2 Programming the ICE Interface

When the ICE FIG-file of the previous section is first saved or the GUI is first run (e.g., by clicking the *Run* button on the Layout Editor's task bar), GUIDE generates a starting *GUI M-file* called ice.m. This file, which can be modified using a standard text editor or MATLAB's M-file editor, determines how the interface responds to user actions. The automatically generated GUI M-file for ICE is as follows:

ice

GUIDE generated starting M-file.

```
function varargout = ice(varargin)
% Begin initialization code - DO NOT EDIT
gui_Singleton = 1;
gui_State = struct('gui_Name',       mfilename, ...
                   'gui_Singleton',  gui_Singleton, ...
                   'gui_OpeningFcn', @ice_OpeningFcn, ...
                   'gui_OutputFcn',  @ice_OutputFcn, ...
                   'gui_LayoutFcn',  [], ...
                   'gui_Callback',   []);
if nargin & ischar(varargin{1})
   gui_State.gui_Callback = str2func(varargin{1});
end

if nargout
   [varargout{1:nargout}] = gui_mainfcn(gui_State, varargin{:});
else
   gui_mainfcn(gui_State, varargin{:});
end
% End initialization code - DO NOT EDIT

function ice_OpeningFcn(hObject, eventdata, handles, varargin)
handles.output = hObject;
guidata(hObject, handles);
% uiwait(handles.figure1);

function varargout = ice_OutputFcn(hObject, eventdata, handles)
varargout{1} = handles.output;

function ice_WindowButtonDownFcn(hObject, eventdata, handles)
function ice_WindowButtonMotionFcn(hObject, eventdata, handles)
function ice_WindowButtonUpFcn(hObject, eventdata, handles)
```

```
function component_popup_Callback(hObject, eventdata, handles)
function smooth_checkbox_Callback(hObject, eventdata, handles)
function reset_pushbutton_Callback(hObject, eventdata, handles)
function slope_checkbox_Callback(hObject, eventdata, handles)
function resetall_pushbutton_Callback(hObject, eventdata, handles)
function pdf_checkbox_Callback(hObject, eventdata, handles)
function cdf_checkbox_Callback(hObject, eventdata, handles)
function mapbar_checkbox_Callback(hObject, eventdata, handles)
function mapimage_checkbox_Callback(hObject, eventdata, handles)
```

This automatically generated file is a useful starting point or prototype for the development of the fully functional `ice` interface. (Note that we have stripped the file of many GUIDE-generated comments to save space.) In the sections that follow, we break this code into four basic sections: (1) the initialization code between the two "`DO NOT EDIT`" comment lines, (2) the figure opening and output functions (`ice_OpeningFcn` and `ice_OutputFcn`), (3) the figure callback functions (i.e., the `ice_WindowButtonDownFcn`, `ice_WindowButtonMotionFcn`, and `ice_WindowButtonUpFcn` functions), and (4) the object callback functions (e.g., `reset_pushbutton_Callback`). When considering each section, completely developed versions of the `ice` functions contained in the section are given, and the discussion is focused on features of general interest to most GUI M-file developers. The code introduced in each section will not be consolidated (for the sake of brevity) into a single comprehensive listing of `ice.m`. It is introduced in a piecemeal manner.

The operation of `ice` was described in Section 6.4. It is also summarized in the following Help text block from the fully developed `ice.m` M-function:

`ice`

Help text block of final version.

```
%ICE Interactive Color Editor.
%
%   OUT = ICE('Property Name', 'Property Value', ...) transforms an
%   image's color components based on interactively specified mapping
%   functions. Inputs are Property Name/Property Value pairs:
%
%      Name               Value
%      --------------     ---------------------------------------------
%      'image'            An RGB or monochrome input image to be
%                         transformed by interactively specified
%                         mappings.
%      'space'            The color space of the components to be
%                         modified. Possible values are 'rgb', 'cmy',
%                         'hsi', 'hsv', 'ntsc' (or 'yiq'), 'ycbcr'. When
%                         omitted, the RGB color space is assumed.
%      'wait'             If 'on' (the default), OUT is the mapped input
%                         image and ICE returns to the calling function
%                         or workspace when closed. If 'off', OUT is the
%                         handle of the mapped input image and ICE
%                         returns immediately.
```

```
%
%  EXAMPLES:
%    ice OR ice('wait', 'off')              % Demo user interface
%    ice('image', f)                        % Map RGB or mono image
%    ice('image', f, 'space', 'hsv')        % Map HSV of RGB image
%    g = ice('image', f)                    % Return mapped image
%    g = ice('image', f, 'wait', 'off');    % Return its handle
%
%  ICE displays one popup menu selectable mapping function at a
%  time. Each image component is mapped by a dedicated curve (e.g.,
%  R, G, or B) and then by an all-component curve (e.g., RGB). Each
%  curve's control points are depicted as circles that can be moved,
%  added, or deleted with a two- or three-button mouse:
%
%     Mouse Button          Editing Operation
%     --------------        ---------------------------------------------
%     Left                  Move control point by pressing and dragging.
%     Middle                Add and position a control point by pressing
%                           and dragging. (Optionally Shift-Left)
%     Right                 Delete a control point. (Optionally
%                           Control-Left)
%
%  Checkboxes determine how mapping functions are computed, whether
%  the input image and reference pseudo- and full-color bars are
%  mapped, and the displayed reference curve information (e.g.,
%  PDF):
%
%    Checkbox              Function
%    ---------------       ---------------------------------------------
%    Smooth                Checked for cubic spline (smooth curve)
%                          interpolation. If unchecked, piecewise linear.
%    Clamp Ends            Checked to force the starting and ending curve
%                          slopes in cubic spline interpolation to 0. No
%                          effect on piecewise linear.
%    Show PDF              Display probability density function(s) [i.e.,
%                          histogram(s)] of the image components affected
%                          by the mapping function.
%    Show CDF              Display cumulative distributions function(s)
%                          instead of PDFs.
%                          <Note: Show PDF/CDF are mutually exclusive.>
%    Map Image             If checked, image mapping is enabled; else
%                          not.
%    Map Bars              If checked, pseudo- and full-color bar mapping
%                          is enabled; else display the unmapped bars (a
%                          gray wedge and hue wedge, respectively).
%
```

```
%   Mapping functions can be initialized via pushbuttons:
%
%      Button          Function
%      -------------   ------------------------------------------------
%      Reset           Init the currently displayed mapping function
%                      and uncheck all curve parameters.
%      Reset All       Initialize all mapping functions.
```

B.2.1 Initialization Code

The opening section of code in the starting GUI M-file (at the beginning of Section B.2) is a standard GUIDE-generated block of initialization code. Its purpose is to build and display ICE's GUI using the M-file's companion FIG-file (see Section B.1) and control access to all internal M-file functions. As the enclosing "DO NOT EDIT" comment lines indicate, the initialization code should not be modified. Each time ice is called, the initialization block builds a structure called gui_State, which contains information for accessing ice functions. For instance, named field gui_Name (i.e., gui_State.gui_Name) contains the MATLAB function mfilename, which returns the name of the currently executing M-file. In a similar manner, fields gui_OpeningFcn and gui_OutputFcn are loaded with the GUIDE generated names of ice's opening and output functions (discussed in the next section). If an ICE GUI object is activated by the user (e.g., a button is pressed), the name of the object's callback function is added as field gui_Callback [the callback's name would have been passed as a string in varargin(1)].

After structure gui_State is formed, it is passed as an input argument, along with varargin(:), to function gui_mainfcn. This MATLAB function handles GUI creation, layout, and callback dispatch. For ice, it builds and displays the user interface and generates all necessary calls to its opening, output, and callback functions. Since older versions of MATLAB may not include this function, GUIDE is capable of generating a stand-alone version of the normal GUI M-file (i.e., one that works without a FIG-file) by selecting **Export...** from the **File** menu. In the stand-alone version, function gui_mainfcn and two supporting routines, ice_LayoutFcn and local_openfig, are appended to the normally FIG-file dependent M-file. The role of ice_LayoutFcn is to create the ICE GUI. In the stand-alone version of ice, it begins with the statement

```
h1 = figure(...
'Units', 'characters',...
'Color', [0.87843137254902 0.874509803921569 0.890196078431373],...
'Colormap', [0 0 0.5625;0 0 0.625;0 0 0.6875;0 0 0.75;...
            0 0 0.8125;0 0 0.875;0 0 0.9375;0 0 1;0 0.0625 1;...
            0 0.125 1;0 0.1875 1;0 0.25 1;0 0.3125 1;0 0.375 1;...
            0 0.4375 1;0 0.5 1;0 0.5625 1;0 0.625 1;0 0.6875 1;...
            0 0.75 1;0 0.8125 1;0 0.875 1;0 0.9375 1;0 1 1;...
            0.0625 1 1;0.125 1 0.9375;0.1875 1 0.875;...
```

```
            0.25 1 0.8125;0.3125 1 0.75;0.375 1 0.6875;...
            0.4375 1 0.625;0.5 1 0.5625;0.5625 1 0.5;...
            0.625 1 0.4375;0.6875 1 0.375;0.75 1 0.3125;...
            0.8125 1 0.25;0.875 1 0.1875;0.9375 1 0.125;...
            1 1 0.0625;1 1 0;1 0.9375 0;1 0.875 0;1 0.8125 0;...
            1 0.75 0;1 0.6875 0;1 0.625 0;1 0.5625 0;1 0.5 0;...
            1 0.4375 0;1 0.375 0;1 0.3125 0;1 0.25 0;...
            1 0.1875 0;1 0.125 0;1 0.0625 0;1 0 0;0.9375 0 0;...
            0.875 0 0;0.8125 0 0;0.75 0 0;0.6875 0 0;0.625 0 0;...
            0.5625 0 0],...
'IntegerHandle', 'off',...
'InvertHardcopy', get(0, 'defaultfigureInvertHardcopy'),...
'MenuBar', 'none',...
'Name', 'ICE - Interactive Color Editor',...
'NumberTitle', 'off',...
'PaperPosition', get(0, 'defaultfigurePaperPosition'),...
'Position', [0.8 65.2307692307693 92.6 30.0769230769231],...
'Renderer', get(0, 'defaultfigureRenderer'),...
'RendererMode', 'manual',...
'WindowButtonDownFcn', 'ice(''ice_WindowButtonDownFcn'', gcbo, [],...
                       guidata(gcbo))',...
'WindowButtonMotionFcn', 'ice(''ice_WindowButtonMotionFcn'', gcbo,...
                       [], guidata(gcbo))',...
'WindowButtonUpFcn', 'ice(''ice_WindowButtonUpFcn'', gcbo, [],...
                       guidata(gcbo))',...
'HandleVisibility', 'callback',...
'Tag', 'ice',...
'UserData', zeros(1,0));
```

to create the main figure window. GUI objects are then added with statements like

uicontrol

Function `uicontrol('PropertyName1', Value1, ...)` *creates a user interface control in the current window with the specified properties and returns a handle to it.*

```
h13 = uicontrol(...
'Parent', h1,...
'Units', 'normalized',...
'Callback', 'ice(''reset_pushbutton_Callback'', gcbo, [],...
            guidata(gcbo))',...
'FontSize', 10,...
'ListboxTop', 0,...
'Position', [0.710583153347732 0.508951406649616...
           0.211663066954644 0.0767263427109974],...
'String', 'Reset',...
'Tag', 'reset_pushbutton');
```

which adds the `Reset` pushbutton to the figure. Note that these statements specify explicitly properties that were defined originally using the Property Inspector of the GUIDE Layout Editor. Finally, we note that the `figure` function was introduced in Section 2.3; `uicontrol` creates a user interface control (i.e., GUI object) in the current figure window based on property name/value pairs (e.g., 'Tag' plus 'reset_pushbutton') and returns a handle to it.

B.2.2 The Opening and Output Functions

The first two functions following the initialization block in the starting GUI M-file at the beginning of Section B.2 are called *opening* and *output functions*, respectively. They contain the code that is executed just before the GUI is made visible to the user and when the GUI returns its output to the command line or calling routine. Both functions are passed arguments hObject, eventdata, and handles. (These arguments are also inputs to the callback functions in the next two sections.) Input hObject is a graphics object handle, eventdata is reserved for future use, and handles is a structure that provides handles to interface objects and any application specific or user defined data. To implement the desired functionality of the ICE interface (see the Help text), both ice_OpeningFcn and ice_OutputFcn must be expanded beyond the "barebones" versions in the starting GUI M-file. The expanded code is as follows:

ice_OpeningFcn

From the final M-file.

```
%-------------------------------------------------------------------%
function ice_OpeningFcn(hObject, eventdata, handles, varargin)
%   When ICE is opened, perform basic initialization (e.g., setup
%   globals, ...) before it is made visible.

% Set ICE globals to defaults.
handles.updown = 'none';              % Mouse updown state
handles.plotbox = [0 0 1 1];          % Plot area parameters in pixels
handles.set1 = [0 0; 1 1];            % Curve 1 control points
handles.set2 = [0 0; 1 1];            % Curve 2 control points
handles.set3 = [0 0; 1 1];            % Curve 3 control points
handles.set4 = [0 0; 1 1];            % Curve 4 control points
handles.curve = 'set1';               % Structure name of selected curve
handles.cindex = 1;                   % Index of selected curve
handles.node = 0;                     % Index of selected control point
handles.below = 1;                    % Index of node below control point
handles.above = 2;                    % Index of node above control point
handles.smooth = [0; 0; 0; 0];        % Curve smoothing states
handles.slope = [0; 0; 0; 0];         % Curve end slope control states
handles.cdf = [0; 0; 0; 0];           % Curve CDF states
handles.pdf = [0; 0; 0; 0];           % Curve PDF states
handles.output = [];                  % Output image handle
handles.df = [];                      % Input PDFs and CDFs
handles.colortype = 'rgb';            % Input image color space
handles.input = [];                   % Input image data
handles.imagemap = 1;                 % Image map enable
handles.barmap = 1;                   % Bar map enable
handles.graybar = [];                 % Pseudo (gray) bar image
handles.colorbar = [];                % Color (hue) bar image

% Process Property Name/Property Value input argument pairs.
wait = 'on';
```

```
if (nargin > 3)
   for i = 1:2:(nargin - 3)
      if nargin - 3 == i
         break;
      end
      switch lower(varargin{i})
      case 'image'
         if ndims(varargin{i + 1}) == 3
            handles.input = varargin{i + 1};
         elseif ndims(varargin{i + 1}) == 2
            handles.input = cat(3, varargin{i + 1}, ...
                                  varargin{i + 1}, varargin{i + 1});
         end
         handles.input = double(handles.input);
         inputmax = max(handles.input(:));
         if inputmax > 255
            handles.input = handles.input / 65535;
         elseif inputmax > 1
            handles.input = handles.input / 255;
         end

      case 'space'
         handles.colortype = lower(varargin{i + 1});
         switch handles.colortype
         case 'cmy'
            list = {'CMY' 'Cyan' 'Magenta' 'Yellow'};
         case {'ntsc', 'yiq'}
            list = {'YIQ' 'Luminance' 'Hue' 'Saturation'};
            handles.colortype = 'ntsc';
         case 'ycbcr'
            list = {'YCbCr' 'Luminance' 'Blue' ...
                  'Difference' 'Red Difference'};
         case 'hsv'
            list = {'HSV' 'Hue' 'Saturation' 'Value'};
         case 'hsi'
            list = {'HSI' 'Hue' 'Saturation' 'Intensity'};
         otherwise
            list = {'RGB' 'Red' 'Green' 'Blue'};
            handles.colortype = 'rgb';
         end
         set(handles.component_popup, 'String', list);
      case 'wait'
         wait = lower(varargin{i + 1});
      end
   end
end

% Create pseudo- and full-color mapping bars (grays and hues). Store
% a color space converted 1x128x3 line of each bar for mapping.
xi = 0:1/127:1;      x = 0:1/6:1;      x = x';
y = [1 1 0 0 0 1 1; 0 1 1 1 0 0 0; 0 0 0 1 1 1 0]';
gb = repmat(xi, [1 1 3]);         cb = interp1q(x, y, xi');
```

```
cb = reshape(cb, [1 128 3]);
if ~strcmp(handles.colortype, 'rgb')
   gb = eval(['rgb2' handles.colortype '(gb)']);
   cb = eval(['rgb2' handles.colortype '(cb)']);
end
gb = round(255 * gb);      gb = max(0, gb);      gb = min(255, gb);
cb = round(255 * cb);      cb = max(0, cb);      cb = min(255, cb);
handles.graybar = gb;      handles.colorbar = cb;

% Do color space transforms, clamp to [0, 255], compute histograms
% and cumulative distribution functions, and create output figure.
if size(handles.input, 1)
   if ~strcmp(handles.colortype, 'rgb')
      handles.input = eval(['rgb2' handles.colortype ...
                           '(handles.input)']);
   end
   handles.input = round(255 * handles.input);
   handles.input = max(0, handles.input);
   handles.input = min(255, handles.input);
   for i = 1:3
      color = handles.input(:, :, i);
      df = hist(color(:), 0:255);
      handles.df = [handles.df; df / max(df(:))];
      df = df / sum(df(:));   df = cumsum(df);
      handles.df = [handles.df; df];
   end
   figure;    handles.output = gcf;
end

% Compute ICE's screen position and display image/graph.
set(0, 'Units', 'pixels');      ssz = get(0, 'Screensize');
set(handles.ice, 'Units', 'pixels');
uisz = get(handles.ice, 'Position');
if size(handles.input, 1)
   fsz = get(handles.output, 'Position');
   bc = (fsz(4) - uisz(4)) / 3;
   if bc > 0
      bc = bc + fsz(2);
   else
      bc = fsz(2) + fsz(4) - uisz(4) - 10;
   end
   lc = fsz(1) + (size(handles.input, 2) / 4) + (3 * fsz(3) / 4);
   lc = min(lc, ssz(3) - uisz(3) - 10);
   set(handles.ice, 'Position', [lc bc 463 391]);
else
   bc = round((ssz(4) - uisz(4)) / 2) - 10;
   lc = round((ssz(3) - uisz(3)) / 2) - 10;
   set(handles.ice, 'Position', [lc bc uisz(3) uisz(4)]);
end
set(handles.ice, 'Units', 'normalized');
graph(handles);    render(handles);
```

```
% Update handles and make ICE wait before exit if required.
guidata(hObject, handles);
if strcmpi(wait, 'on')
   uiwait(handles.ice);
end

%-------------------------------------------------------------------%
function varargout = ice_OutputFcn(hObject, eventdata, handles)
%  After ICE is closed, get the image data of the current figure
%  for the output. If 'handles' exists, ICE isn't closed (there was
%  no 'uiwait') so output figure handle.

if max(size(handles)) == 0
   figh = get(gcf);
   imageh = get(figh.Children);
   if max(size(imageh)) > 0
      image = get(imageh.Children);
      varargout{1} = image.CData;
   end
else
   varargout{1} = hObject;
end
```

`ice_OutputFcn`

From the final M-file.

Rather than examining the intricate details of these functions (see the code's comments and consult Appendix A or the index for help on specific functions), we note the following commonalities with most GUI opening and output functions:

1. The `handles` structure (as can be seen from its numerous references in the code) plays a central role in most GUI M-files. It serves two crucial functions. Since it provides handles for all the graphic objects in the interface, it can be used to access and modify object properties. For instance, the `ice` opening function uses

   ```
   set(handles.ice, 'Units', 'pixels');
   uisz = get(handles.ice, 'Position');
   ```

   to access the size and location of the ICE GUI (in pixels). This is accomplished by setting the `Units` property of the `ice` figure, whose handle is available in `handles.ice`, to `'pixels'` and then reading the `Position` property of the figure (using the `get` function). The `get` function, which returns the value of a property associated with a graphics object, is also used to obtain the computer's display area via the `ssz = get(0, 'Screensize')` statement near the end of the opening function. Here, `0` is the handle of the computer display (i.e., root figure) and `'Screensize'` is a property containing its extent.

   In addition to providing access to GUI objects, the `handles` structure is a powerful conduit for sharing application data. Note that it holds the default values for twenty-three global `ice` parameters (ranging from the mouse state in `handles.updown` to the entire input image in `handles.input`). They must survive every call to ice and are added to handles at the start of ice_OpeningFcn. For instance, the handles.set1 global is created by the statement

```
handles.set1 = [0 0; 1 1]
```

where set1 is a named field containing the control points of a color mapping function to be added to the handles structure and [0 0; 1 1] is its default value [curve endpoints (0,0) and (1,1)]. Before exiting a function in which handles is modified,

```
guidata(hObject, handles)
```

Function guidata (H, DATA) *stores the specified data in the figure's application data.* H *is a handle that identifies the figure—it can be the figure itself, or any object contained in the figure.*

must be called to store variable handles as the application data of the figure with handle hObject.

2. Like many built-in graphics functions, ice_OpeningFcn processes input arguments (except hObject, eventdata, and handles) in property name and value pairs. When there are more than three input arguments (i.e., if nargin > 3), a loop that skips through the input arguments in pairs [for i = 1:2:(nargin – 3)] is executed. For each pair of inputs, the first is used to drive the switch construct,

```
switch lower(varargin{i})
```

which processes the second parameter appropriately. For case 'space', for instance, the statement

```
handles.colortype = lower(varargin{i + 1});
```

sets named field colortype to the value of the second argument of the input pair. This value is then used to setup ICE's color component popup options (i.e., the String property of object component_popup). Later, it is used to transform the components of the input image to the desired mapping space via

```
handles.input = eval(['rgb2' ...
    handles.colortype '(handles.input)']);
```

where built-in function eval(s) causes MATLAB to execute string s as an expression or statement (see Section 12.4.1 for more on function eval). If handles.input is 'hsv', for example, eval argument ['rgb2' 'hsv' '(handles.input)'] becomes the concatenated string 'rgb2hsv(handles.input)', which is executed as a standard MATLAB expression that transforms the RGB components of the input image to the HSV color space (see Section 6.2.3).

3. The statement

```
% uiwait(handles.figure1);
```

in the starting GUI M-file is converted into the conditional statement

```
if strcmpi(wait, 'on') uiwait(handles.ice); end
```

in the final version of ice_OpeningFcn. In general,

```
uiwait(fig)
```

blocks execution of a MATLAB code stream until either a uiresume is executed or figure fig is destroyed (i.e., closed). [With no input arguments, uiwait is the same as uiwait(gcf) where MATLAB function gcf returns the handle of the current figure]. When ice is not expected to return a mapped version of an input image, but return immediately (i.e., before the ICE GUI is closed), an input property name/value pair of 'wait'/'off' must be included in the call. Otherwise, ICE will not return to the calling routine or command line until it is closed. That is, until the user is finished interacting with the interface (and color mapping functions). In this situation, function ice_OutputFcn can not obtain the mapped image data from the handles structure, because it does not exist after the GUI is closed. As can be seen in the final version of the function, ICE extracts the image data from the CData property of the surviving mapped image output figure. If a mapped output image is not to be returned by ice, the uiwait statement in ice_OpeningFcn is not executed, ice_OutputFcn is called immediately after the opening function (long before the GUI is closed), and the handle of the mapped image output figure is returned to the calling routine or command line.

Finally, we note that several internal functions are invoked by ice_OpeningFcn. These—and all other ice internal functions—are listed next. Note that they provide additional examples of the usefulness of the handles structure in MATLAB GUIs. For instance, the

```
nodes = getfield(handles, handles.curve)
```

and

```
nodes = getfield(handles, ['set' num2str(i)])
```

statements in internal functions graph and render, respectively, are used to access the interactively defined control points of ICE's various color mapping curves. In its standard form,

```
F = getfield(S,'field')
```

returns to F the contents of named field 'field' from structure S.

ice

Final M-file internal functions.

```
%-------------------------------------------------------------------%
function graph(handles)
%   Interpolate and plot mapping functions and optional reference
%   PDF(s) or CDF(s).

nodes = getfield(handles, handles.curve);
c = handles.cindex;       dfx = 0:1/255:1;
colors = ['k' 'r' 'g' 'b'];

% For piecewise linear interpolation, plot a map, map + PDF/CDF, or
% map + 3 PDFs/CDFs.
if ~handles.smooth(handles.cindex)
   if (~handles.pdf(c) & ~handles.cdf(c)) | ...
         (size(handles.df, 2) == 0)
      plot(nodes(:, 1), nodes(:, 2), 'b-', ...
           nodes(:, 1), nodes(:, 2), 'ko', ...
           'Parent', handles.curve_axes);
   elseif c > 1
      i = 2 * c - 2 - handles.pdf(c);
      plot(dfx, handles.df(i, :), [colors(c) '-'], ...
           nodes(:, 1), nodes(:, 2), 'k-', ...
           nodes(:, 1), nodes(:, 2), 'ko', ...
           'Parent', handles.curve_axes);
   elseif c == 1
      i = handles.cdf(c);
      plot(dfx, handles.df(i + 1, :), 'r-', ...
           dfx, handles.df(i + 3, :), 'g-', ...
           dfx, handles.df(i + 5, :), 'b-', ...
           nodes(:, 1), nodes(:, 2), 'k-', ...
           nodes(:, 1), nodes(:, 2), 'ko', ...
           'Parent', handles.curve_axes);
   end

% Do the same for smooth (cubic spline) interpolations.
else
   x = 0:0.01:1;
   if ~handles.slope(handles.cindex)
      y = spline(nodes(:, 1), nodes(:, 2), x);
   else
      y = spline(nodes(:, 1), [0; nodes(:, 2); 0], x);
   end
   i = find(y > 1);        y(i) = 1;
   i = find(y < 0);        y(i) = 0;

   if (~handles.pdf(c) & ~handles.cdf(c)) | ...
         (size(handles.df, 2) == 0)
      plot(nodes(:, 1), nodes(:, 2), 'ko', x, y, 'b-', ...
           'Parent', handles.curve_axes);
   elseif c > 1
      i = 2 * c - 2 - handles.pdf(c);
      plot(dfx, handles.df(i, :), [colors(c) '-'], ...
            nodes(:, 1), nodes(:, 2), 'ko', x, y, 'k-', ...
```

```
                'Parent', handles.curve_axes);
        elseif c == 1
            i = handles.cdf(c);
            plot(dfx, handles.df(i + 1, :), 'r-', ...
                 dfx, handles.df(i + 3, :), 'g-', ...
                 dfx, handles.df(i + 5, :), 'b-', ...
                 nodes(:, 1), nodes(:, 2), 'ko', x, y, 'k-', ...
                 'Parent', handles.curve_axes);
      end
   end
   % Put legend if more than two curves are shown.
   s = handles.colortype;
   if strcmp(s, 'ntsc')
      s = 'yiq';
   end
   if (c == 1) & (handles.pdf(c) | handles.cdf(c))
      s1 = ['-- ' upper(s(1))];
      if length(s) == 3
         s2 = ['-- ' upper(s(2))];       s3 = ['-- ' upper(s(3))];
      else
         s2 = ['-- ' upper(s(2)) s(3)];   s3 = ['-- ' upper(s(4)) s(5)];
      end
   else
      s1 = '';     s2 = '';     s3 = '';
   end
   set(handles.red_text, 'String', s1);
   set(handles.green_text, 'String', s2);
   set(handles.blue_text, 'String', s3);
   %-----------------------------------------------------------------------%
   function [inplot, x, y] = cursor(h, handles)
   %   Translate the mouse position to a coordinate with respect to
   %   the current plot area, check for the mouse in the area and if so
   %   save the location and write the coordinates below the plot.

   set(h, 'Units', 'pixels');
   p = get(h, 'CurrentPoint');
   x = (p(1, 1) - handles.plotbox(1)) / handles.plotbox(3);
   y = (p(1, 2) - handles.plotbox(2)) / handles.plotbox(4);
   if x > 1.05 | x < -0.05 | y > 1.05 | y < -0.05
      inplot = 0;
   else
      x = min(x, 1);      x = max(x, 0);
      y = min(y, 1);      y = max(y, 0);
      nodes = getfield(handles, handles.curve);
      x = round(256 * x) / 256;
      inplot = 1;
      set(handles.input_text, 'String', num2str(x, 3));
      set(handles.output_text, 'String', num2str(y, 3));
```

```
end
set(h, 'Units', 'normalized');

%-------------------------------------------------------------------%
function y = render(handles)
%   Map the input image and bar components and convert them to RGB
%   (if needed) and display.

set(handles.ice, 'Interruptible', 'off');
set(handles.ice, 'Pointer', 'watch');
ygb = handles.graybar;        ycb = handles.colorbar;
yi = handles.input;           mapon = handles.barmap;
imageon = handles.imagemap & size(handles.input, 1);

for i = 2:4
   nodes = getfield(handles, ['set' num2str(i)]);
   t = lut(nodes, handles.smooth(i), handles.slope(i));
   if imageon
      yi(:, :, i - 1) = t(yi(:, :, i - 1) + 1);
   end
   if mapon
      ygb(:, :, i - 1) = t(ygb(:, :, i - 1) + 1);
      ycb(:, :, i - 1) = t(ycb(:, :, i - 1) + 1);
   end
end
t = lut(handles.set1, handles.smooth(1), handles.slope(1));
if imageon
   yi = t(yi + 1);
end
if mapon
   ygb = t(ygb + 1);      ycb = t(ycb + 1);
end

if ~strcmp(handles.colortype, 'rgb')
   if size(handles.input, 1)
      yi = yi / 255;
      yi = eval([handles.colortype '2rgb(yi)']);
      yi = uint8(255 * yi);
   end
   ygb = ygb / 255;       ycb = ycb / 255;
   ygb = eval([handles.colortype '2rgb(ygb)']);
   ycb = eval([handles.colortype '2rgb(ycb)']);
   ygb = uint8(255 * ygb);      ycb = uint8(255 * ycb);
else
   yi = uint8(yi);     ygb = uint8(ygb);     ycb = uint8(ycb);
end

if size(handles.input, 1)
   figure(handles.output);      imshow(yi);
end
ygb = repmat(ygb, [32 1 1]);       ycb = repmat(ycb, [32 1 1]);
axes(handles.gray_axes);           imshow(ygb);
axes(handles.color_axes);          imshow(ycb);
figure(handles.ice);
```

```
set(handles.ice, 'Pointer', 'arrow');
set(handles.ice, 'Interruptible', 'on');

%------------------------------------------------------------------------%
function t = lut(nodes, smooth, slope)
%  Create a 256 element mapping function from a set of control
%  points. The output values are integers in the interval [0, 255].
%  Use piecewise linear or cubic spline with or without zero end
%  slope interpolation.

t = 255 * nodes;     i = 0:255;
if ~smooth
   t = [t; 256 256];   t = interp1q(t(:, 1), t(:, 2), i');
else
   if ~slope
      t = spline(t(:, 1), t(:, 2), i);
   else
      t = spline(t(:, 1), [0; t(:, 2); 0], i);
   end
end
t = round(t);     t = max(0, t);     t = min(255, t);

%------------------------------------------------------------------------%
function out = spreadout(in)
% Make all x values unique.

% Scan forward for non-unique x's and bump the higher indexed x--
% but don't exceed 1. Scan the entire range.
nudge = 1 / 256;
for i = 2:size(in, 1) - 1
   if in(i, 1) <= in(i - 1, 1)
      in(i, 1) = min(in(i - 1, 1) + nudge, 1);
   end
end

% Scan in reverse for non-unique x's and decrease the lower indexed
% x -- but don't go below 0. Stop on the first non-unique pair.
if in(end, 1) == in(end - 1, 1)
   for i = size(in, 1):-1:2
      if in(i, 1) <= in(i - 1, 1)
         in(i - 1, 1) = max(in(i, 1) - nudge, 0);
      else
         break;
      end
   end
end

% If the first two x's are now the same, init the curve.
if in(1, 1) == in(2, 1)
   in = [0 0; 1 1];
end
out = in;
```

```
%-------------------------------------------------------------------%
function g = rgb2cmy(f)
%   Convert RGB to CMY using IPT's imcomplement.

g = imcomplement(f);

%-------------------------------------------------------------------%
function g = cmy2rgb(f)
%   Convert CMY to RGB using IPT's imcomplement.

g = imcomplement(f);
```

B.2.3 Figure Callback Functions

The three functions immediately following the ICE opening and closing functions in the starting GUI M-file at the beginning of Section B.2 are *figure callbacks* `ice_WindowButtonDownFcn`, `ice_WindowButtonMotionFcn`, and `ice_WindowButtonUpFcn`. In the automatically generated M-file, they are *function stubs*—that is, MATLAB `function` definition statements without supporting code. Fully developed versions of the three functions, whose joint task is to process mouse events (clicks and drags of mapping function control points on ICE's `curve_axes` object), are as follows:

`ice`
Figure Callbacks

```
%-------------------------------------------------------------------%
function ice_WindowButtonDownFcn(hObject, eventdata, handles)
%   Start mapping function control point editing. Do move, add, or
%   delete for left, middle, and right button mouse clicks ('normal',
%   'extend', and 'alt' cases) over plot area.

set(handles.curve_axes, 'Units', 'pixels');
handles.plotbox = get(handles.curve_axes, 'Position');
set(handles.curve_axes, 'Units', 'normalized');
[inplot, x, y] = cursor(hObject, handles);
if inplot
   nodes = getfield(handles, handles.curve);
   i = find(x >= nodes(:, 1));    below = max(i);
   above = min(below + 1, size(nodes, 1));
   if (x - nodes(below, 1)) > (nodes(above, 1) - x)
      node = above;
   else
      node = below;
   end
   deletednode = 0;

   switch get(hObject, 'SelectionType')
   case 'normal'
      if node == above
         above = min(above + 1, size(nodes, 1));
      elseif node == below
         below = max(below - 1, 1);
      end
      if node == size(nodes, 1)
```

```
            below = above;
         elseif node == 1
            above = below;
         end
         if x > nodes(above, 1)
            x = nodes(above, 1);
         elseif x < nodes(below, 1)
            x = nodes(below, 1);
         end
         handles.node = node;     handles.updown = 'down';
         handles.below = below;   handles.above = above;
         nodes(node, :) = [x y];
      case 'extend'
         if ~length(find(nodes(:, 1) == x))
            nodes = [nodes(1:below, :); [x y]; nodes(above:end, :)];
            handles.node = above;    handles.updown = 'down';
            handles.below = below;   handles.above = above + 1;
         end
      case 'alt'
         if (node ~= 1) & (node ~= size(nodes, 1))
            nodes(node, :) = [];   deletednode = 1;
         end
         handles.node = 0;
         set(handles.input_text, 'String', '');
         set(handles.output_text, 'String', '');
      end
      handles = setfield(handles, handles.curve, nodes);
      guidata(hObject, handles);
      graph(handles);
      if deletednode
         render(handles);
      end
   end
```

setfield

Functions S = setfield(S, 'field', V) *sets the contents of the specified field to value* V. *The changed structure is returned.*

```
%-------------------------------------------------------------------------%
function ice_WindowButtonMotionFcn(hObject, eventdata, handles)
%  Do nothing unless a mouse 'down' event has occurred. If it has,
%  modify control point and make new mapping function.

if ~strcmpi(handles.updown, 'down')
   return;
end
[inplot, x, y] = cursor(hObject, handles);
if inplot
   nodes = getfield(handles, handles.curve);
   nudge = handles.smooth(handles.cindex) / 256;
   if (handles.node ~= 1) & (handles.node ~= size(nodes, 1))
      if x >= nodes(handles.above, 1)
         x = nodes(handles.above, 1) - nudge;
      elseif x <= nodes(handles.below, 1)
         x = nodes(handles.below, 1) + nudge;
      end
```

```
   else
      if x > nodes(handles.above, 1)
         x = nodes(handles.above, 1);
      elseif x < nodes(handles.below, 1)
         x = nodes(handles.below, 1);
      end
   end
   nodes(handles.node, :) = [x y];
   handles = setfield(handles, handles.curve, nodes);
   guidata(hObject, handles);
   graph(handles);
end
%-------------------------------------------------------------------%
function ice_WindowButtonUpFcn(hObject, eventdata, handles)
%   Terminate ongoing control point move or add operation. Clear
%   coordinate text below plot and update display.
update = strcmpi(handles.updown, 'down');
handles.updown = 'up';     handles.node = 0;
guidata(hObject, handles);
if update
   set(handles.input_text, 'String', '');
   set(handles.output_text, 'String', '');
   render(handles);
end
```

In general, figure callbacks are launched in response to interactions with a figure object or window—not an active `uicontrol` object. More specifically,

- The `WindowButtonDownFcn` is executed when a user clicks a mouse button with the cursor in a figure but not over an enabled `uicontrol` (e.g., a pushbutton or popup menu).
- The `WindowButtonMotionFcn` is executed when a user moves a depressed mouse button within a figure window.
- The `WindowButtonUpFcn` is executed when a user releases a mouse button, after having pressed the mouse button within a figure but not over an enabled `uicontrol`.

The purpose and behavior of `ice`'s figure callbacks are documented (via comments) in the code. We make the following general observations about the final implementations:

1. Because the `ice_WindowButtonDownFcn` is called on all mouse button clicks in the `ice` figure (except over an active graphic object), the first job of the callback function is to see if the cursor is within `ice`'s plot area (i.e., the extent of the `curve_axes` object). If the cursor is outside this area, the mouse should be ignored. The test for this is performed by internal function `cursor`, whose listing was provided in the previous section. In `cursor`, the statement

```
p = get(h, 'CurrentPoint');
```

returns the current cursor coordinates. Variable `h` is passed from `ice_WindowButtonDownFcn` and originates as input argument `hObject`. In all figure callbacks, `hObject` is the handle of the figure requesting service. Property `'CurrentPoint'` contains the position of the cursor relative to the figure as a two-element row vector [x y].

2. Since `ice` is designed to work with two- and three-button mice, `ice_WindowButtonDownFcn` must determine which mouse button causes each callback. As can be seen in the code, this is done with a `switch` construct using the figure's `'SelectionType'` property. Cases `'normal'`, `'extent'`, and `'alt'` correspond to the left, middle, and right button clicks on three-button mice (or the left, shift-left, and control-left clicks of two-button mice), respectively, and are used to trigger the add control point, move control point, and delete control point operations.

3. The displayed ICE mapping function is updated (via internal function `graph`) each time a control point is modified, but the output figure, whose handle is stored in `handles.output`, is updated on *mouse button releases only*. This is because the computation of the output image, which is performed by internal function `render`, can be time-consuming. It involves mapping separately the input image's three color components, remapping each by the "all-component" curve, and converting the mapped components to the RGB color space for display. Note that without adequate precautions, the mapping function's control points could be modified inadvertently during this lengthy output mapping process.

   To prevent this, `ice` controls the interruptibility of its various callbacks. All MATLAB graphics objects have an `Interruptible` property that determines whether their callbacks can be interrupted. The default value of every object's `'Interruptible'` property is `'on'`, which means that object callbacks can be interrupted. If switched to `'off'`, callbacks that occur during the execution of the *now* noninterruptible callback are either ignored (i.e., cancelled) or placed in an *event queue* for later processing. The disposition of the interrupting callback is determined by the `'BusyAction'` property of the object being interrupted. If `'BusyAction'` is `'cancel'`, the callback is discarded; if `'queue'`, the callback is processed after the noninterruptible callback finishes.

   The `ice_WindowButtonUpFcn` function uses the mechanism just described to suspend temporarily (i.e., during output image computations) the user's ability to manipulate mapping function control points. The sequence

```
set(handles.ice, 'Interruptible', 'off');
set(handles.ice, 'Pointer', 'watch');
                    :
set(handles.ice, 'Pointer', 'arrow');
set(handles.ice, 'Interruptible', 'on');
```

   in internal function `render` sets the `ice` figure window's `'Interruptible'` property to `'off'` during the mapping of the output image and pseudo- and full-color bars. This prevents users from modifying mapping function control points while a mapping is being performed. Note also that the figure's 'Pointer' property is set to 'watch' to indicate visually that ice is busy and reset to 'arrow' when the output computation is completed.

B.2.4 Object Callback Functions

The final nine lines of the starting GUI M-file at the beginning of Section B.2 are *object callback* function stubs. Like the automatically generated figure callbacks of the previous section, they are initially void of code. Fully developed versions of the functions follow. Note that each function processes user interaction with a different ice uicontrol object (pushbutton, etc.) and is named by concatenating its Tag property with string '_Callback'. For example, the callback function responsible for handling the selection of the displayed mapping function is named the component_popup_Callback. It is called when the user activates (i.e., clicks on) the popup selector. Note also that input argument hObject is the handle of the popup graphics object—not the handle of the ice figure (as in the figure callbacks of the previous section). ICE's object callbacks involve minimal code and are self-documenting.

ice
Object Callbacks

```
%-------------------------------------------------------------------%
function component_popup_Callback(hObject, eventdata, handles)
%   Accept color component selection, update component specific
%   parameters on GUI, and draw the selected mapping function.

c = get(hObject, 'Value');
handles.cindex = c;
handles.curve = strcat('set', num2str(c));
guidata(hObject, handles);
set(handles.smooth_checkbox, 'Value', handles.smooth(c));
set(handles.slope_checkbox, 'Value', handles.slope(c));
set(handles.pdf_checkbox, 'Value', handles.pdf(c));
set(handles.cdf_checkbox, 'Value', handles.cdf(c));
graph(handles);

%-------------------------------------------------------------------%
function smooth_checkbox_Callback(hObject, eventdata, handles)
%   Accept smoothing parameter for currently selected color
%   component and redraw mapping function.

if get(hObject, 'Value')
   handles.smooth(handles.cindex) = 1;
   nodes = getfield(handles, handles.curve);
   nodes = spreadout(nodes);
   handles = setfield(handles, handles.curve, nodes);
else
   handles.smooth(handles.cindex) = 0;
end
guidata(hObject, handles);
set(handles.ice, 'Pointer', 'watch');
graph(handles);      render(handles);
set(handles.ice, 'Pointer', 'arrow');
```

```
%-------------------------------------------------------------------------%
function reset_pushbutton_Callback(hObject, eventdata, handles)
%   Init all display parameters for currently selected color
%   component, make map 1:1, and redraw it.

handles = setfield(handles, handles.curve, [0 0; 1 1]);
c = handles.cindex;
handles.smooth(c) = 0;   set(handles.smooth_checkbox, 'Value', 0);
handles.slope(c) = 0;    set(handles.slope_checkbox, 'Value', 0);
handles.pdf(c) = 0;      set(handles.pdf_checkbox, 'Value', 0);
handles.cdf(c) = 0;      set(handles.cdf_checkbox, 'Value', 0);
guidata(hObject, handles);
set(handles.ice, 'Pointer', 'watch');
graph(handles);       render(handles);
set(handles.ice, 'Pointer', 'arrow');
%-------------------------------------------------------------------------%
function slope_checkbox_Callback(hObject, eventdata, handles)
%   Accept slope clamp for currently selected color component and
%   draw function if smoothing is on.

if get(hObject, 'Value')
   handles.slope(handles.cindex) = 1;
else
   handles.slope(handles.cindex) = 0;
end
guidata(hObject, handles);
if handles.smooth(handles.cindex)
   set(handles.ice, 'Pointer', 'watch');
   graph(handles);       render(handles);
   set(handles.ice, 'Pointer', 'arrow');
end
%-------------------------------------------------------------------------%
function resetall_pushbutton_Callback(hObject, eventdata, handles)
%   Init display parameters for color components, make all maps 1:1,
%   and redraw display.

for c = 1:4
   handles.smooth(c) = 0;          handles.slope(c) = 0;
   handles.pdf(c) = 0;             handles.cdf(c) = 0;
   handles = setfield(handles, ['set' num2str(c)], [0 0; 1 1]);
end
set(handles.smooth_checkbox, 'Value', 0);
set(handles.slope_checkbox, 'Value', 0);
set(handles.pdf_checkbox, 'Value', 0);
set(handles.cdf_checkbox, 'Value', 0);
guidata(hObject, handles);
set(handles.ice, 'Pointer', 'watch');
graph(handles);       render(handles);
set(handles.ice, 'Pointer', 'arrow');
```

```
%-------------------------------------------------------------------------%
function pdf_checkbox_Callback(hObject, eventdata, handles)
%   Accept PDF (probability density function or histogram) display
%   parameter for currently selected color component and redraw
%   mapping function if smoothing is on. If set, clear CDF display.

if get(hObject, 'Value')
   handles.pdf(handles.cindex) = 1;
   set(handles.cdf_checkbox, 'Value', 0);
   handles.cdf(handles.cindex) = 0;
else
   handles.pdf(handles.cindex) = 0;
end
guidata(hObject, handles);    graph(handles);
%-------------------------------------------------------------------------%
function cdf_checkbox_Callback(hObject, eventdata, handles)
%   Accept CDF (cumulative distribution function) display parameter
%   for selected color component and redraw mapping function if
%   smoothing is on. If set, clear CDF display.

if get(hObject, 'Value')
   handles.cdf(handles.cindex) = 1;
   set(handles.pdf_checkbox, 'Value', 0);
   handles.pdf(handles.cindex) = 0;
else
   handles.cdf(handles.cindex) = 0;
end
guidata(hObject, handles);    graph(handles);
%-------------------------------------------------------------------------%
function mapbar_checkbox_Callback(hObject, eventdata, handles)
%   Accept changes to bar map enable state and redraw bars.

handles.barmap = get(hObject, 'Value');
guidata(hObject, handles);    render(handles);
%-------------------------------------------------------------------------%
function mapimage_checkbox_Callback(hObject, eventdata, handles)
%   Accept changes to the image map state and redraw image.

handles.imagemap = get(hObject, 'Value');
guidata(hObject, handles);    render(handles);
```

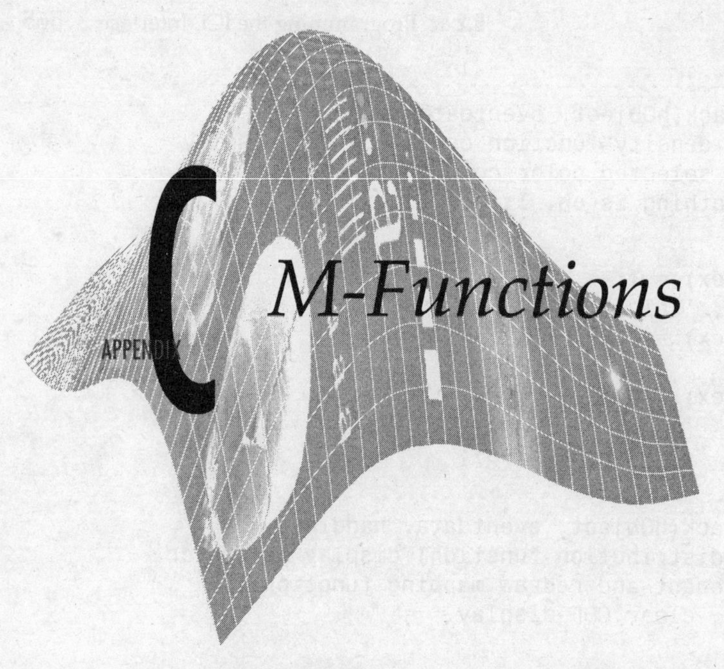

M-Functions

Preview

This appendix contains a listing of all the M-functions that are not listed earlier in the book. The functions are organized alphabetically. The first two lines of each function are typed in bold letters as a visual cue to facilitate finding the function and reading its summary description.

A

```
function f = adpmedian(g, Smax)
%ADPMEDIAN Perform adaptive median filtering.
%   F = ADPMEDIAN(G, SMAX) performs adaptive median filtering of
%   image G.  The median filter starts at size 3-by-3 and iterates up
%   to size SMAX-by-SMAX.  SMAX must be an odd integer greater than 1.

% SMAX must be an odd, positive integer greater than 1.
if (Smax <= 1) | (Smax/2 == round(Smax/2)) | (Smax ~= round(Smax))
   error('SMAX must be an odd integer > 1.')
end
[M, N] = size(g);

% Initial setup.
f = g;
f(:) = 0;
alreadyProcessed = false(size(g));

% Begin filtering.
for k = 3:2:Smax
   zmin = ordfilt2(g, 1, ones(k, k), 'symmetric');
   zmax = ordfilt2(g, k * k, ones(k, k), 'symmetric');
   zmed = medfilt2(g, [k k], 'symmetric');
```

```
    processUsingLevelB = (zmed > zmin) & (zmax > zmed) & ...
        ~alreadyProcessed;
    zB = (g > zmin) & (zmax > g);
    outputZxy  = processUsingLevelB & zB;
    outputZmed = processUsingLevelB & ~zB;
    f(outputZxy) = g(outputZxy);
    f(outputZmed) = zmed(outputZmed);

    alreadyProcessed = alreadyProcessed | processUsingLevelB;
    if all(alreadyProcessed(:))
        break;
    end
end

% Output zmed for any remaining unprocessed pixels.  Note that this
% zmed was computed using a window of size Smax-by-Smax, which is
% the final value of k in the loop.
f(~alreadyProcessed) = zmed(~alreadyProcessed);
```

B

```
function rc_new = bound2eight(rc)
%BOUND2EIGHT Convert 4-connected boundary to 8-connected boundary.
%   RC_NEW = BOUND2EIGHT(RC) converts a four-connected boundary to an
%   eight-connected boundary.  RC is a P-by-2 matrix, each row of
%   which contains the row and column coordinates of a boundary
%   pixel.  RC must be a closed boundary; in other words, the last
%   row of RC must equal the first row of RC.  BOUND2EIGHT removes
%   boundary pixels that are necessary for four-connectedness but not
%   necessary for eight-connectedness.  RC_NEW is a Q-by-2 matrix,
%   where Q <= P.

if ~isempty(rc) & ~isequal(rc(1, :), rc(end, :))
    error('Expected input boundary to be closed.');
end

if size(rc, 1) <= 3
    % Degenerate case.
    rc_new = rc;
    return;
end

% Remove last row, which equals the first row.
rc_new = rc(1:end - 1, :);

% Remove the middle pixel in four-connected right-angle turns.  We
% can do this in a vectorized fashion, but we can't do it all at
% once. Similar to the way the 'thin' algorithm works in bwmorph,
% we'll remove first the middle pixels in four-connected turns where
% the row and column are both even; then the middle pixels in all
% the remaining four-connected turns where the row is even and the
% column is odd; then again where the row is odd and the column is
% even; and finally where both the row and column are odd.
```

```
remove_locations = compute_remove_locations(rc_new);
field1 = remove_locations & (rem(rc_new(:, 1), 2) == 0) & ...
         (rem(rc_new(:, 2), 2) == 0);
rc_new(field1, :) = [];

remove_locations = compute_remove_locations(rc_new);
field2 = remove_locations & (rem(rc_new(:, 1), 2) == 0) & ...
         (rem(rc_new(:, 2), 2) == 1);
rc_new(field2, :) = [];

remove_locations = compute_remove_locations(rc_new);
field3 = remove_locations & (rem(rc_new(:, 1), 2) == 1) & ...
         (rem(rc_new(:, 2), 2) == 0);
rc_new(field3, :) = [];

remove_locations = compute_remove_locations(rc_new);
field4 = remove_locations & (rem(rc_new(:, 1), 2) == 1) & ...
         (rem(rc_new(:, 2), 2) == 1);
rc_new(field4, :) = [];

% Make the output boundary closed again.
rc_new = [rc_new; rc_new(1, :)];

%-------------------------------------------------------------------%
function remove = compute_remove_locations(rc)

% Circular diff.
d = [rc(2:end, :); rc(1, :)] - rc;

% Dot product of each row of d with the subsequent row of d,
% performed in circular fashion.
d1 = [d(2:end, :); d(1, :)];
dotprod = sum(d .* d1, 2);

% Locations of N, S, E, and W transitions followed by
% a right-angle turn.
remove = ~all(d, 2) & (dotprod == 0);

% But we really want to remove the middle pixel of the turn.
remove = [remove(end, :); remove(1:end - 1, :)];

if ~any(remove)
   done = 1;
else
   idx = find(remove);
   rc(idx(1), :) = [];
end
```

```
function rc_new = bound2four(rc)
%BOUND2FOUR Convert 8-connected boundary to 4-connected boundary.
%   RC_NEW = BOUND2FOUR(RC) converts an eight-connected boundary to a
%   four-connected boundary.  RC is a P-by-2 matrix, each row of
%   which contains the row and column coordinates of a boundary
%   pixel.  BOUND2FOUR inserts new boundary pixels wherever there is
%   a diagonal connection.

if size(rc, 1) > 1
   % Phase 1: remove diagonal turns, one at a time until they are all gone.
```

```
    done = 0;
    rc1 = [rc(end - 1, :); rc];
    while ~done
       d = diff(rc1, 1);
       diagonal_locations = all(d, 2);
       double_diagonals = diagonal_locations(1:end - 1) & ...
          (diff(diagonal_locations, 1) == 0);
       double_diagonal_idx = find(double_diagonals);
       turns = any(d(double_diagonal_idx, :) ~= ...
                  d(double_diagonal_idx + 1, :), 2);
       turns_idx = double_diagonal_idx(turns);
       if isempty(turns_idx)
          done = 1;
       else
          first_turn = turns_idx(1);
          rc1(first_turn + 1, :) = (rc1(first_turn, :) + ...
                                    rc1(first_turn + 2, :)) / 2;
          if first_turn == 1
             rc1(end, :) = rc1(2, :);
          end
       end
    end
    rc1 = rc1(2:end, :);
end

% Phase 2: insert extra pixels where there are diagonal connections.

rowdiff = diff(rc1(:, 1));
coldiff = diff(rc1(:, 2));

diagonal_locations = rowdiff & coldiff;
num_old_pixels = size(rc1, 1);
num_new_pixels = num_old_pixels + sum(diagonal_locations);
rc_new = zeros(num_new_pixels, 2);

% Insert the original values into the proper locations in the new RC
% matrix.
idx = (1:num_old_pixels)' + [0; cumsum(diagonal_locations)];
rc_new(idx, :) = rc1;

% Compute the new pixels to be inserted.
new_pixel_offsets = [0 1; -1 0; 1 0; 0 -1];
offset_codes = 2 * (1 - (coldiff(diagonal_locations) + 1)/2) + ...
    (2 - (rowdiff(diagonal_locations) + 1)/2);
new_pixels = rc1(diagonal_locations, :) + ...
    new_pixel_offsets(offset_codes, :);

% Where do the new pixels go?
insertion_locations = zeros(num_new_pixels, 1);
insertion_locations(idx) = 1;
insertion_locations = ~insertion_locations;

% Insert the new pixels.
rc_new(insertion_locations, :) = new_pixels;
```

```
function B = bound2im(b, M, N, x0, y0)
%BOUND2IM Converts a boundary to an image.
%   B = BOUND2IM(b) converts b, an np-by-2 or 2-by-np array
%   representing the integer coordinates of a boundary, into a binary
%   image with 1s in the locations defined by the coordinates in b
%   and 0s elsewhere.
%
%   B = BOUND2IM(b, M, N) places the boundary approximately centered
%   in an M-by-N image. If any part of the boundary is outside the
%   M-by-N rectangle, an error is issued.
%
%   B = BOUND2IM(b, M, N, X0, Y0) places the boundary in an image of
%   size M-by-N, with the topmost boundary point located at X0 and
%   the leftmost point located at Y0. If the shifted boundary is
%   outside the M-by-N rectangle, an error is issued. X0 and X0 must
%   be positive integers.

[np, nc] = size(b);
if np < nc
   b = b'; % To convert to size np-by-2.
   [np, nc] = size(b);
end

% Make sure the coordinates are integers.
x = round(b(:, 1));
y = round(b(:, 2));

% Set up the default size parameters.
x = x - min(x) + 1;
y = y - min(y) + 1;
B = false(max(x), max(y));
C = max(x) - min(x) + 1;
D = max(y) - min(y) + 1;

if nargin == 1
   % Use the preceding default values.
elseif nargin == 3
   if C > M | D > N
      error('The boundary is outside the M-by-N region.')
   end
   % The image size will be M-by-N. Set up the parameters for this.
   B = false(M, N);
   % Distribute extra rows approx. even between top and bottom.
   NR = round((M - C)/2);
   NC = round((N - D)/2); % The same for columns.
   x = x + NR; % Offset the boundary to new position.
   y = y + NC;
elseif nargin == 5
   if x0 < 0 | y0 < 0
      error('x0 and y0 must be positive integers.')
   end
   x = x + round(x0) - 1;
   y = y + round(y0) - 1;
```

```
   C = C + x0 - 1;
   D = D + y0 - 1;
   if C > M | D > N
      error('The shifted boundary is outside the M-by-N region.')
   end
   B = false(M, N);
else
   error('Incorrect number of inputs.')
end

B(sub2ind(size(B), x, y)) = true;
```

```
function B = boundaries(BW, conn, dir)
%BOUNDARIES Trace object boundaries.
%   B = BOUNDARIES(BW) traces the exterior boundaries of objects in
%   the binary image BW.  B is a P-by-1 cell array, where P is the
%   number of objects in the image. Each cell contains a Q-by-2
%   matrix, each row of which contains the row and column coordinates
%   of a boundary pixel.  Q is the number of boundary pixels for the
%   corresponding object.  Object boundaries are traced in the
%   clockwise direction.
%
%   B = BOUNDARIES(BW, CONN) specifies the connectivity to use when
%   tracing boundaries.  CONN may be either 8 or 4.  The default
%   value for CONN is 8.
%
%   B = BOUNDARIES(BW, CONN, DIR) specifies the direction used for
%   tracing boundaries.  DIR should be either 'cw' (trace boundaries
%   clockwise) or 'ccw' (trace boundaries counterclockwise).  If DIR
%   is omitted BOUNDARIES traces in the clockwise direction.

if nargin < 3
   dir = 'cw';
end

if nargin < 2
   conn = 8;
end

L = bwlabel(BW, conn);

% The number of objects is the maximum value of L.  Initialize the
% cell array B so that each cell initially contains a 0-by-2 matrix.
numObjects = max(L(:));
if numObjects > 0
   B = {zeros(0, 2)};
   B = repmat(B, numObjects, 1);
else
   B = {};
end

% Pad label matrix with zeros.  This lets us write the
% boundary-following loop without worrying about going off the edge
% of the image.
Lp = padarray(L, [1 1], 0, 'both');
```

```
% Compute the linear indexing offsets to take us from a pixel to its
% neighbors.
M = size(Lp, 1);
if conn == 8
   % Order is N NE E SE S SW W NW.
   offsets = [-1, M - 1, M, M + 1, 1, -M + 1, -M, -M-1];
else
   % Order is N E S W.
   offsets = [-1, M, 1, -M];
end

% next_search_direction_lut is a lookup table.  Given the direction
% from pixel k to pixel k+1, what is the direction to start with when
% examining the neighborhood of pixel k+1?
if conn == 8
   next_search_direction_lut = [8 8 2 2 4 4 6 6];
else
   next_search_direction_lut = [4 1 2 3];
end

% next_direction_lut is a lookup table.  Given that we just looked at
% neighbor in a given direction, which neighbor do we look at next?
if conn == 8
   next_direction_lut = [2 3 4 5 6 7 8 1];
else
   next_direction_lut = [2 3 4 1];
end

% Values used for marking the starting and boundary pixels.
START    = -1;
BOUNDARY = -2;

% Initialize scratch space in which to record the boundary pixels as
% well as follow the boundary.
scratch = zeros(100, 1);

% Find candidate starting locations for boundaries.
[rr, cc] = find((Lp(2:end-1, :) > 0) & (Lp(1:end-2, :) == 0));
rr = rr + 1;

for k = 1:length(rr)
   r = rr(k);
   c = cc(k);
   if (Lp(r,c) > 0) & (Lp(r - 1, c) == 0) & isempty(B{Lp(r, c)})
      % We've found the start of the next boundary.  Compute its
      % linear offset, record which boundary it is, mark it, and
      % initialize the counter for the number of boundary pixels.
      idx = (c-1)*size(Lp, 1) + r;
      which = Lp(idx);

      scratch(1) = idx;
      Lp(idx) = START;
      numPixels = 1;
      currentPixel = idx;
      initial_departure_direction = [];
```

```
done = 0;
next_search_direction = 2;
while ~done
   % Find the next boundary pixel.
   direction = next_search_direction;
   found_next_pixel = 0;
   for k = 1:length(offsets)
      neighbor = currentPixel + offsets(direction);
      if Lp(neighbor) ~= 0
         % Found the next boundary pixel.

         if (Lp(currentPixel) == START) & ...
                isempty(initial_departure_direction)
            % We are making the initial departure from
            % the starting pixel.
            initial_departure_direction = direction;

         elseif (Lp(currentPixel) == START) & ...
                (initial_departure_direction == direction)
            % We are about to retrace our path.
            % That means we're done.
            done = 1;
            found_next_pixel = 1;
            break;
         end
         % Take the next step along the boundary.
         next_search_direction = ...
              next_search_direction_lut(direction);
         found_next_pixel = 1;
         numPixels = numPixels + 1;
         if numPixels > size(scratch, 1)
            % Double the scratch space.
            scratch(2*size(scratch, 1)) = 0;
         end
         scratch(numPixels) = neighbor;

         if Lp(neighbor) ~= START
            Lp(neighbor) = BOUNDARY;
         end

         currentPixel = neighbor;
         break;
      end

      direction = next_direction_lut(direction);
   end

   if ~found_next_pixel
      % If there is no next neighbor, the object must just
      % have a single pixel.
      numPixels = 2;
      scratch(2) = scratch(1);
      done = 1;
```

```
            end
         end
         % Convert linear indices to row-column coordinates and save
         % in the output cell array.
         [row, col] = ind2sub(size(Lp), scratch(1:numPixels));
         B{which} = [row - 1, col - 1];
      end
   end

   if strcmp(dir, 'ccw')
      for k = 1:length(B)
         B{k} = B{k}(end:-1:1, :);
      end
   end
```

```
function [s, su] = bsubsamp(b, gridsep)
%BSUBSAMP Subsample a boundary.
%   [S, SU] = BSUBSAMP(B, GRIDSEP) subsamples the boundary B by
%   assigning each of its points to the grid node to which it is
%   closest.  The grid is specified by GRIDSEP, which is the
%   separation in pixels between the grid lines. For example, if
%   GRIDSEP = 2, there are two pixels in between grid lines. So, for
%   instance, the grid points in the first row would be at (1,1),
%   (1,4), (1,6), ..., and similarly in the y direction. The value
%   of GRIDSEP must be an even integer. The boundary is specified by
%   a set of coordinates in the form of an np-by-2 array.  It is
%   assumed that the boundary is one pixel thick.
%
%   Output S is the subsampled boundary. Output SU is normalized so
%   that the grid separation is unity.  This is useful for obtaining
%   the Freeman chain code of the subsampled boundary.

% Check input.
[np, nc] = size(b);
if np < nc
   error('B must be of size np-by-2.');
end
if gridsep/2 ~= round(gridsep/2)
   error('GRIDSEP must be an even integer.')
end

% Some boundary tracing programs, such as boundaries.m, end with
% the beginning, resulting in a sequence in which the coordinates
% of the first and last points are the same. If this is the case
% in b, eliminate the last point.
if isequal(b(1, :), b(np, :))
   np = np - 1;
   b = b(1:np, :);
end

% Find the max x and y spanned by the boundary.
xmax = max(b(:, 1));
ymax = max(b(:, 2));
```

```
% Determine the number of grid lines with gridsep points in
% between them that can fit in the intervals [1,xmax], [1,ymax],
% without any points in b being left over. If points are left
% over, add zeros to extend xmax and ymax so that an integral
% number of grid lines are obtained.
% Size needed in the x-direction:
L = gridsep + 1;
n = ceil(xmax/L);
T = (n - 1)*L + 1;

% Zx is the number of zeros that would be needed to have grid
% lines without any points in b being left over.
Zx = abs(xmax - T - L);
if Zx == L
   Zx = 0;
end
% Number of grid lines in the x-direction, with L pixel spaces
% in between each grid line.
GLx = (xmax + Zx - 1)/L + 1;

% And for the y-direction:
n = ceil(ymax/L);
T = (n - 1)*L + 1;
Zy = abs(ymax - T - L);
if Zy == L
   Zy = 0;
end
GLy = (ymax + Zy - 1)/L + 1;

% Form vectors of x and y grid locations.
I = 1:GLx;
% Vector of grid line locations intersecting x-axis.
X(I) = gridsep*I + (I - gridsep);

J = 1:GLy;
% Vector of grid line locations intersecting y-axis.
Y(J) = gridsep*J + (J - gridsep);

% Compute both components of the cityblock distance between each
% element of b and all the grid-line intersections.  Assign each
% point to the grid location for which each comp of the cityblock
% distance was less than gridsep/2. Because gridsep is an even
% integer, these assignments are unique. Note the use of meshgrid to
% optimize the code.
DIST = gridsep/2;
[XG, YG] = meshgrid(X, Y);
Q = 1;
for k=1:np
   [I,J] = find(abs(XG - b(k, 1)) <= DIST & abs(YG - b(k, 2)) <= ...
               DIST);
   IL = length(I);
   ord = k*ones(IL, 1); % To keep track of order of input coordinates
```

```
   K = Q + IL - 1;
   d1(Q:K, :) = cat(2, X(I), ord);
   d2(Q:K, :) = cat(2, Y(J), ord);
   Q = K + 1;
end

% d is the set of points assigned to the new grid with line
% separation of gridsep. Note that it is formed as d=(d2,d1) to
% compensate for the coordinate transposition inherent in using
% meshgrid (see Chapter 2).
d = cat(2, d2(:, 1), d1); % The second column of d1 is ord.

% Sort the points using the values in ord, which is the last col in
% d.
d = fliplr(d); % So the last column becomes first.
d = sortrows(d);
d = fliplr(d); % Flip back.

% Eliminate duplicate rows in the first two components of
% d to create the output. The cw or ccw order MUST be preserved.
s = d(:, 1:2);
[s, m, n] = unique(s, 'rows');

% Function unique sorts the data--Restore to original order
% by using the contents of m.
s = [s, m];
s = fliplr(s);
s = sortrows(s);
s = fliplr(s);
s = s(:, 1:2);

% Scale to unit grid so that can use directly to obtain Freeman
% chain code.  The shape does not change.
su = round(s./gridsep) + 1;
```

C

```
function image = changeclass(class, varargin)
%CHANGECLASS changes the storage class of an image.
%   I2 = CHANGECLASS(CLASS, I);
%   RGB2 = CHANGECLASS(CLASS, RGB);
%   BW2 = CHANGECLASS(CLASS, BW);
%   X2 = CHANGECLASS(CLASS, X, 'indexed');

%   Copyright 1993-2002 The MathWorks, Inc.  Used with permission.
%   $Revision: 1.2 $  $Date: 2003/02/19 22:09:58 $

switch class
case 'uint8'
   image = im2uint8(varargin{:});
case 'uint16'
   image = im2uint16(varargin{:});
```

```
case 'double'
   image = im2double(varargin{:});
otherwise
   error('Unsupported IPT data class.');
end
```

```
function [VG, A, PPG]= colorgrad(f, T)
%COLORGRAD Computes the vector gradient of an RGB image.
%   [VG, VA, PPG] = COLORGRAD(F, T) computes the vector gradient, VG,
%   and corresponding angle array, VA, (in radians) of RGB image
%   F. It also computes PPG, the per-plane composite gradient
%   obtained by summing the 2-D gradients of the individual color
%   planes. Input T is a threshold in the range [0, 1]. If it is
%   included in the argument list, the values of VG and PPG are
%   thresholded by letting VG(x,y) = 0 for values <= T and VG(x,y) =
%   VG(x,y) otherwise. Similar comments apply to PPG.  If T is not
%   included in the argument list then T is set to 0. Both output
%   gradients are scaled to the range [0, 1].

if (ndims(f) ~= 3) | (size(f, 3) ~= 3)
   error('Input image must be RGB.');
end

% Compute the x and y derivatives of the three component images
% using Sobel operators.
sh = fspecial('sobel');
sv = sh';
Rx = imfilter(double(f(:, :, 1)), sh, 'replicate');
Ry = imfilter(double(f(:, :, 1)), sv, 'replicate');
Gx = imfilter(double(f(:, :, 2)), sh, 'replicate');
Gy = imfilter(double(f(:, :, 2)), sv, 'replicate');
Bx = imfilter(double(f(:, :, 3)), sh, 'replicate');
By = imfilter(double(f(:, :, 3)), sv, 'replicate');

% Compute the parameters of the vector gradient.
gxx = Rx.^2 + Gx.^2 + Bx.^2;
gyy = Ry.^2 + Gy.^2 + By.^2;
gxy = Rx.*Ry + Gx.*Gy + Bx.*By;
A = 0.5*(atan(2*gxy./(gxx - gyy + eps)));
G1 = 0.5*((gxx + gyy) + (gxx - gyy).*cos(2*A) + 2*gxy.*sin(2*A));

% Now repeat for angle + pi/2.  Then select the maximum at each point.
A = A + pi/2;
G2 = 0.5*((gxx + gyy) + (gxx - gyy).*cos(2*A) + 2*gxy.*sin(2*A));
G1 = G1.^0.5;
G2 = G2.^0.5;
% Form VG by picking the maximum at each (x,y) and then scale
% to the range [0, 1].
VG = mat2gray(max(G1, G2));

% Compute the per-plane gradients.
RG = sqrt(Rx.^2 + Ry.^2);
GG = sqrt(Gx.^2 + Gy.^2);
```

```
BG = sqrt(Bx.^2 + By.^2);
% Form the composite by adding the individual results and
% scale to [0, 1].
PPG = mat2gray(RG + GG + BG);

% Threshold the result.
if nargin == 2
   VG = (VG > T).*VG;
   PPG = (PPG > T).*PPG;
end
```

```
function I = colorseg(varargin)
%COLORSEG Performs segmentation of a color image.
%   S = COLORSEG('EUCLIDEAN', F, T, M) performs segmentation of color
%   image F using a Euclidean measure of similarity. M is a 1-by-3
%   vector representing the average color used for segmentation (this
%   is the center of the sphere in Fig. 6.26 of DIPUM). T is the
%   threshold against which the distances are compared.
%
%   S = COLORSEG('MAHALANOBIS', F, T, M, C) performs segmentation of
%   color image F using the Mahalanobis distance as a measure of
%   similarity. C is the 3-by-3 covariance matrix of the sample color
%   vectors of the class of interest. See function covmatrix for the
%   computation of C and M.
%
%   S is the segmented image (a binary matrix) in which 0s denote the
%   background.

% Preliminaries.
% Recall that varargin is a cell array.
f = varargin{2};
if (ndims(f) ~= 3) | (size(f, 3) ~= 3)
   error('Input image must be RGB.');
end
M = size(f, 1); N = size(f, 2);
% Convert f to vector format using function imstack2vectors.
[f, L] = imstack2vectors(f);
f = double(f);
% Initialize I as a column vector.  It will be reshaped later
% into an image.
I = zeros(M*N, 1);
T = varargin{3};
m = varargin{4};
m = m(:)'; % Make sure that m is a row vector.

if length(varargin) == 4
   method = 'euclidean';
elseif length(varargin) == 5
   method = 'mahalanobis';
else
   error('Wrong number of inputs.');
end
```

```
switch method
case 'euclidean'
   % Compute the Euclidean distance between all rows of X and m. See
   % Section 12.2 of DIPUM for an explanation of the following
   % expression. D(i) is the Euclidean distance between vector X(i,:)
   % and vector m.
   p = length(f);
   D = sqrt(sum(abs(f - repmat(m, p, 1)).^2, 2));
case 'mahalanobis'
   C = varargin{5};
   D = mahalanobis(f, C, m);
otherwise
   error('Unknown segmentation method.')
end

% D is a vector of size MN-by-1 containing the distance computations
% from all the color pixels to vector m. Find the distances <= T.
J = find(D <= T);

% Set the values of I(J) to 1.  These are the segmented
% color pixels.
I(J) = 1;

% Reshape I into an M-by-N image.
I = reshape(I, M, N);
```

```
function c = connectpoly(x, y)
%CONNECTPOLY Connects vertices of a polygon.
%   C = CONNECTPOLY(X, Y) connects the points with coordinates given
%   in X and Y with straight lines. These points are assumed to be a
%   sequence of polygon vertices organized in the clockwise or
%   counterclockwise direction. The output, C, is the set of points
%   along the boundary of the polygon in the form of an nr-by-2
%   coordinate sequence in the same direction as the input. The last
%   point in the sequence is equal to the first.

v = [x(:), y(:)];

% Close polygon.
if ~isequal(v(end, :), v(1, :))
   v(end + 1, :) = v(1, :);
end

% Connect vertices.
segments = cell(1, length(v) - 1);
for I = 2:length(v)
   [x, y] = intline(v(I - 1, 1), v(I, 1), v(I - 1, 2), v(I, 2));
   segments{I - 1} = [x, y];
end

c = cat(1, segments{:});
```

D

```
function s = diameter(L)
%DIAMETER Measure diameter and related properties of image regions.
%   S = DIAMETER(L) computes the diameter, the major axis endpoints,
```

```
%   the minor axis endpoints, and the basic rectangle of each labeled
%   region in the label matrix L. Positive integer elements of L
%   correspond to different regions. For example, the set of elements
%   of L equal to 1 corresponds to region 1; the set of elements of L
%   equal to 2 corresponds to region 2; and so on.  S is a structure
%   array of length max(L(:)). The fields of the structure array
%   include:
%
%     Diameter
%     MajorAxis
%     MinorAxis
%     BasicRectangle
%
%   The Diameter field, a scalar, is the maximum distance between any
%   two pixels in the corresponding region.
%
%   The MajorAxis field is a 2-by-2 matrix.  The rows contain the row
%   and column coordinates for the endpoints of the major axis of the
%   corresponding region.
%
%   The MinorAxis field is a 2-by-2 matrix.  The rows contain the row
%   and column coordinates for the endpoints of the minor axis of the
%   corresponding region.
%
%   The BasicRectangle field is a 4-by-2 matrix.  Each row contains
%   the row and column coordinates of a corner of the
%   region-enclosing rectangle defined by the major and minor axes.
%
%   For more information about these measurements, see Section 11.2.1
%   of Digital Image Processing, by Gonzalez and Woods, 2nd edition,
%   Prentice Hall.

s = regionprops(L, {'Image', 'BoundingBox'});

for k = 1:length(s)
    [s(k).Diameter, s(k).MajorAxis, perim_r, perim_c] = ...
        compute_diameter(s(k));
    [s(k).BasicRectangle, s(k).MinorAxis] = ...
        compute_basic_rectangle(s(k), perim_r, perim_c);
end

%-------------------------------------------------------------------%
function [d, majoraxis, r, c] = compute_diameter(s)
%   [D, MAJORAXIS, R, C] = COMPUTE_DIAMETER(S) computes the diameter
%   and major axis for the region represented by the structure S. S
%   must contain the fields Image and BoundingBox.  COMPUTE_DIAMETER
%   also returns the row and column coordinates (R and C) of the
%   perimeter pixels of s.Image.

% Compute row and column coordinates of perimeter pixels.
[r, c] = find(bwperim(s.Image));
r = r(:);
```

```
c = c(:);
[rp, cp] = prune_pixel_list(r, c);

num_pixels = length(rp);
switch num_pixels
case 0
   d = -Inf;
   majoraxis = ones(2, 2);

case 1
   d = 0;
   majoraxis = [rp cp; rp cp];

case 2
   d = (rp(2) - rp(1))^2 + (cp(2) - cp(1))^2;
   majoraxis = [rp cp];

otherwise
   % Generate all combinations of 1:num_pixels taken two at at time.
   % Method suggested by Peter Acklam.
   [idx(:, 2) idx(:, 1)] = find(tril(ones(num_pixels), -1));
   rr = rp(idx);
   cc = cp(idx);

   dist_squared = (rr(:, 1) - rr(:, 2)).^2 + ...
      (cc(:, 1) - cc(:, 2)).^2;
   [max_dist_squared, idx] = max(dist_squared);
   majoraxis = [rr(idx,:)' cc(idx,:)'];

   d = sqrt(max_dist_squared);

   upper_image_row = s.BoundingBox(2) + 0.5;
   left_image_col = s.BoundingBox(1) + 0.5;

   majoraxis(:, 1) = majoraxis(:, 1) + upper_image_row - 1;
   majoraxis(:, 2) = majoraxis(:, 2) + left_image_col - 1;
end
%-------------------------------------------------------------------%
function [basicrect, minoraxis] = compute_basic_rectangle(s, ...
                                                perim_r, perim_c)
%   [BASICRECT,MINORAXIS] = COMPUTE_BASIC_RECTANGLE(S, PERIM_R,
%   PERIM_C) computes the basic rectangle and the minor axis
%   end-points for the region represented by the structure S.  S must
%   contain the fields Image, BoundingBox, MajorAxis, and
%   Diameter. PERIM_R and PERIM_C are the row and column coordinates
%   of perimeter of s.Image. BASICRECT is a 4-by-2 matrix, each row
%   of which contains the row and column coordinates of one corner of
%   the basic rectangle.

% Compute the orientation of the major axis.
theta = atan2(s.MajorAxis(2, 1) - s.MajorAxis(1, 1), ...
              s.MajorAxis(2, 2) - s.MajorAxis(1, 2));

% Form rotation matrix.
T = [cos(theta) sin(theta); -sin(theta) cos(theta)];
```

```
% Rotate perimeter pixels.
p = [perim_c perim_r];
p = p * T';

% Calculate minimum and maximum x- and y-coordinates for the rotated
% perimeter pixels.
x = p(:, 1);
y = p(:, 2);
min_x = min(x);
max_x = max(x);
min_y = min(y);
max_y = max(y);

corners_x = [min_x max_x max_x min_x]';
corners_y = [min_y min_y max_y max_y]';

% Rotate corners of the basic rectangle.
corners = [corners_x corners_y] * T;

% Translate according to the region's bounding box.
upper_image_row = s.BoundingBox(2) + 0.5;
left_image_col = s.BoundingBox(1) + 0.5;

basicrect = [corners(:, 2) + upper_image_row - 1, ...
             corners(:, 1) + left_image_col - 1];

% Compute minor axis end-points, rotated.
x = (min_x + max_x) / 2;
y1 = min_y;
y2 = max_y;
endpoints = [x y1; x y2];

% Rotate minor axis end-points back.
endpoints = endpoints * T;

% Translate according to the region's bounding box.
minoraxis = [endpoints(:, 2) + upper_image_row - 1, ...
             endpoints(:, 1) + left_image_col - 1];

%-------------------------------------------------------------------%
function [r, c] = prune_pixel_list(r, c)
%   [R, C] = PRUNE_PIXEL_LIST(R, C) removes pixels from the vectors
%   R and C that cannot be endpoints of the major axis.  This
%   elimination is based on geometrical constraints described in
%   Russ, Image Processing Handbook, Chapter 8.

top = min(r);
bottom = max(r);
left = min(c);
right = max(c);

% Which points are inside the upper circle?
x = (left + right)/2;
y = top;
radius = bottom - top;
inside_upper = ( (c - x).^2 + (r - y).^2 ) < radius^2;
```

```
% Which points are inside the lower circle?
y = bottom;
inside_lower = ( (c - x).^2 + (r - y).^2 ) < radius^2;

% Which points are inside the left circle?
x = left;
y = (top + bottom)/2;
radius = right - left;
inside_left = ( (c - x).^2 + (r - y).^2 ) < radius^2;

% Which points are inside the right circle?
x = right;
inside_right = ( (c - x).^2 + (r - y).^2 ) < radius^2;

% Eliminate points that are inside all four circles.
delete_idx = find(inside_left & inside_right & ...
                  inside_upper & inside_lower);

r(delete_idx) = [];
c(delete_idx) = [];
```

F

```
function  c = fchcode(b, conn, dir)
%FCHCODE Computes the Freeman chain code of a boundary.
%   C = FCHCODE(B) computes the 8-connected Freeman chain code of a
%   set of 2-D coordinate pairs contained in B, an np-by-2 array. C
%   is a structure with the following fields:
%
%      c.fcc    = Freeman chain code (1-by-np)
%      c.diff   = First difference of code c.fcc (1-by-np)
%      c.mm     = Integer of minimum magnitude from c.fcc (1-by-np)
%      c.diffmm = First difference of code c.mm (1-by-np)
%      c.x0y0   = Coordinates where the code starts (1-by-2)
%
%   C = FCHCODE(B, CONN) produces the same outputs as above, but
%   with the code connectivity specified in CONN. CONN can be 8 for
%   an 8-connected chain code, or CONN can be 4 for a 4-connected
%   chain code. Specifying CONN=4 is valid only if the input
%   sequence, B, contains transitions with values 0, 2, 4, and 6,
%   exclusively.
%
%   C = FHCODE(B, CONN, DIR) produces the same outputs as above, but,
%   in addition, the desired code direction is specified. Values for
%   DIR can be:
%
%      'same'     Same as the order of the sequence of points in b.
%                  This is the default.
%
%      'reverse'  Outputs the code in the direction opposite to the
%                  direction of the points in B.  The starting point
%                  for each DIR is the same.
```

```
%
%   The elements of B are assumed to correspond to a 1-pixel-thick,
%   fully-connected, closed boundary. B cannot contain duplicate
%   coordinate pairs, except in the first and last positions, which
%   is a common feature of boundary tracing programs.
%
%   FREEMAN CHAIN CODE REPRESENTATION
%   The table on the left shows the 8-connected Freeman chain codes
%   corresponding to allowed deltax, deltay pairs. An 8-chain is
%   converted to a 4-chain if (1) if conn = 4; and (2) only
%   transitions 0, 2, 4, and 6 occur in the 8-code.  Note that
%   dividing 0, 2, 4, and 6 by 2 produce the 4-code.
%
%      -----------------------  ----------------
%      deltax | deltay | 8-code  corresp 4-code
%      -----------------------  ----------------
%        0       1       0           0
%       -1       1       1
%       -1       0       2           1
%       -1      -1       3
%        0      -1       4           2
%        1      -1       5
%        1       0       6           3
%        1       1       7
%      -----------------------  ----------------
%
%   The formula z = 4*(deltax + 2) + (deltay + 2) gives the following
%   sequence corresponding to rows 1-8 in the preceding table: z =
%   11,7,6,5,9,13,14,15. These values can be used as indices into the
%   table, improving the speed of computing the chain code. The
%   preceding formula is not unique, but it is based on the smallest
%   integers (4 and 2) that are powers of 2.

% Preliminaries.
if nargin == 1
   dir = 'same';
   conn = 8;
elseif nargin == 2
   dir = 'same';
elseif nargin == 3
   % Nothing to do here.
else
   error('Incorrect number of inputs.')
end
[np, nc] = size(b);
if np < nc
   error('B must be of size np-by-2.');
end

% Some boundary tracing programs, such as boundaries.m, output a
% sequence in which the coordinates of the first and last points are
```

```
% the same. If this is the case, eliminate the last point.
if isequal(b(1, :), b(np, :))
   np = np - 1;
   b = b(1:np, :);
end

% Build the code table using the single indices from the formula
% for z given above:
C(11)=0; C(7)=1; C(6)=2; C(5)=3; C(9)=4;
C(13)=5; C(14)=6; C(15)=7;

% End of Preliminaries.

% Begin processing.
x0 = b(1, 1);
y0 = b(1, 2);
c.x0y0 = [x0, y0];

% Make sure the coordinates are organized sequentially:
% Get the deltax and deltay between successive points in b. The
% last row of a is the first row of b.
a = circshift(b, [-1, 0]);

% DEL = a - b is an nr-by-2 matrix in which the rows contain the
% deltax and deltay between successive points in b. The two
% components in the kth row of matrix DEL are deltax and deltay
% between point (xk, yk) and (xk+1, yk+1).  The last row of DEL
% contains the deltax and deltay between (xnr, ynr) and (x1, y1),
% (i.e., between the last and first points in b).
DEL = a - b;

% If the abs value of either (or both) components of a pair
% (deltax, deltay) is greater than 1, then by definition the curve
% is broken (or the points are out of order), and the program
% terminates.
if any(abs(DEL(:, 1)) > 1) | any(abs(DEL(:, 2)) > 1);
   error('The input curve is broken or points are out of order.')
end

% Create a single index vector using the formula described above.
z = 4*(DEL(:, 1) + 2) + (DEL(:, 2) + 2);

% Use the index to map into the table. The following are
% the Freeman 8-chain codes, organized in a 1-by-np array.
fcc = C(z);

% Check if direction of code sequence needs to be reversed.
if strcmp(dir, 'reverse')
   fcc = coderev(fcc); % See below for function coderev.
end

% If 4-connectivity is specified, check that all components
% of fcc are 0, 2, 4, or 6.
if conn == 4
   val = find(fcc == 1 | fcc == 3 | fcc == 5 | fcc ==7 );
   if isempty(val)
```

```
        fcc = fcc./2;
    else
        warning('The specified 4-connected code cannot be satisfied.')
    end
end

% Freeman chain code for structure output.
c.fcc = fcc;

% Obtain the first difference of fcc.
c.diff = codediff(fcc,conn); % See below for function codediff.

% Obtain code of the integer of minimum magnitude.
c.mm = minmag(fcc); % See below for function minmag.

% Obtain the first difference of fcc
c.diffmm = codediff(c.mm, conn);

%-------------------------------------------------------------------%
function cr = coderev(fcc)
%   Traverses the sequence of 8-connected Freeman chain code fcc in
%   the opposite direction, changing the values of each code
%   segment. The starting point is not changed. fcc is a 1-by-np
%   array.

% Flip the array left to right.  This redefines the starting point
% as the last point and reverses the order of "travel" through the
% code.
cr = fliplr(fcc);

% Next, obtain the new code values by traversing the code in the
% opposite direction. (0 becomes 4, 1 becomes 5, ... , 5 becomes 1,
% 6 becomes 2, and 7 becomes 3).
ind1 = find(0 <= cr & cr <= 3);
ind2 = find(4 <= cr & cr <= 7);
cr(ind1) = cr(ind1) + 4;
cr(ind2) = cr(ind2) - 4;

%-------------------------------------------------------------------%
function z = minmag(c)
%MINMAG Finds the integer of minimum magnitude in a chain code.
%   Z = MINMAG(C) finds the integer of minimum magnitude in a given
%   4- or 8-connected Freeman chain code, C. The code is assumed to
%   be a 1-by-np array.

% The integer of minimum magnitude starts with min(c), but there
% may be more than one such value. Find them all,
I = find(c == min(c));
% and shift each one left so that it starts with min(c).
J = 0;
A = zeros(length(I), length(c));
for k = I;
    J = J + 1;
    A(J, :) = circshift(c,[0 -(k-1)]);
end
```

```
% Matrix A contains all the possible candidates for the integer of
% minimum magnitude. Starting with the 2nd column, succesively find
% the minima in each column of A. The number of candidates decreases
% as the seach moves to the right on A.  This is reflected in the
% elements of J.  When length(J)=1, one candidate remains.  This is
% the integer of minimum magnitude.
[M, N] = size(A);
J = (1:M)';
for k = 2:N
   D(1:M, 1) = Inf;
   D(J, 1) = A(J, k);
   amin = min(A(J, k));
   J = find(D(:, 1) == amin);
   if length(J)==1
      z = A(J, :);
      return
   end
end
%-------------------------------------------------------------------%
function d = codediff(fcc, conn)
%CODEDIFF Computes the first difference of a chain code.
%   D = CODEDIFF(FCC) computes the first difference of code, FCC. The
%   code FCC is treated as a circular sequence, so the last element
%   of D is the difference between the last and first elements of
%   FCC.  The input code is a 1-by-np vector.
%
%   The first difference is found by counting the number of direction
%   changes (in a counter-clockwise direction) that separate two
%   adjacent elements of the code.

sr = circshift(fcc, [0, -1]); % Shift input left by 1 location.
delta = sr - fcc;
d = delta;
I = find(delta < 0);

type = conn;
switch type
case 4 % Code is 4-connected
   d(I) = d(I) + 4;
case 8 % Code is 8-connected
   d(I) = d(I) + 8;
end
```

G

```
function g = gscale(f, varargin)
%GSCALE Scales the intensity of the input image.
%   G = GSCALE(F, 'full8') scales the intensities of F to the full
%   8-bit intensity range [0, 255].  This is the default if there is
```

```
%   only one input argument.
%
%   G = GSCALE(F, 'full16') scales the intensities of F to the full
%   16-bit intensity range [0, 65535].
%
%   G = GSCALE(F, 'minmax', LOW, HIGH) scales the intensities of F to
%   the range [LOW, HIGH]. These values must be provided, and they
%   must be in the range [0, 1], independently of the class of the
%   input. GSCALE performs any necessary scaling. If the input is of
%   class double, and its values are not in the range [0, 1], then
%   GSCALE scales it to this range before processing.
%
%   The class of the output is the same as the class of the input.

if length(varargin) == 0 % If only one argument it must be f.
   method = 'full8';
else
   method = varargin{1};
end

if strcmp(class(f), 'double') & (max(f(:)) > 1 | min(f(:)) < 0)
   f = mat2gray(f);
end

% Perform the specified scaling.
switch method
case 'full8'
   g = im2uint8(mat2gray(double(f)));
case 'full16'
   g = im2uint16(mat2gray(double(f)));
case 'minmax'
   low = varargin{2}; high = varargin{3};
   if low > 1 | low < 0 | high > 1 | high < 0
      error('Parameters low and high must be in the range [0, 1].')
   end
   if strcmp(class(f), 'double')
      low_in = min(f(:));
      high_in = max(f(:));
   elseif strcmp(class(f), 'uint8')
      low_in = double(min(f(:)))./255;
      high_in = double(max(f(:)))./255;
   elseif strcmp(class(f), 'uint16')
      low_in = double(min(f(:)))./65535;
      high_in = double(max(f(:)))./65535;
   end
   % imadjust automatically matches the class of the input.
   g = imadjust(f, [low_in high_in], [low high]);
otherwise
   error('Unknown method.')
end
```

I

```
function [X, R] = imstack2vectors(S, MASK)
%IMSTACK2VECTORS Extracts vectors from an image stack.
%   [X, R] = imstack2vectors(S, MASK) extracts vectors from S, which
%   is an M-by-N-by-n stack array of n registered images of size
%   M-by-N each (see Fig. 11.24). The extracted vectors are arranged
%   as the rows of array X. Input MASK is an M-by-N logical or
%   numeric image with nonzero values (1s if it is a logical array)
%   in the locations where elements of S are to be used in forming X
%   and 0s in locations to be ignored. The number of row vectors in X
%   is equal to the number of nonzero elements of MASK. If MASK is
%   omitted, all M*N locations are used in forming X.  A simple way to
%   obtain MASK interactively is to use function roipoly. Finally, R
%   is an array whose rows are the 2-D coordinates containing the
%   region locations in MASK from which the vectors in S were
%   extracted to form X.

% Preliminaries.
[M, N, n] = size(S);
if nargin == 1
   MASK = true(M, N);
else
   MASK = MASK ~= 0;
end

% Find the set of locations where the vectors will be kept before
% MASK is changed later in the program.
[I, J] = find(MASK);
R = [I, J];

% Now find X.

% First reshape S into X by turning each set of n values along the third
% dimension of S so that it becomes a row of X. The order is from top to
% bottom along the first column, the second column, and so on.
Q = M*N;
X = reshape(S, Q, n);

% Now reshape MASK so that it corresponds to the right locations
% vertically along the elements of X.
MASK = reshape(MASK, Q, 1);

% Keep the rows of X at locations where MASK is not 0.
X = X(MASK, :);

function [x, y] = intline(x1, x2, y1, y2)
%INTLINE Integer-coordinate line drawing algorithm.
%   [X, Y] = INTLINE(X1, X2, Y1, Y2) computes an
%   approximation to the line segment joining (X1, Y1) and
%   (X2, Y2) with integer coordinates.  X1, X2, Y1, and Y2
%   should be integers.  INTLINE is reversible; that is,
%   INTLINE(X1, X2, Y1, Y2) produces the same results as
%   FLIPUD(INTLINE(X2, X1, Y2, Y1)).
```

```
%   Copyright 1993-2002 The MathWorks, Inc. Used with permission.
%   $Revision: 5.11 $  $Date: 2002/03/15 15:57:47 $

dx = abs(x2 - x1);
dy = abs(y2 - y1);

% Check for degenerate case.
if ((dx == 0) & (dy == 0))
   x = x1;
   y = y1;
   return;
end

flip = 0;
if (dx >= dy)
   if (x1 > x2)
      % Always "draw" from left to right.
      t = x1; x1 = x2; x2 = t;
      t = y1; y1 = y2; y2 = t;
      flip = 1;
   end
   m = (y2 - y1)/(x2 - x1);
   x = (x1:x2).';
   y = round(y1 + m*(x - x1));
else
   if (y1 > y2)
      % Always "draw" from bottom to top.
      t = x1; x1 = x2; x2 = t;
      t = y1; y1 = y2; y2 = t;
      flip = 1;
   end
   m = (x2 - x1)/(y2 - y1);
   y = (y1:y2).';
   x = round(x1 + m*(y - y1));
end

if (flip)
   x = flipud(x);
   y = flipud(y);
end
```

```
function phi = invmoments(F)
%INVMOMENTS Compute invariant moments of image.
%   PHI = INVMOMENTS(F) computes the moment invariants of the image
%   F. PHI is a seven-element row vector containing the moment
%   invariants as defined in equations (11.3-17) through (11.3-23) of
%   Gonzalez and Woods, Digital Image Processing, 2nd Ed.
%
%   F must be a 2-D, real, nonsparse, numeric or logical matrix.

if (ndims(F) ~= 2) | issparse(F) | ~isreal(F) | ~(isnumeric(F) | ...
                                                  islogical(F))
   error(['F must be a 2-D, real, nonsparse, numeric or logical '...
          'matrix.']);
end
```

```
F = double(F);

phi = compute_phi(compute_eta(compute_m(F)));

%-------------------------------------------------------------------%
function m = compute_m(F)

[M, N] = size(F);
[x, y] = meshgrid(1:N, 1:M);

% Turn x, y, and F into column vectors to make the summations a bit
% easier to compute in the following.
x = x(:);
y = y(:);
F = F(:);

% DIP equation (11.3-12)
m.m00 = sum(F);
% Protect against divide-by-zero warnings.
if (m.m00 == 0)
   m.m00 = eps;
end
% The other central moments:
m.m10 = sum(x .* F);
m.m01 = sum(y .* F);
m.m11 = sum(x .* y .* F);
m.m20 = sum(x.^2 .* F);
m.m02 = sum(y.^2 .* F);
m.m30 = sum(x.^3 .* F);
m.m03 = sum(y.^3 .* F);
m.m12 = sum(x .* y.^2 .* F);
m.m21 = sum(x.^2 .* y .* F);

%-------------------------------------------------------------------%
function e = compute_eta(m)

% DIP equations (11.3-14) through (11.3-16).

xbar = m.m10 / m.m00;
ybar = m.m01 / m.m00;

e.eta11 = (m.m11 - ybar*m.m10) / m.m00^2;
e.eta20 = (m.m20 - xbar*m.m10) / m.m00^2;
e.eta02 = (m.m02 - ybar*m.m01) / m.m00^2;
e.eta30 = (m.m30 - 3 * xbar * m.m20 + 2 * xbar^2 * m.m10) / m.m00^2.5;
e.eta03 = (m.m03 - 3 * ybar * m.m02 + 2 * ybar^2 * m.m01) / m.m00^2.5;
e.eta21 = (m.m21 - 2 * xbar * m.m11 - ybar * m.m20 + ...
          2 * xbar^2 * m.m01) / m.m00^2.5;
e.eta12 = (m.m12 - 2 * ybar * m.m11 - xbar * m.m02 + ...
          2 * ybar^2 * m.m10) / m.m00^2.5;

%-------------------------------------------------------------------%
function phi = compute_phi(e)

% DIP equations (11.3-17) through (11.3-23).
```

```
phi(1) = e.eta20 + e.eta02;
phi(2) = (e.eta20 - e.eta02)^2 + 4*e.eta11^2;
phi(3) = (e.eta30 - 3*e.eta12)^2 + (3*e.eta21 - e.eta03)^2;
phi(4) = (e.eta30 + e.eta12)^2 + (e.eta21 + e.eta03)^2;
phi(5) = (e.eta30 - 3*e.eta12) * (e.eta30 + e.eta12) * ...
         ( (e.eta30 + e.eta12)^2 - 3*(e.eta21 + e.eta03)^2 ) + ...
         (3*e.eta21 - e.eta03) * (e.eta21 + e.eta03) * ...
         ( 3*(e.eta30 + e.eta12)^2 - (e.eta21 + e.eta03)^2 );
phi(6) = (e.eta20 - e.eta02) * ( (e.eta30 + e.eta12)^2 - ...
                                 (e.eta21 + e.eta03)^2 ) + ...
         4 * e.eta11 * (e.eta30 + e.eta12) * (e.eta21 + e.eta03);
phi(7) = (3*e.eta21 - e.eta03) * (e.eta30 + e.eta12) * ...
         ( (e.eta30 + e.eta12)^2 - 3*(e.eta21 + e.eta03)^2 ) + ...
         (3*e.eta12 - e.eta30) * (e.eta21 + e.eta03) * ...
         ( 3*(e.eta30 + e.eta12)^2 - (e.eta21 + e.eta03)^2 );
```

M

```
function [x, y] = minperpoly(B, cellsize)
%MINPERPOLY Computes the minimum perimeter polygon.
%   [X, Y] = MINPERPOLY(F, CELLSIZE) computes the vertices in [X, Y]
%   of the minimum perimeter polygon of a single binary region or
%   boundary in image B. The procedure is based on Slansky's
%   shrinking rubber band approach. Parameter CELLSIZE determines the
%   size of the square cells that enclose the boundary of the region
%   in B. CELLSIZE must be a nonzero integer greater than 1.
%
%   The algorithm is applicable only to boundaries that are not
%   self-intersecting and that do not have one-pixel-thick
%   protrusions.

if cellsize <= 1
   error('CELLSIZE must be an integer > 1.');
end

% Fill B in case the input was provided as a boundary. Later
% the boundary will be extracted with 4-connectivity, which
% is required by the algorithm. The use of bwperim assures
% that 4-connectivity is preserved at this point.
B = imfill(B, 'holes');
B = bwperim(B);
[M, N] = size(B);

% Increase image size so that the image is of size K-by-K
% with (a) K >= max(M,N) and (b)  K/cellsize = a power of 2.
K = nextpow2(max(M, N)/cellsize);
K = (2^K)*cellsize;

% Increase image size to nearest integer power of 2, by
% appending zeros to the end of the image. This will allow
% quadtree decompositions as small as cells of size 2-by-2,
```

```
% which is the smallest allowed value of cellsize.
M = K - M;
N = K - N;
B = padarray(B, [M N], 'post'); % f is now of size K-by-K

% Quadtree decomposition.
Q = qtdecomp(B, 0, cellsize);

% Get all the subimages of size cellsize-by-cellsize.
[vals, r, c] = qtgetblk(B, Q, cellsize);

% Get all the subimages that contain at least one black
% pixel. These are the cells of the wall enclosing the boundary.
I = find(sum(sum(vals(:, :, :)) >= 1));
x = r(I);
y = c(I);

% [x', y'] is a length(I)-by-2 array.  Each member of this array is
% the left, top corner of a black cell of size cellsize-by-cellsize.
% Fill the cells with black to form a closed border of black cells
% around interior points. These cells are the cellular complex.
for k = 1:length(I)
   B(x(k):x(k) + cellsize-1, y(k):y(k) + cellsize-1) = 1;
end

BF = imfill(B, 'holes');

% Extract the points interior to the black border. This is the region
% of interest around which the MPP will be found.
B = BF & (~B);

% Extract the 4-connected boundary.
B = boundaries(B, 4, 'cw');
% Find the largest one in case of parasitic regions.
J = cellfun('length', B);
I = find(J == max(J));
B = B{I(1)};

% Function boundaries outputs the last coordinate pair equal to the
% first.  Delete it.
B = B(1:end-1,:);

% Obtain the xy coordinates of the boundary.
x = B(:, 1);
y = B(:, 2);

% Find the smallest x-coordinate and corresponding
% smallest y-coordinate.
cx = find(x == min(x));
cy = find(y == min(y(cx)));

% The cell with top leftmost corner at (x1, y1) below is the first
% point considered by the algorithm.  The remaining points are
% visited in the clockwise direction starting at (x1, y1).
x1 = x(cx(1));
y1 = y(cy(1));
```

```
% Scroll data so that the first point is (x1, y1).
I = find(x == x1 & y == y1);
x = circshift(x, [-(I - 1), 0]);
y = circshift(y, [-(I - 1), 0]);

% The same shift applies to B.
B = circshift(B, [-(I - 1), 0]);

% Get the Freeman chain code.  The first row of B is the required
% starting point. The first element of the code is the transition
% between the 1st and 2nd element of B, the second element of
% the code is the transition between the 2nd and 3rd elements of B,
% and so on.  The last element of the code is the transition between
% the last and 1st elements of B. The elements of B form a cw
% sequence (see above), so we use 'same' for the direction in
% function fchcode.
code = fchcode(B, 4, 'same');
code = code.fcc;

% Follow the code sequence to extract the Black Dots, BD, (convex
% corners) and White Dots, WD, (concave corners). The transitions are
% as follows: 0-to-1=WD; 0-to-3=BD; 1-to-0=BD; 1-to-2=WD; 2-to-1=BD;
% 2-to-3=WD; 3-to-0=WD; 3-to-2=dot.  The formula t = 2*first - second
% gives the following unique values for these transitions: -1, -3, 2,
% 0, 3, 1, 6, 4.  These are applicable to travel in the cw direction.
% The WD's are displaced one-half a diagonal from the BD's to form
% the half-cell expansion required in the algorithm.

% Vertices will be computed as array "vertices" of dimension nv-by-3,
% where nv is the number of vertices. The first two elements of any
% row of array vertices are the (x,y) coordinates of the vertex
% corresponding to that row, and the third element is 1 if the
% vertex is convex (BD) or 2 if it is concave (WD). The first vertex
% is known to be convex, so it is black.
vertices = [x1, y1, 1];
n = 1;
k = 1;
for k = 2:length(code)
   if code(k - 1) ~= code(k)
      n = n + 1;
      t = 2*code(k-1) - code(k); % t = value of formula.
      if t == -3 | t == 2 | t == 3 | t == 4 % Convex: Black Dots.
         vertices(n, 1:3) = [x(k), y(k), 1];
      elseif t == -1 | t == 0 | t == 1 | t == 6 % Concave: White Dots.
         if t == -1
            vertices(n, 1:3) = [x(k) - cellsize, y(k) - cellsize,2];
         elseif t==0
            vertices(n, 1:3) = [x(k) + cellsize, y(k) - cellsize,2];
         elseif t==1
            vertices(n, 1:3) = [x(k) + cellsize, y(k) + cellsize,2];
         else
            vertices(n, 1:3) = [x(k) - cellsize, y(k) + cellsize,2];
         end
```

```
        else
            % Nothing to do here.
        end
    end
end

% The rest of minperpoly.m processes the vertices to
% arrive at the MPP.

flag = 1;
while flag
    % Determine which vertices lie on or inside the
    % polygon whose vertices are the Black Dots. Delete all
    % other points.
    I = find(vertices(:, 3) == 1);
    xv = vertices(I, 1); % Coordinates of the Black Dots.
    yv = vertices(I, 2);
    X = vertices(:, 1); % Coordinates of all vertices.
    Y = vertices(:, 2);
    IN = inpolygon(X, Y, xv, yv);
    I = find(IN ~= 0);
    vertices = vertices(I, :);

    % Now check for any Black Dots that may have been turned into
    % concave vertices after the previous deletion step. Delete
    % any such Black Dots and recompute the polygon as in the
    % previous section of code. When no more changes occur, set
    % flag to 0, which causes the loop to terminate.
    x = vertices(:, 1);
    y = vertices(:, 2);
    angles = polyangles(x, y); % Find all the interior angles.
    I = find(angles > 180 & vertices(:, 3) ==1);
    if isempty(I)
        flag = 0;
    else
        J = 1:length(vertices);
        for k = 1:length(I)
            K = find(J ~= I(k));
            J = J(K);
        end
        vertices = vertices(J, :);
    end
end

% Final pass to delete the vertices with angles of 180 degrees.
x = vertices(:, 1);
y = vertices(:, 2);
angles = polyangles(x, y);
I = find(angles ~= 180);

% Vertices of the MPP:
x = vertices(I, 1);
y = vertices(I, 2);
```

P

```
function B = pixeldup(A, m, n)
%PIXELDUP Duplicates pixels of an image in both directions.
%   B = PIXELDUP(A, M, N) duplicates each pixel of A M times in the
%   vertical direction and N times in the horizontal direction.
%   Parameters M and N must be integers.  If N is not included, it
%   defaults to M.

% Check inputs.
if nargin < 2
   error('At least two inputs are required.');
end
if nargin == 2
   n = m;
end

% Generate a vector with elements 1:size(A, 1).
u = 1:size(A, 1);

% Duplicate each element of the vector m times.
m = round(m); % Protect against nonintergers.
u = u(ones(1, m), :);
u = u(:);

% Now repeat for the other direction.
v = 1:size(A, 2);
n = round(n);
v = v(ones(1, n), :);
v = v(:);
B = A(u, v);
```

```
function angles = polyangles(x, y)
%POLYANGLES Computes internal polygon angles.
%   ANGLES = POLYANGLES(X, Y) computes the interior angles (in
%   degrees) of an arbitrary polygon whose vertices are given in
%   [X, Y], ordered in a clockwise manner.  The program eliminates
%   duplicate adjacent rows in [X Y], except that the first row may
%   equal the last, so that the polygon is closed.

% Preliminaries.
[x y] = dupgone(x, y); % Eliminate duplicate vertices.
xy = [x(:) y(:)];
if isempty(xy)
   % No vertices!
   angles = zeros(0, 1);
   return;
end
if size(xy, 1) == 1 | ~isequal(xy(1, :), xy(end, :))
   % Close the polygon
   xy(end + 1, :) = xy(1, :);
end

% Precompute some quantities.
d = diff(xy, 1);
```

```
v1 = -d(1:end, :);
v2 = [d(2:end, :); d(1, :)];
v1_dot_v2 = sum(v1 .* v2, 2);
mag_v1 = sqrt(sum(v1.^2, 2));
mag_v2 = sqrt(sum(v2.^2, 2));

% Protect against nearly duplicate vertices; output angle will be 90
% degrees for such cases. The "real" further protects against
% possible small imaginary angle components in those cases.
mag_v1(~mag_v1) = eps;
mag_v2(~mag_v2) = eps;
angles = real(acos(v1_dot_v2 ./ mag_v1 ./ mag_v2) * 180 / pi);

% The first angle computed was for the second vertex, and the
% last was for the first vertex. Scroll one position down to
% make the last vertex be the first.
angles = circshift(angles, [1, 0]);

% Now determine if any vertices are concave and adjust the angles
% accordingly.
sgn = convex_angle_test(xy);

% Any element of sgn that's -1 indicates that the angle is
% concave. The corresponding angles have to be subtracted
% from 360.
I = find(sgn == -1);
angles(I) = 360 - angles(I);

%-------------------------------------------------------------------%
function sgn = convex_angle_test(xy)
%   The rows of array xy are ordered vertices of a polygon. If the
%   kth angle is convex (>0 and <= 180 degress) then sgn(k) =
%   1. Otherwise sgn(k) = -1. This function assumes that the first
%   vertex in the list is convex, and that no other vertex has a
%   smaller value of x-coordinate. These two conditions are true in
%   the first vertex generated by the MPP algorithm. Also the
%   vertices are assumed to be ordered in a clockwise sequence, and
%   there can be no duplicate vertices.
%
%   The test is based on the fact that every convex vertex is on the
%   positive side of the line passing through the two vertices
%   immediately following each vertex being considered.  If a vertex
%   is concave then it lies on the negative side of the line joining
%   the next two vertices. This property is true also if positive and
%   negative are interchanged in the preceding two sentences.

% It is assumed that the polygon is closed.  If not, close it.
if size(xy, 1) == 1 | ~isequal(xy(1, :), xy(end, :))
   xy(end + 1, :) = xy(1, :);
end

% Sign convention: sgn = 1 for convex vertices (i.e, interior angle > 0
% and <= 180 degrees), sgn = -1 for concave vertices.
```

```
% Extreme points to be used in the following loop.  A 1 is appended
% to perform the inner (dot) product with w, which is 1-by-3 (see
% below).
L = 10^25;
top_left = [-L, -L, 1];
top_right = [-L, L, 1];
bottom_left = [L, -L, 1];
bottom_right = [L, L, 1];

sgn = 1; % The first vertex is known to be convex.

% Start following the vertices.
for k = 2:length(xy) - 1
   pfirst= xy(k - 1, :);
   psecond = xy(k, :); % This is the point tested for convexity.
   pthird = xy(k + 1, :);
   % Get the coefficients of the line (polygon edge) passing
   % through pfirst and psecond.
   w = polyedge(pfirst, psecond);

   % Establish the positive side of the line w1x + w2y + w3 = 0.
   % The positive side of the line should be in the right side of the
   % vector (psecond - pfirst).  deltax and deltay of this vector
   % give the direction of travel. This establishes which of the
   % extreme points (see above) should be on the + side. If that
   % point is on the negative side of the line, then w is replaced by -w.

   deltax = psecond(:, 1) - pfirst(:, 1);
   deltay = psecond(:, 2) - pfirst(:, 2);
   if deltax == 0 & deltay == 0
      error('Data into convexity test is 0 or duplicated.')
   end
   if deltax <= 0  & deltay >= 0 % Bottom_right should be on + side.
      vector_product = dot(w, bottom_right); % Inner product.
      w = sign(vector_product)*w;
   elseif deltax <= 0 & deltay <= 0 % Top_right should be on + side.
      vector_product = dot(w, top_right);
      w = sign(vector_product)*w;
   elseif deltax >= 0 & deltay <= 0  % Top_left should be on + side.
      vector_product = dot(w, top_left);
      w = sign(vector_product)*w;
   else % deltax >= 0 & deltay >= 0, so bottom_left should be on + side.
      vector_product = dot(w, bottom_left);
      w = sign(vector_product)*w;
   end
   % For the vertex at psecond to be convex, pthird has to be on the
   % positive side of the line.
   sgn(k) = 1;
   if (w(1)*pthird(:, 1) + w(2)*pthird(:, 2) + w(3)) < 0
      sgn(k) = -1;
   end
end
```

```
%-------------------------------------------------------------------%
function w = polyedge(p1, p2)
%   Outputs the coefficients of the line passing through p1 and
%   p2. The line is of the form w1x + w2y + w3 = 0.

x1 = p1(:, 1);  y1 = p1(:, 2);
x2 = p2(:, 1);  y2 = p2(:, 2);
if x1==x2
   w2 = 0;
   w1 = -1/x1;
   w3 = 1;
elseif y1==y2
   w1 = 0;
   w2 = -1/y1;
   w3 = 1;
elseif x1 == y1 & x2 == y2
   w1 = 1;
   w2 = 1;
   w3 = 0;
else
   w1 = (y1 - y2)/(x1*(y2 - y1) - y1*(x2 - x1) + eps);
   w2 = -w1*(x2 - x1)/(y2 - y1);
   w3 = 1;
end
w = [w1, w2, w3];
%-------------------------------------------------------------------%
function [xg, yg] = dupgone(x, y)
% Eliminates duplicate, adjacent rows in [x y], except that the
% first and last rows can be equal so that the polygon is closed.

xg = x;
yg = y;
if size(xg, 1) > 2
   I = find((x(1:end-1, :) == x(2:end, :)) & ...
         (y(1:end-1, :) == y(2:end, :)));
   xg(I) = [];
   yg(I) = [];
end
```

R

```
function [xn, yn] = randvertex(x, y, npix)
%RANDVERTEX Adds random noise to the vertices of a polygon.
%   [XN, YN] = RANDVERTEX[X, Y, NPIX] adds uniformly distributed
%   noise to the coordinates of vertices of a polygon. The
%   coordinates of the vertices are input in X and Y, and NPIX is the
%   maximum number of pixel locations by which any pair (X(i), Y(i))
%   is allowed to deviate. For example, if NPIX = 1, the location of
%   any X(i) will not deviate by more than one pixel location in the
%   x-direction, and similarly for Y(i). Noise is added independently
%   to the two coordinates.
```

```
% Convert to columns.
x = x(:);
y = y(:);

% Preliminary calculations.
L = length(x);
xnoise = rand(L, 1);
ynoise = rand(L, 1);
xdev = npix*xnoise.*sign(xnoise - 0.5);
ydev = npix*ynoise.*sign(ynoise - 0.5);

% Add noise and round.
xn = round(x + xdev);
yn = round(y + ydev);

% All pixel locations must be no less than 1.
xn = max(xn, 1);
yn = max(yn, 1);
```

S

```
function [st, angle, x0, y0] = signature(b, varargin)
%SIGNATURE Computes the signature of a boundary.
%   [ST, ANGLE, X0, Y0] = SIGNATURE(B) computes the
%   signature of a given boundary, B, where B is an np-by-2 array
%   (np > 2) containing the (x, y) coordinates of the boundary
%   ordered in a clockwise or counterclockwise direction. The
%   amplitude of the signature as a function of increasing ANGLE is
%   output in ST. (X0,Y0) are the coordinates of the centroid of the
%   boundary. The maximum size of arrays ST and ANGLE is 360-by-1,
%   indicating a maximum resolution of one degree. The input must be
%   a one-pixel-thick boundary obtained, for example, by using the
%   function boundaries. By definition, a boundary is a closed curve.
%
%   [ST, ANGLE, X0, Y0] = SIGNATURE(B) computes the signature, using
%   the centroid as the origin of the signature vector.
%
%   [ST, ANGLE, X0, Y0] = SIGNATURE(B, X0, Y0) computes the boundary
%   using the specified (X0, Y0) as the origin of the signature
%   vector.

% Check dimensions of b.
[np, nc] = size(b);
if (np < nc | nc ~= 2)
   error('B must be of size np-by-2.');
end

% Some boundary tracing programs, such as boundaries.m, end where
% they started, resulting in a sequence in which the coordinates
% of the first and last points are the same. If this is the case,
% in b, eliminate the last point.
if isequal(b(1, :), b(np, :))
   b = b(1:np - 1, :);
```

```
   np = np - 1;
end

% Compute parameters.
if nargin == 1
   x0 = round(sum(b(:, 1))/np); % Coordinates of the centroid.
   y0 = round(sum(b(:, 2))/np);
elseif nargin == 3
   x0 = varargin{1};
   y0 = varargin{2};
else
   error('Incorrect number of inputs.');
end

% Shift origin of coord system to (x0, y0)).
b(:, 1) = b(:, 1) - x0;
b(:, 2) = b(:, 2) - y0;

% Convert the coordinates to polar.  But first have to convert the
% given image coordinates, (x, y), to the coordinate system used by
% MATLAB for conversion between Cartesian and polar cordinates.
% Designate these coordinates by (xc, yc). The two coordinate systems
% are related as follows:  xc = y and yc = -x.
xc = b(:, 2);
yc = -b(:, 1);
[theta, rho] = cart2pol(xc, yc);

% Convert angles to degrees.
theta = theta.*(180/pi);

% Convert to all nonnegative angles.
j = theta == 0; % Store the indices of theta = 0 for use below.
theta = theta.*(0.5*abs(1 + sign(theta)))...
        - 0.5*(-1 + sign(theta)).*(360 + theta);
theta(j) = 0; % To preserve the 0 values.

temp = theta;
% Order temp so that sequence starts with the smallest angle.
% This will be used below in a check for monotonicity.
I = find(temp == min(temp));

% Scroll up so that sequence starts with the smallest angle.
% Use I(1) in case the min is not unique (in this case the
% sequence will not be monotonic anyway).
temp = circshift(temp, [-(I(1) - 1), 0]);

% Check for monotonicity, and issue a warning if sequence
% is not monotonic. First determine if sequence is
% cw or ccw.
k1 = abs(temp(1) - temp(2));
k2 = abs(temp(1) - temp(3));
if k2 > k1
   sense = 1; % ccw
elseif k2 < k1
   sense = -1; % cw
```

```
else
   warning(['The first 3 points in B do not form a monotonic ' ...
            'sequence.']);
end
% Check the rest of the sequence for monotonicity. Because
% the angles are rounded to the nearest integer later in the
% program, only differences greater than 0.5 degrees are
% considered in the test for monotonicity in the rest of
% the sequence.
flag = 0;
for k = 3:length(temp) - 1
   diff = sense*(temp(k + 1) - temp(k));
   if diff < -.5
      flag = 1;
   end
end
if flag
   warning('Angles do not form a monotonic sequence.');
end
% Round theta to 1 degree increments.
theta = round(theta);
% Keep theta and rho together.
tr = [theta, rho];
% Delete duplicate angles.  The unique operation
% also sorts the input in ascending order.
[w, u, v] = unique(tr(:, 1));
tr = tr(u,:); % u identifies the rows kept by unique.
% If the last angle equals 360 degrees plus the first
% angle, delete the last angle.
if tr(end, 1) == tr(1) + 360
   tr = tr(1:end - 1, :);
end
% Output the angle values.
angle = tr(:, 1);
% The signature is the set of values of rho corresponding
% to the angle values.
st = tr(:, 2);
```

```
function [srad, sang, S] = specxture(f)
%SPECXTURE Computes spectral texture of an image.
%   [SRAD, SANG, S] = SPECXTURE(F) computes SRAD, the spectral energy
%   distribution as a function of radius from the center of the
%   spectrum, SANG, the spectral energy distribution as a function of
%   angle for 0 to 180 degrees in increments of 1 degree, and S =
%   log(1 + spectrum of f), normalized to the range [0, 1]. The
%   maximum value of radius is min(M,N), where M and N are the number
%   of rows and columns of image (region) f. Thus, SRAD is a row
%   vector of length = (min(M, N)/2) - 1; and SANG is a row vector of
%   length 180.
```

```
% Obtain the centered spectrum, S, of f. The variables of S are
% (u, v), running from 1:M and 1:N, with the center (zero frequency)
% at [M/2 + 1, N/2 + 1] (see Chapter 4).
S = fftshift(fft2(f));
S = abs(S);
[M, N] = size(S);
x0 = M/2 + 1;
y0 = N/2 + 1;

% Maximum radius that guarantees a circle centered at (x0, y0) that
% does not exceed the boundaries of S.
rmax = min(M, N)/2 - 1;

% Compute srad.
srad = zeros(1, rmax);
srad(1) = S(x0, y0);
for r = 2:rmax
   [xc, yc] = halfcircle(r, x0, y0);
   srad(r) = sum(S(sub2ind(size(S), xc, yc)));
end

% Compute sang.
[xc, yc] = halfcircle(rmax, x0, y0);
sang = zeros(1, length(xc));
for a = 1:length(xc)
   [xr, yr] = radial(x0, y0, xc(a), yc(a));
   sang(a) = sum(S(sub2ind(size(S), xr, yr)));
end

% Output the log of the spectrum for easier viewing, scaled to the
% range [0, 1].
S = mat2gray(log(1 + S));

%-------------------------------------------------------------------%
function [xc, yc] = halfcircle(r, x0, y0)
%   Computes the integer coordinates of a half circle of radius r and
%   center at (x0,y0) using one degree increments.
%
%   Goes from 91 to 270 because we want the half circle to be in the
%   region defined by top right and top left quadrants, in the
%   standard image coordinates.

theta=91:270;
theta = theta*pi/180;
[xc, yc] = pol2cart(theta, r);
xc = round(xc)' + x0; % Column vector.
yc = round(yc)' + y0;

%-------------------------------------------------------------------%
function [xr, yr] = radial(x0, y0, x, y);
%   Computes the coordinates of a straight line segment extending
%   from (x0, y0) to (x, y).
%
```

```
%   Based on function intline.m. xr and yr are
%   returned as column vectors.

 [xr, yr] = intline(x0, x, y0, y);
```

```
function [v, unv] = statmoments(p,  n)
%STATMOMENTS Computes statistical central moments of image histogram.
%   [W, UNV] = STATMOMENTS(P, N) computes up to the Nth statistical
%   central moment of a histogram whose components are in vector
%   P. The length of P must equal 256 or 65536.
%
%   The program outputs a vector V with V(1) = mean, V(2) = variance,
%   V(3) = 3rd moment, . . . V(N) = Nth central moment. The random
%   variable values are normalized to the range [0, 1], so all
%   moments also are in this range.
%
%   The program also outputs a vector UNV containing the same moments
%   as V, but using un-normalized random variable values (e.g., 0 to
%   255 if length(P) = 2^8). For example, if length(P) = 256 and V(1)
%   = 0.5, then UNV(1) would have the value UNV(1) = 127.5 (half of
%   the [0 255] range).

Lp = length(p);
if (Lp ~= 256) & (Lp ~= 65536)
   error('P must be a 256- or 65536-element vector.');
end
G = Lp - 1;

% Make sure the histogram has unit area, and convert it to a
% column vector.
p = p/sum(p); p = p(:);

% Form a vector of all the possible values of the
% random variable.
z = 0:G;

% Now normalize the z's to the range [0, 1].
z = z./G;

% The mean.
m = z*p;

% Center random variables about the mean.
z = z - m;

% Compute the central moments.
v = zeros(1, n);
v(1) = m;
for j = 2:n
   v(j) = (z.^j)*p;
end

if nargout > 1
   % Compute the uncentralized moments.
   unv = zeros(1, n);
   unv(1)=m.*G;
```

```
   for j = 2:n
      unv(j) = ((z*G).^j)*p;
   end
end
```

```
function [t] = statxture(f, scale)
%STATXTURE Computes statistical measures of texture in an image.
%   T = STATXURE(F, SCALE) computes six measures of texture from an
%   image (region) F. Parameter SCALE is a 6-dim row vector whose
%   elements multiply the 6 corresponding elements of T for scaling
%   purposes. If SCALE is not provided it defaults to all 1s.  The
%   output T is 6-by-1 vector with the following elements:
%      T(1) = Average gray level
%      T(2) = Average contrast
%      T(3) = Measure of smoothness
%      T(4) = Third moment
%      T(5) = Measure of uniformity
%      T(6) = Entropy
if nargin == 1
   scale(1:6) = 1;
else % Make sure it's a row vector.
   scale = scale(:)';
end
% Obtain histogram and normalize it.
p = imhist(f);
p = p./numel(f);
L = length(p);

% Compute the three moments. We need the unnormalized ones
% from function statmoments. These are in vector mu.
[v, mu] = statmoments(p, 3);

% Compute the six texture measures:
% Average gray level.
t(1) = mu(1);
% Standard deviation.
t(2) = mu(2).^0.5;
% Smoothness.
% First normalize the variance to [0 1] by
% dividing it by (L-1)^2.
varn = mu(2)/(L - 1)^2;
t(3) = 1 - 1/(1 + varn);
% Third moment (normalized by (L - 1)^2 also).
t(4) = mu(3)/(L - 1)^2;
% Uniformity.
t(5) = sum(p.^2);
% Entropy.
t(6) = -sum(p.*(log2(p + eps)));

% Scale the values.
t = t.*scale;
```

X

```
function [B, theta] = x2majoraxis(A, B, type)
%X2MAJORAXIS Aligns coordinate x with the major axis of a region.
%   [B2, THETA] = X2MAJORAXIS(A, B, TYPE) aligns the x-coordinate
%   axis with the major axis of a region or boundary. The y-axis is
%   perpendicular to the x-axis.  The rows of 2-by-2 matrix A are the
%   coordinates of the two end points of the major axis, in the form
%   A = [x1 y1; x2 y2]. On input, B is either a binary image (i.e.,
%   an array of class logical) containing a single region, or it is
%   an np-by-2 set of points representing a (connected) boundary. In
%   the latter case, the first column of B must represent
%   x-coordinates and the second column must represent the
%   corresponding y-coordinates. On output, B contains the same data
%   as the input, but aligned with the major axis. If the input is an
%   image, so is the output; similarly the output is a sequence of
%   coordinates if the input is such a sequence. Parameter THETA is
%   the initial angle between the major axis and the x-axis. The
%   origin of the xy-axis system is at the bottom left; the x-axis is
%   the horizontal axis and the y-axis is the vertical.
%
%   Keep in mind that rotations can introduce round-off errors when
%   the data are converted to integer coordinates, which is a
%   requirement.  Thus, postprocessing (e.g., with bwmorph) of the
%   output may be required to reconnect a boundary.

% Preliminaries.
if islogical(B)
   type = 'region';
elseif size(B, 2) == 2
   type = 'boundary';
   [M, N] = size(B);
   if M < N
      error('B is boundary. It must be of size np-by-2; np > 2.')
   end
   % Compute centroid for later use. c is a 1-by-2 vector.
   % Its 1st component is the mean of the boundary in the x-direction.
   % The second is the mean in the y-direction.
   c(1) = round((min(B(:, 1)) + max(B(:, 1))/2));
   c(2) = round((min(B(:, 2)) + max(B(:, 2))/2));

   % It is possible for a connected boundary to develop small breaks
   % after rotation. To prevent this, the input boundary is filled,
   % processed as a region, and then the boundary is re-extracted. This
   % guarantees that the output will be a connected boundary.
   m = max(size(B));
   % The following image is of size m-by-m to make sure that there
   % there will be no size truncation after rotation.
   B = bound2im(B,m,m);
   B = imfill(B,'holes');
```

```
else
   error('Input must be a boundary or a binary image'.)
end

% Major axis in vector form.
v(1) = A(2, 1) - A(1, 1);
v(2) = A(2, 2) - A(1, 2);
v = v(:);  % v is a col vector

% Unit vector along x-axis.
u = [1; 0];

% Find angle between major axis and x-axis. The angle is
% given by acos of the inner product of u and v divided by
% the product of their norms. Because the inputs are image
% points, they are in the first quadrant.
nv = norm(v);
nu = norm(u);
theta = acos(u'*v/nv*nu);
if theta > pi/2
   theta = -(theta - pi/2);
end
theta = theta*180/pi;  % Convert angle to degrees.

% Rotate by angle theta and crop the rotated image to original size.
B = imrotate(B, theta, 'bilinear', 'crop');

% If the input was a boundary, re-extract it.
if strcmp(type, 'boundary')
   B = boundaries(B);
   B = B{1};
   % Shift so that centroid of the extracted boundary is
   % approx equal to the centroid of the original boundary:
   B(:, 1) = B(:, 1) - min(B(:, 1)) + c(1);
   B(:, 2) = B(:, 2) - min(B(:, 2)) + c(2);

end
```

Index

A

B

C

D

G

J

K

L

N

S

T